D1594230

Microbial Phylogeny and Evolution

Microbial Phylogeny and Evolution
Concepts and Controversies

Edited by
Jan Sapp

2005

Oxford University Press, Inc., publishes works that further
Oxford University's objective of excellence
in research, scholarship, and education.

Oxford New York
Auckland Cape Town Dar es Salaam Hong Kong Karachi
Kuala Lumpur Madrid Melbourne Mexico City Nairobi
New Delhi Shanghai Taipei Toronto

With offices in
Argentina Austria Brazil Chile Czech Republic France Greece
Guatemala Hungary Italy Japan Poland Portugal Singapore
South Korea Switzerland Thailand Turkey Ukraine Vietnam

Published by Oxford University Press, Inc.
198 Madison Avenue, New York, New York 10016

www.oup.com

Library of Congress Cataloging-in-Publication Data
Microbial phylogeny and evolution : concepts and controversies /
edited by Jan Sapp.
p. cm.
Includes bibliographical references.
ISBN-13: 978-0-19-516877-1
ISBN 0-19-516877-1 (alk. paper)
1. Microorganisms—Evolution. I. Sapp, Jan.
[DNLM: 1. Bacteria—genetics. 2. Evolution, Molecular.
3. Genetics, Microbial. 4. Phylogeny. 5. Sequence Analysis.
QW 51 M6269 2004]
QR13.M527 2004
579'.138—dc22 2004000565

9 8 7 6 5 4 3 2 1

Printed in the United States of America
on acid-free paper

For Joshua Lederberg

Foreword

Postludian Remarks—Phylogeny versus Evolution

JOSHUA LEDERBERG

As I was not a primary participant in the symposium, these remarks stem from my reading the early manuscript, at the invitation of Jan Sapp, the organizer and editor. His introductory chapter already practices the exegetical function, and I would be unqualified to emulate or improve on his labors.

What an illumination the reading has been for me. I will confess to having had something of an agnostic position about the possibility of verifying our speculations about the major kingdoms, and especially the origin of Eukarya. I dare not admit to fixed conclusions on my own part, but now having had the experience of the finely spun arguments, at least I can say that I have a better understanding of the question. In that regard, I anticipate being in the company of a horde of biologists who have been bewildered by the controversies and have had no prior opportunity for the direct confrontation of views represented in this volume. One may have to look back to the nineteenth century and before to locate examples of equally weighty disputations in biology.

Besides the larger canvas of evolutionary drama, every chapter has a revelation of natural history. How thrilling to be reminded of the connections of the apical complexes seen in hemoparasites like malaria, with the plastids of red algae. In turn that has inspired Bob Haselkorn to seek antimicrobials for Toxoplasma among familiar herbicides directed against the chloroplasts of common weeds.

Autobiographically, my own fascination with symbiogenesis was sparked by Luigi Provasoli's report 55 years ago that Euglena could be "cured" of its chloroplast by the application of streptomycin. Later, this was corroborated in spinach seedlings. Within a few years, we had the reports from Boris Ephrussi about similar responses—now the marker was oxidative metabolism (read mitochondria)—of yeast to acriflavine. From then on, I could see no definitive boundary between infection and (extranuclear) heredity.

This motivated my coining the term "plasmid" in 1952 to stress the vacuity of the argument over whether a particle was a virus, a symbiont, or a gene. It also made me more receptive to then-marginalized speculations that mitochondria and chloroplasts were derived from free-living microbes gone obligate symbionts.

Not until prodded by Jan Sapp did I really understand how symbiogenesis was the foundation for a major paradigm shift from arboreal modes of ascent—the Darwinian paradigm—to a mode where major convergences of prior evolutionary histories engendered sudden innovations, just like lichens from alga and fungus obvious beneath our feet. The book harbors deeply felt contrary convictions about the prevalence and importance of LGT (lateral gene transfer). The least controversial examples are the migration of genes from mitochondrial and plastid symbionts to the nuclear genomes of their hosts. Of course, it was not until the early 1950s that the full panoply of genetic mechanisms was uncovered that could justify thoughts of widespread sharing of genetic information among diverse taxa of microbes. Early examples were the generation of new serovars of Salmonella (then given Linnean names as if species) by transduction of flagellar markers. Then around 1960, Lou Baron showed that Salmonella could be hybridized with *Escherichia coli*, mediated by the conjugal plasmid F. Few might doubt that these laboratory phenomena played some role in natural evolution, but that participation was hard to prove prior to the availability of genomic sequences.

The core arguments of the book relate to the selection of genes to serve as molecular clocks. For phylogenetic mensuration they should be highly conserved, uniform in mutation rates, ubiquitous, inert to natural selection; to wit, devoid of any interesting phenotype or adaptation to new habitats. These desiderata are pertinent to phylogenetic resolution: They leave out almost everything that relates to evolutionary diversification. There should be some discomfort when the kingdoms comprising multicellulars all together project as a handprint on the wall map of the major categories, such as Archaea, Bacteria, and Eukarya. They sandwich *Homo sapiens* in a narrow confine between corn and mushrooms. That may be a phylogenetic reality; perhaps we need a better terminology to contrast that placement with biological evolution.

When we look, say, within the animal or plant world, we have to recognize that genomes evolve along more complex pathways than the accumulation of single-nucleotide substitutions. Block deletions, duplications, inversions, and transfers of chromosome segments and the accumulation of repeated sequences, intragenomic recombination, hybridization, and heteroploidy all play a role in issuing adaptive phenotypes, and are all highly constrained by natural selection. Nor is the assumption of mutation rates, uniform in time or place, a reasonable one. Mutators—often by relaxation of corrective editing of error-prone replication—are well documented, and exogenous transposable elements play a leading role in so-called spontaneous mutation in *Drosophila*. Finally, many signatures of relationship are confounded by productive aberrations of lateral gene transfer and symbiosis. There is an essential tension between phylogenetic analysis and evolutionary description.

The murkiest swamp is doubtless the shore of Darwin's warm little pond, the chemical drama of the very first stages of life on earth. There are many attractions to the "RNA world model," at some early stage depending on RNA ribozymia to provide the catalytic functions for molecular self-replication. The provenience of the concentrated phosphorylated and chemically reactive precursors for RNA synthesis poses problems of another order. We would be greatly heartened if we could find some vestige of an RNA-dominated (actually a DNA-free) premicrobe. The RNA viruses do not quite qualify as they all depend on DNA-rich hosts. As the microbial world is constantly presenting new surprises, perhaps the RNA-cyte will still be fished out of some depths.

As an alternative option, or rather a more primitive stage, I have suggested we gamble on the heretical possibility that some polypeptide sequences may have a faint, template-directed capacity to orient assembly so as to constitute a primitive self-replication. There are barely more than hints that this actually happens, but who has looked? The warm pond would not lack for such monomeric precursors, whether populated by Urey-Miller sparks in a reducing atmosphere or by cosmic infall, or whether the earth's primary composition is now revealed by deep vents, and inorganic systems will catalyze the condensation into polymers. Aboriginal proteins might then be the springboard for evolutionary perfection and elaboration, in due course concentrating and recruiting other monomers, like the nucleic acids. Whether polypeptide, polynucleotide, or who knows what other primitive polymer, one category at a time would seem a wise counsel for our speculations. Most important is how these categories motivate further experimentation on the minimum conditions for template-driven assembly.

In the symposium, there was some suggestion that autotrophs had to precede heterotrophs, in order to provide the specific nutrients. That is likely true in more recent times, but we hardly want to invoke the anabolically sufficient (and therefore highly complex) autotroph at the threshold. Years ago, Horowitz pointed out that as nutrients were gradually depleted from the pond by biological utilization, there would be powerful natural selection for the elaboration of synthetic pathways to make up the deficits, one by one, in reverse order of synthesis. As Darwin foresaw, the earliest environments would be the richest (up to the advent of photosynthesis).

Briefly, to some other points under discussion: We are quizzical about the fitness of nature's experiments in LGT or more farreaching admixture of genomes. It is understood that alien intrusions may be incompatible with complex networks, but we have contrary evidence in the far-reaching promiscuity of many plasmids, whose host opportunities transcend the kingdoms. A well-adapted example is the rhizobial symbiosis, and its neighbor *Agrobacterium tumefaciens*, with a plasmid normally hosted by a bacterium introducing itself to the plant genome. We can see the F plasmid, *E. coli*'s sex factor, growing in yeast cells.

We cannot simply address fitness by growth rates in the laboratory. If we did we would have to assign zero fitness to the vast majority of taxa in the soil that are uncultivable at all with current techniques. Exponential growth is vanishingly rare in natural settings, as the bugs themselves quickly exhaust available nutrients—cattle rumen may be a notable exception. Competitive survival in fluctuating natural environments entails much more than rapid growth in the cocoon of the laboratory.

Are the metabolic innovations of Archaea reactive to antibiotics produced by gram-positive bacteria? That is more credible when they share temperate habitats. The extreme scalding and briny habitats would already drive many natural experiments for biochemical adaptations.

Is phagotrophy a precondition for the acquisition of microbial symbionts? Many invasive bacteria that we would not customarily describe as phagotrophic have developed their own mechanisms for access to host epithelial cells. Within microbiota, contemporary parasitic Bdellovibrio will penetrate at least the outer membrane, living in the periplasm of gram-negative bacterial hosts. Bacteriophages have evolved their own inoculation needles, and their exit is facilitated by lysozymes that digest host walls. Any of these mechanisms might be hijacked by other bacteria as a route of access, and do not overlook bacterial hosts that are genetically (Mycoplasma) or transiently (L-forms) devoid of walls, the latter particularly animated by lysozymes, or B-lactam antibiotics, which are, after all, natural products. Indeed, bacterial lysozymes have been reported, although they may well stem from prophages integrated in the bacterial chromosome. Similar wall stretching can be induced in obligate halophiles planted in hypotonic media. There is likely to be far more quasi-phagotrophy among bacteria than meets the eye. The rate of such gobbling experiments may be attenuated, but the biosphere is a powerful, big place.

Acknowledgments

My sincere thanks to the authors of this volume for their generosity of spirit, hard work, and enthusiasm. This volume has its origin in a three-day conference held at the Université du Québec à Montréal in October 2002, sponsored by the Canada Research Chair in the History of the Biological Sciences. The success of that conference relied not only on outstanding participants but also on the exceptional organizational abilities of Lucie Comeau, whose tireless efforts—and those of her assistant, Frédérick Mongrain—were appreciated by everyone.

Preface

Molecular studies of bacterial phylogeny have effected a profound revolution in biology. Beginning in the 1970s, 16s ribosomal RNA phylogenetics by Carl Woese and his collaborators led to a natural classification of bacteria and to the construction of a universal phylogenetic tree, with three Domains or Urkingdoms: the Archaea, the Eubacteria, and the Eucarya. Studies of molecular evolution have also led to new concepts for understanding evolutionary change, and to the confirmation of old concepts, including the paramount role of symbiosis in the origin of the eukaryotic cell from which all plants and animals have sprung. The origin of mitochondria from alpha-proteobacteria and the origin of chloroplasts from cyanobacteria are well established.

Neo-Darwinian evolutionary theory, based on gene mutations and recombination between individuals of a species, did not include bacteria (*sensu lato*, or in the broad sense); nor could it. The evolutionary synthesis was constructed in the 1930s and early 1940s, before bacterial genetics was established, before the bacterium's mechanisms of heredity were elucidated, and before bacterial phylogenetics was considered possible. My own historical introduction highlights the controversies over microbial classification from Pasteur to the present. It illustrates the difficulties of constructing a microbial phylogeny before the advent of the field of molecular evolution, the revolution in microbial evolution by rRNA phylogenies, how the current controversies have arisen with the rise of genomics, and how the general concepts of microbial evolution contrast with classical neo-Darwinian theory.

Microbial evolutionary biology has been undergoing another transformation, with the birth of bacterial genomics, since the mid-1990s. These new data have led to several conceptual modifications and to wholly new issues. The importance of lateral gene transfer (between "species") across the bacterial phylogenetic spectrum is a subject of heated debate. Even the standard symbiotic scenario for evolution is being reconsidered, and the complex twists and turns in chloroplast evolutionary history have been revealed. Although the origin and evolution of mitochondria and plastids continue to be rich veins of research, several researchers have proposed further that the eukaryotic cell nucleus also evolved from a symbiotic event of some kind. The scope and significance of hereditary symbiosis among animals also remain subjects of controversy.

This volume brings together leading microbial evolutionists, often with markedly different viewpoints, to discuss these issues. Does the phylogenetic "tree" based on rRNA phylogenies still stand in the age of genomics? What is the scope and significance of lateral gene transfer across the bacterial phylogenetic spectrum? Can one trace bacterial genealogies at all in the face of lateral gene transfer? Is the course of the first 2 billion years or so of evolution unknowable? These issues are discussed in the first ten chapters. A full range of views are expressed in this volume.

Pace, Ludwig, and Schleifer (in chaps. 2 and 3) articulate the significance of rRNA-based microbial phylogenies. Woese (chap. 4) discusses why our understanding of the cell can never be separated from its evolution, the conceptual problems with the eukaryote–prokaryote dichotomy, the importance of lateral gene transfer in cell evolution, and the nature of the translation apparatus. Pace (chap. 2) provides a tour of the macrostructure of the phylogenetic tree and emphasizes how understanding of the tree has expanded through recent molecular studies of microbial diversity in the environment. Ludwig and Schleifer (chap. 3) compare the weighty data based on rRNA phylogenies to evidence based on other gene phylogenies.

While these authors elaborate on the importance of rRNA phylogenies, others qualify such studies as providing a definitive tree. A range of alternative views are offered here. At one pole, Doolittle (chap. 5) argues that lateral gene transfer may be so pervasive as to completely erase bacterial phylogenetic tracks. Martin (chap. 6) proposes that lateral gene transfer does not completely dash hopes for tracing the early evolution of life, but that one has to supplement genomics and rRNA histories with other data from biochemical evolution and from the fossil record. Morowitz and his collaborators (chap. 7) offer arguments for the robustness of metabolic phenotype compared to a noisy changing bacterial genotype. Gupta (chap. 8) and Lake and his collaborators (chap. 9) agree that it is indeed possible to trace genealogies, but they also suggest modifications to the model based on rRNA. On the opposite pole, Kurland (chap. 10) argues that the actual evidence for evolutionary significant lateral gene transfer is weak and that its phylogenetic importance has been grossly exaggerated.

Gray (chap. 11) offers us an overview of the central questions and controversies in research on mitochondrial evolution: What was the nature of the proto-mitochondria? Did mitochondria arise only once? How has the mitochondrial proteome evolved? Kurland (chap. 10) also considers some modifications to the standard symbiotic scenario for mitochondria, and Martin (chap. 6) argues for a radical change to theory, one that joins the symbiotic origin for mitochondria with that of nucleus. Archibald and Keeling (chap. 12) provide a lucid overview of the recent issues in chloroplast evolution: What was the nature of the proto-chloroplasts? Have chloroplasts evolved only once? Although all plastids may have descended from a single common ancestor, photosynthesis has spread horizontally among unrelated protist groups by symbioses involving two protists (secondary symbiosis). In still other cases, such secondary plastids have been replaced with those of an unrelated alga.

Although there have been several propositions that the nucleus is also a chimera of sorts, there is no consensus among scientists. Gupta (chap. 8) argues that the nucleus originated from a fusion between a diderm (gram negative) and monoderm (gram positive) bacteria, which he believes is the fundamental dichotomy among bacteria. Lake and his colleagues (chap. 9) consider that the nucleus involved an engulfment of one kind of bacteria by another, whereas Martin (chap. 6) has proposed that the ancestor of the mitochondria was involved in the origin of the nucleus. Margulis and her collaborators (chap. 13) maintain, instead, that the nucleus involved the merger of a spirochete-like eubacterium and an archaebacterium, which would account for the eukaryotic cell cytoskeleton and protist motility. The need to account for the origin of eukaryotic cell cytoskeleton is discussed by Dolan (chap. 14).

Werren's overview (chap. 15) of the importance of hereditary bacterial symbiosis in insects, shrimp, spiders, and worms, detected by techniques of molecular screening, reminds us that the scope of microbial symbiosis extends far and wide. This chapter effectively contradicts the assertions of leading neo-Darwinian theorists that hereditary symbiosis among animals is a rare occurrence.

Contents

Contributors

John M. Archibald
Assistant Professor
Department of Biology
Dalhousie University
Halifax, Nova Scotia
Canada B3H 4H7

Daniel Broyles
Research Consultant
Krasnow Institute
Mail Stop 2A1
George Mason University
Fairfax, VA 22030, USA

Michael F. Dolan
Lecturer
Department of Biology
University of Massachusetts
Amherst, MA 01003-5820, USA

W. Ford Doolittle
Professor
Department of Biochemistry and Molecular Biology
Tupper Medical Building
Dalhousie University
Halifax, Nova Scotia
Canada B3H 4H7

Michael W. Gray
Professor
Department of Biochemistry and Molecular Biology
Tupper Medical Building
Dalhousie University
Halifax, Nova Scotia
Canada B3H 4H7

Radhey Gupta
Professor
Department of Biochemistry
Health Sciences Centre, Room 4N59
McMaster University
Hamilton, Ontario
Canada L8N 325

Patrick J. Keeling
Associate Professor
Department of Botany
University of British Columbia
3529-6270 University Boulevard
Vancouver, British Columbia
V6T 1Z4 Canada

Charles Kurland
Professor of Molecular Biology, emeritus
Lunds University
Microbial Ecology
Ekologihuset
223 62 Lund, Sweden

James Lake
Professor
Molecular Biology Institute
University of California, Los Angeles
Los Angeles, CA 90095-1570, USA

Howard Lasus
Research Consultant
Krasnow Institute
George Mason University
Fairfax, VA 22030, USA

Joshua Lederberg
President Emeritus and University Professor
The Rockefeller University
New York, NY, 10021, USA

Wolfgang Ludwig
Lehrstuhl für Mikrobiologie
Technische Universität München
Am Hochanger 4
85350 Freising, Germany

Lynn Margulis
Distinguished Professor
Department of Geosciences
University of Massachusetts
Amherst, MA 01003-5820, USA

William Martin
Professor
Institut für Botanik
Universität Düsseldorf
ED-40225 Düsseldorf, Germany

Hannah Melnitsky
Department of Geosciences
University of Massachusetts
Amherst, MA 01003-5820, USA

Jonathan E. Moore
Molecular Biology Institute
University of California, Los Angeles
Boyer Hall, Box 951570
Los Angeles, CA 90095-1570, USA

Harold J. Morowitz
Clarence J. Robinson Professor of Biology and Natural Philosophy
Krasnow Institute
Mail Stop 2A1
George Mason University
Fairfax, VA 22030, USA

Norman R. Pace
Professor
Department of Molecular Cellular and Developmental Biology
University of Colorado
Boulder, CO 80309-0347, USA

Frederick A. Rainey
Professor
Department of Biology
Lousiana State University
Baton Rouge, LA 70803, USA

Maria C. Rivera
Molecular Biology Institute
University of California, Los Angeles
Boyer Hall, Box 951570
Los Angeles, CA 90095-1570, USA

Jan Sapp
Professor
Department of Biology
York University
4700 Keele St.
Toronto, Ontario
Canada M3J 1P3

Karl-Heinz Schleifer
Professor
Lehrstuhl für Mikrobiologie
Technische Universität München
Am Hochanger 4
85350 Freising, Germany

Anne B. Simonson
Eiserling Graduate Fellow
Molecular Biology Institute
University of California, Los Angeles
Los Angeles, CA 90095-1570, USA

John H. Werren
Professor
Biology Department
University of Rochester
Rochester, NY, 14627, USA

Carl Woese
Professor
Department of Microbiology
University of Illinois at Urbana-Champaign
601 South Goodwin Avenue
B103 Chemical and Life Sciences Laboratory
Urbana, IL 61801-3709, USA

Microbial Phylogeny and Evolution

1

The Bacterium's Place in Nature

JAN SAPP

The neo-Darwinian synthesis is about plants and animals. Gene mutations and genetic recombination between individuals of a species are held to provide the fuel for evolution by natural selection, but this is effectively a sterile view of evolution, lacking bacteria. Certainly, the authors of that synthesis in the 1930s and 1940s assumed that plants and animals evolved from the "lower" or "primitive" organisms that bacteria were conceived to be. Still, bacteria lacked biological definition, and they remained outside any general evolutionary or phylogenetic framework. The relationships of bacteria to each other and to other forms of life, as well as their mechanisms of inheritance, have been subjects of perennial discussion and debate. My aim in this chapter is to outline the issues from the nineteenth century to the present. I discern four phases in the study of bacteria, each one characterized by a range of possibilities defined by the current theories and beliefs about bacteria and by the way of observing and discussing them.

The first phase, occurring from 1862 to 1945, is characterized by the rise of the germ theory of disease and the recognition of the physiological diversity of bacteria. A natural, phylogenetic classification of bacteria confronted concepts of pleomorphorism and bacterial physiological adaptation, on the one hand, and a lack of morphological traits on the other. Debates centered on which kind of classification, natural or artificial, was most useful and doable. A scheme for a natural classification was constructed, based on the primary importance of physiological traits, but it was criticized for being founded on unreliable traits and on erroneous assumptions about the evolution of physiological complexity. Supported by diverse practical interests, nonphylogenetic classification was seen as the most realistic and appropriate classification. During this period, it was widely assumed by bacteriologists that bacteria possessed no species as such, that production was solely by division, and that bacterial heredity and evolution involved a vague Lamarckian mechanism. A phylogenetic order was generally thought to be unachievable. Bacteria were thus objects—germs—without a natural history, that were to be exploited in industry, to be avoided as agents of killer diseases, and to be hunted and killed.

A second phase, taking place from 1946 to 1977, is characterized by the rise of bacterial genetics and molecular biology. Bacteria were shown to possess genes and to exhibit genetic recombination; Lamarckian mechanisms were rejected, but some bacterial genes were shown to be transmitted between some bacterial types by various means. Although these mechanisms of lateral gene transfer were exploited in molecular biology and biotechnology, they were virtually ignored in evolutionary theory. Based on electron microscopic imagery, bacteria were labeled "prokaryotes" and were sharply distinguished from nucleated cells, "eukaryotes," as representing the greatest discontinuity in nature. Mitochondria and chloroplasts were shown to possess genetic systems, and theories about their symbiotic origin reemerged and were debated as highly speculative.

In a third phase, taking place from 1977 to 1994, molecular evolutionary studies of bacteria based on sequence comparisons of small subunit ribosomal RNA (SUU rRNA) successfully challenged artificial classification of bacteria, and a universal genealogical tree of life was created. Comparisons of SSU rRNA, cell wall structure, and translation machinery distinguished three domains or Urkingdoms: Archaea, Bacteria and Eucarya. The symbiotic origins of chloroplasts and mitochondria were substantiated and systematically investigated.

A fourth phase, occurring from 1995 to the present, is marked by the rise of bacterial genomics. Genes other than those for ribosomal RNA often indicated different phylogenies, and methodological debates arose over which molecules were the most reliable for tracing organismic genealogies. Genomic data also indicated the ubiquity of lateral gene transfer across the bacterial taxonomic spectrum, and the issue of whether it is impossible to trace bacterial phylogeny reemerged. Investigations of symbiosis shifted toward understanding the complex manner in which mitochondria and chloroplasts were acquired, the nature of the host that acquired organellar symbionts, and the chimeric nature of the nucleus.

The First Era

Germ Theory

The germ theory of disease is one of the esteemed icons of modern science. The concept of germs as the agents of killer diseases, against which one could potentially act, represented the late nineteenth-century transition to modern medicine.[1] The demonstration that invisible organisms were inducers of putrefaction and fermentation is also one of the great hallmarks of general biology. Life did not result from decayed organic matter: It was its cause. This was the conclusion of the debates and experiments over "spontaneous generation," which Louis Pasteur (1822–1895) brought to a head in 1862.[2] Pasteur subsequently showed that undesired fermentation could be prevented in wine and beer by heating it to 57°C for a few minutes. He argued that every fermentation process is associated with one specific germ: one to break sugar down to make alcohol, another when sugar forms lactic acid. He supported Robert Koch's (1843–1910) claim of 1876 that anthrax was alway caused by the anthrax bacillus, though it was sometimes obscured because the organism formed spores that were difficult to identify.

The microbe hunters of the late nineteenth century were compared with the heros of Greek mythology as microbial agents were discovered for many infectious diseases. New methods were developed for detecting these fearful enemies and for learning how they were transported, how they multiplied, and how they could be arrested. Bacteriologists in pathology, agriculture, and industry were admired for their detailed studies of the physiological properties of bacteria and for ascertaining their great importance for life and disease, as well as for their refined methods of culture, preparation, staining, and observations with ever-improving microscopes.

Pathological and public health laboratories expanded and multiplied, while pathologists became occupied with the development of antitoxins and antisera and with the identification and development of vaccines, exemplified by Pasteur's achievement in 1885 of a preventative vaccine against rabies. In 1888, the Pasteur Institute was founded, and an international network of forty other Pasteur Institutes was subsequently established. In 1891, the German government furnished Koch with a research institute, and in that year, in London, the British Institute for Preventive Medicine was founded, which, in 1903, changed its name to the Lister Institute to honor its champion of germ theory, surgeon Joseph Lister (1827–1912). Professors of pathology in the universities taught practical classes in bacteriology to medical, veterinary, and hygiene students and to candidates for diplomas in public health and in sanitary and food inspection.[3] Studies of bacteria were also important for research on soil fertility, nitrogen-fixation, dairy products, plant diseases, and industrial fermentation (primarily brewing).

Despite these brilliant results, few specialists attempted to understand bacteria in light of evolution. They had been referred to as Infusoria because they were readily found in infusions of decaying organic matter, but it soon became common place to call them "germs." Lister first used the expression "theory of germs" in a letter in 1874 to Pasteur.[4] In 1878, the term "microbe" was introduced by Charles Sédillot and was used interchangeably with "germ." The word "bacteria" (from the Greek meaning "little rod" or "staff") was also frequently employed, beginning in the 1870s, as a general term to embrace the smallest of germs, "all those minute, rounded, ellipsoid, rod-shaped, thread-like or spiral forms."[5]

Nothing was known of bacterial evolution: how they varied, how they were related to one another, and how they were related to other microscopic organisms, plants, and animals. Bacteria had a past with no history. In this regard, there had not been advance since Antony van Leeuwenhoek (1632–1723), chamberlain of the Council-Chamber of Delft, observed bacteria among the very little animalcules in sea water and pepper-water and described them in his letters to the Royal Society of London in 1676.[6] In the nineteenth century and most of the twentieth century, the study of bacterial natural history was largely the domain of botanists and a few professors of bacteriology in universities who classified bacteria as plants.

Chaos or Kingdom?

In his *Systema Naturae* of 1735, Carolus Linnaeus (1707–1778) arranged all "natural bodies" into a hierarchical ordering of three kingdoms: Mineral Kingdom (*Regnum*

Lapideum); Vegetable Kingdom (*Regnum Vegetabile*); next closest to God; and Animal Kingdom (*Regnum Animale*), closest to God. In his *System Vegitabilum* (1774), Linnaeus proposed the generic term "chaos" to microscopic life that he could not order and classify.[7] In 1812, Georges Cuvier (1769–1832) classified the Infusoria in the order of worms as zoophytes, but "in which all is reduced to a homogeneous pulp."[8] In 1838, Christian Gottlieb Ehrenberg (1795–1876) denied that the Infusoria were homogeneous blobs and reported that they (including the bacteria) had organs as complex as any "higher animal." Ehrenberg's views were contradicted by Ferdinand Cohn (1828–1898), who in 1872 reclassified the bacteria as plants and asserted that some of the structures Ehrenberg had reported were simply imaginary. Indeed, the strongest magnifying lens of the 1870s, the immersion system of Hartnack, gave a magnifying power of just 3000 to 4000. Most bacteria were at the diffraction limits of resolution, and one could not clearly observe their interiors or the details of their reproduction.

Charles Darwin (1809–1882) said virtually nothing about the Infusoria, but post-Darwinian botanists usually spoke of the "bacteria" as belonging to the class of lower fungi; that is, "as one-celled fungi reproducing by simple fission only."[9] They were referred to as "the fission fungi" (Schizomycetes), as Carl von Naegeli (1817–1891) had named them in 1857, or more generally as fission-plants (Schizophyta). Still, there were proposals to place the bacteria, along with other microbes, in a new Kingdom because, it was argued, they constituted the main trunk of the tree of life and therefore belonged equally to both of its main branches: plants and animals. Some microbes possessed locomotion similar to animals, yet they had modes of living that were more like plants than animals. Thus, it was supposed that in the realm of the microscopic were the not-quite-animal and not-quite-plant ancestors of all living things. In 1860, famed paleontologist Richard Owen (1804–1892) called the new kingdom, "the Protozoa."[10] The same year, John Hogg renamed " the fourth kingdom" the "Primigenum" so as not to limit them to animal ancestors, as implied by Owen's neologistic Protozoa."[11] The Primigenal kingdom would embrace all the primary beings or "Protoctista," including the Protozoa and the Protophyta (the ancestors of the plants). In his *Generelle Morphologie der Organismen* (1866), Ernst Haeckel (1834–1919) designated the third living kingdom, the "Protista," as the first living creatures. These creatures included the "Protozoa" and the "Protophyta" as well as the "Protista Neutralia," or those that were not ancestral to plant or animal.

Haeckel placed the bacteria in the order Moneres (later Monera) at "the lowest stage of the protist kingdom." Bacteria were unique, he argued, because unlike other protists, they possessed no nucleus. They were as different from nucleated cells as "a hydra was from a vertebrate," or "a simple alga from a palm."[12] Haeckel had postulated the existence of such nonnucleated primitive life as part of his "monist philosophy" of life in terms of physical laws that break down the explanatory barrier between life and nonlife.[13] Later in his *Wonders of Life* of 1904, he included the Cyanophyceae or "Chromacea" (blue-green algae) among the Moneras, along with the bacteria. Although they were usually classified as a class of algae, Haeckel asserted that the blue-green algae lacked a nucleus and that the only real comparison between them and plants was with the chromatophores (chromatella or chloroplasts). He thus suggested that the plant cell evolved as "a symbiosis between a plasmodomonous green and plasmophagus not-green companions."[14] Such ideas had not been uncommon since the 1880s, although they were only systematically investigated a century later, when suitable techniques became available.[15]

Despite Haeckel's modern vision, botanists continued to refer to the bacteria as plants: Schizomycetes comparable to the Schizophyceae (fission algae). This was not simply the result of conservativism in nomenclature. Whether or not bacteria (and Cyanophyceae) actually possessed a real nucleus, and whether they divided in the manner of other microbes, remained controversial well into the twentieth century. In 1911, British protozoologist Clifford Dobell reviewed in detail the work of forty-nine authors and concluded that bacteria definitely possess nuclei and, often, a complex life cycle: "All bacteria which have been adequately investigated are—like all other protista—nucleated cells. . . . There is no evidence that enucleated Bacteria exist. The Bacteria are in no way a group of simple organisms, but rather a group displaying a high degree of morphological differentiation coupled in many cases with a life-cycle of considerable complexity."[16] Still others maintained that although bacteria had no nuclei and exhibited no true mitotic division, blue-green algae did contain a nucleus-like body that may undergo a simple form of mitotic division.[17] The definition by Williams and Park in 1929 was representative:

> Bacteria may be defined as extremely minute, simple, unicellular microorganisms, which reproduce themselves under suitable conditions with exceeding rapidity, usually by transverse division, and grow without aid of chlorophyl. They have no morphological nucleus, but contain nuclear material which is generally diffused throughout the cell body in the form of larger or smaller granules.[18]

Microbes remained divided among botanists and zoologists as "lower plants" and "lower animals." On the basis of "agreed usage," the green flagellates from which plants were thought to have descended, as were the bacteria, were the domain of botanists, and the colorless flagellata—the protozoa, Heliozoa, Foraminifera, and Infusoria—were the subject of zoologists. By the third decade of the twentieth century, some botanists expressed the opinion that the Schizophyta or Monera did not belong in the plant kingdom. American botanist Edwin Bingham Copeland (1873–1964) argued in 1927 that a plant kingdom that included the bacteria was "no more natural than a kingdom of the stones."[19] Copeland reassured botanists that botany was "still the most convenient place to study them," but added that, "It is however, very important that common sense, consistency, reasonableness never be ignored. There is no other one thing so important in systematic biology as the fact that the grouping of organisms reflects and expresses their true relationships. It is inconsistent and unreasonable to begin the course in botany by doing violence to this basic principle."[20]

In 1938, Edwin Copeland's son, Herbert Faulkner Copeland (b. 1902) wrote a more detailed paper and then, in 1956, a book proposing that Haeckel's subdivision "Monera" be granted its own kingdom. He did so on the basis of two assumptions, one evolutionary and one morphological: "that they are the comparatively little modified descendants of whatever single form of life appeared on earth, and that they are sharply distinguished from other organisms by the absence of nuclei."[21] H.F. Copeland argued that there were thus four kingdoms of life: Monera, Protista, Plantae, and Animalia. Copeland's arguments relied on a natural classification or genealogy in accordance with evolutionary precepts. However, leading bacteriologists were most reluctant to assign bacteria to their own kingdom, and they had abandoned any hope of classifying bacteria based on genealogy. Debates over how to order and classify the bacteria had persisted ever since Darwin.

Natural versus Artificial

Naturalists before Darwin who had grouped plants and animals into species, genera and families had believed their natural system revealed "the plan of the Creator," but their order of things assumed a new significance in the light of evolutionary theory. Darwin wrote in the *Origin* that: "All true classification is genealogical; that community of descent is the hidden bond which naturalists have been unconsciously seeking, and not some unknown plan of creation, or the enunciation of general propositions, and the mere putting together and separating objects more or less alike."[22] All the innumerable species, genera, and families of beings descended, each within its own class or group, from common parents, and all have been modified in the course of descent. He inferred further from analogy that, "all plants and animals have descended from one common prototype" and that "probably all organic beings which have ever lived on this earth have descended from some one primordial form, into which life was first breathed."[23]

When a large number of organisms had multiple resemblances, conforming to a single common structural pattern (e.g., the vertebrate pattern or the arthropod pattern in the animal kingdom), this was because they belonged to a single evolutionary stem and had a closer genetic relationship to each other than they did to other plants or animals that were constructed according to a different plan. Classification became, in principle, the art of grouping organisms in the manner that expresses best the degree of their evolutionary relatedness. A classification on these principles is known as a natural or phylogenetic system. Take all individuals that appear to be alike and group them as species, gather all the species that appear sufficiently similar and place them into the genus, and lump together the related genera in families, similar families in orders, orders in classes, classes in phyla, and phyla into kingdoms. When arrangement of these divisions is done to represent "blood" relationships, genealogies, they would express the course of evolution.

Post-Darwinian classification of plants and animals was based on comparative anatomy, life histories of the organisms, comparative embryology, differences in reproduction, and paleontological evidence dating back to the Cambrian explosion of 570 million years ago. Solid evidence from Precambrian paleontology did not come until the middle of the twentieth century, and as far as most observers could discern, bacteria possessed little signs of a life cycle, no developmental history, and no true reproduction from an egg, except for spore formation in some types. The life history of bacteria seemed to be limited to alternate processes of elongation and transverse fission.

Bacteria showed morphological diversity only in size and shape, such as, spherical, rod shaped, or spiral. Even then, one could not know whether this was really morphological diversity or only different stages in the development of the one organism. Indeed, many early workers believed that bacteria were highly variable and pleomorphic and that the many chemical and morphological changes observed in a culture only reflected transformations undergone by a single kind of organism. Famed botanist Carl von Naegeli was known for his belief in bacterial pleomorphism (derived from the Greek doctrine of many shapes), expounded in his book of 1877. As he commented, "For ten years I have been investigating thousands of different forms of dividing yeast and really could not say that even division into two different species is compelling."[24]

Ferdinand Cohn's "Untersuchungen über Bacterien" (1872) is often taken today as marking the starting-point of modern bacteriology.[25] Cohn was convinced that bacteria could be separated in just as good and distinct species as other "lower plants and animals" and that it was only their extraordinary smallness that made it difficult to distinguish species, which he argued, lived together in mixed cultures. The inability to distinguish between species was thus a technical problem, not a natural one. Techniques were lacking for isolating bacteria, making pure cultures, and observing them under different conditions so as to distinguish age and verify whether the observed diversity was the result of developmental states or of mixed cultures of different species of bacteria.[26]

Though bacteria had only a few basic shapes, they were known for their great physiological diversity, reflecting their varied environments, and their products. The question was whether and how one could or should use physiological differences to classify them. Cohn considered morphological differences to be primary; they would be used to determine genera and higher groups. Physiological differences would indicate only varieties or races; they originated from the same germ, but through constant natural or artificial culture under the same conditions, became hereditary.[27] He named four tribes and six genera—all on the basis of morphology:

Tribe I

Sphaerobacteria (sphere bacteria)

Genus 1. *Micrococcus*

Tribe II Microbacteria (rod bacteria)

Genus 2. *Bacterium* (filament bacteria)

Tribe III Desmobacteria

Genus 3. *Bacillus*

Genus 4. *Vibrio*

Tribe IV Spirobacteria

Genus 5. *Spirillum*

Genus 6. *Spirochaeta.*

During the last two decades of the nineteenth century, with ever-improving microscopic techniques, more morphological differences were disclosed in groups previously regarded as homogeneous. The formation of special aggregates, such as chains, clump tetrad, or packets, came to be accepted as an important additional criterion for characterizing spherical bacteria. The arrangement of organs of locomotion in the case of motile bacteria, as well as the ability to produce spores, was included in characterizations of rod-shaped bacteria. Thus, accepted genera or species were split into new ones. New staining techniques revealing differences in the chemical composition of otherwise similar organisms were also used. In 1884 Danish bacteriologist Hans Christian Gram (1853–1938), working in Berlin, published the famous procedure he used to distinguish *pneumococci* from *Klebsiella pneumoniae.* When appropriately stained and placed in alcohol, the color was discharged from certain bacteria, but retained in others.

More physiological or biochemical differences were also discerned. Industrial bacteriologists searched for specific microbes that were capable of decomposing certain organic substances or causing spoilage of food products, or that were responsible for the production of special chemicals. This led to the isolation and study of more or less well defined physiological groups and their nutritional physiology. This added to the number of species and genera while extending and refining criteria for characterizations.

Many physiological traits were also discovered by soil microbiologists. Beginning in the mid 1880s Sergei Winogradsky (1856–1953) at Strasbourg (and later in Zurich and in St. Petersburg,) discovered autotrophic bacteria (those that obtain their energy by the oxidation of inorganic elements and compounds), sulfur-metabolizing bacteria, iron bacteria, and later, nonsymbiotic nitrifying soil bacteria.[28] In 1888, Martins Willem Beijerinck (1851–1931), at the Netherlands Yeast and Alcohol Manufactory in Delft (and after 1895 at the Delft Polytechnic School), demonstrated the presence of nitrogen-fixing bacteria in legumes by isolated nitrogen-fixing root-nodule bacteria in pure cultures, which later greatly affected agriculture. He studied plant galls, contributed to knowledge of the tobacco mosaic virus, and worked on bacteria responsible for butyric alcohol fermentation, as well as those lactose ferments living in symbiosis with fermenting yeast in sour-milk preparations such as yogurt and Kefir.[29] Beijernick's studies of microbial ecology mark the beginning of "The Delft School," and pioneering work on comparative biochemistry of microbes showing their great metabolic diversity was subsequently carried out by Albert Jan Kluyver (1888–1956).[30]

Diagnostic techniques developed for pathogens were extended to nonpathogenic bacteria. Koch had published new techniques for isolating pure bacterial cultures in 1883. He isolated the tubercle bacillus, his colleague Friedrich Loeffer isolated the diphtheria bacillus, and Georg Gaffky did the same for typhoid. These scientists established the concept of bacterial specificity: typhoid germs descended from typhoid germs and tubercle bacilli from tubercle bacilli. Medical bacteriologists noted that the agents of disease possessed a variety of physiological characteristics that were not usually correlated with any specific morphological characteristic. For example, the colon bacillus and the typhoid germ were morphologically indistinguishable: one is a normal intestinal resident, the other a cause of severe disease. It was obvious to pathologists that classification and identification of bacteria could not be based entirely on morphology.

Thus, by the end of the century, bacteria were distinguished by a combination of criteria: morphological characteristics, cultural traits, biochemical activity, pathogenicity for plants and animals, and serum reactions. There were the Nitrosomonas, Nitrobacter, Azotobacter, Aerobacter, Thiobacillus, Photobacterium, Granulobacter, and many others. In addition to these "biochemical genera" and Cohn's "form genera," there were "color genera," "disease genera," and "nutritional genera," such as Flavobacterium, Rodococcus, Chromobacterium, Rhodospirilum, Phytomonas, Pneumococcus, and Haemiphilus. All of these genera were useful and expedient: It was useful to put vinegar bacteria, nitrogen-fixing bacteria, or luminous bacteria into groups. However, none of this held any phylogenetic meaning, and, because the proposed new genera were recognized by diverse criteria, organizing them into a natural phylogeny and uniting them into higher taxonomic entities was virtually impossible. Moreover, any natural classification based on physiology was considered dubious.

The problem with using physiological characters was that so many physiological properties seemed to be directly adaptive according to environmental circumstances. Similar adaptive characters were likely to arise in different groups under the influence of similar environmental conditions and therefore would be misleading as to true phylogenetic relationships. The same physiological trait may have been acquired relatively recently in one case and long ago in another, and one could not tell the difference.

Walter Migula (1863–1938) elaborated on Cohn's classification, based on morphology, when he compiled and integrated most of the new acquired knowledge in his two-volume *System der Bakterien*, published in 1897–1900.[31] He increased the number of morphological groups (genera) because additional characters such as type of flagella and sporulation had come to light. He also used some physiological criteria to distinguish bacteria into two orders, the Eubacteria, or true bacteria, and the Thiobacteria, the sulfur bacteria, a morphological diverse group found in places where hydrogen sulfide is present.[32] However, critics argued that Migula's classification made no phylogenetic sense. By using a single, apparently arbitrary, character such as flagella as a criterion for classification, for example, he grouped together widely different types of bacteria that were alike in no single other respect except that they have flagella. As one commentator put it, "These genera based on motility are on par with a division of animals into those with wings and those without, which would place bats and birds and flying fish and bees in one group and cats and ordinary fishes and worker ants in another."[33] Migula's classification was fruitless even if one accepted the motility genera: His scheme left over one-third of all known bacteria in the genus *Bacillus*.

Revolt from Morphology

Although Migula's classification was discouraging, some bacteriologists turned to physiological traits to develop a natural classification. They argued that imitating botanical and zoological classifications had paralyzed and distorted a true natural understanding of the bacteria. After all, whereas plants and animals developed complex structural modifications to obtain food materials of certain limited kinds, the bacteria had maintained themselves by acquiring the power of assimilating simple and abundant foods of various sorts. Evolution had developed gross structure in one case without altering metabolism, and it had produced diverse metabolism in the other case without altering gross structure.

The first attempt at developing a phylogenetic system based principally on physiological characters was made by Danish mycologist Sigurd Orla-Jensen (1870–1949) in a series of papers in 1909.[34] His phylogenetic system was rooted in part in theoretical considerations about the origin of life. He surmised that the first organisms on earth must have developed in an environment in which there was neither light nor organic matter. Because the only organisms known to be capable of that were the chemosynthetic bacteria, he thus assumed that they were the first organisms on earth; he based the rest of his system on an evaluation of increasing physiological complexity, reflecting nutrient requirements and the natural habitat of various groups of bacteria. He also used morphological characters, arguing that the postulated succession of bacterial types could also be derived from consideration of complexity in structure. His emphasis on physiology resulted in a new nomenclature that was simple and rational. Various genera were designated by names that concisely described the physiological and morphological properties of the organisms.

Though received with circumspection and criticism, Orla-Jensen's scheme was highly influential. Those who applauded it thought it was high time that the diverse physiological traits played an important role in classification, because that is how bacteria have attained their most remarkable diversity. As Yale bacteriologist Charles-Edward Winslow (1877–1957), editor-in-chief of the *Journal of Bacteriology* (1916–1944) commented in 1914, "There is as wide a difference in metabolism between pneumococci and the nitrifying bacteria as there is in structure between a liverwort and an oak."[35]

Still, Winslow and many others were as wary of physiological traits as Cohn had been. When comparing two organisms with the same physiological trait, how could one tell whether the trait in question was old in both organisms or whether one of the organisms had recently acquired the trait? One way to avoid the error of confusing convergent adaptive characters with those that indicate real community of descent was to use several independent characters that occur in correlation.[36] Whatever criticisms Orla-Jensen's system received, no one had advanced a better explanation for the gradual development of life on earth, and his system continued to be highly regarded by some until the mid-1940s. As Winslow commented, "I believe that no one who has thought seriously about bacterial relationships can study it carefully without feeling that it is by far the most successful attempt yet made at a real biological classification of the group and that future progress will probably consist in its modification and extension rather than in any profound reversal of its basic principles."[37]

Much of the opposition to Orla-Jensen's system arose from the threat that it posed to the stability of existing nomenclature.[38] Bacterial species were often shifted from one genus to another, and differences between species and genera were being split more and more finely. Old species were made into genera, and genera into families. These changes, informed by evolutionary speculation, caught the ire of those bacteriologists who wanted a classification and nomenclature to be stable. After all, for many scientists, that stability was the very reason for having a classification in the first place. Many bacteriologists had given up hope of a natural system, and others actively opposed constructing one, arguing that it was not necessary or even desirable.

Compromise and Confusion

Bacterial systematics remained in a confused state. R.E. Buchanan, a former president of the Society of American Bacteriologists, commented in 1925 that, "if science is to be defined as a system of *classified* knowledge, the subject of bacteriology is laboring under a serious handicap."[39] As Robert Breed noted in 1928, "A review of the literature will show that the most popular term that has been used to describe systematic bacteriology is 'chaos'; and this irrespective of the period of history under consideration."[40] The aims of bacterial systematics were clear: present a conception of the natural relationships of bacteria, provide a greater degree of stability for the names used for groups, and prevent nomenclatural confusion. There was power in naming as F.W. Andrewes remarked in 1930: "Solomon, we are told, knew the names of all the spirits, and, having their names, he held them subject to his will. The parable may be applied to bacteriology, as to many other sciences."[41] However, systematic bacteriology was comparable to the Tower of Babel.

Part of the problem was that bacteriologists looked at naming from different points of view. For phylogenetically minded bacteriologists, no classification was accurate or complete if it did not record or imply all the life phenomena of the organism, including its pedigree. However, the vast majority of those scientists concerned with microbial life were preoccupied with immediately practical problems, and most of them had been trained for their work from the standpoint of some practical art—medicine, veterinary science, sanitary engineering, or agriculture—rather than from the more general and fundamental standpoint of the biologist. Pathologists, hygienists, brewers, and chemists regarded the organism simply as an object to be named for convenience because it brought about certain changes in tissues, waters, and other media with which they were concerned.[42] They were interested in a reliable and stable scheme of classification that was useful, not one that was based on ever-changing phylogenetic understanding. The aims of the evolutionist were thus far removed from those who sought a reliable (artificial) system of classification for practical purposes.

Phylogeneticists and those who wanted a practical classification also looked for different things. The former looked for similarities and differences between genera and species, yet for practical purposes, the characters that would not rank as species characters by systematists could be of paramount importance. Consider the pathologists or medical bacteriologists treating patients by the administration of an antiserum. The pathologists and bacteriologists must know the extent of type specificity exhibited by the pathogen. Two bacteria may be morphologically and culturally identical and, for phylogenetic reasons, placed in the same species. However, one of them may be pathogenic, the other not, or one may differ in their immunity reactions. For example, two strains of meningococci may be isolated from two different cases of epidemic meningitis that appear to be identical under the microscope and in all artificial cultures. Yet an immune serum against one will not agglutinate the other, nor can a therapeutic serum against one be used successfully in the treatment of an infection by the other. These physiological effects may not be of sufficient importance to warrant creating a new species, but they are of considerable importance in the treatment of disease. Consider the industrial bacteriologists who attempt to manufacture vinegar by use of pure cultures of acetic acid bacteria, or solvents by the bacterial fermentation of carbohydrates. Although they might agree with a specific distinction between *Acetobacter aceti* and *Acetobacter rancens*, it is possible they would prefer classification on the basis of such properties as the ability to grow on media with high alcohol concentrations or to produce certain specified concentrations of acetic acid. Or the bacteriologists could point out that *Clostridium butylicium* is too broad a group to help them in selecting desired cultures for acetone or butyl-alcohol production.

It was difficult to maintain a stable classification because new forms were being continuously discovered. Thousands of "species" had been identified by the 1920s—only a tiny fraction of the number thought to exist because bacteriologists had only studied those habitats that were of some practical importance. Determining boundaries of various categories—species, genus, family—was confusing. For animals and plants this was less a problem. There was no difficulty in understanding that humans are a species. There is a gap between humans and the nearest other living species, the anthropoid apes, with no living intermediate types, no continuous graduation between the two. Even so, drawing lines between species was a matter settled by arbitration and compromise between authorities. Among bacteria, however, such gaps became less and less apparent, and the gradations between species were more subtle and more continuous. Although the species concept was hardly settled for plants and animals, it was obvious to some bacteriologists that "bacterial species" was an artificial division that did not actually exist.[43]

Despite the differences in criteria for bacterial classification arising from divergent interests, ideas, education, and training, it was evident that there must be developed and recognized some code of rules or laws for creating a single classification that would serve all ends. Such "Codes of Nomenclature" had been developed both in botany and in zoology and had been discussed at their respective international congresses, but bacteriology lacked both codes and international congresses. The Society of American Bacteriologists had been founded in 1899, both in protest to the "necessary but dangerous specialization" among bacteriologists and to bring together workers in all fields for a consideration of their problems in light of the underlying, unifying principles of bacteriology as a member of the group of the biologic sciences.[44]

In his presidential address of 1914, C.E.A. Winslow called for a court of appeal on matters of systematic bacteriology, a court to which all suggested classification, past and future, would be referred for official acceptance or rejection, in whole or in part. It would be based on an international committee, like that of botany.[45] Constructing an acceptable classification even of plants and animals so that different biologists would employ the same names for the same kinds of organisms, he argued, had always involved some sort of compromise. He called for an international organization for bacteriologist comparable to those maintained by botanists and zoologists to bring about the necessary taxonomic changes.[46] He suggested that fifteen bacteriologists from the principal scientific countries be invited to act as an international commission on the characterization and classification of bacterial types.[47]

However, there was still another problem. Medical concerns were leading bacteriology to become even less defined. With the discovery that many diseases of human animals were caused by "protozoa," bacteriologists tended to refer to themselves as "microbiologists" because in many cases the disease-causing microbes were best studied by methods developed in bacteriological laboratories. This led to further taxonomic muddle. It proved somewhat confusing to use the botanical code for the plant forms (bacteria) and the zoological code for the protozoa.[48]

The Society of American Bacteriologists (renamed in 1960 The American Society for Microbiology because of its broadened scope) subsequently appointed a committee to work out some sort of internationally acceptable classification that might be followed by all bacteriologists. The committee, chaired by Winslow, made a final report in 1920.[49] It agreed to follow the codes of both botanists and zoologists insofar as they might be applicable and appropriate. Accordingly, a committee of the Society of American Bacteriologists, headed by David Bergey (1845–1932), prepared a *Manual of Determinative Bacteriology* in 1923. The book became famous as *Bergey's Manual*, and it grew through many editions. As it did, it became increasingly more voluminous and complex. The first three editions were sponsored by the society, with the following statement on the flyleaf:

> Published at the direction of the Society. In publishing this Manual the Society of American bacteriologists disclaims any responsibility for the system of classification followed. The classification given has not been formally approved by the Society, and is in no sense *official* or *standard*.[50]

What resulted from the committee was at best a compromise between the most divergent ideas that had been expressed in the course of time. It was a merger, using all kinds of properties as it suited—morphological, physiological, nomenclatural, utilitarian, cultural, and pathogenic—to name groups and build up of an arbitrary classification system. *Bergey's Manual* only added fuel to the controversy over how physiological and morphological characteristics should be used in classification.

Realists versus Idealists: Return to Morphology

Although in botany and zoology only a natural, phylogenetic classification was considered satisfactory, bacteriologists continued to show a strong aversion to genealogy. Those who wished to have a natural classification subjected *Bergey's Manual* to fierce criticism, arguing that its inconsistent and arbitrary criteria for classification resulted in a profoundly confused order of things. The result was a complete lack of homology in the various groups.[51]

Disputants in the United States characterized their differences as being between "realists" and "idealists." Idealists wanted a natural classification that would serve all ends: All organisms would be grouped into a system that would provide both a plan of their family tree and a key to their identification. Members of the school at Delft, Albert Jan Kluyver and his former student Cornelis B. van Niel (1897–1985), who moved to Hopkins Marine laboratory in 1929, and van Niel's former student Roger Stanier (1916–1982) denounced *Bergey's Manual* as inadequate and effectively incompetent and admonished its editors for ignoring various criticisms.[52] Realists, however, argued that natural relationships were obscure and that a natural system would be highly speculative and constantly changing. As Robert S. Breed, one of the editorial board members of *Bergey's Manual*, commented in 1929: "Realistic workers have on their side been impatient with idealists who have introduced many . . . unjustified speculations regarding relationships between the various groups of bacteria."[53] He commented, "Why should we care whether one or the other type of character is more useful in defining a genus, or whether we should use one type of characters in defining a genus and another in defining a species?"[54]

Stanier and van Niel insisted in 1941 that the "realistic" attitude of pessimism was not justified, and that some relationships could be recognized and incorporated into a system of classification. Even granting that "the true course of evolution can never be known and that any phylogenetic system has to be based to some extent on hypothesis," there was still good reason to prefer an imperfect natural system to a pure empirical one. A phylogenetic system, they reasoned, could be altered and improved as new facts come to light, and its very weaknesses would point to the kind of experimental work necessary for improvement. However, they also argued that

> an empirical system was largely unmodifiable because the differential characters employed are arbitrarily chosen and usually cannot be altered to any great extent without disrupting the whole system. Its sole ostensible advantage is its greater immediate practical utility; but if the differential characters used are not mutually exclusive (and such mutual exclusiveness may be difficult to attain when the criteria employed are purely arbitrary) even this advantage disappears. The wide separation of closely related groups caused by the use of arbitrary differential characters naturally enough shocks "idealists," but when these characters make it impossible to tell with certainty in what order a given organism belongs, an empirical system loses its value even for 'realists."[55]

Bergey's Manual illustrated the weakness of the empirical approach, beginning with its definition of bacteria. Defining the bacteria so as to distinguish the group from other groups of microbes was a difficult task. Even so, Stanier and van Niel scoffed, "A more inadequate definition than that given by Bergey would be hard to conceive."[56]

Stanier and van Niel offered a new method for classification based on a rational weighting of morphological and physiological characters. They argued that morphological characters only should be used to distinguish larger units of classification (tribes and families), whereas both physiological and morphological traits should be used for genera. Like others before them, they cautioned that one had to be wary of using physiological characteristics "because of the wide range of adaptation which is manifested in bacteria." They also unearthed theoretical assumptions underlying Orla-Jensen's scheme, which pointed to foundational problems inherent in any phylogenetic system based primarily on physiology: "the chief stumbling block of drawing up a phylogenetic system on a primarily physiological basis is the necessity of making a large number of highly speculative assumptions as to what constitute primitive and advanced metabolic types."[57] Orla-Jensen had assumed that chemosynthetic bacteria were the most primitive group because they could live in complete absence of organic matter and hence were independent of other living forms (autotrophs). However, the ability of an organism to synthesize all its cellular constituents using carbon dioxide as the only carbon source would require a highly developed enzymatic apparatus, and it was hard to imagine how such an apparatus could have originated that by any mechanism in an inorganic world. Of course, there was the hypothesis of panspermia, which postulated that organisms on earth were transported here from another planet or solar system by one means or another. There were several objections to this hypothesis, however: such an organism would have had to withstand the heat of atmospheric entry, and during interstellar transit the germs would be exposed to ultraviolet irradiation of such intensity and for such periods of time that it seemed inconceivable that any germ could reach the earth without being killed. Even then, the problem of the origin of life would not be solved—it would merely be shifted from one part of the universe to another.

Orla-Jensen's assumption that organisms had arisen when the earth was devoid of organic matter was effectively challenged by Aleksandr Ivanovich Oparin (1894–1980) in his 1924 book *Origin of Life*, translated into English in 1938.[58] Oparin postulated that life did not originate in an environment devoid of organic matter, but that many organic compounds had emerged from strictly chemical reactions before organisms appeared. Without microbes to decompose them, they could have persisted for long periods of time, and as a consequence more and more complex molecules or molecular aggregates could have arisen through chemical interactions. With the emergence of such complexes, a fortuitous combination of circumstances might have yielded systems with properties of self-propagation. This system would be characterized by a minimum of synthetic ability, emerging gradually, step by step. As a consequence, the earliest living forms would have been heterotrophs; the development of autotrophism would be a later adaptation to an environment in which organic materials had become scarce through the activities of heterotrophs.[59]

This was easier to imagine than a model in which organisms appeared already equipped with a complete set of synthetic mechanisms, as implied in Orla-Jensen's phylogenetic system.[60] Furthermore, to construct another system based on increased metabolic complexity, using Oparin's model as a starting point, would be foolhardy, Stanier and van Niel argued, because André Lwoff had shown that many of the differences in power of synthesis that were exhibited in related groups were the result of losses. Therefore, an increase in synthetic ability may not necessarily reflect phylogenetic trends.[61]

In 1941, van Niel and Stanier upheld the scheme Kluyver and van Niel had proposed 5 years earlier—that only morphological properties be used for the demarcation of larger units, tribes, families, and so forth, whereas both morphological and biochemical characters would be used to define genera.[62] van Niel and Stanier insisted that there were enough discernable morphological differences to do the job: "Clearly paramount is the structure of the individual vegetative cell, including such points as the nature of the cell wall, the presence and location of chromatin material, the functional structures (e.g. locomotion), the method of cell division, and the shape of the cell. A closely allied character is the type of organization of cells into larger structures. In addition, the nature and structure of reproductive or resting cells or cell masses deserve due consideration."[63]

Like that of Orla-Jensen, Kluyver and van Niel's model was based on fundamental assumptions about an evolutionary trajectory toward increased complexity from "lower" to "higher" forms. The only difference was that the trend would be toward increased morphological complexity. Their primary assumption was that the simplest shape, the sphere, was the original shape of all bacteria. Evolution proceeded toward increasingly intricate form, toward more complex life cycles, "in the direction from unicellularity to multicellularity. The highest developmental stage in the group of spherical organisms is in all probability displayed by the cocci able to form endospores. . . . Endospore forming rods with pritrichous flagella present a higher stage of development in these groups . . . further development of these universally immotile bacteria can have given rise to the mycobacteria which apparently form the connecting link with the simpler actinomycetes."[64] Thus, Kluyver and van Niel argued that a rational and a more enduring natural classification should replace an arbitrary nomenclature.

A Failed Revolution

When van Niel returned to the issue of bacterial classification in 1946 at a famous Cold Spring Harbor Symposia on "Heredity and Variation in Microorganisms," his support for the scheme he and Kluyver had proposed 10 years earlier and his zeal for a natural system had dwindled.[65] He conceded that bacteriologists' "fragmentary knowledge of bacterial phylogeny is far from sufficient to construct anything like a complete system. Even for a general outline along phylogenetic lines, the available information is entirely inadequate. Much of this is, of course, the result of the paucity of characteristics, especially those of a developmental nature."[66] His paper had a tone of despair. He had painted himself into a corner. He insisted that, "we cannot yet use physiological or biochemical characters as a sound guide for the development of a 'natural system' of classification for bacteria." Only morphology reflected phylogeny. Yet, he admitted that "in the vast majority of cases," morphology "was not any more useful than physiology." Nonetheless, he avowed, "the search for a basis upon which a 'natural system' can be constructed must continue."[67] By 1955, van Niel, following Winogradsky, finally gave up all hope for a natural phylogeny. Winogradsky commented that the principle of phylogenetic classification "is impossible to apply to bacteria."[68] He reminded bacteriologists that the naming of things in *Bergey's Manual* of species, genera, tribes, families, and orders was only a facade. To avoid the delusion that it represented a natural ordering, Winogradsky and van Niel suggested using "biotypes" instead of "species" and using common names such as "sulfur bacteria," "photosynthetic bacteria," and "nitrogen-fixing bacteria" instead of Latin names, with their phylogenetic implications.[69]

Although van Niel could not report on progress in the phylogenetic classification of bacteria at the Cold Spring Harbor Symposia, that meeting marked a great revolution in understanding mechanisms of bacterial heredity and variation. Bacteria, long thought not to contain genes and to reproduce solely asexually by division, were shown to have genes that recombined during conjugation.

The Second Era

Birth of Bacterial Genetics

During the 1930s and early 1940s, the architects of "the evolutionary synthesis" united Mendelian laws of heredity and Darwinian selection theory. Gene mutations and recombination within species provided the fuel for evolution by natural selection. Although the principles of the evolutionary synthesis were assumed to extend to nucleated protists, bacteria were projected to remain a (rather large) exception. The conventional wisdom was that bacteria adapt to their chemical and physical environment in a direct "Lamarckian" fashion. Bacteriologists insisted that changes of this sort were very familiar to them, as, for example, the increase in virulence on passage through susceptible animals or in the converse process of attenuation by exposure to unfavorable environmental conditions, as in the production of vaccines for anthrax by Pasteur.[70]

There was also a reported polymorphism of bacteria, a supposed phenomenon called "cyclogeny," accounts of which had persisted for decades.[71] Variants of the old idea of pleomorphism were widespread between 1916 and 1935. Cyclogeny was not a return to the extreme views about the existence of only one species of bacteria, but it contradicted assumptions about simplicity of the bacterial life cycle (monomorphism) with reported demonstrations that bacteria had life cycle complexities comparable to the "higher fungi." Then, in the 1920s, there were further reports that diverse kinds of bacteria could pass through a submicroscopic stage that was so minute as to pass through bacterial filters, and were later able to reproduce the original type of cell from which they were derived. Stanford bacteriologist W.H. Manwaring commented in 1934:

> About the only conventional law of genetics and organic evolution that is not definitely challenged by current bacteriologists is the nineteenth century denial of the possibility of spontaneous generation of a bacterial cell. Even this is questioned by certain recent theorists in their hypothetical transformation of certain normal enzymes into "pathogenic genes"or "filterable viruses", and in their apparently successful synthesis of "Twort genes" by the chemical oxidation of certain heat-sterilized organic products.
>
> Whether or not future refinements in immuno-chemical technique can or will bridge the gap between the apparent Lamarckian world of bacteriology and the presumptive Darwinian world of higher biological science is beyond current prophecy.[72]

That bacterial species pass sequentially through specific stages of development relied on the claim that a number of physiological, serological, morphological, and cultural traits varied together. Researchers later, in the 1940s, demonstrated that the bacterial characters of a culture were capable of independent variation and were not regenerated by submicroscopic filtrable particles.[73]

That bacteria reproduce solely asexually by fission was the most common assumption. In *The Evolution of Genetic Systems*, in 1939, British cytogeneticist C.D. Darlington referred to "asexual bacteria without gene recombination"[74] and "genes which are still undifferentiated in viruses and bacteria." Julian Huxley summarized what everyone "knew" about bacteria in 1942:

> Bacteria (and *a fortiori* viruses if they can be considered to be true organisms), in spite of occasional reports of a sexual cycle, appear to be not only wholly asexual but pre-mitotic. Their hereditary constitution is not differentiated into specialized parts with different func-

tions. They have no genes in the sense of accurately quantized portions of hereditary substances; and therefore they have no need for accurate division of the genetic system which is accomplished by mitosis. The entire organism appears to function as soma and germplasm, and evolution must be a matter of alteration in the reaction-system as a whole. That occasional "mutations" occur we know, but there is no ground for supposing that they are similar in nature to those of higher organisms, nor since they are usually reversible according to conditions, that they play the same part in evolution. We must, in fact, expect that the processes of variation and evolution in bacteria are quite different from the corresponding processes in multicellular organisms. But their secret has not yet been unraveled.[75]

In 1946, Joshua Lederberg (b. 1925) launched a new field of bacterial genetics when he experimentally demonstrated genetic recombination in *Escherichia coli*.[76] He and Edward Tatum announced the news at the Cold Spring Harbor meeting that year[77]: Genes could be transmitted to bacteria by viruses as well. The study of bacteria and its viruses was developed by Max Delbrück, Alfred Hershey, and Savador Luria as well as Lederberg and his collaborators in the United States and by André Lwoff, Elie Wollman, François Jacob, and Jacques Monod in France. Hershey and Delbrück reported on genetic recombination in bacteriophage in 1946. In the early 1950s, Lwoff synthesized the work to clarify the mysterious phenomenon of lysogeny, by which the genome of a bacteriophage is integrated into the chromosome of its bacterial host—the genome and related infecting phages are prevented from reproducing—a subject of considerable importance for immunity.[78] The prophage genome remains integrated within the lysogenic bacterium's chromosome unless the bacterium is exposed to certain stimuli, when the prophage genome is excised from the bacterial chromosome and viruses are reproduced and kill their host.

Bacterial viruses were studied as "naked genes." The very properties that had hitherto excluded bacteria from classical genetics made them valuable to emerging molecular biology. Small in size and rapid in growth, bacteria double in number every 30 minutes (some *Vibrios* are known today to do it in 9 minutes), and one could grow enormous populations of billions of microbes in a few cubic centimeters. Studies of genetic mutations and mechanisms of genetic regulation in bacteria, led by Jacob and Monod during the 1950s, indicated that the previous Lamarckian interpretations resulted in part from mixing up reversible physiological adaptations (induced enzyme formation) with more or less irreversible hereditary "mutations."[79] However, bacterial geneticists did show that bacteria had various mechanisms of heredity not known to be operational in plants and animals. Bacteria possessed several mechanisms for transmitting and acquiring genes between taxonomic groups.

Conjugation: Bacteria possessed genes arranged along structures similar to the chromosomes of flies, guinea pigs, or humans. In addition to a main circular "chromosome," bacteria possessed smaller rings of genes, which Lederberg named "plasmids" in 1952. In bacterial conjugation, genetic material of the "male" plasmids, and sometimes small to large parts of the main chromosome, are transferred to the "female" recipient, and some genes may recombine with the female's chromosome.
Transformations: Certain kinds of bacteria can absorb and incorporate into their own chromosome the DNA released by dead bacteria.
Transduction: Viruses or bacteriophage can act as vehicles to transfer genes between bacteria.[80]

Lederberg extended the concept of heredity in 1952 to embrace what he called "infective heredity." He coined the term "plasmid" in the context of debates over the scope and significance of cytoplasmic heredity and over the concept of "plasmagenes." The debates go back to the first decade of the twentieth century. While geneticists, led by T.H. Morgan in the United States upheld the exclusive or predominant role of nuclear chromosomal genes in heredity, others, especially embryologists and European geneticists, argued that the cytoplasm also played a role in heredity. During the 1940s, evidence for cytoplasmic inheritance was revived in the United States with evidence of "plasmagenes" in microbes. Some of the cytoplasmic particles such as *kappa* in *Paramecium aurelia* were showed to be infectious under certain laboratory conditions. On this basis, leading classical geneticists including H.J. Muller dismissed cytoplasmic genetic agents as parasites of little significance for heredity.[81]

Lederberg introduced "plasmid"as a generic term for any cytoplasmic genetic element, whether infectious or not. He suggested that many genes may have been incorporated into bacterial cells and that infectious heredity is a macromechanism of heredity comparable to hybridization in its effects. He also reconsidered the notion that fully integrated organelles such as mitochondria and chloroplasts might have evolved as symbionts—an idea that had lingered on the margins of biology since the nineteenth century.[82]

Bacteria were biologized from a genetic perspective, and their structural organization was becoming well understood, but their mechanisms of heredity were investigated primarily for understanding genes and their regulation. They were studied in reference to the genetics of plants and animals to provide a unifying concept of heredity at the molecular level. Transformations and viral infections were important for identifying the genetic material as DNA. These mechanisms were also highly valued for the subsequent development of biotechnology, but their general significance for bacterial evolution was not recognized until almost 50 years later. From a phylogenetic point of view, little more was known about bacteria than in the days of Pasteur.

Viruses, Prokaryotes, and Eukaryotes

The molecular biology of the gene and the deployment of the electron microscope (increasing resolving power a thousand-fold over the light microscope) after the Second World War led to refinement in the concept of a bacterium, both biochemically morphologically. That articulation began with a new concept of the virus. The word virus (latin for slimy liquid, poison, offensive odor or taste) had changed its meaning over the centuries. In the sixteenth century, it was used in its original sense to refer to poison or venom, as in the story of Cleopatra pouring the virus of an asp into a wound made in her arm by her own teeth. In the eighteenth century, one referred to virus in the sense of contagious pus. In the nineteenth century, "viruses" were considered as microbes. Bacteria were isolated using filters, permeable to toxins but impermeable to the bacteria, but there were anomalies. Some infectious agents were so small that they could pass through a bacterial filter. These were called "filterable viruses." Other small, obligate parasitic bacteria of the rickettsia type, barely resolvable by the light microscope, were often thought to be transitional between the filterable virus and the typical bacterium. Thus, bacteria were thought to range from the size of some algae to the size of filterable viruses. As late as 1958, *Bergey's Manual* suggested a new kingdom, "protophytes," which would include both bacteria and viruses.[83]

A year earlier, however, a major distinction between viruses and bacteria had been articulated, based on molecular structure and physiology, in a now-classic paper by André

Lwoff: "The Concept of Virus." The virus contained either RNA or DNA enclosed in a coat of protein, and it possessed few if any enzymes, except those concerned with attachment to and penetration into the host cell. The virus was not organized like a cell, and it did not reproduce by division like a cell. Its replication occurred only within a susceptible cell, which always contains both DNA and RNA—an array of different proteins endowed with enzymatic functions that are mainly concerned with the generation of ATP and the synthesis of varied organic constituents of the cell from chemical compounds in the environment. "Viruses should be treated as viruses," Lwoff concluded, "because viruses are viruses."[84] Only the word "virulence," for poisonousness, was retained for both bacterial and viral diseases. There were no biological entities that could properly be described as transitional between a virus and a cellular organism, and the differences between them were of such a nature that it was indeed difficult to visualize any kind of intermediate organization.

Five years later, Stanier and van Niel wrote a sister paper to Lwoff's, entitled, "The Concept of a Bacterium." Although they were no longer willing to defend the classification scheme they had defended in 1941, Stainer and van Niel were able to offer a definition of bacteria that would distinguish them both from viruses and from other protists. "Any good biologist finds it intellectually distressing to devote his life to the study of a group that cannot be readily and satisfactorily defined in biological terms; and the abiding intellectual scandal of bacteriology has been the absence of a clear concept of a bacterium."[85] Certainly many microscopists since the days of Haeckel had recognized a difference between bacteria (and blue-green algae), which lacked a true nucleus, and other protists that contained a nucleus. In the early 1920s, Lwoff's mentor Edouard Chatton (1883–1947), at the University of Strasbourg, referred to Cyanophyceae, Bacteriacae, and Spirochaetaceae as "procaryotes" (Greek before karyon or nucleus) to distinguish them from the protozoa, which he called "eucaryotes" (Greek: eu-karyon; true nucleus)."[86]

At Lwoff's suggestion, Stanier and van Niel adopted these terms, asserting that the distinctive property of bacteria and blue-green algae is the procaryotic nature of their cells. "The principle distinguishing features of the procaryotic cell are: 1 absence of internal membranes which separate the resting nucleus from the cytoplasm, and isolate the enzymatic machinery of photosynthesis and of respiration in specific organelles; 2 nuclear division by fission, not by mitosis, a character possibly related to the presence of a single structure which carries all the genetic information of the cell; and 3 the presence of a cell wall which contains a specific mucopeptide as its strengthening element."[87] In effect the bacteria were defined negatively in terms of what they lacked. Eukaryotes had a membrane-bound nucleus, a cytoskeleton, an intricate system of internal membranes, mitochondria that perform respiration, and in the case of plants, chloroplasts. Bacteria (prokaryotes) were smaller and lacked all of these structures.

Just as there were no transitional forms between virus and bacteria, Stanier and his collaborators insisted, there were no transitional forms between bacteria and all other organisms. Stanier, Michael Doudoroff, and Edward Adelberg declared in their famed book *The Microbial World* of 1963 that, "In fact, this basic divergence in cellular structure, which separates the bacteria and blue-green algae from all other cellular organisms, represents the greatest single evolutionary discontinuity to be found in the present-day world."[88]

During the 1950s, palaeobiologists extended the fossil record back to indicate the presence of eukaryotic fossils in rock that was some 2000 million years old.

Paleontologists who marveled at the great burst in plant and animal diversity of 500 million years ago referred to it as "the Cambrian explosion," but those who studied cellular organization insisted that the real "big bang" of biology occurred some 1500 million years earlier, when the eukaryote arose. With its membrane-bound nucleus and all the associated features, such as mitosis, meiosis, and multiple chromosomes to package up to tens of thousands of genes per cell, the eukaryote provided the organismic conditions for the differentiation of tissues, organs, and organ systems of plants and animals.

Stanier, Doudoroff and Adelberg asserted that there was a common origin for bacteria in the remote evolutionary past, at the same time they insisted that bacteria simply could not be arranged phylogenetically. Only four principle groups were able to be discerned: blue-green algae, myxobacteria, spirochetes, and eubacteria; "Beyond this point, however, any systematic attempt to construct a detailed scheme of natural relationships becomes the purest speculation, completely unsupported by any sort of evidence."[89] There was only one possible conclusion: "the ultimate scientific goal of biological classification cannot be achieved in the case of bacteria."[90]

Symbiosis Revisited

Beginning in 1967, Lynn Margulis (b. 1938) promoted the argument that the large gap separating bacteria and eukaryotes could have occurred by leaps and bounds, by a series of events in which bacteria invaded a primitive "amoeboid microbe"[91] She suggested that mitochondria had been first acquired as food by predatory microbes, but they resisted digestion and proved to be of benefit to their host because the primitive atmosphere had increased in oxygen (with the advent of photosynthetic cyanobacteria), which otherwise would be toxic to their hosts; the engulfed mitochondrial ancestor provided ATP to its protist host through respiration. As Margulis put it much later, "the release of ATP to its host would be analogous to throwing cash into the streets."[92] Chloroplasts would have originated subsequently when a protist engulfed cyanobacteria. Their selective advantage to their host is obvious: Protists that once needed a constant food supply henceforth could thrive on nothing more than light together with air, water, and a few dissolved minerals.

By the 1960, the electron microscope revealed the structural similarities between mitochondria, chloroplasts, and bacteria. DNA had been discovered in these organelles; the organelle genomes were circular, as is bacterial DNA, and both organelles possessed ribosomes and a full protein synthesis apparatus.[93] Genetic research programs had emerged, dissecting chloroplasts and mitochondrial genomes, led by Ruth Sager, Nicholas Gillham, and others.[94] Genetic studies of *kappa* in Paramecium showed how chloroplasts and mitochondria could have originated as bacterial symbionts. Margulis also extended the reach of symbiosis to account for the origin of centrioles/kinetosomes and, therefore, for the origin of cell motility and mitosis. Since the nineteenth century, centrioles had been thought to divide by fission, to spin out spindles, and to play a crucial role in cell division by mitosis in many kinds of organisms. This idea was strengthened in the late 1960s with suggestive evidence, at that time, that centrioles might possess their own DNA (this evidence was later refuted).

Although the evidence had changed, the image of cytoplasmic organelles as quasi-independent organisms had been discussed since the late nineteenth century.[95] That chloroplasts originated as symbionts had been suggested by Andreas Schimper in 1883, discussed by Haeckel in 1905, and was developed most prominently by Russian botanist Constantin Merezhkowsky (1855–1921) at Kazan University in 1905. In 1909/1910, Merezhkowsky coined the word "symbiogenesis" for the synthesis of new organisms by symbiosis.[96] In Geneva in 1918, he wrote an elaborate paper, which he considered his most important work, arguing that chloroplasts had originated from symbiotic blue-green algae (cyanobacteria) in the remote past. He also maintained, as had others before him, that nucleus and cytoplasm had originated as a symbiosis of two different kinds of microbes.[97]

In his book *Les Symbiotes* (1918), French biologist Paul Portier (1866–1962) at the Institut Océanographique de Monaco, developed an elaborate theory of symbiosis as a fundamental aspect of life. He argued that mitochondria originated as symbionts and that they had been transformed over eons by their intracellular existence. Thus, he declared that, "All living beings, all animals from Amoeba to Man, all plants from Cryptogams to Dicotyledons are constituted by an association, the *'emboitement'* of two different beings. Each living cell contains in its cytoplasm formations which histologists call 'mitochondria'. These organelles are, for me, nothing other than symbiotic bacteria, which I call 'symbiotes.'"[98]

In the mid-1920s, French-Canadian Félix d'Herelle (1873–1949) discussed the perpetuation of mixed cultures of bacteria and their viruses (he named them "bacteriophages") in terms of symbiosis.[99] He referred to the bacteria that harbor viruses ("lysogenic bacteria") as "microlichens." The morphological and physiological changes resulting from symbiosis led d'Herelle to assert, in 1926, that "symbiosis is in large measure responsible for evolution."[100] The idea of virus-harboring bacteria was rejected as unbelievable for decades before it was revitalized by bacterial geneticists of the 1950s.[101]

In the United States, during the 1920s, Ivan Wallin (1883–1969) at the University of Colorado argued that mitochondria originated as symbiotic bacteria. He also emphasized the importance of bacterial symbiosis as a generator of new tissues and new organs. In his book *Symbionticism and the Origin of Species* (1927), Wallin proposed that bacterial symbionts were the source of new genes and the primary mechanism for the origin of species. He also claimed that he had cultured mitochondria to prove their actual bacterial nature.[102] Wallin's book met with virtual silence in the United States.[103]

Although these ideas are often seen in light of today's views of organellar symbiosis, we should also emphases the differences: Merezhkowsky denied that mitochondria were symbionts, and those who promoted the idea that mitochondria were symbionts did so with the widely held belief in bacterial pleomorphism. When Portier argued that mitochondria were symbionts, he argued that there was one symbiote in nature with a very extensive morphological and physiological plasticity. When Wallin argued that mitochondria were symbiotic bacteria, he argued that they were also the source of other organelles, including chloroplasts and the organelles of motility centriole/kinetosomes. Similarly, Félix d'Herelle argued that there was only one kind of bacteriophage, capable of wide variation based on adaptive powers. We should also note that for d'Herelle, the bacteriophage, or filterable virus, was an organism, and as much as an infective organism in bacteria, as was the bacteria *Cholera vibrio* within their human hosts.[104]

No matter how they were conceived, however, the evolutionary effects of interspecies integration resulting from microbial infections remained close to the margins of "polite biological society." A series of disciplinary aims and doctrines confronted the study of symbiosis in evolution.[105] The emphasis on conflict and competition in nature, which many critics argued then and now, is merely a reflection of dominant views of human social progress. The overwhelming interest in the disease-causing effects of some microbes means that most biologists were concerned with killing the microbes or with using them as technologies for chemistry and genetics rather than with understanding their evolution and ecology. Studies of hereditary symbiosis conflicted with the doctrine that the chromosomes in the cell nucleus were the predominant—if not exclusive—vehicles of heredity. A view that emphasized the prevalence of heredity symbiosis confronted the evolutionary synthesis based on Mendelian gene mutations and recombination as the basis of evolution, and challenged the idea that the mechanisms of macroevolution were the same as those for microevolution. Finally, studies of the role of symbiosis in cell origins struggled under the charge of speculation and metascience.

Crucial evidence was also lacking in the 1960s and 1970s. One way to demonstrate that an organelle was a symbiont was to culture it outside the cell, in a test tube or on a petri dish. Initially, Margulis imagined, as had Wallin before her, that biologists might indeed learn to culture chloroplasts, mitochondria, and centrioles. However, it soon became evident that these organelles were highly integrated into the nuclear genetic system: Only a small fraction of the genes needed for mitochondrial and chloroplast functions were actually located in the organelles themselves.[106] This integration lent support to the alternative theory that these organelles had arisen by "direct filiation"; that is, they had developed gradually from within the nucleated cell.

The symbiotic theory was underdetermined by data, and leading cell biologists argued that theories about eukaryotic cell origins were unscientific because they could not be proven. Roger Stanier spoke for many in 1970 when he commented that, "Evolutionary speculation constitutes a kind of metascience, which has the same fascination for some biologists that metaphysical speculation possessed for some medieval scholastics. It can be considered a relatively harmless habit, like eating peanuts, unless it assumes the form of an obsession; then it becomes a vice."[107] This assertion was short-lived, however. By the end of that decade, the field of "molecular evolution" had emerged. One could reconstruct bacterial phylogeny by comparative molecular morphology and extend that knowledge to investigate symbiosis and the origins of eukaryotes.

The Third Era

Molecular Evolution

Comparative molecular morphology for taxonomic purposes was predicted by Francis Crick (b. 1916) in 1958, a few years before the genetic code was cracked:

> Biologists should realize that before long we shall have a subject which might be called "protein taxonomy"—the study of amino acid sequences of proteins of an organism and the comparison of them between species. It can be argued that these sequences are the most delicate expression possible of the phenotype of an organism and that vast amounts of evolutionary information may be hidden away within them.[108]

The field of molecular evolution began with methods developed by another famed British chemist, Frederick Sanger (b. 1918). He succeeded in developing new techniques for amino acid sequencing and used them to deduce the complete sequence of insulin, for which he was awarded the Nobel Prize in chemistry in 1958.[109] In 1980, he shared half a Nobel Prize in chemistry with Walter Gilbert "for their contributions concerning the determination of base sequences in nucleic acids."

In 1965, Linus Pauling (1901–1994) and Emile Zuckerkandl articulated how phylogenies might be reconstructed from comparisons of molecular structures.[110] Instead of comparative anatomy and physiology, one could base family trees on differences in the order or sequence of the building blocks in selected genes as well as proteins. The approach was eminently logical. Individual genes comprising unique sequences of the nucleotides adenine, guanine, cytosine, and thymine (AGCT) would typically serve as the blueprints for the primary structure of proteins, which consist of strings of specifically ordered amino acids. As genes mutate, the simplest change would be to replace one base for another; for example, replacing the G in the sequence ". . . AAG . . ." to C, yielding ". . . AAC . . ," which in turn could change the amino acid glutamic acid (AAG) to aspartic acid (AAC) at a certain position in some protein. Some of these small changes would have little effect, but others could drastically change the encoded protein: Small changes can have large effects. (Consider, e.g., that DNA sequences of humans differ by only 0.1%, and between humans and chimpanzees they differ only by 2%).

Genetic mutations that either have no effect or that improve protein function would accumulate over time. As two species diverge from an ancestor, the sequences of the genes they share also diverge, and as time advances, the genetic divergence will increase. One could therefore reconstruct the evolutionary past of species and make phylogenetic trees by assessing the sequence divergence of genes or of proteins isolated from those organisms.[111] Pauling and Zuckerkandl thus introduced what they called "the molecular clock." Their idea was of a rate constancy at the molecular level: that changes in the amino acid sequence of a protein from different species should be "approximately proportional in number to evolutionary time."[112] In other words, molecules may not evolve in the same (irregular) way that the morphological features of an organism did.

This was not the first announcement of molecular evolution, nor of the idea behind a molecular clock. In 1963, Emmanuel Margoliash (b. 1920) and his colleagues compared similarities and differences in amino acid sequences of cytochrome *c* molecules from horses, humans, pigs, rabbits, chickens, tuna, and baker's yeast to infer phylogenetic relationships.[113] Zuckerkandl and Pauling had also pioneered the use of amino acid sequence comparisons to infer evolutionary relationships in primate phylogeny with data from hemoglobin sequences, but their paper of 1965 crystallized the idea of molecular evolution for many who entered this field.

In light of the molecular revolution, Stanier had a slight change of heart about the possibility of bacterial phylogeny in 1971, when he commented that

> we have at our disposal a variety of methods for ascertaining (within certain limits) *relationships* among the bacteria; and that where relationship can be firmly established, it affords a more satisfactory basis for the construction of taxa than does mere resemblance. As the philosopher G.C. Lichtenberg remarked 200 years ago, there is significant difference between *still* believing something and believing it *again*. It would be obtuse still to believe in the desirability of basing bacterial classification on evolutionary considerations. However, there may be solid grounds for believing it again, in the new intellectual and experimental climate which has been produced by the molecular biological revolution.[114]

Techniques for sequencing RNA and DNA dramatically improved in the 1970s and 1980s.[115] To compare the nucleotide sequences of genes, one needed the means to clone DNA; that is, to make many copies of sequences from minute samples. The invention of the polymerase chain reaction (PCR) solved this problem. PCR revolutionized many aspects of biology. In criminal investigations, DNA "fingerprints" could be prepared from cells in a tiny speck of dried blood or at the base of a single human hair. Gene amplification was also crucial to the development of the human genome project and its promised new era of gene therapy. Sequencing to understand microbial evolutionary history developed in parallel, but at a comparatively modest scale.

During the 1970s, Peter Sneath and his colleagues in the United Kingdom called for a classification of bacteria based on the computer-assisted numerical taxonomy he had developed since the 1950s, and using new molecular evidence based on "nucleic acid pairing, protein sequences, and nucleic acid sequences."[116] Numerical taxonomy was the most widely discussed approach to the classification in microbiology textbooks of the 1960s and 1970s.[117]

Evolution Revolution

A revolution in the study of bacterial evolution was launched by Carl Woese (b. 1928) at the University of Illinois (Urbana). He developed an empirical framework for the natural classification of microbes based on comparisons of ribosomal RNA sequences. In doing so, he offered a radical revision of the history of life on earth. Educated in biophysics at Yale, Woese had two complementary interests: comparing genetic systems to unravel microbial genealogies, and using genealogies to understand the evolution of the genetic system itself.[118] His aim was to understand how the complex mechanism for translating nucleic acids into the amino acid sequences of proteins had evolved. Chemists led by Stanley Miller's experiments of 1953 showed how certain amino acids might have been synthesized in an abiotic soup billions of years ago. How the amino acid sequences of proteins and the nucleotide sequences of DNA and RNA, which specify those sequences, had evolved and came together in the appropriate way to make a functional protein was more difficult to fathom. How did the codons evolve for specific amino acids, and how did they come to be arranged in such a manner to make the primary structure of a protein? This problem, the evolution of the genetic code, interested Woese.

Research on the origins of life increased dramatically when the National Aeronautics and Space Administration (NASA) became a major sponsor in the 1960s. Its original charter included a mission for the agency to search for extraterrestrial life, or to at least understand how life might have arisen on earth. Its funding was also crucial for many of the major advances in microbial evolutionary biology.

Francis Crick and others had suggested that the evolution of the code could have been some sort of historical quirk, a "frozen accident." In other words, it evolved merely because the cell had a need to have a relationship between nucleic acid and protein structure, not because there were relationships between nucleic acids and amino acids that forced the genetic code to evolve.[119] However, Woese suspected that the translation machinery had to have evolved in steps, with selection acting on the mechanism's speed and accuracy,[120] and this is what led him to microbial phylogeny. He supposed that one might be able to follow the translation machinery's early evolution, before cells reached their present sophisticated complexity. To do so meant that one had to construct a universal evolutionary framework, a deep phylogeny, a universal tree that would, in effect, encompass all organisms.[121] The universal tree would therefore be far more than the ultimate ordering of life on Earth; it held the secret to its existence as well. Thus, Woese set out with the hope of tracing cell life back to universal ancestors, "the progenote," which might not possess a modern translation machinery.

The first step was to construct a natural classification of bacteria; one that ordered them in terms of their evolutionary history, their genealogy. Not surprisingly, he focused on ribosomes, more particularly ribosomal RNAs (rRNAs). All cells need rRNAs to construct proteins, and therefore their similarities and differences could be used to track every lineage of life from bacteria to elephants. Ribosomal RNA is also abundant in cells, so that it was easy to extract. Ribosomes served at the core of an organism, and because they interacted with at least 100 proteins, Woese suspected that their molecular sequences would change so slowly that the sequences would hardly differ between species. Ribosomes would thus be among the most "conserved" elements in all organisms, and would therefore make excellent recorders of life's long evolutionary past. Ribosomes are composed of two pieces or subunits, with a smaller one slightly cupped inside a larger one. Woese chose to compare sequences of the small subunit RNA SSU rRNA. He believed that SSU rRNA sequences (or more precisely the genes encoding them) would change so slowly over evolutionary time that they would retain traces of ancestral patterns from billions of years ago at the deep roots of the phylogenetic tree. SSU rRNA would thus serve as a "universal molecular chronometer."

His methods were at first indirect and tedious,[122] but by the mid 1970s, Woese and his collaborators sequenced the SSU rRNA from about sixty kinds of bacteria and arranged them by genetic similarity.[123] Their results contradicted the standard classification based on morphological similarities of bacteria. *Bergey's Manual* distinguished the gliding bacteria, the sheathed bacteria, the appendaged bacteria, the spiral and curved bacteria, the rickettsias, *Flavobacterium*, and *Pseudomonas*, but Woese argued that these groups had no biological or evolutionary meaning; they were really paraphyletic, or polyphyletic; that is, they were not genealogically coherent groups.

By the late 1980s, the study of microbial phylogeny by rRNA sequences attracted many biologists, who classified several thousands of bacterial "species" so as to sketch an outline of a universal tree of life. With a universal evolutionary tree, biologists could begin to understand bacteria, as they do the rest of life, as organisms with histories and evolutionary relationships to one another and to all other organisms. Molecular phylogenies based on ribosomal RNAs were also applied to animals and to protists,[124] but no studies caused more controversy and attracted more interest than those of Woese and his colleagues.

A Trilogy of Life

In 1977, Woese and George Fox (working on methanogens in collaboration with Ralph Wolfe) announced that they had discovered a new form of life: a group of bacteria-like organisms that was genetically and historically very different. A student of David Nanney's sugggested the name "archaebacteria" to distinguish the newly discovered group of organisms from true bacteria or eubacteria.[125] The methane-generating organisms, methanogens, the first organisms they assigned to the archaebacteria, suggested the antiquity of the group. As Woese and Fox wrote in 1977, "The apparent antiquity of the methanogenic phenotype plus the fact that it seems well suited to the type of environment presumed to exist on earth 3–4 billion years ago lead us tentatively to name this urkingdom the archaebacteria. Whether or not other biochemically distinct phenotypes exist in this kingdom is clearly an important question upon which may turn our concept of the nature and ancestry of the first prokaryotes."[126]

The archaebacteria had other traits in common. They lived in extreme environments. Woese and colleagues identified as archaebacteria the salt-loving halophiles found in brines five times as salty as the oceans, and the thermophiles found in geothermal environments that would cook other organisms, and they found archaebacteria in anaerobic habitats where even trace amounts of oxygen prove lethal. Still other phenotypic features corroborated the conclusion from the RNA data about the uniqueness of the group. Their cell membranes are made up of unique lipids that are quite distinct in their physical properties, and the structures of the proteins responsible for several crucial cellular processes such as transcription and translation are different from those of their counterparts in bacteria.

For the press and the public, the discovery of the archaebacteria was a momentous event: It touched on the age old concern of where we came from. As the first organisms, however, this "universal ancestor" contradicted biochemists' assumptions about the conditions in which life first emerged on earth. The leading theory had long held that life began when lightning activated molecules in the atmosphere, which then reacted chemically with one another. The atmosphere then deposited those compounds in the oceans, where they continued to react to produce a warm soup of organic molecules. Darwin had said little about the origins of life, except this famous note in a letter to Joseph Hooker in 1871, "But if (and oh! what a big if!) we could conceive in some warm little pond with all sorts of ammonia and phosphoric salts, light, heat, electricity and etc., present that a protein compound was chemically formed, ready to undergo still more complex changes."[127] The SSU rRNA tree seemed to point to a strikingly different theoretical geochemical context, one that was emerging from other research and theorizing of the late 1970s. If the first organisms were archaebacterial-like creatures, instead of evolving in a mild soup of organic molecules, they may have been born in boiling, sulfurous pools or hot, mineral-laden, deep-sea volcanic vents.

Comparative studies of rRNA sequences also offered a novel conceptual scheme for the first billion years of life on earth. Formerly, it had been supposed that life's history was straightforward and progressive: from simple bacteria, to more complex bacteria to cells with nuclei, from which came plants and animals. However, the rRNA phylogenies indicated that life divided into major independent lineages much sooner than biologists had ever imagined. The stunning implication of this branching was that the prokaryotic world was richly diversified beyond what anyone had imagined. Plants and animals were but recent twigs on what amounted to a great microbial tree of life.

The postulation of a tripartite division of life, eubacteria, archaebacteria, and eukaryotes, contradicted the established belief in the basic dichotomy in the living world championed by van Niel, Stanier, and their colleagues: the dichotomy of the eukaryote and the prokaryote. By the 1970s, biologists, led by Robert H. Whittaker and Margulis, generally agreed that eukaryotes embraced four kingdoms: Planta, Animalia, Protista, and Fungi. Prokaryotes were classified as a fifth kingdom.[128] As Woese and his collaborators saw it, the prokaryote–eukaryote dichotomy had no phylogenetic meaning, and furthermore, it was a hindrance to understanding bacterial evolution.

The rRNA approach to bacterial phylogeny, and the three-domain proposal, was developed by many biologists including Otto Kandler and Wolfram Zillig and Kandler's former students, Karl Stetter and Karl-Heinz Schleifer in Germany, and Mitchel Sogin, Gary Olsen, James Lake, and Norman Pace in the United States. In the early 1980s, Linda Bonen, then Woese's technician, transferred the technology to Canada, where programs emerged led by Ford Doolittle and Michael Gray, who developed microbial phylogeny as a special program within the Canadian Institute for Advanced Research. The work on SSU rRNA led to an upheaval in bacterial systematics and a major revision of texts in regard to the universal genealogical tree.

Microbes were largely ignored by most biologists and virtually unknown to the public except in contexts of disease and rot, or bread, cheese, beer, wine, and biowarfare. Biology textbooks still teach biodiversity almost exclusively in terms of animals and plants; insects usually top the count of species, with about a half-million described to date. Plants and animals obviously showed far greater and more elaborate morphological differences, but bacteria were far more diverse biochemically. Those who worked on the evolution and phylogeny of Eubacteria and Archaebacteria suggested that they possessed greater biological diversity than plants and animals combined.[129] As Pace remarked, within one insect species one can find hundreds or thousands of distinct microbial "species."[130] A handful of soil contains billions of them—so many different types that accurate numbers remained unknown. Most life in the ocean is microbial. Bacteria can live in an incredible variety of conditions from well below freezing to above the normal boiling temperature of water. Extreme halophiles thrive in brines so saturated that they would pickle other life. Other microbes live in the deep border of the trench at the bottom of the Red Sea in hot saline loaded with toxic heavy metal ions. They are also found growing in oil deposits, deep underground. Microbial researchers argued that they had barely scratched the surface of microbial diversity. The entire surface of this planet down to a depth of at least several kilometers may be a habitat for eubacteria and archaebacteria. Bacteria, it was announced in 1998, constituted the greatest biomass on earth.[131]

There were technical problems of studying bacterial diversity. Technically, knowledge of microorganisms (and their niches) depends mainly on studies of pure cultures in the laboratory. However, those who studied microbial diversity estimated that more than 99% of the organisms seen microscopically could so far not be cultivated by routine techniques.[132] Beginning in the 1980s, Pace and his collaborators developed means to get around these limitations.[133] He reasoned that an inventory of microbes in a niche could be taken by sequencing rRNA genes obtained from DNA isolated directly from the habitat itself. Arguing that biologists' understanding of the makeup of the microbial world is rudimentary, in 1998 Pace called for a representative survey of the Earth's microbiodiversity with the use of automated sequencing technology.[134]

The tree of life was widely branching and had deep roots. In 1989, two groups led by Peter Gogarten and Naoyuki Iwabe independently used ancient gene duplications to root the tree of life by means of an outgroup.[135] Though they used different molecular markers and different algorithms, the two studies reached the same conclusion: the root of the universal tree appeared to be located between the eubacteria on the one side and the archaebacteria and eukaryotes on the other. Archaebacteria were more closely related to eukaryotes than they were to Eubacteria. To emphasize archaebacteria's difference from eubacteria, in 1990, Woese and collaborators shortened the name to "Archaea," and they proposed three formal "domains," above the level of kingdoms: the Bacteria, the Archaea, and the Eucarya.[136] In their view, Archaea were more different from Bacteria than humans were from plants. Though widely endorsed, this proposed rooting of the tree was soon refuted in the 1990s with evidence of lateral gene transfer,[137] whose ubiquity became widely recognized with a new approach based on whole genomes, with the rise of microbial genomics.

The Fourth Era

Bacterial Genomics

In the mid-1990s, the U.S. Department of Energy instituted a "Microbial Genome Initiative" as an offshoot of the "Human Genome Project" it had initiated with the National Institutes of Health 5 years earlier. In 1995, researchers at The Institute for Genomic Research (TIGR) headed by Craig Venter, published the sequence of *Haemophilus influenzae*, and the following year Woese and Olsen, together with researchers at TIGR, published the complete sequence of the first archaebacterium: *Methanococcus jannaschi*.[138] The Human Genome Project was rationalized in terms of its medical benefits, and microbial genomics had been justified similarly—each microbe for a specific practical purpose: medical, agricultural, or industrial[139]—but Woese and his colleagues also saw a deeper and more fundamental rationale. Humans were stressing the biosphere, and there would soon come a day when a deep knowledge of the biosphere and its capacity to adapt will be critical. Bacteria are largely responsible for the overall state of the biosphere: Our oxygen atmosphere exists (directly or indirectly) because of them, and they are vital to the regulation of the planet's surface temperature through their roles in carbon dioxide turnover and methane production and utilization. Thus, microbial genomics was needed to explore microbial diversity, to understand the interaction between microorganisms and their environments, and to reveal their evolutionary dynamics.

Can only God make a tree?

By the late 1990s, just when the three-domain proposal and the outlines of a "universal phylogenetic tree" were becoming well established, the microbial order based on rRNAs was challenged by data from complete genome analysis of bacteria. Phylogenies based on genes other than those for rRNA often indicated different genealogies, and indeed a somewhat chaotic order. The new genomic data also indicated that archaebacteria and bacteria had many genes in common; perhaps they were not that different after all. Classical evolutionist Ernst Mayr led the attack on rRNA phylogenies and Woese's three-domain proposal in 1998, and a few disaffected molecular phylogenicists soon followed.

That one could actually construct phylogenies, whether of bacteria or primates, based on comparisons of one or a few molecules had already been a fiercely contested issue. Classical evolutionists such as Mayr, Theodosius Dobzhansky, and George Gaylord Simpson were opposed to the whole field of "molecular evolution" from the very beginning.[140] There were several aspects to their resistance. At the most general level, there were basic institutional issues between molecular biology on the one hand—well-funded and rapidly growing—and traditional evolutionary studies on the other. The swift rise of molecular biology was perceived to be in direct competition with the aims and interests of evolutionary biologists.[141] In 1963, Dobzhansky reminded readers of *Science* that there was stimulating research going on in "organismic as well as molecular genetics."[142] His note was followed by a commentary from Mayr requesting "more financial and moral support for the classical areas."[143] When "molecular evolution" and the idea of a "molecular clock" emerged, the architects of the synthesis resented the "intrusion" into their domain; they argued that evolution was an "affair of phenotypes," one simply could not reduce it to comparative molecular morphology. Not only did classical evolutionists reject molecular methods for studying evolution but they also rejected their answers.

Attempts to stop the molecular clocks had come to a head when molecular evolutionists argued that the amino acid sequences of proteins did not evolve by adaptation and natural selection. Many changes simply had no effect on the protein structure, and therefore, they had no adaptive value. The evolution of proteins without selection was dubbed "the neutral theory" by Motoo Kimura, and "non-Darwinian evolution" by Jack King and Thomas Jukes in the late 1960s.[144] Mayr and Simpson found it incredible that molecular and morphological evolution could be different in mechanism and rate. In any case, they argued, the only evolution that mattered operated by natural selection.[145] Changes at the molecular level that did not affect the phenotype were really of "no interest for organismal biologists as they are not involved in the evolution of whole organisms."[146] Though the debate between panselectionists and neutralists has fizzled in recent years, the scope and significance of the neutral theory in molecular evolution remains unsettled to this day, despite the fact that the entire technology of molecular evolution for classification purposes is based on the assumptions of the neutral theory.[147]

From the outset there were those who argued that one could not classify microbes on the basis of rRNA phylogenies.[148] In 1990, Mayr sent a note to *Nature* protesting that separating the Archaea from Bacteria and claiming they formed a "domain," "superkingdom," or "empire" was grossly misleading.[149] Mayr's offensive on the rRNA phylogenies 8 years later came when the three-domain proposal was getting into the biology textbooks and when microbial evolutionists reported that bacteria contained a diversity that rivalled, and indeed surpassed, that of all of the macrobiological world. As Mayr saw it, Woese's three-domain proposal was absurd—prokaryotes did not possess a degree of diversity even remotely comparable to the eukaryotic world—and it was preposterous to compare the molecular genetic differences between Bacteria and Archaea to the huge morphological differences between eukaryotes and prokaryotes. He reasserted that evolution was "an affair of phenotypes," and, on this basis, he insisted that "all archaebacteria are nearly indistinguishable"; even if one took prokaryotes as a whole, he argued, it "does not reach anywhere the size and diversity of eukaryotes."[150] Microbial phylogenists had so far described only about 200 archaebacterial species and only about 10,000 eubacterial species, whereas Mayr suspected that within eukaryotes there were more than 30 million species. There were 10,000 species of birds alone and of course hundreds of thousands of insect species.[151] Mayr remarked,

> the eukaryote genome is larger than the prokaryote genome by several orders of magnitude. And it is precisely this part of the eukaryote genome that is most characteristic for the eukaryotes. This includes not only the genetic program for the nucleus and mitosis, but the capacity for sexual reproduction, meiosis, and the ability to produce the wonderful organic diversity represented by jellyfish, butterflies, dinosaurs, hummingbirds, yeasts,

> giant kelp, and giant sequoias. To sweep all this under the rug and claim that difference between the two kinds of bacteria is of the same weight as the difference between the prokaryotes and the extraordinary world of the eukaryotes strikes me as incomprehensible.[152]

Woese responded that the difference between himself and Mayr was not simply a matter of molecular versus organismic approach. Mayr looked at evolution from the top down, from the present to the past, observing the great phenotypic diversity of plants and animals evolved over the previous 500 million years of evolution. Woese looked from the bottom up; his concern was to understand evolutionary processes over the first 3 billion years of evolution, based on observing differences in molecules and genes. From his perspective, bacteria could not be defined negatively and in opposition to eukaryotes or in terms of the kingdoms to which some of them later gave rise. Bacteria had to be understood in their own terms, and from a historical perspective. As Woese commented, "The science of biology is very different from these two perspectives and its future even more so."[153] As he also noted, however, Mayr's critique had come at an opportune time, when the three-domain proposal seemed to be under considerable strain from molecular studies of whole genomes over the previous 3 years.

Defection grew from within the ranks of molecular evolutionists during the late 1990s. Several leading microbial phylogenicists saw in Mayr's critique much that they considered to be true, as central features of the Archaeal story of the 1980s were challenged. First, analyses of whole genomes (more than 70 had been sequenced by 2003) had shown that Archaebacteria and Eubacteria possessed numerous genes in common; they shared a rich biochemical complexity. These data did seem to contradict the hypothesis that the Archaea were so very different from Bacteria because the two groups diverged when life was quite new. Second, comparisons of genes for other functions seemed to contradict the phylogenetic lineages deduced from rRNA sequences. For example, although comparisons of SSU rRNA placed the microsporidia low on the phylogenetic tree, comparisons of the gene for the enzyme RNA polymerase placed the microsporidia higher on the tree, with the fungi. Therefore, some suggested that one could not use the SUU rRNA trees to trace microbial life.[154]

There was a third fundamental issue. Not only did the phylogenies from the new genomic studies disagree with the traditional rRNA-based phylogenies but the new genome data also conflicted among themselves. Comparisons of individual gene phylogenies (other than those concerned with the translation machinery) often indicated different organismic genealogies. Phylogeneticists suspected that the mix-up was caused by evolutionary mechanisms whose scope and significance they may have severely underestimated: gene transfer between groups.

Lateral Gene Transfer

In addition to gene transfer from parent to offspring, or "vertical transfer," there is also transmission of genes between distinct evolutionary lineages—"horizontal gene transfer," or "lateral gene transfer" (LGT). A bacterium of one strain may have acquired one or several genes from a completely unrelated organism. Therefore, similarities and differences in some genes may not be a measure of genealogical relationship.[155] For example, if organism type A and organism type B carry the same gene for a protein, it may not be because they both belong to the same taxonomic group, but because one of them acquired that gene (by "infection" or passive uptake) from a third type of organism, which is not ancestral to them. LGT could in principle blur the genetic record.

It had been well known since the mid-1950s that bacteria possess several mechanisms for transmitting genes between unrelated groups—through transformations, viral transduction, and conjugation. However, the importance of these mechanisms in bacterial evolution was usually overlooked.[156] Bacterial geneticists tended to view bacterial heredity in the way classical geneticists had that of plants and animals, in terms of gene mutations and gene exchanges between related strains. When the cause of bacterial antibiotic resistance was debated among bacteriologists in the late 1940s, just 8 years after their great success, discussions centered on about whether acquired resistance was the result of environmentally induced adaptive hereditary changes or mutation and natural selection. LGT as the cause of widespread antibiotic resistance was not considered until a decade later, when Japanese researchers made a case for it.

Although sanitary conditions were considered very good in Japan, bacillary dysentery was one of the most important infectious diseases, and shortly after the Second World War, a high incidence of antibiotic resistance appeared. Beginning in 1957, Japanese researchers isolated Shigella strains with multiple antibiotic resistances. In 1960, they reported that antibiotic resistance could be easily transferred between *E. coli* and Shigella in the intestines of humans.[157] That year, studies of mixed cultures indicated that cell-to-cell contact or conjugation was essential, and that the resistant factors were located on an "episome," the term introduced by François Jacob and Eli Wollman in 1958 for a genetic particle that may or may not become associated with the main bacterial "chromosome."[158] Thus, acquired antibiotic resistance represented what Lederberg had called "infective heredity." As Tsutomu Watanabe warned in 1963, "The medical importance of infective drug resistance, especially multiple drug resistance, is apparently limited to Japan at present, but it could become a serious world-wide problem in the future."[159] The complacency of the pharmaceutic industry about the threat of antibiotic resistance led to the synthesis of large numbers of antibiotics over the next three decades, which by the 1990s resulted in "the crisis in antibiotic resistance."[160]

By the end of the twentieth century, lateral gene transfer among bacteria was recognized as an urgent medical problem in regard to antibiotic resistance. At the same time, analyses of complete genome sequences led to suggestions by bacterial evolutionists that the extent of lateral gene transfer was far greater than was previously appreciated. Bacteria can adapt to new environmental conditions such as antibiotic resistance by acquiring DNA in several ways: inheritance of a plasmid that may either remain autonomous replicons or recombine into the chromosome; integration of a lysogenic phage into the chromosome; or insertion of a linear DNA fragment into the chromosome (usually by transposition or recombination with flanking homologous sequences).[161] LGT could scramble the phylogenetic record.

The significance of LGT for bacterial phylogeny was not a completely new issue with genomic studies. It had been noted at the outset of the new molecular morphology for bacterial phylogenetics.[162] Molecular comparisons of the amino acid sequences of the protein enzyme cytochrome had been shown to correlate well with what was known of the phylogeny of animals from morphological and paleontological evidence, but in the late 1970s, these sequences did not correlate well with any of the recognized species of the photosynthetic bacteria *Rhodospirillaceae* as recognized in *Bergey's Manual*. Instead of questioning the taxonomy in *Bergey's Manual*, groups led by Meyer and

Kamen at the University of Edinburgh and at University of California, San Diego, suggested that gene transfer between species could be responsible[163]:

> Our observations are consistent with the hypothesis that bacterial evolution proceeds both by the assimilation of genes for single functions or for whole metabolic pathways from other organisms as well as by mutation and selection. We have no information about the frequency of successful gene assimilation or about the timescale of protein evolution and speciation in bacteria. We have no evidence at all to suggest that either gene product (such as cytochrome *c*) or an assemblage of genes (such as *Rhodospirillum rubrum*) should be stable through geological time.[164]

There were two responses to this argument. First, Richard Dickerson at the California Institute of Technology emphasized that the disagreement with *Bergey's Manuel* was not a problem, because the classification therein did not reflect genealogies, and that although "proteins that were extrinsically useful to a bacterium" would not be useful for deciphering bacterial phylogeny because of LGT and consequential scrambling of the genetic record, this would not be true for highly integrated proteins "with an attendant metabolic setting" such as cytochrome *c*.[165] Woese and colleagues responded with the same point about *Bergey's Manual*; and added that "if comparative analysis of several unrelated macromolecules yields essentially the same phylogenetic tree, then that pattern is extremely unlikely to reflect the lateral transfer of genes."[166] On this basis, they argued that the 16S rRNA data was in agreement with the cytochrome *c* data, which would be virtually impossible to explain by lateral transfer of genes. In 1986, Meyer, Kamen and their collaborators insisted again that one could not use amino acid sequences as indicators of phylogeny. This time, however, they made no mention of LGT but argued instead that convergent mutations and back mutations would blur the evolutionary record. Paraphrasing the poet Joyce Kilmer, they concluded: "Only God can make a tree."[167]

Lateral gene transfer had been generally thought to affect metabolic genes for accessory functions, not genes at the "heart" of the organism, such as those for rRNA, which interact with so many proteins. However, announcements of increased cases of LGT occurred after 1995 with the rise of bacterial genomics.[168] In 1998, Howard Ochmann and Jeffrey Lawrence reported that about 10% of the *E. coli* genome consists of genes that had been acquired in more than 200 events of LGT, following the divergence of *E. coli* and Salmonella some 100 million years ago.[169] About 18% of the *E. coli* genome entailed relatively recent acquisitions. These data indicated that LGT would have a profound effect on evolutionary genome comparisons.[170]

Recognition of the pervasiveness of LGT entailed modifications to two seminal views about microbial evolution and the course of early evolution: their treelike branching genealogies, and its hierarchical nature. In the Darwinian order of things, you sort plants and animals into species. Then you sort species resembling one another into genera, genera into families, families into orders, and so on. Each species belongs to one and only one genus, each genus to one and only one family, and so on. This hierarchical order is based on common descent from an ever-decreasing number of ancestors, leading back to the origin of life, the common ancestor of all living things. This was fine as long as there were barriers to gene transfer between species.

Evidence for gene transfer between bacterial groups contradicted this view of evolution. Bacteria are composites; they have acquired and integrated genes from diverse taxa.

Thus, instead of the branching genealogical tree that Darwin imagined for plants and animals, the pattern of early evolution (at least) is reticulated and involves the inheritance of acquired characteristics, but by mechanisms quite different from those attributed to Lamarck. LGTs between taxa make it resemble more a web than a tree. Accordingly, the resulting evolutionary order may be nonhierarchical.[171]

LGT blurs the boundaries between "species." The ease with which genes are interchanged among bacteria reinforced long-standing views that, "the biological species concept" (in the general sense of a reproductively isolated group) does not apply to bacteria.[172] Isolating mechanisms segregating Mendelian populations do not apply, and a test of hybrid sterility is irrelevant. Certainly, many bacteriologists of the 1950s and 1960s had also recognized that the concept of species did not apply to bacteria. This was not because bacteria could exchange genes between distantly related groups, but because laboratory studies indicated that sexual reproduction was a rare event for bacteria—as it was for most microorganisms.[173] As Samuel T. Cowan concluded in 1962, "the microbial species does not exist; it is impossible to define except in terms of nomenclatural type; and it is one of the greatest myths of microbiology."[174]

There had been few reflections about the importance of lateral gene transfer before the rise of bacterial genomics, although speculations of its ubiquity had led some theorists such as Sorin Sonea in Montreal to conceive of the entire bacterial world as a superorganism.[175] Speculations about the nature and intensity of lateral gene transfer led some *fin de siècle* bacterial phylogenicists to fear, as had some cytochrome *c* researchers two decades earlier, that the whole enterprise of classification may be insolvent, that a natural phylogeny of bacteria may be impossible. Bacterial (*sensu lato*) phylogeny may lay beyond the chronicles of history. This was the most extreme message that some researchers had gleaned from comparisons of complete genomes since the mid-1990s. Biologists who had once been chief advocates of SUU rRNA phylogenies, chief among them W. Ford Doolittle at Dalhousie University in Canada, became most skeptical of ever constructing a universal phylogeny (see chapter 5).[176]

Other microbial phylogeneticists remained confident in molecular phylogeny. Some offered alternative models based on other molecular data. James Lake and Maria Rivera at the University of California, Los Angeles, continued to support an argument they had made in the early 1990s, based on cladistic arguments and on ribosomal structural differences, that the Archaea was not a monophyletic group. They distinguished, within the Archaea, a group they called the Eocytes (early cells), which, they argued, was more closely related to Eukaryotes than to Archaea (see chap. 9).[177] Radhey Gupta at McMaster's University argued, based on studies of several protein phylogenies, that the bacteria world exhibited a fundamental dichotomy between what he called monoderms (bacteria possessing a single membrane) and diderms (those with a double membrane) corresponding to gram-positive and gram-negative bacteria (see chap. 8).[178]

Still others, including Karl-Heinz Schleifer in Germany, Charles Kurland in Sweden, and Woese and his collaborator Gary Olsen, upheld microbial phylogenies based on some 20,000 rRNA molecules catalogued in public databases (see chaps. 2, 3, 4, and 10). Because the SSU rRNA gene is at the core of the cell's most complex machinery and interacted with so many proteins, they argued, it would be unlikely to be transferable between phylogenetic groups without disrupting core cellular systems.[179] Therefore, rRNA comparisons would be one of the few reliable means for tracking bacterial lineages. The central question for all microbial phylogeneticists was whether or not genes for rRNA are exchanged frequently between groups.[180] Thus, microbial evolutionists called for further studies in comparative genomics to establish the principles governing LGT across the bacterial phylogenetic spectrum.[181]

In the meantime, Woese reconsidered an idea he had begun to develop in 1982 (see also chap. 4).[182] He interpreted the new genomic evidence indicated many shared genes between the Archaea, the Bacteria, and the Eucarya in terms of intense LGT occurring before the groups emerged as distinct fundamental domains. The evidence could be understood in terms of his long-searched-for transitional stages in the evolution of the translation apparatus. Ever since Darwin, biologists had assumed that all life on earth arose from a single ancestral cell,[183] but Woese disagreed with the canon of a single ancestral mother of all cells. He speculated that instead of the expected "first cell," the progenote was a population of precellular entities with underdeveloped and error-prone replication, and with translation machinery. Before the development of the modern translation apparatus, evolution would be driven by a different mode and tempo. At this early time, there were no individual lineages that could be distinguished as such because of so much gene mutation and intense LGT. These processes would generate enormous diversity very quickly. Primitive systems would be modular and exchange parts freely, but as the translation machinery evolved, becoming refractory to lateral gene transfer, so too did definable lineages. This was the great Darwinian divide, when the three domains, Archaea, Bacteria, and Eucarya, emerged out of the chaos.

Woese likened the emergence of the three domains to physical annealing: There would first be a period of intense genetic "heat" (high mutation rates and intense gene transfer between lineages that would have short histories) when cellular entities were simple and information systems were inaccurate. It would be impossible to discern organismic genealogies. This intense period would be followed by genetic "cooling," with the development of the modern cell with a sophisticated translation apparatus resulting in the emergence of genealogically recognized domains and taxa. Thus, as Woese concluded:

> The universal ancestor is not a discrete entity. It is, rather, a diverse community of cells that survives and evolves as a biological unit. This communal ancestor has a physical history but not a genealogical one. Over time, this ancestor refined into a smaller number of increasingly complex cell types with the ancestors of the three primary groupings of organisms arising as a result.[184]

Extending and Modifying Symbiosis

LGT applies to eukaryotes as well. By the mid-1990s it became clear to all phylogeneticists that the eukaryotic cell was also fundamentally chimeric. That the cytoplasmic organelles mitochondria and chloroplast were symbionts had been agreed to earlier. Comparing ribosomal RNAs of chloroplast, mitochondrial, and nuclear origin with each other and with different kinds of bacteria provided the rigor and closed the main controversy about their origin.[185] Based on SSU rRNA comparisons, genes encoded in those organellar genomes indicated that mitochondria and chloroplasts were of eubacterial origin (alpha-proteobacteria and cyanobacteria, respectively). Reviewing the evidence in 1982, Gray and Doolittle considered the matter resolved (see also chaps. 11 and 12).[186] There were no comparable organellar data to test Margulis's theory that centrioles/kinetosomes arose as symbionts. The evidence for DNA in centrioles had been on-again, off-again since the 1960s, but it was effectively refuted in the 1990s by evidence from electron microscopy and molecular hybridization, which indicated that genes affecting centriolar/flagellar function are located in the nucleus.[187] However, it was still possible that they were transferred there from symbionts.

At the end of the twentieth century, the origin of the eukaryotic-cell nucleus had emerged as one of the most pressing problems of microbial evolution.[188] By that time, genomic comparisons of ancient genes indicated that the archaebacterial genome and the main eukaryotic genome had many genes in common, and that several properties of the transcriptional and translational apparatus of the Eukaryote could be found in archaebacteria but not in eubacteria.[189] There were also nuclear genes of eubacteria ancestry thought to be transferred there from chloroplasts or mitochondrial genomes as expected. However, many ancient eubacterial genes that were not obviously transferred from organelles were also found in the nucleus.[190] Several interpretations were compatible with the available molecular phylogenetic evidence.

First, the nucleus of the eukaryote may possess eubacterial genes other than those that were transported from the organelles, because lateral gene transfer had been rampant before the rise of the three domains, as Woese had suggested. The issue here is twofold: the timing of events, and establishing whether the eubacteria genes in eukaryote are traceable to one taxonomic group of bacteria or to many. One aim is to sort among the data for individual gene transfers and for clusters of genes traceable to one potential symbiont.

Second, all or most of the eubacteria genes in the nucleus may actually derive from the mitochondrial ancestor.[191] Although the standard models propose that the nucleus emerged before the mitochondria, others have suggested that the nucleus emerged after the mitochondrial symbiosis. This interpretation was compatible with one of the most discussed models: the "hydrogen hypothesis" proposed by William Martin and Miklós Müller in 1998, according to which the nucleus may have originated from a merger between an Archaebacterial organism and the Eubacterial ancestor of the mitochondria (see also chap. 6).[192] According to this hypothesis, an anaerobic, autotrophic, and hydrogen-dependent host, living in an environment scarce in hydrogen, established a tight physical relationship with heterotrophic bacteria that were able to produce molecular H_2 through anaerobic fermentation. The engulfed symbiont, an anaerobic alphaproteobacterium, would initially supply its methanogenic archaebacterial host genes for glycolytic carbohydrate metabolism. Selection to feed the symbiont carbohydrates favored the transfer of genes from mitochondrial genome to host genome.[193] Thus, this hypothesis argues that the initial advantage of this symbiotic association was not ATP export from a symbiont into the cytosol of a eukaryotic host through respiration, as proposed previously by the traditional oxygen hypothesis for the acquisition of mitochondria, but rather the excretion of molecular H_2 produced by the symbionts in an archaebacterial host. An obvious mitochondrial origin of the many eubacterial genes in the nucleus could have been obscured by mutation or by LGT.

The nature of the host for the mitochondria had been a simmering issue since the 1970s and 1980s. The principal idea of that time was that only eukaryotic cells could accomplish phagocytosis and engulf symbionts. Therefore, the missing host for mitochondria would have been some kind of eukaryote that itself could not have arisen by bacterial symbiosis. In 1983, Tom Cavalier-Smith gave this missing link the name of Archezoa.[194] It would be a subkingdom of eukaryotes that diverged before the acquisition of chloroplasts and mitochondria. Such amitochondriate eukaryotes were found to exist, but it was not certain whether these organisms once had mitochondria and subsequently lost them. If so, relic mitochondrial genes would be present. Special searches for relic mitochondrial genes in amitochondriate protists were conducted to test the Archezoa hypothesis during the 1990s. Many amitochondriate candidates turned out to have the signature "mitochondrial gene" relics: mitochondria may have been lost many times in the course of evolution. To Martin and others, it seemed reasonable to suggest that there were no primordial amitochondriate eukaryotes; the original host cell for the mitochondria was not a eukaryote as in the traditional serial endosymbiosis theory, but, rather, an archaebacterium.[195]

Third, Margulis and her collaborators have searched in the new data for evidence of the symbiotic origin of the cytoskeleton mitotic and motility apparatus.[196] With evidence that the centriolar genome is located in the nucleus and with evidence of ancient eubacteria genes of nonmitochondrial origin, Margulis and her colleagues have suggested that the nucleus developed from a symbiosis between thermoacidophilic Archaebacteria and motile Eubacteria that led to the mitotic spindle and motility features of the cell (see chaps. 13 and 14).

Fourth, nonmitochondrial genes in the nucleus may have been transferred from peroxisomes: organelles as widely distributed throughout nature as mitochondria, but much more simple in structure and composition. In 1982, Christian de Duve at the Rockefeller Institute suggested that peroxisomes arose from aerobic bacteria that were adopted as endosymbionts before mitochondria.[197] Although today they carry out various metabolic activities, de Duve suggested that the original benefit of peroxisomes was to rescue their anaerobic hosts from the toxic effects of oxygen, which greatly accumulated in the primitive atmosphere some 2 billion years ago after photosynthetic cyanobacteria arose.[198] Unlike mitochondria and chloroplasts, peroxisomes have no remnants of an independent genetic system.

Fifth, nonmitochondrial eubacterial genes in the nucleus could have been acquired from gene transfers from symbionts acquired and lost after the primordial eukaryotic cell was formed.[199] Although the existence of mitochondrial genes in many, if not all, extant amitochondriate protists seemed to rule out the Archezoa hypothesis and is consistent with the view that the mitochondria came before the nucleus, other scientists have emphasized that one cannot take the data at face value. They argue instead that those genes thought to be of mitochondrial descent may actually be the result of other bacterial gene acquisition and losses. After all, acquired bacterial symbionts are common among protists.

Sixth, it was possible that the eukaryotic cell, that is, its nucleus, may have been formed from an ancient symbiosis or fusion of some kind before the emergence of other organelles. The idea that the nucleus may be have evolved as a microbial symbiont living in a primitive host cell is an old one.[200] Japanese biologist Shôsaburô Watasē suggested at a Woods Hole lecture in 1893 that centrioles, as well as nucleus and cytoplasm, may have also arisen as symbionts, and this idea was later mentioned by others.[201] In 1903, Theodor Boveri discussed the idea that the nucleus was a symbiont, and between 1905 and 1918, Constantin Merezhkowsky developed this view as part of his theory of "symbiogenesis."[202] These early suggestions were based on cytological evidence of sharply stained chromatin in the nucleus suspended in the cytoplasm with different chemical properties, as well as on evidence of symbioses such as lichens and nitrogen-fixing bacteria. For Merezhkowsky and those who preceded him, the nucleus was a colony of primitive microbes (chromatin), and the cytoplasm was the body of another kind of host microbe.

Over the last decade, several models have been proposed, according to which the nucleus emerged from some kind of symbiosis independent of any other organelles. In the 1980s, James Lake and Maria Rivera argued that the nucleus evolved from an engulfed (Eocyte) Archaebacterial symbiont of a Eubacterial host (see also chap. 9).[203] Arguing that the genes affecting the cytoskeleton, which allowed phagocytosis, are found in no existing bacterial lineages, Mitchel Sogin, Wolfram Zillig and coworkers, Russell Doolittle, Moreia, and López-García have suggested that the eukaryotic nucleus was formed from the cellular fusion of an archaebacterium and a gram-negative eubacterium.[204] A similar idea has been proposed by Hyman Hartman and Alexie Federoff.[205] Radhey Gupta has argued that the nucleus resulted from a fusion event between a diderm and a monoderm bacterium. In his scheme, the context is oxygen and antibiotic warfare; one partner (an oxygen-tolerant diderm eubacteria) provided protection against oxygen, and the other partner (an archaebacteria monoderm) provided antibiotic protection (see also chap. 8).[206]

The origin of the eukaryote remains unsettled, as disputants marshal various kinds of data in support of their favored model. All possible theoretic positions seem to be filled. The larger paradigm of thinking in terms of LGTs and symbiosis is shared by all, and this sharply distinguishes research on bacterial evolution from classical thinking. Indeed, molecular studies of microbial phylogeny have dramatically transformed thinking about microbial evolution. The acquisition of genes and of whole genomes contradicts the traditional Darwinian conceptions.[207] Symbiosis, and LGT, were, and continue to be, trivialized or ignored by leading evolutionists. Stephen Gould regarded the symbiotic origin of mitochondria and chloroplasts as "entering the quirky and incidental side" of evolution.[208] Based on theoretical assumptions about the evolution of cooperation and gene-based selection, leading neo-Darwinian theorists have also insisted that the inheritance of acquired bacteria is a rare exceptional phenomenon in plants and animals. Thus, John Maynard Smith and Eors Szathmáry asserted in 1999 that "transmission of symbionts through the host egg is unusual."[209] Despite such assertions, hereditary symbiosis is prevalent in animals too.[210]

Surveys based on molecular phylogenetic techniques for screening have, so far, found bacteria of the Genus *Wolbachia* in more than 16% of all known insect species (it may be present in as many as 80%), including each of the major insect orders (see chap. 15).[211] These bacteria are thought to be the most common hereditary infection on Earth, rampant throughout the invertebrate world, infecting shrimp, spiders, and parasitic worms, as well as insects.[212] Their complete distribution in arthropods and other phyla are yet to be determined. *Wolbachia* are α-proteobacteria, like mitochondria, and they appear to have evolved as specialists in manipulating reproduction and development of their hosts. They cause a number of profound reproductive alterations in insects, including cytoplasmic incompatibility between strains and related species, parthenogenesis induction, and femininization: They can convert genetic males into reproductive females (and produce intersexes). Sometimes, as in the case of weevils (one of the most notorious pests of stored grain), *Wolbachia* are inherited together with other bacterial symbionts that allow the animal better adaptation to the environment by providing vitamins and energy, and by enhancing the insect's ability to fly.[213] *Wolbachia* have considerable evolutionary interest, especially as a mechanism for rapid speciation. Studies of hereditary symbionts of insects are well-funded today, not because of their evolutionary significance, but because of their potential for pest control in agriculture and as a mechanism for modifying arthropod vectors of human disease.[214]

Recapitulation

The bacterium has always had an odd place in the life sciences—outside or close to the margins of evolutionary biology. From the nineteenth century to the present, there have been repeated claims that bacteria defy natural classification based on genealogy, or that they do not possess clearly marked species in the sense of plants and animals, but the arguments for the absence of bacterial species or an indecipherable evolutionary past have changed.

In a first phase, from 1860 to 1940, typified by the rise of the germ theory of fermentation and disease, a natural classification that reflected genealogy was considered by most bacteriologists not to be possible, essential, or even necessarily beneficial. Bacteria were classified for practical purposes, not for their natural history and evolution. Those who searched for a natural classification faced lack of morphological diversity on the one hand and great physiological diversity and extreme plasticity on the other. Before the rise of bacterial genetics, the inapplicability of a general concept of species was framed between a perceived pleomorphism and physiological adaptation based on vague Lamarckian mechanisms on the one hand, and a lack of sexuality on the other. Bacteria were thought to reproduce solely by binary fission; there was no discernable sexual recombination.

In a second phase, after the Second World War, when genetic recombination was demonstrated in bacteria, the focus of bacterial geneticists was on the molecular biology of genes: their regulation, immunity, and disease. Non-Mendelian mechanisms were discerned. A concept of infective heredity was constructed, the possibility that mitochondria and chloroplasts evolved as bacterial symbionts was reconsidered, and by the 1960s the importance of LGT was recognized for antibiotic resistance in the war against disease. Those bacteriologists who had insisted on a natural classification based on morphology finally admitted that a natural phylogeny of bacteria may be impossible. Instead, they articulated an all-encompassing morphological dichotomy distinguishing bacteria (prokaryotes) from all other organisms (eukaryotes) as representing the greatest discontinuity in evolutionary history.

In a third phase, beginning in the 1970s, new techniques based on comparative molecular morphology revolutionized bacterial taxonomy, providing a natural phylogeny based on rRNA comparisons deep within the genetic machinery of the organism. Changes in rRNA sequences were conceived of as a universal chronometer. Ribosomal RNA phylogenies resulted in the elaboration of a universal phylogenetic tree, a tripartite division of life—the Archaea, Bacteria, and Eucarya—and they provided molecular evidence for the conjecture that mitochondria and chloroplasts had evolved from bacterial symbionts.

In a fourth phase after the mid-1990s, with the development of genomics, the hitherto unappreciated ubiquity of LGT was postulated to explain many gene histories other than those for rRNA. The species concept was again considered to be inapplicable to bacteria, not because of the absence of genetic recombination, as long thought, but because there seemed to be so little barrier to it. Doubts about the inability to construct bacterial genealogies arose anew because of the scrambling of the genetic record from LGT. While debates continue over which molecules (if any) provide the most reliable phylogenetic guide, so too do debates over the origin of the eukaryotic cell nucleus and over the inheritance of acquired bacterial genomes.

Notes

1. Not surprisingly, the history of bacteriology is seen almost exclusively from the perspective of pathology and is virtually a history of germ theory and practice. See, for example, P. De Kruif, *Microbe Hunters* (New York: Harcourt, Brace and Co., 1926), and W. Bulloch, *The History of Bacteriology* (New York: Oxford University Press, 1938). See also N.J. Tomes and J.H. Warner, "Introduction to Special Issue on Rethinking the Reception of the Germ Theory of Disease: Comparative Perspectives," *Journal of the History of Medicine and Allied Sciences* 52 (1997): 7–16.

2. See, for example, J. Farley, *The Spontaneous Generation Controversy from Descartes to Oparin* (Princeton, NJ: Princeton University Press, 1974), and G.L. Geison, *The Private Science of Louis Pasteur* (Princeton, NJ: Princeton University Press, 1995). See also J. Strick, *Sparks of Life; Darwinism and the Victorian Debate over Spontaneous Generation* (Chicago: University of Chicago Press, 2000), and H. Harris, *Things Come to Life. Spontaneous Generation Revisited* (Oxford: Oxford University Press, 2002).

3. K. Vernon, "Pus, Sewage, Beer and Milk: Microbiology in Britain, 1870–1940," *History of Science* 28 (1990): 290–325.

4. K. Codell Carter, "The Development of Pasteur's Concept of Disease Causation and the Emergence of Specific Causes in Nineteenth-Century Medicine," *Bulletin of the History of Medicine* 65 (1991): 528–548; 530.

5. See, for example, G.S. Woodhead, *Bacteria and Their Products* (London: Walter Scott, 1891), 24.

6. C. Dobell, *Anthony Leeuwenhoek and his "Little Animals"* (New York: Russel and Russel, 1958), 131–133.

7. C. Linné, *Systema Vegitabilum*, 13th edition (Göttingen: Dietrich, 1774). See R.S. Breed, "The Present Status of Systematic Bacteriology," *Journal of Bacteriology* 15 (1928): 143–163; 143.

8. Georges Cuvier, "Sur un nouveau rapprochement à établir entre les classes qui composent le règne animal," *Annales Muséum d'Histoire Naturelle* 19 (1812): 73–84; 83.

9. A.T. Henrici, *The Biology of Bacteria. An Introduction to General Microbiology*, 2nd edition (Boston: D.C. Heath and Co., 1939), 82.

10. Richard Owen, *Paleontology or a Systematic Summary of Extinct Animals and Their Geological Relations* (Edinburgh: A. and C. Black 1860). See also L.J. Rothschild, "Protozoa, Protista, Protoctista: What's in A Name?" *Journal of the History of Biology* 22 (1989): 277–305, and M. Regan, "A Third Kingdom of Eukaryotic Life: History of an Idea," *Archiv. Protistenkd* 148 (1997): 225–243.

11. J. Hogg, "On the Distinction of a Plant and an Animal, and on a Fourth Kingdom of Nature," *Edinburgh New Philosophical Journal*, N.S. 12 (1860): 216–225.

12. E. Haeckel, *Generelle Morphologic der Organismen*. 2 vols. (Berlin: George Reimer, 1866), 205.

13. As microscopic methods improved, Haeckel recognized that many of the organisms he had assigned to Monera were either nonexistent or turned out to have a nucleus. Among his original examples, only *Vibrio* was true to the definition. Ernst Haeckel, *The Wonders of life. A Popular Study of Biological Philosophy*, translated by Joseph McCabe (New York: Harper and Brothers, 1905).

14. Ibid., 195–196.

15. J. Sapp, *Evolution by Association. A History of Symbiosis* (New York: Oxford University Press, 1994).

16. C.C. Dobell, "Contributions to the Cytology of the Bacteria," *Quarterly Journal of Microscopical Sciences* 56 (1911): 395–506; 488, 489.

17. B.F. Lutman, *Microbiology* (NewYork: McGraw-Hill, 1929).

18. William Park, Anna Williams, and Charles Krumwiede, *Pathogenic Microorganisms. A Practical Manual for Students, Physicians and Health Officers*, 9th edition (Philadelphia: Lea and Febiger, 1929), 28.

19. E.B.Copeland, "What is a Plant?" *Science* 65 (1927): 388–390; 390.

20. Ibid.

21. H.F. Copeland, "The Kingdoms of Organisms" *The Quarterly Review of Biology* 13 (1938): 383–420; 386. Idem, *The Classification of Lower Organisms* (Palo Alto, CA: Pacific Books, 1956).

22. Charles Darwin, *On the Origin of Species*, with an introduction by Ernst Mayr. Facsimile edition of 1859 (Cambridge, MA: Harvard University Press, 1964), 420.

23. Ibid., 484.

24. C. Naegeli, *Die niederen Pilze in ihren. Bezihungen zu den Infectionskrankheiten und der gesundheitspflege*. (Munich: Oldenburg, 1877), 20. Quoted in Christoph Gradmann, "Isolation, Contamination, and Pure Culture: Monomorphism of Pathogenic Micro-Organisms as Research Problem 1860–1880," *Perspectives on Science* 9 (2000): 147–171; 151.

25. See Ferdinand Cohn, "Uber die Bacterien, die Kleinstein Lebenden Wesen," *Samsung Gemeinverständlicher Wissenschaftlicher Vorträge* Berlin: np, 1872), vol. 7. Idem., *Bacteria: The Smallest of living Organisms*, Translated by Charles S. Dolley (1881), Introduction by Morris C. Leikind (Baltimore, MD: The Johns Hopkins Press, 1939), 4. This paper was also one of the earliest works in bacteriology to be published in English.

26. F. Cohn, "Untersuchungen über Bacterien," *Beiträge zur Biologie der Pflanzen* 1 (2) (1875): 127–222. Idem, "Untersuchungen über Bacterien II," *Beiträge zur Biologie der Pflanzen* 1 (3) (1875): 141–207. Cohn's belief in bacterial diversity was based on observations of large bacteria (especially the spirilla) that maintained the same form under different conditions and that showed no intermediate forms. In 1875, he grouped the blue-green algae within the bacteria.

27. Cohn defined the bacteria as "cells without chlorophyll, spherical, oblong, or cylindrical, containing also twisted or curved forms, which reproduce exclusively by transverse fission, and are either single or in families of cells." He suggested that, "bacteria which cause different chemical and pathological processes consist of a small number of individual species which have developed into a large number of natural and cultural races, which, since they only reproduce by asexual means, are able to maintain their physiological characteristics with more firmness." He argued that, because all bacteria reproduce only by asexual methods, such as budding or fission, such a fixation of a race characteristic was easier to accept. In the same way that summer rye cannot be used as winter rye, although both races are of the same origin and could be returned to the same race through continued culture over a long period of time, top yeast cannot be used for the preparation of Bavarian beer (bottom yeast), and almost every wine or beer producer has his or her own yeast, so that it seemed probable that many alcohol producing yeasts are only a large number of cultural races of the same species.

28. S.A. Waksman, *Sergie N. Winogradsky. His Life and Work*, (New Brunswick, NJ: Rutgers University Press, 1953).

29. G. Van Iverson, L.E. den Dooren de Jong and A.J. Kluyver, *Martinus Beijerinck. His life and Work* (The Hague: Martinus Nijhoff, 1940).

30. A.J. Kluyver, *The Chemical Activities of Microorganisms* (London: University Press, 1931).

31. W. Migula, *System der Bakterien. Handbuch der Morphologie, Entwickelungsgeschichte und Systematik der Bakterien*. 2 vols. (Jena: G. Fischer, 1897–1900).

32. Winogradsky, in 1888–1887, had shown that hydrogen sulfide and sulfur fulfill the same function that organic substances generally perform in other organisms: acting as an energy source. Sulfur bacteria provided the first example of a new kind of physiology, and they were the prototype of the group later identified as chemosynthetic bacteria, for those that derive energy from

the oxidation of inorganic substances organic substances. This grouping was the first application of physiological characters for the creation of a large systematic entity.

33. C.E.A. Winslow, "The Characterization and Classification of Bacterial Types," *Science* 39 (1914): 77–91; 84.

34. S. Orla-Jensen, "Die Hauptlinien des natürlichen Bakteriensystems nebst einer Uebersicht der Gärungsphenomene," *Zentralbl. F. Bakt. Parasitenk.* II Bd 22 (1909): 305–346. See also R.E. Buchanan, *General Systematic Bacteriology* (Baltimore, MD: Williams and Wilkins, 1925). Although Orla-Jensen discarded purely morphological classification, especially that based on presence or absence of flagella, he did include arrangement of flagella.

35. Winslow, "The Characterization and Classification of Bacterial Types," 85.

36. Ibid.

37. Ibid., 89.

38. Buchanan, *General Systematic Bacteriology.* See also, S. Orla-Jensen, "The Main Lines of the Natural Bacterial System," *Journal of Bacteriology* 6 (1921): 263–273.

39. Ibid., 9.

40. Breed, "The Present Status of Systematic Bacteriology," 143.

41. F.W. Andrewes, *A System of Bacteriology* (London: Stationary Office, 1930). Quoted in C.H. Andrewes, "Classification and Nomenclature of Viruses," *Annual Review of Microbiology* 6 (1952): 119–138.

42. Buchanan, *General Systematic Bacteriology*, 11.

43. See O. Rahn, "Contributions to the Classification of Bacteria, 1–IV." *Zentralblatt für Bakteriologie*, 78 (1929): 1–21. Ibid., 79: 321–343. O. Rahn, "New Principles for the Classification of Bacteria," *Zentralblatt für Bakteriologie*, 96 (1937): 273–286. See also C.B. van Niel, "The Classification and Natural Relationships of Bacteria," *Cold Spring Harbor Symposia on Quantitative Biology* 11 (1946): 185–301; 297.

44. Winslow, "The Characterization and Classification of Bacterial Types," 77.

45. Ibid., 89.

46. Such commissions on classification and nomenclature were established in older biological sciences, as for example, the International Commission on Zoological Nomenclature appointed by the Third International Zoological Congress in 1895 and made permanent at the fourth congress in 1898. Its work was more along the line of precise legal definitions and determination of priority in terminology than would be the case with a similar commission in bacteriology. Bacteriology did not have an international congress to which such a commission could report on all phases of its work. Although for one group, the colon-typhoid group, a commission on systematic relations was created by the International Congress on Hygiene and Demography.

47. Winslow, "The Characterization and Classification of Bacterial Types," 90.

48. See Buchanan, *General Systematic Bacteriology*, 109.

49. See C.-E.A. Winslow et al. Committee of the Society of American Bacteriologists on Characterization and Classification of Bacterial Types, 1917. Preliminary Report. "The Families and Genera of Bacteria," *Journal of Bacteriology* 2 (1917): 505–566. Idem, "The Families and Genera of Bacteria. Final Report of the Committee of the Society of American Bacteriologists on Characterization and Classification of Bacterial Types," *Journal of Bacteriology* 5 (1920): 191–229.

50. Quoted in Henrici, *The Biology of Bacteria*, 317. After the publication of the third edition in 1934, the Society of American Bacteriologists relinquished sponsorship for *Bergey's Manual.* It transferred all of the rights and interests in the manual to Bergey, who would create an educational trust (designating himself, Robert S. Reed, and E.G.D. Murray as trustees) for the purpose of preparing, editing, and publishing revisions and successive editions of the manual. Since the creation of the trust, the trustees have published, successively, the fourth, fifth, sixth, seventh, and eighth editions of the Manual (dated 1934, 1939, 1948, 1957, and 1974). In 1977 the trust published an abbreviated version of the eighth edition, called *The Shorter Bergey's Manual of Determinative Bacteriology*. This volume contained the outline classification of the bacteria, the descriptions of all genera and higher taxa, all of the keys and tables for the diagnosis of species, all of the illustrations, and two of the introductory chapters; however, it did not contain the detailed species descriptions, most of the taxonomic comments, the etymology of names, and references to authors.

51. A.R. Prévot, "Etudes de sytématique bactérienne, I,II," *Annales des sciences naturelles botanique* 10e Sér., 15 (1933): 223–261. A.J. Kluyver and C.B. van Niel, "Prospects for a Natural System of Classification of Bacteria," *Zentralblatt für Bakteriologie*, Bd 94 (1936): 369–402; 377.

52. Kluyver and van Niel, "Prospects for a Natural System," 377. R.Y. Stanier and C.B. van Niel, "The Main Outlines of Bacterial Classification," *Journal of Bacteriology* 42 (1941): 437–466; 443–444.

53. R.S. Breed in *Bergey's Manual of Determinative Bacteriology*, 5th edition (Baltimore, MD: Williams and Wilkins, 1939), 39.

54. Breed, "The Present Status of Systematic Bacteriology," 145.

55. Stanier and van Niel, "The Main Outlines of Bacterial Classification," 437–438.

56. Ibid., 439.

57. Ibid., 443. See also Kluyver and van Niel, "Prospects for a Natural System" 377.

58. A.I. Oparin, *Origin of Life*, translated with annotations by Sergius Morgulis (New York: Dover, 1938).

59. Stanier and van Niel, "The Main Outlines of Bacterial Classification," 444.

60. Van Niel, "The Classification and Natural Relationships of Bacteria," 289.

61. Ibid., 290.

62. Kluyver and van Niel, "Prospects for a Natural System," 293. Kluyver, Stanier, and van Niel recognized one major exception to their suggestion that exclusively morphological characters should be used for classification above the genera (physiology and morphology for genera), and this pertained to the blue-green algae. Because this classification did not distinguish blue-green algae, Kluyver, Stainer, and van Niel argued that the primary use of physiological characters was essential in that case. Van Niel wrote in 1946, "No matter how rational this solution may have appeared at the time, its application leads to some very serious difficulties, especially when it comes to a separation of the bacteria from certain of the blue-green algae. And no one will challenge the assertion that such a separation is both desirable and concerned with the delineation of large groups. Here as Stanier and van Niel (1940) have pointed out, the primary use of physiological characters is essential. That obstacles might be encountered, which would render the restriction to morphology for the diagnosis of larger units unsatisfactory, was recognized by Kluyver and van Niel"; "The Classification and Natural Relationships of Bacteria," 293. For a critique of the system proposed by Kluyver and van Niel, see P.B. White, " Remarks on Bacterial Taxonomy," *Zentralblatt für Bakteriologie*, 96 (1937): 145–149. See also, E.G. Pringsheim, "The Relationship Between Bacteria and Myxophyceae," *Bacteriological Reviews* 13 (1949): 47–98.

63. Stanier and van Niel, "The Main Outlines of Bacterial Classification," 444.

64. Kluyver and van Niel, "Prospects for a Natural System," 387–388.

65. "Whether the further elaboration of a rational nomenclature along the lines laid down by Orla-Jensen, and further expanded by Kluyver and van Niel would prove adequate or whether it might even be preferable to drop the use of Latin names with their taxonomic implications, is a matter for future developments And while I am fully in agreement with the opinion that stability in nomenclature is of great importance, I must once more insist that, in the long run, it may turn out to be easier to gain adherence to a more rational, modernized system than to the current one." van Niel, "The Classification and Natural Relationships of Bacteria," 297–298.

66. Ibid., 290.

67. Ibid.

68. S. Winogradsky, "Sur la classification des bactéries," *Annales de l'Institut Pasteur* 82 (1952): 125–131; 131. He held the same to be true for viruses which he believed would always be identified only by their destructive effects. See also, Andrewes, "Classification and Nomenclature of Viruses."

69. Ibid. van Niel, "Classification and Taxonomy of the Bacteria and Blue Green Algae," in *A Century of Progress in the Natural Sciences* 1853–1953. (San Francisco: California Academy of Sciences, 1955), 89–114.

70. Winslow, "The Characterization and Classification of Bacterial Types," 80. C.N. Hinshelwood, "Bacterial Growth," *Biological Reviews* 19 (1944): 150–163.

71. See Olga Amsterdamska, "Stabilizing Instability: The Controversy over Cylogenic Theo-

ries of Bacterial Variation during the Interwar period," *Journal of the History of Biology* 24 (1991): 191–222.

72. W.H. Manwaring, "Environmental Transformation of Bacteria," *Science* 79 (1934): 466–470.

73. See W. Braun, "Bacterial Dissociation: A Critical Review of a Phenomenon of Bacterial Variation," *Bacteriological Reviews* 11 (1947): 75–114. van Niel viewed the claims about bacterial life cycles to be "generally disregarded" by 1946. See van Niel,"The Classification and Natural Relationships of Bacteria," 290.

74. C.D. Darlington, *The Evolution of Genetics Systems* (Cambridge: Cambridge University Press, 1939), 70.

75. J. Huxley, *Evolution: The Modern Synthesis* (London: Allen and Unwin, 1942), 131–132. As Lederberg wrote in one of his first reviews on microbial genetics, "The lack of outward differentiation of bacteria and viruses does give the appearance of holo-cellular propagation and of identity between direct transmission and inheritance. Geneticists and bacteriologists alike have . . . shown justifiable hesitation in accepting unanalyzed genetic variations as gene mutations." Joshua Lederberg, "Problems in Microbial Genetics," *Heredity* 11 (1948):145–198; 153.

76. J. Lederberg and E. Tatum, "Gene Recombination in *Escherichia coli*," *Nature* 158 (1946): 558.

77. J. Lederberg and E. Tatum, "Novel Genotypes in Mixed Cultures of Biochemical Mutants of Bacteria," *Cold Spring Harbor Symposia on Quantitative Biology* 11 (1946): 113–114.

78. A. Lwoff, "Lysogeny," *Bacteriological Reviews* 17 (1953): 269–337.

79. See J. Sapp, *Genesis. The Evolution of Biology* (New York: Oxford University Press, 2003). Beyond the Gene: *Cytoplasmic Inheritance and the Struggle for Authority in Genetics* (New York: Oxford University Press, 1987). E. Jablonka and M. Lamb, *Epigenetic Inheritance and Evolution—The Lamarckian Dimension* (Oxford: Oxford University Press, 1994).

80. J. Lederberg, "Cell Genetics and Hereditary Symbiosis," *Physiological Reviews* 32 (1952): 403–430. Idem, "Genetic Transduction," *American Scientist* 44 (1956): 264–280.

81. Sapp, *Beyond the Gene.*

82. See J. Sapp, *Evolution by Association.* Idem, *Genesis.*

83. R.S. Breed, E.G.D. Murray, and A. Parker Hitchens, *Bergey's Manual of Determinative Bacteriology*, 6th edition (Baltimore, MD: Williams and Wilkins, 1948).

84. A. Lwoff, "The Concept of Virus," *Journal of General Microbiology*, 17 (1957): 239–253; 252.

85. R.Y. Stanier and C.B. van Niel, "The Concept of a Bacterium," *Archiv für Mikrobiologie* 42 (1962): 17–35; 17.

86. E. Chatton, "Pansporella Perplexa. Reflexions sur la Biologie et la Phylogénie des Protozoaires," *Ann. des Sciences naturelles Zoologique 10e serie*, 7, (1925): 1–84; 76. Edouard Chatton, "Titre and Travaux Scientifique (1906–1937) de Edouard Chatton," (Sottano, Italy: Sette, 1938).

87. Stanier and van Niel, "The Concept of a Bacterium," 21.

88. R. Stanier, M. Douderoff, and E. Adelberg, *The Microbial World*, 2nd edition (Engelwood Cliffs, NJ: Prentice-Hall, 196

89. Stanier, Doudoroff, and Adelberg, *The Microbial World*, 2nd edition, 409.

90. Ibid.

91. L. Sagan, "On the Origin of Mitosing Cells," *Journal of Theoretical Biology* 14 (1967): 225–274.

92. L. Margulis, *Symbiosis in Cell Evolution*, 2nd edition, (New York: W.H. Freeman, 1981), 208.

93. H. Ris and W. Plaut, "Ultrastructure of DNA-Containing Areas in the Chloroplast of Chlamydomonas," *Journal of Cell Biology* 13 (1962): 383–391, 388–390. Sylan Nass and Margit M.K. Nass, "Intramitochondrial Fibers with DNA characteristics," *Journal of Cell Biology* 19 (1963): 613–628. F.J.R. Taylor, "Implications and Extensions of the Serial Endosymbiosis Theory of the Origin of Eukaryotes," *Taxon* 23 (1974): 229–258.

94. See R. Sager, *Cytoplasmic Genes and Organelles* (New York: Academic Press, 1972). N. Gillham, *Organelle Heredity* (New York: Raven Press, 1978).

95. See L. Margulis, *Symbiosis in Cell Evolution* (San Francisco: Freeman, 1982).

96. See C. Merezhkowsky, "Über Natur und Ursprung der Chromatophoren im Pflanzenreiche,"

Biologishes Centralblatt 25 (1905): 593–604. English translation in W. Martin and K.V. Kowallik, "Annoted English Translation of Merezhkowsky's 1905 paper 'Über Natur und Ursprung der Chromatophoren im Pflanzenreiche,'" *European Journal of Phycology* 34 (1999): 287–295. C. Mérejkovsky, "La Plante considére'e comme un complexe symbiotique," *Bulletin de la Société Naturelles* 6 (1920): 17–98. For a account of Merezhkowsky's life, see Jan Sapp, Francisco Carrapiço, and Mikhail Zolotonosov, "The Hidden Face of Constantin Merezhkowsky," *History and Philosophy of the Life Sciences* 24 (2002): 413–440.

97. C. Merezhkowsky, "Theorie der zwei Plasmaarten als Grundlage der Symbiogenesis, einer neuen Lehre von der Entstehung der Organismen," *Biologisches Centralblatt*, 30 (1910), 277–303, 321–347; 353–367.

98. P. Portier, *Les Symbiotes* (Paris: Masson, 1918), vii.

99. William Summers, *Félix d'Herelle and the Origins of Molecular Biology* (New Haven, CT: Yale University Press, 1999).

100. F. d'Herelle, *The Bacteriophage and Its Behavior*. Translated by George H. Smith (Baltimore, MD: Williams and Wilkins, 1926), 320.

101. See Summers, *Félix d'Herelle;* Sapp, *Evolution by Association.*

102. I.E. Wallin, *Symbionticism and the Origin of Species* (Baltimore, MD: Williams and Wilkins, 1927).

103. Sapp, *Evolution by Association.*

104. Summers, *Félix d'Herelle*, 90–91. T. van Helvoort, "The Construction of Bacteriophage as Bacterial Virus: Linking Endogenous and Exogenous Thought Styles," *Journal of the History of Biology* 27 (1994): 91–139.

105. Sapp, *Evolution by Association.*

106. See, for example, Michael Gray, Gertraud Burger, and B. Franz Lang, "Mitochondrial Evolution," *Science* 283 (1999): 1476–1481.

107. R. Stanier, "Some Aspects of the Biology of Cells and Their Possible Evolutionary Significance," in H.P. Charles and B.C. Knight, eds., *Organization and Control in Prokaryotic Cells. Twentieth Symposium of the Society for General Microbiology* (Cambridge, MA: Cambridge University Press, 1970), 1–38, 31.

108. F.H.C. Crick, 1958. "The Biological Replication of Macromolecules," *Symposia of the Society for Experimental Biology* 12 (1958): 138–163; 142.

109. See F. Sanger, "Chemistry of Insulin," *Science* 129 (1959): 1340–1344.

110. E. Zuckerkandl and L. Pauling, "Molecules as Documents of Evolutionary History," *Journal of Theoretical Biology* 8 (1965): 357–66. See also G.L. Morgan, "Emile Zuckerkandl, Linus Pauling, and the Molecular Evolutionary Clock, 1959–1965," *Journal of the History of Biology* 31 (1998): 155–178, and M. Dietrich, "The Origins of the Neutral Theory of Molecular Evolution," *Journal of the History of Biology* 27 (1994): 21–59. Idem, "Paradox and Persuasion: Negotiating the Place of Molecular Evolution within Evolutionary Biology," *Journal of the History of Biology* 31 (1998): 85–111.

111. Zuckerkandl and Pauling, "Molecules as Documents."

112. E. Zuckerkandl and L. Pauling, "Evolutionary Divergence and Convergence in Proteins," in V. Bryson and H. Vogel, eds., *Evolving Genes and Proteins* (New York: Academic Press, 1965), 97–166, 148.

113. See, for example, E. Margoliash, "Primary Structure and Evolution in Cytochrome *c*," *Procedings of the National Academy of Sciences USA* 50 (1963): 672–679.

114. R.Y. Stanier, "Toward an Evolutionary Taxonomy of the Bacteria," *Recent Advances in Microbiology*, edited by Perez-Miravete and Dionisio Peláez, Mexico: D. F. Asociatió Mexicana de Microbiologia, (1971): 595–604; 595.

115. A.M. Maxam and W. Gilbert, "A New Method for Sequencing DNA," *Proceedings of the National Academy of Sciences USA* 74 (1977): 1258. K.F. Mullis, et al., "Specific Enzymatic Amplification of DNA in vitro: The Polymerase Chain Reaction," *Cold Spring Harbor Symposia on Quantitative Biology*, 51 (1987): 263–273. See also H.F .Judson," A History of the Science and Technology Behind Gene Mapping and Sequencing," in D.J. Kevles and L. Hood, eds., *The Code of Codes: Scientific and Social Issues in the Human Genome Project* (Cambridge, MA: Harvard University Press, 1992), 37–80. P. Rabinow, *Making PCR. A Story of Biotechnology* (Chicago: University of Chicago Press, 1996).

116. See, for example, P.H.A. Sneath, "Phylogeny of Micro-Organisms," *Symposia of the Society for General Microbiology* 24 (1974): 1–39.

117. See J. Hagen, "The Introduction of Computers into Systematic Research in the United States during the 1960s." *Studies in Philosophy of Biological and Biomedical Sciences* 32 (2001): 291–314. See also, M.T. MacDonnel and R.R. Colwell, "The Contribution of Numerical Taxonomy to the Systematics of Gram-Negative Bacteria," In M. Goodfellow, D. Jones, and F.G. Priest, eds., *Computer Assisted Bacteria Systematics* (Orlando, FL: Academic Press, 1985).

118. C.R. Woese, *The Genetic Code: The Molecular Basis of Genetic Expression* (Harper and Row, New York, 1967). See also, Lily E. Kay, *Who Wrote the Book of Life? A History of the Genetic Code*. (Stanford, CA: Stanford University Press, 2000).

119. See Woese, *The Genetic Code*, 179–195.

120. Ibid., 194.

121. Ibid., 176.

122. Woese relied on a tedious, labor-intensive technique known as oligonucleotide cataloging. In this method, an rRNA molecule, which is a long string of four nucleotides (adenine, cytosine, uracil, and guanine, or A, C, U, G) was broken into small fragments by cutting it at every G residue. Each of these fragments or oligonucleotides was then broken into subfragments with enzymes that sliced at different residues. This allowed one to reconstruct the sequence of the original rRNA fragment.

123. C.R. Woese, G.E. Fox, L. Zablen, T. Uchida, L. Bonen, K. Pechman, B.J. Lewis, and D. Stahl, "Conservation of Primary Structure in 16s Ribosomal RNA," *Nature* 254 (1975): 83–86; C.R. Woese and G.E. Fox, "The Concept of Cellular Evolution," *Journal of Molecular Evolution* 10 (1977): 1–6; C.R. Woese, "Bacterial Evolution," *Microbial Reviews* 51 1987): 221–271.

124. See M.L. Sogin, "Evolution of Eukaryotic Microorganisms and their Small Subunit Ribosomal RNAs," *American Zoologist* 29 (1989): 487–499.

125. C.R. Woese and G.E. Fox, "Phylogenetic Structure of the Prokaryotic Domain: The Primary Kingdoms," *Proceedings of the National Academy of Sciences USA* 74 (1977): 5088–5090; 5089.

126. Woese and Fox, "Phylogenetic Structure," 5089.

127. C. Darwin to J. Hooker, 1871 in F. Darwin ed., *Life and letters of Charles Darwin*, vol. 3 (London: John Murray, 1887), 18.

128. R.H. Whittaker, "New Concepts of Kingdoms of Organisms," *Science* 163 (1969): 150–160. R.H. Whittaker and L. Margulis, "Protist Classification and the Kingdoms of Organisms, *Biosystems* 10 (1978): 3–18. L. Margulis and K.V. Schwartz, *Five Kingdoms and Illustrated Guide to the Phyla of Life on Earth* (San Francisco: W.H. Freeman, 1988).

129. G. Olsen, C. Woese, "Lessons from an Archael Genome: What are we Learning from *Methanococcus jannaschii*?" *Trends in Genetics* 12 (1996): 377–379.

130. N. Pace, "A Molecular View of Microbial Diversity and the Biosphere," *Science* 276 (1997): 734–740, 734.

131. W.B. Whitman, D.C. Coleman, and W.J. Wiebe, "Prokaryotes: The Unseen Majority," *Proceedings of the National Academy of Sciences USA* 95 (1998): 6578–6583.

132. Pace, "A Molecular View."

133. N.R. Pace, D.A. Stahl, D.J. Lane, and G.J. Olsen, "Analyzing Natural Microbial Populations by rRNA Sequences," *American Society of Microbiology News* 51 (1985): 4–12.

134. Pace, "A Molecular View."

135. J.P. Gogarten et al., "Evolution of the Vacuolar H+ -ATPase: Implications for the Origin of Eukaryotes," *Proceedings of the National Academy of Sciences USA* 86 (1989): 6661–6665. N. Iwabe et al., "Evolutionary Relationships of Archaebacteria, Eubacteria and Eukaryotes Inferred from Phylogenetic Trees of Duplicated Genes," *Proceedings of the National Academy of Sciences USA* 86 (1989): 9355–9359.

136. C.R. Woese, O. Kandler, and M.L. Wheelis, "Towards a Natural System of Organisms: Proposal for the Domains Archaea, Bacteria, and Eucarya, "*Proceedings of the National Academy of Sciences USA* 87 (1990): 4576–4579.

137. E. Hilario and P. Gogarten, "Horizontal transfer of ATPase genes—The Tree of Life Becomes a Net of Life," *BioSystems* 31 (1993): 111–119.

138. R.D. Fleischmann and 35 others, "Whole-Genome Random Sequencing and Assembly

of *Haemophilus influenzae* Rd.," *Science* 269 (1995): 468–470. D.J. Bult, O. White, G.J. Olsen, and C.R. Woese, "Complete Genome sequence of the Methanogenic Archaeon, *Methanococcus jannaschii*," *Science* 273 (1996): 1058–1073. Olsen and Woese, "Lessons from an Archaeal Genome, 377–379.

139. C. Woese, "A Manifesto for Microbial Genomics," *Current Biology* 8 (1999): R781–R783.

140. See Dietrich, "Paradox and Persuasion," 87–111.

141. See P. Abir-Am, "The Politics of Macromolecules: Molecular Biologists, Biochemists, and Rhetoric, *Osiris* 7 (1992): 164–191.

142. T. Dobzhansky, "Evolutionary and Population Genetics," *Science* 142 (1963): 1131–1135.

143. E. Mayr, "The New Versus the Classical in Science," *Science*, 141 (1963): 763.

144. See M. Kimura, *The Neutral Theory of Molecular Evolution* (Cambridge: Cambridge University Press, 1983). Morgan, "Emile Zuckerkandl, Linus Pauling, and the Molecular Evolutionary Clock," 155–178. See also Dietrich, "The Origins of the Neutral Theory of Molecular Evolution," 21–59. Idem, "Paradox and Persuasion." W. Provine, "The Neutral Theory of Molecular Evolution in Historical Perspective," in B. Takahata and J. Crow, eds., *Population Biology of Genes and Molecules* (Tokyo: Baifukan, 1990), 17–31.

145. Simpson and Mayr protested that evolution worked on whole phenotypes and could not single out genes unless they had phenotypic effects separable both phenotypically and genetically from other genes. See Dietrich, "Paradox and Persuasion," 97. William Provine has suggested that underlying the confrontation there was substantial confusion about what neutral theorists were arguing. Neutral theorists were not arguing that functional proteins emerged without selection, but only that one could detect amino acid sequence changes unaffected by selection. In other words, drift and natural selection were not mutually exclusive alternatives. Provine,"The Neutral Theory of Molecular Evolution in Historical Perspective."

146. G.G. Simpson, *Concession to the Improbable: An Unconventional Autobiography* (New Haven, CT: Yale University Press, 1978), 269.

147. M. Nei, *Molecular Evolutionary Genetics* (New York: Columbia University Press, 1987).

148. See C. Woese, "The Archaea: their History and Significance," in M. Kates, D.J. Kushner and A.T. Matheson, eds., *The Biochemistry of Archae* (*Archaebacteria*) (Dordrecht: Elsevier, 1993), vii–xxxix, vii. See also, Virginia Morell, "Microbiology's Scarred Revolutionary" *Science* 276 (1997): 699–702.

149. See E. Mayr, "A Natural System of Organisms," *Nature* 348 (1990): 491. See also, T. Cavalier-Smith, "Bacteria and Eukaryotes," *Nature* 356 (1992): 570.

150. E. Mayr, "Two Empires or Three?"*Proceedings of the National Academy of Sciences USA* 95 (1998): 9720–9723; 9722.

151. Ibid., 9722. Protistologists counted 200,000 species of protists. See John Corliss, "Toward a Nomenclatural Protist Perspective," in L. Margulis, H. I. McKhann, L. Olendzenski, eds., *Illustrated Glossary of Protoctista* (Boston: Johns and Bartlett, 1993), xxvii–xxxii.

152. Mayr, "Two Empires or Three," 9723.

153. C.R. Woese, "Default Taxonomy; Ernst Mayr's View of the Microbial World," *Proceedings of the National Academy of Sciences USA* 95 (1998): 11043–11046; 11045.

154. See W.F. Doolittle, "Phylogenetic Classification and the Universal Tree," *Science* 284 (1999): 21124–21128.

155. For an overview of the biology of lateral gene transfer, see F. Bushman, *Lateral DNA Transfer. Mechanisms and Consequences* (New York: Cold Spring Harbor Laboratory Press, 2002).

156. See, however, E.S. Anderson, "Possible Importance of Transfer Factors in Bacterial Evolution," *Nature* 209 (1966): 637–638, and D. Reanney, "Extrachromosomal Elements as Possible Agents of Adaptation and Development," *Bacteriological Reviews* 40 (1976): 552–590.

157. T. Watanabe, "Infective Heredity of Multiple Drug Resistance in Bacteria," *Bacteriological Reviews* 27 (1963): 87–115.

158. F. Jacob and E.L. Wollman, "Les épisomes, élements génétiques ajoutés," *Comptes Rendus des Académie des Sciences* 247 (1958): 75–92.

159. Watanabe, "Infective Heredity," 108.

160. H.C. Neu, "The Crisis in Antibiotic Resistance," *Science* 257 (1992): 1064–1073.

161. See J. Claverys, M. Prudhomme, I. Mortier-Barriere, and B. Martin, "Adaptation to the Environment: *Streptococcus pneumoniae*, a Paradigm for Recombination-Mediated Genetic Plasticity?" *Molecular Microbiol*ogy 35 (2000): 251–259. J. Davison, "Genetic Exchange Between Bacteria in the Environment" *Plasmid* 42 (1999): 73–91. J. Lawrence and J. Roth, "Evolution of Coenzyme B12 Synthesis Among Enteric Bacteria: Evidence for Loss and Reacquisition of a Multigene Complex," *Genetics* 142 (1997): 11–24. M. Maiden, "Horizontal Genetic Exchange, Evolution, and Spread of Antibiotic Resistance in Bacteria," *Clinical Infectious Diseases* 27 (1998): S12–S20. K. Sowers and H. Schreier, "Gene Transfer Systems for the Archaea," *Trends in Microbiol*ogy 7 (1999): 212–219.

162. Roger Stanier had recognized this problem in 1971 when he wrote that, "genetic studies on procaryote are complicated by a phenomenon no known to exist among eucaryote. A bacterium may be a genetic chimera, some of its phenotypic traits being determined by episomes that are transferable among (and expressed in) a considerable range of species, having markedly different chromosomal genomes. It is therefore conceivable that false inferences concerning the relatedness of a series of bacteria could be reached by the study of one or more shared characters determined by episomal genes." Stanier, "Toward an Evolutionary Taxonomy of the Bacteria," 597. See also, Sneath, "Phylogeny of Micro-Organisms."

163. T.E. Meyer, R.G. Bartsch, and M.D. Kamen, "Cytochrome *c2* sequence variation among the recognised species of purple nonsulphur photo synthetic bacteria," *Nature* 278 (1979): 659–660. R.P. Ambler, T.E. Meyer, and M.D. Kamen, "Anomalies in amino acid sequence of small cytochrome *c* and cytochromes *c'* from two species of purple photo synthetic bacteria," *Nature* 278 (1979): 661–662.

164. Meyer, Bartsch, and Kamen, "Cytochrome *c2* Sequence Variation," 660.

165. Richard Dickerson, "Evolution and Gene Transfer in Purple Photosynthetic Bacteria," *Nature* 283 (1980): 210–212.

166. C.R Woese, J. Gibson, and G.E. Fox, "Do Genealogical Patterns in Purple Photosynthetic Bacteria Reflect Interspecific Gene Transfer?" *Nature* 283(1980): 212–214.

167. T.E. Meyr, M.A. Cusanovich, and M.D. Kamen, "Evidence Against Use of Bacterial Amino Acid Sequence Data for Construction of All-Inclusive Phylogenetic Trees," *Proceedings of the National Academy of Sciences USA* 83 (1986): 217–220; 220. Like van Niel four decades earlier, they maintained "a wary optimism that a natural classification for bacteria will eventually emerge." p. 220.

168. J. Raymond, O. Zhaxybayeva, J.P. Gogarten, S.Y. Gerdes, and Re. E. Blankenship, "Whole-Genome Analysis of Photosynthetic Prokaryotes," *Science* 298 (2002): 1616–1620.

169. J.G. Lawrence and H. Ochman, "Molecular Archaeology of the *Escherichia coli* genome," *Proceedings of the National Academy of Sciences USA* 95 (1998): 9413–9417. See also H. Ochamn, J.G. Lawrence, and E.A. Grolsman, "Lateral Gene Transfer and the Nature of Bacterial Innovation," *Nature* 405 (2000): 299–304.

170. See William Martin, "Mosaic Bacterial Chromosomes: A Challenge en route to a Tree of Genomes," *BioEssays* 21 (1999): 99–104.

171. Doolittle, "Phylogenetic Classification."

172. See H. Ochman, Jeffery G. Lawrence, and Eduardo Grolsman, "Lateral Gene Transfer and the Nature of Bacterial Innovation," *Nature* 405 (2000): 299–304. Jonathan Eisen, "Horizontal gene Transfer Among Microbial Genomes: New Insights from Complete Genome Analysis," *Current Opinion in Genetics and Development* 10 (2000): 606–611.

173. See A. Lwoff, "La notion d'espèce bactérienne à la lumière des découvertes recentes. L'espèce bactérienne" *Annales Institutes Pasteur* 94 (1958): 137. P. Shaeffer, "La notion d'espèce après les recherches récente de génétique bactérienne," *Annales of the Institutes Pasteur* 94 (1958): 167–178.

174. S.T. Cowan, "The Microbial Species—A Macromyth," in *Microbial Classification*, 12th Symposium of the Society for General Microbiology (Cambridge: Cambridge University Press, 1962), 433–455; 451.

175. S. Sonea and P. Panisset, *The New Bacteriology* (Boston: Jones and Bartlett, 1983). S. Sonea and L.G. Mathieu, *Prokaryotology* (Montreal: Les Presses de l'Université de Montréal, 2000).

176. W. Martin, "Mosaic Bacterial Chromosomes." W.F. Doolittle, "Uprooting the Tree of Life," *Scientific American* February (2000), 90–95. D.M. Faguy and W.F. Doolittle, "Horizontal Transfer of Catalase-Peroxidase Genes Between Archaea and Pathogenic Bacteria." *Trends in Genetics* 16(2000): 196–197.

177. M. Rivera and J. Lake, "Evidence that Eukaryotes and Eocyte Prokaryotes are Immediate Relatives," *Science* 257 (1992): 74–76.

178. R. Gupta, "Protein Phylogenies and Signature Sequences: A Reappraisal of Evolutionary Relationships among Archaebacteria, Eubacteria, and Eukaryotes," *Microbiology and Molecular Biological Reviews* 62 (1998): 1435–1491. See also, E.V. Koonin A.R. Musegian, M.Y. Galperin, and D.R. Walker, "Comparison of Archaeal and Bacterial Genomes: Computer Analysis of Protein Sequences Predicts Novel Functions and Suggests a Chimeric Origin for the Archaea," *Molecular Microbio*logy 25 (1997): 619–637.

179. G.E. Fox, K.R. Pechman, and C.R. Woese, "Comparative Cataloging of 165 Ribosomal Ribonucleic Acid: Molecular Approach to Procaryotic Systematics, *International Journal of Systematica, 27* (1977): 44–47. See also R. Jain, M.C. Rivera, and J.A. Lake, "Horizontal Gene Transfer Among Genomes: the Complexity Hypothesis," *Proceedings of the National Academy of Sciences USA* 99 (1999): 3801–3806.

180. As William Martin has noted, there is an apparent circular element to this logic because if the groups are defined by their rRNA, there can be no lateral exchange of rRNA by definition. See, however, T. Asai, D. Zaporojets, C. Squires, and C.L. Squires, "An *Esherichia coli* Strain With All Chromosomal rRNA Operons Inactivated: Complete Exchange of rRNA Genes Between Bacteria," *Proceedings of the National Academy of Sciences USA* 96 (1999): 1971–1976.

181. W. Martin, "Mosaic Bacterial Chromosomes." Doolittle, "Phylogenetic Classification and the Universal Tree."

182. C. Woese, "The Universal Ancestor," *Proceedings of the National Academy of Sciences USA* 95 (1998): 6854–6859. See also, O. Kandler, "The Early Diversification of Life and the Origin of the Three Domains: A Proposal," in Juergen Wiegel and Michael Adams, eds., *Thermophiles: The Keys to Molecular Evolution and the Origin of Life?* (New York, Taylor and Francis, 1998), 19–31.

183. Darwin had reasoned that, "probably all organic beings which have ever lived on this earth have descended from one primordial form, into which life was first breathed." C. Darwin, *On the Origin of Species*, with an introduction by E. Mayr, facsimile edition of 1859 (Cambridge, MA: Harvard University Press, 1964), 484.

184. Woese, "The Universal Ancestor," 6854.

185. C.R. Woese and G.E. Fox, "Phylogenetic Structure of the Prokaryote Domain," 5088–5090. C.R. Woese, "Endosymbionts and Mitochondrial origins," *Journal of Molecular Evolution* 10 (1977): 93–96.

186. M.W. Gray and W.F. Doolittle, "Has the Endosymbiont Hypothesis Been Proven?" Microbial Reviews 46 (1982): 1–42. See also Michael Gray, "The Endosymbiont Hypothesis Revisited," *International Review of Cytology* 141 (1992): 233–257.

187. J.L. Hall and D.J. Luck, "Basal Body-Associated DNA: *In situ* studies in *Chlamydomonas reinhardtii," Proceedings of the National Academy of Sciences USA* 92 (1995): 5129–5133. One of the biological lessons from this is that cell structures do not have to have nucleic acids to be inherited. See also, Jan Sapp, "Freewheeling Centrioles," *History and Philosophy of the Life Sciences* 20 (1998): 255–290.

188. In 1974, Jeremy Pickett-Heaps supported the idea the nucleus was a symbiont by noting that some dinoflagellates have two nuclei, "one characteristic of dinoflagellates, and the other a more typical eukaryotic nucleus." J. Pickett-Heaps, "The Evolution of Mitosis and the Eukaryotic Condition," *BioSystems* 6 (1974): 37–48.

189. S.D. Bell and S.P. Jackson, "Transcription and Translation in Archae: A Mosaic of Eukaryal and Bacterial Features," *Trends in Microbiology* 6 (1997): 222–228. W.F. Doolittle, "Some Aspects of the Biology of Cells and their Evolutionary Significance," *in* D.M. Roberts, P. Sharp, G. Alderson, and M.A. Collins, eds., *Evolution of Microbial Life: 54th Symposium of the Society for General Microbiology* (Cambridge: Cambridge University Press, 1996), 1–26.

190. J.R. Brown and W.F. Doolittle, "Archaea and the Prokaryote-to-Eukaryote Transition," *Microbiology and Molecular Biology Reviews* 61 (1997): 456–502. D.F. Feng, G. Co, and R.F.

Doolittle, "Determining Divergence Times and a Protein Cock: Update and Re-Evaluation," *Proceedings of the National Academy of Sciences USA* 94 (1997): 13028–13033.

191. W. Martin and M. Muller, "The Hydrogen Hypothesis for the First Eukaryote," *Nature* (London) 392 (1998): 37–41. W. Martin and R.G. Hermann, "Gene transfer from Organelles to the Nucleus: How Much, What Happens and Why?" *Plant Physiology* 118 (1998): 9–17. A. Roger, "Reconstructing Early Events in Eukaryotic Evolution," *The American Naturalist* 154 (1999): S146–S163.

192. See Martin and Muller, "The Hydrogen Hypothesis," 37–41. T. Marin Embley and William Martin, "A Hydrogen-Producing Mitochondrion," *Nature* 396 (1998): 517–519.

193. Kurland and Andersson have argued that the main benefit initially derived from mitochondrial acquisition was oxygen detoxification of its host, not ATP production. According to Kurland and Andersson's idea, certain anaerobic archaebacteria might have survived in environments that were becoming oxidized by respiring proteobacteria. The engulfed symbionts using O_2 for respiration may have decreased the intracellular oxygen concentration of host cells. In parallel, the symbiont could get into a new niche. but rather clearing from the host O_2 and its free radicals. C.G. Kurland and S.G.E. Andersson, "Origin and Evolution of the Mitochondrial Proteome," *Microbiology and Molecular Biology Reviews* 64 (2000): 786–820.

194. T. Cavalier-Smith, "Endosymbiotic Origin of the Mitochondrial envelope in W. Schwemmler and H.E.A. Schenk, eds., *Endocytobiology II. Intracellular Space as an Oligogenetic Ecosystem*, (Berlin: De Gutyer, 1983), 265–280. Idem, "A Six-Kingdom Classified and Unified Phylogeny," Ibid., 1027–1034.

195. Roger, "Reconstructing Early Events in Eukaryotic Evolution," 153. He remarked, "At this stage, it seems reasonable to suggest the data indicate that mitochondrial endosymbiosis was established in a common ancestor of most, if not all, known living eukaryotes. One of the most astonishing implications of this view is that classical mitochondria appear to have been lost 13–16 times during the course of eukaryotic evolution."

196. Michael Chapman, Michael Dolan, and Lynn Margulis, "Centrioles and Kinetosomes: Form, Function, and Evolution." *Quarterly Review of Biology* 75 (2000): 409–429.

197. See C. de Duve, "The Birth of Complex Cells," *Scientific American* (1996): 50–57.

198. More recently Kurland and Anderson have suggested that mitochondrial may have actually first benefitted their host in this manner.

199. Doolittle, "Some Aspects of the Biology of Cells." M.L. Sogin, "History Assignment: When was the Mitochondrion Founded?" *Current Opinion in Genetics and Development* 7 (1997): 792–799.

200. See Sapp, *Evolution by Association.*

201. S. Watasē. "On the Nature of Cell Organization," *Woods Hole Biological Lectures*, 1893, 83–103. See also, Sapp, *Evolution by Association.*

202. S. Watasē. "On the Nature of Cell Organization." Theodor Boveri, "Ergennisse über die Konstitution der chromatischen Kernsubstance," Jena (1904), 90. Merezhkowsky, "Theories der Zwei Plasmaarten," 277–303, 321–347, 353–367. For historical discussion, see Sapp, *Evolution by Association.*

203. James Lake, E. Henderson, M.W. Clark, and A.T. Matheson, "Mapping Evolution with Ribosome Structure: Intralineage Constancy and Interlineage Variation," *Proceedings of the National Academy of Sciences USA* 79 (1982): 5948–5952, 5951; James Lake and Maria Rivera, "Was the Nucleus the First Symbiont?" *Proceedings of the National Academy of Sciences USA* 91 (1994): 2880–2881. See also, T. Horiike, K. Hamada, S. Kanaya, and T. Shinozawa, "Origin of Eukaryotic Cell Nuclei by Symbiosis of Archaea in Bacteria is Revealed by Homology-Hits Analysis," *Nature Cell Biology* 3 (2001): 210–214. M. Takemura, "Poxviruses and the Origin of the Eukaryotic Nucleus," *Journal of Molecular Evolution* 52(2001): 419–425.

204. M.L. Sogin "Early Evolution and the Origin of Eukaryotes," *Current Opinion in Genetics and Development* 4 (1991): 457–463. W. Zillig, Palm, P., Klenk, H.P., Pühler, G. Groop, F., and C. Schliper, "Phylogeny of DNA-dependent RNA-Polymerases: Testimony for the Origin of Eukaryotes," in F. Rodriguez-Valera ed., *General and Applied Aspects of Halophilic Microorganisms* (New York: Plenum Press, 1991), 321–332. R.F. Doolittle, "The Origins and Evolution of Eukaryotic Proteins," *Philosophical Transactions of the Royal Society Science* 349 (1995): 235–240. D. Moreira and P. Lopez-Garcia, "Symbiosis Between Methogenic Archaea and

alpha-Proteobactera as the Origin of Eukaryotes: The Syntrophic Hypothesis," *Journal of Molecular Evolution*. 47 (1998): 517–530.

205. H. Hartmann and A. Fedorov, "The Origin of the Eukaryotic Cell: A Genomic Investigation," *Proceedings of the National Academy of Sciences USA* 99 (2002): 1420–1425.

206. See R. Gupta and G.B. Golding, "The Origin of the Eukaryotic Cell," *Trends in Biology* 21 (1996): 166–170; R. Gupta, "Protein Phylogenies and Signature Sequences: A Reappraisal of Evolutionary Relationships Among Archaebacteria, Eubacteria and Eukaryotes," *Microbiology and Molecular Biology Reviews* 62 (1998): 1435–1491.

207. See Jan Sapp, *Genesis*.

208. S. Jay Gould, *Wonderful Life. The Burgess Shale and the Nature of History* (London: Hutchison Radius, 1989), 310. Symbiosis is not mentioned at all, and bacteria are allotted three pages out of 1200, in Stephen Jay Gould's treatise, *The Structure of Evolutionary Thought* (Cambridge, MA: Harvard University Press, 2002).

209. J. Maynard Smith and E. Szathmáry, *The Origins of Life. From the Birth of Life to the Origin of Language* (New York: Oxford University Press, 1999), 107. See also Idem, *The Great Transitions in Life* (New York: W.H. Freeman, 1995), 195.

210. See Jan Sapp, *Genesis*.

211. J.H. Werren, "Biology of Wolbachia," *Annual Review of Entomology* 42 (1997): 587–609. See also, James Higgins and Abdu F. Azad, "Use of Polymerase Chain Reaction to Detect Bacteria in Arthropods: A Review," *Journal of Medical Entomology* 32 (1995): 13–22. Wenseelers, F. Ito, S. van Taorm, R. Huybrechts, F. Volckhaert, and J. Billen. "Widespread Occurrence of the Micro-Organism Wolbachia in Ants," *Proceedings of the Royal Society of London B* 265 (1998): 1447–1452. J.H. Werren and D. Windsor, "Wolbachia Infection Frequencies in Insects: Evidence of A Global Equilibrium," *Proceedings of the Royal Society of London R* 267 (2000): 1977–1995.

212. See C. Zimmer, "Wolbachia: A Tale of Sex and Survival," *Science* 292 (2001): 1093–1096.

213. H. Abdelaziz et al., "Four Intracellar Genomes Direct Weevil Biology: Nuclear, Mitochondrial, Principal Endosymbiont, and *Wolbachia*," *Proceedings of the National Academy of Sciences USA* 96 (1999): 6814–6819.

214. C.B. Beard, R.V. Durvasula, and F.F. Richards, "Bacterial Symbiosis in Arthropods and the Control of Disease Transmission," *Emerging Infectious Diseases* 4 (1998): 581–591.

2

The Large-Scale Structure of the Tree of Life

NORMAN R. PACE

The development of molecular phylogeny opened an entirely new perspective on the diversity and history of life. Before the comparison of gene sequences, relationships of fossils provided the main view of the course of macroevolution. The traits used to relate fossil remnants were primarily morphological, however, and could not be applied to microbial organisms. Thus, the evolution of microbial lineages, which constitute most cellular diversity, could not be traced. With the techniques of molecular phylogenetics, contemporary organisms are relatable quantitatively, in terms of DNA sequence differences, regardless of morphology. Variation in sequences thus is a measure of the extent of biodiversity. Gene sequences also can be used for the inference of maps of the history of evolution, in the form of phylogenetic trees. The results are illuminating and provide grist for conjecture and controversy over the evolutionary process. The purpose of this chapter is to tour the emerging large-scale structure of the phylogenetic tree of life. I emphasize how our understanding of the extent of the tree has expanded because of recent molecular studies of microbial diversity in the environment.

Molecular Phylogeny—Inference of Phylogenetic Trees

The basic notion of molecular phylogeny is simple. Sequences of orthologous genes (genes with common ancestry and function) from different organisms are aligned so that corresponding DNA bases can be compared. Differences between pairs of sequences are calculated and treated as some measure of the evolutionary distance that has separated the pairs of organisms. Just as geographical maps can be constructed from distances between land features, evolutionary maps ("phylogenetic trees") can be inferred from evolutionary distances (sequence changes) between homologous genes. Calculations of the paths of evolution are fraught with statistical uncertainties, however.

The process of inferring the best relatedness trees from pairwise sequence counts is complex and depends on the models of evolution used to calculate such trees.[1] One complexity that vexes attempts to infer the deeper relationships in the universal phylogenetic tree with certainty is that the actual number of sequence changes was greater than the observed number. This is because of the probabilities of back mutations, where no change is seen, and multiple past mutations, which are counted as only one change. Numbers of probable mutational events per observed mutation can be estimated statistically, but then a significant amount of the information used to build trees becomes inferential, not directly observed. The mathematics of estimating probable changes from observed change are such that deeper branch points in phylogenetic trees are accompanied by greater statistical uncertainty as to their position. Still another complexity is that different lines of descent have evolved at different rates, which confuses tree-building algorithms.

Current advanced methods for inference of phylogenetic relationships are well developed collectively to cope with the problems mentioned, and with others, but statistical vagaries are inescapable. The methods in common use are dependent on different models for reconstructing relationships, and this can influence the topological outcome of phylogenetic calculations. Popular methods for inferring phylogenetic trees from sequence relationships include evolutionary distance (ED), maximum parsimony (MP), and maximum likelihood (ML). ED uses corrected sequence differences directly as distances to calculate the pattern of ancestral connections. MP presumes that the fewest changes make the best trees, so optimal relatedness patterns are estimated by the minimum number of changes required to generate the topology. ML is a statistical method that calculates the likelihood of a particular topology given the sequence differences. In each case, statistical uncertainties in the calculations render any particular result questionable. As a consequence, nodes in trees are tested many times using the same method and with subsets of the sequence collection: so-called "bootstrap analysis." The reliability of a particular result, for instance, a branch-point in a tree or the composition of a relatedness group, is tested by the frequency with which the result occurs in the set of bootstrap trees. At the current state of their development, the different methods for calculating phylogenetic trees give generally comparable results. Nonetheless, intrinsic uncertainties in any tree must be acknowledged, particularly those dealing with the placement of deeper branches.

What Gene for Deep Phylogeny?

Any collection of orthologous gene sequences can be used to infer phylogenetic relationships among those genes. Genes used to infer the overall structure of evolution—a universal phylogenetic tree that depicts relationships between genetic lines of descent and not simply genes—have special constraints on their properties.[2] One is that the gene must occur in all forms of life, so that all can be related to one another: The hemoglobin gene, for instance, would not be useful for large-scale phylogeny because most organisms do not contain the gene. A second constraint is that the gene must have resisted, over the ages, lateral transfer between genetic lines of descent. Genomic studies have shown clearly that many kinds of genes have experienced extensive lateral transfer during the course of their evolution.[3,4] Such genes cannot be used to track the cellular lines of descent because gene history is not concordant with organismic history. A third constraint on genes for inference of global phylogenetic trees is content of sufficient information—numbers of homologous nucleotides—so that relationships can be established with the best statistical reliability. There are not many genes that meet all these requirements. Most genes occur in only a limited diversity of organisms, and many have undergone lateral transfer. The most generally accepted large-scale phylogenetic results are based on the use of ribosomal RNA (rRNA) gene sequences—those of the large (LSU) and small (SSU) subunits of rRNAs. Ribosomes are present in all cells and major organelles, and phylogenetic trees inferred with these gene sequences are congruent with trees constructed using other elements of the cellular nucleic acid–based, information-processing machinery. Thus, changes in the rRNA sequences seem to reflect the evolutionary path of the genetic machinery.

SSU rRNA sequences were first used for phylogenetic studies by Carl Woese, even before it was possible to determine gene sequences rapidly. Woese prepared radioactive rRNAs from many diverse organisms, mostly microbes, and compared their content of short patches, of sequences (fragments called oligonucleotides). The prevailing notion of life's evolutionary diversity at the time was framed in the context of two kinds of organisms, prokaryotes and eukaryotes. As a consequence, it was surprising that the rRNA sequences from diverse organisms fell into three, not two, fundamentally distinct groups.[5] There had to be three primary lines of evolutionary descent, three "domains" of life. These are now termed Archaea (formerly Archaebacteria), (eu)Bacteria, and Eucarya (eukaryotes).[6] Woese's 1977 paper reporting the discovery of Archaea sparked publicity and controversy. The concept of three primary relatedness groups of life touched off a flurry of refutations defending the prokaryote–eukaryote or the five-kingdoms notions to account for biological organization. These familiar notions had never previously been tested, however, and the analysis of rRNA sequences proved them to be fundamentally incorrect. The shift in public and textbook treatment of the large organization of life is ongoing.

The Three Phylogenetic Domains of Life

Figure 2.1 is derived from a maximum likelihood tree calculated using the particular set of rRNA sequences.[7] The figure is a rough map of the course of evolution of the genetic core of cells (the collection of genes that propagates replication and gene expression). The dimension along the lines is sequence change, not time. Estimated evolutionary change that separates contemporary sequences (organisms) is read along line segments. The "root" of the universal tree, the point of origin for modern lineages, cannot be established using sequences of only one type of molecule. However, phylogenetic studies of gene families that resulted from gene duplications before the last common ancestor of the three domains place the root on the bacterial line.[8,9] This means that Eucarya and Archaea had a common history that excluded the descendants of the bacterial line. This period of evolutionary history shared by Eucarya and Archaea was an important time in the evolution of cells, during which the refinement of the primordial information-processing mechanisms occurred.[10] Thus, modern representatives of

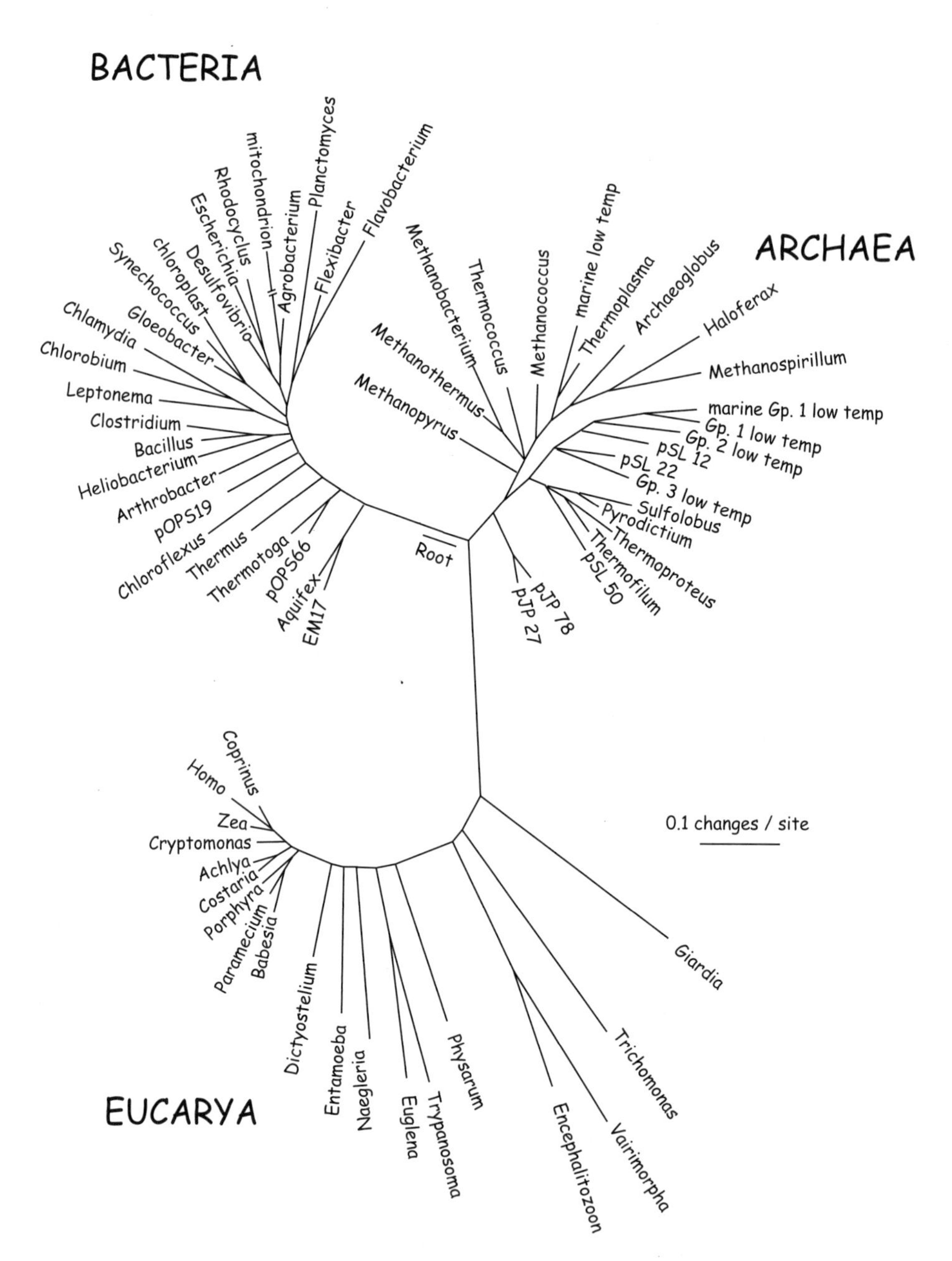

Figure 2.1. Universal tree based on small-subunit ribosomal RNA sequences. Sixty-four rRNA sequences representative of all known phylogenetic domains were aligned, and a tree was produced with a maximum-liklihood method.[7] That tree was modified, resulting in the composite one shown, by trimming and adjusting branch points to incorporate the results of other analyses. The scale bar corresponds approximately to 0.1 changes per nucleotide.

Eucarya and Archaea share many properties that differ from bacterial cells in fundamental ways. One example of the similarities and differences is in the nature of the transcription machinery. The RNA polymerases of Eucarya and Archaea resemble each other far more than either resembles the bacterial type of polymerase. Moreover, whereas all bacterial cells use sigma factors to regulate the initiation of transcription, eucaryal and archaeal cells use TATA-binding proteins.[11,12] The shared evolutionary history of Eucarya and Archaea indicates that we may be able to recognize fundamental elements of our own cells through study of the far simpler archaeal version.

The specific relationship of the phylogenetic domains Eucarya and Archaea also means that the time-honored grouping of "prokaryotes" is no longer intellectually tenable. The term has two meanings, both proven fundamentally incorrect by the topology of the universal tree. One meaning of "procaryote" is "noneukaryote," to distinguish organisms that lack a nuclear membrane from those that do. Logically, however, lack of a quality cannot be a distinguishing property that would specifically relate Bacteria and Archaea. No information is no information. The other, formal, meaning of "prokaryote" would indicate "predecessor of eukaryote." Instead, the molecular trees show that the eukaryotic nuclear line of descent extends into the precellular period. The nuclear component of the modern eukaryotic cell could not have derived from an ancient bacterial or archaeal symbiosis, because those lines of descent were themselves not established at the time of emergence of the eucaryal line. Molecular trees based on rRNA and other reliable genes show unequivocally that the eukaryotic nuclear line of descent is as old as the archaeal line.

The rRNA sequence information, along with other molecular data, solidly confirms the century-old notion that mitochondria and chloroplasts are derived from bacterial symbionts.[13] Sequence comparisons establish that mitochondria are representatives of the Proteobacteria, the group indicated by *Escherichia* and *Agrobacterium* in figure 2.1. Chloroplasts are derived from cyanobacteria, represented by *Synechococcus* and *Gloeobacter* in figure 2.1. Thus, both the respiratory and photosynthetic capacities of eukaryotic cells were obtained from bacterial symbionts. Because mitochondria and chloroplasts diverge from peripheral branches in molecular trees (fig. 2.1), their incorporation into the eukaryotic cell must have occurred relatively late in evolutionary history, after the radiation that gave rise to the main bacterial groups. Moreover, the most deeply divergent eukaryotes in phylogenetic trees even lack mitochondria. These latter kinds of organisms, little-studied but sometimes troublesome anaerobic creatures such as *Giardia* and *Trichomonas*, nonetheless contain at least a few bacterial-type genes.[14] These genes may be evidence of an earlier symbiosis that was lost, or perhaps a gene transfer event between the evolutionary domains.

It's a Microbial World

A sobering aspect of large-scale phylogenetic trees such as shown in figure 2.1 is the graphical realization that most of our knowledge in biological sciences has focused on but a small slice of biological diversity. Thus, the organisms most represented in our textbooks of biology, animals (*Homo* in Fig. 2.1), plants (*Zea*) and fungi (*Coprinus*), constitute peripheral branches of eukaryotic cellular diversity. Life's diversity is mainly microbial in nature. Although the biosphere is absolutely dependent on the activities of microorganisms, our understanding of the makeup and natural history of microbial ecosystems is, at best, rudimentary. One reason for the paucity of information is that microbial biologists traditionally have relied on laboratory cultures for the detection and identification of microbes. Yet more than 99% of environmental microbes are not cultured using standard techniques. As a consequence, the makeup of the natural microbial world remains largely unknown.[15]

The development of cloning and sequencing technology, coupled with the relational perspective afforded by phylogenetic trees, made it possible to identify environmental microbes without the requirement for culture.[15] The phylogenetic types of microbes in natural communities can be assessed by sequencing rRNA genes cloned directly from environmental DNA. This molecular approach to the analysis of microbial ecosystems sidesteps the need to culture organisms to learn something about them. The sequences are incisive identifiers of the organisms and can be used as the basis for procedures with which to study the organisms in their natural habitats. As diagrammed in figure 2.2, rRNA genes are amplified from environmental DNA and sequenced. The result is a rough census of the phylogenetic kinds of organisms that make up a community. The sequences then can be used to design tools such as fluorescently labeled hybridization probes and specific primers for polymerase chain reactions with which to visualize and study particular organisms in their natural settings and to aid in their culture.

A sequence-based phylogenetic assessment of an uncultured organism can provide insight into many of the properties of the organism through comparison with its studied relatives. Many of the phylotypes encountered in the environment have no close rela-

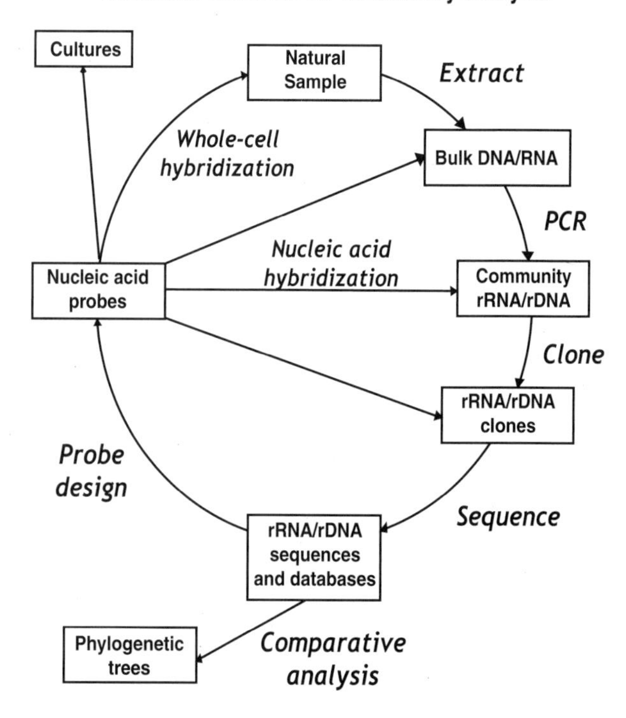

Figure 2.2. Molecular methods applied to microbial community analysis. Through the use of the suite of molecular methods described in the diagram, environmental microbes can be identified, visualized, and counted. The methods make possible the study of microbial natural history, previously generally intractable. (Modified from reference 35.

tives in the culture collections. As a consequence, little can be inferred about the properties of the organisms that correspond to the sequences. Regardless of the properties of the organisms they represent, however, environmental rRNA sequences have provided additional perspective on the topology of the universal tree.

Bacteria

Most of our knowledge of bacteria has derived from the study of only a few types, mainly cultured organisms and in the context of disease or industrial products. Any general census of bacteria that make up natural microbial communities was not possible until the development of the molecular methods that identify rRNA sequence-based phylotypes without culture. Although studies of environmental microbes have only begun, it is already evident that culture-based techniques did not address the main breadth of microbial diversity. Nonetheless, as rRNA sequences have accumulated, the large pattern of bacterial evolution has emerged.

The phylogenetic tree shown in figure 2.1 is based on a calculated result, with the sequences included. Trees inferred with such a diversity of sequences can accurately portray relationships between the domains, but the order of branches within the domains is likely to be inaccurate because of the small number of taxa selected for the analysis. A summary of the results of tree calculations with different methods and different suites of bacterial rRNA sequences is diagrammed in figure 2.3.[16] The wedges indicate the radiations of the major clades, the major relatedness groups that are known so far. These are termed "phylogenetic divisions," or "phyla." The number of known bacterial divisions has expanded substantially in recent years. The first compilation of bacterial molecular diversity, formed by Woese in 1987, included only about 12 divisions. About 40 such deeply related groups of bacteria have now been identified by rRNA sequences. Only about two-thirds of the bacterial divisions have cultured representatives (filled wedges in figure 2.3). The remaining divisions (open wedges) have only been detected in molecular surveys of environmental rRNA genes. Organisms that belong to these bacterial divisions without cultured members sometimes are abundant in their respective environments, and therefore, their activities are likely significant in the local biogeochemistry. Sequences that identify members of the WS6 division, for instance, are conspicuous in hydrocarbon bioremediation sites and thus are likely to be important for that process.[17] OP11 sequences, first detected in a Yellowstone hot spring,[18] commonly are abundant in anoxic environments.[19] The environmental rRNA sequences thus point to areas for investigation by microbial biologists.

Phylogenetic analyses of available molecular sequences, rRNA and protein, have failed to resolve convincingly any specific branching orders of the bacterial divisions. Trees produced using rRNA sequences often indicate (e.g., figs. 2.1 and 2.3) that a few of the division lineages (e.g., Aquificales, Thermotogales) branch more deeply than the main radiation, but this is possibly an artifact of the high-temperature nature of those organisms and their rRNAs. The base of the bacterial tree is best seen as a polytomy, an expansive radiation that is not resolved with the current data. It is likely that future studies will draw together some of the groups that now seem to constitute division-level diversity. It also is likely that still more division-level groups will be discovered. An important direction for establishing the pattern of bacterial evolution is the accumulation of

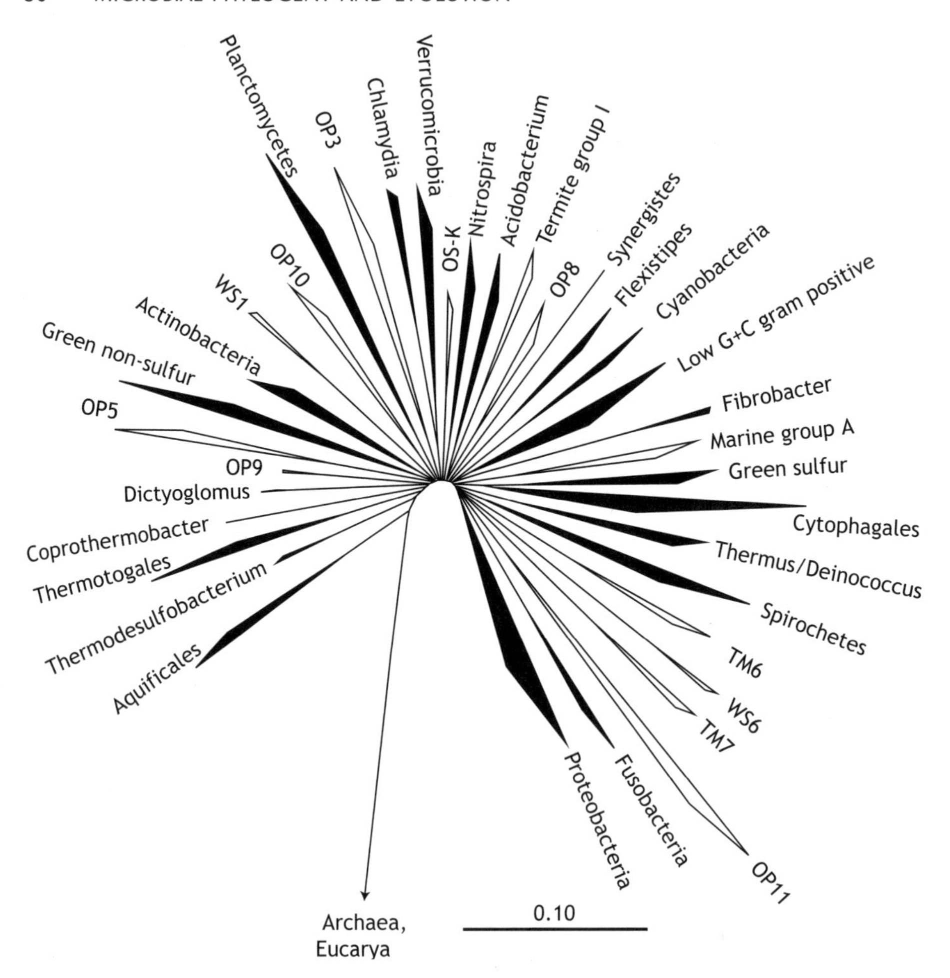

Figure 2.3. Diagrammatic representation of the phylogenetic divisions of Bacteria. Phylogenetic trees containing sequences from the indicated organisms or groups of organisms, chosen to represent the broad diversity of Bacteria, were used as the basis of the figure. Wedges indicate that several representative sequences fall within the indicated depth of branching. Filled wedges represent bacterial divisions with at least one cultured representative. Open wedges represent bacterial divisions represented only by environmental sequences and are named after ribosomal RNA gene clone libraries (OP, WS, TM, OS). The smaller or larger areas of the sectors correspond to smaller or larger numbers of sequences available. The scale corresponds approximately to 0.1 changes per nucleotide.[16]

additional sequences to represent the entire diversity of the bacterial divisions. Broad taxon representation of sequences is required to produce the most accurate phylogenetic trees.[20] At present, most rRNA sequences are from only a few of the bacterial divisions. Further environmental surveys with molecular methods will be the most efficient way, and possibly the only way, to gather a broader information base on bacterial diversity. It is likely, as well, that genomic studies will contribute to the resolution of the bacterial tree. For instance, the common occurrence of gene families could be evidence for a specific relationship between divisions that are not convincingly relatives as determined with the accuracy of the rRNA trees. Although the understanding of the fine structure of the bacterial tree will improve, the current picture of the base of that tree, as an expansive radiation of independent lines of genetic descent, is unlikely to change.

The large-scale structure of the bacterial phylogenetic tree (Fig. 2.3), a line of descent with no (surviving) branches and then a burst of diversifying genetic lineages, is intriguing. This evolutionary radiation surely was one of the great landmarks in biology. The consequences of that diversification included profound modification of this planet through the metabolic activities of the resulting organisms. What could have sparked such a spectacular radiation in the bacterial tree? One possibility is that the expansive genetic differentiation resulted when early life developed sufficient sophistication that stable, independent lines of descent could be established. Before that, the rudimentary nature of biochemical processes probably precluded the establishment of independent genetic lineages. Woese has suggested that early genes would have been shared by communities of replicating entities, and he likens the establishment of the cellular lineages to an annealing process.[4,21] Initially, mutation rates and lateral transfer would have been high. As increasingly complex and specific genetic machineries arose, both mutation rates and lateral transfer would have tapered off, and discrete genetic lines of descent could be established.

Archaea

In 1977, when Archaea were recognized as being fundamentally distinct from both bacteria and eukaryotes, only a few representatives had been cultured and studied. The properties of these organisms seemed uniformly unusual. Some of the cultured species were highly anaerobic methanogens that used molecular hydrogen as an energy source and respired with carbon dioxide, making methane. Others thrived in saturated brine, for instance, the Dead Sea, and produced a rhodopsin-like pigment akin to that in our own eyes. A third kind of was acidophilic thermophiles, found in acidic geothermal springs. Most examples of archaea that have been cultured since their recognition also have had those properties. As a consequence, archaea popularly have been considered restricted to environments that are "extreme" by human standards. Molecular studies have shown, however, that this perception is seriously distorted. Archaeal rRNA genes belonging to uncultured organisms are widely distributed in environments that are not necessarily extreme; for instance, soils and marine and fresh waters. Our understanding of the structure of the archaeal phylogenetic tree rests on only about 1000 rRNA sequences, less than half of which are from cultured organisms, and the others of which are from environmental surveys of rRNA genes. Relatively few environments have been analyzed for archaea, however, so the extent of diversity that makes up that phylogenetic domain surely is far broader than we know.

Figure 2.4 is a diagram of the known phylogenetic makeup of the domain Archaea. There are two main relatedness groups, Euryarchaeota and Crenarchaeota. A potential third deeply divergent lineage of archaea, Korarchaeota, is represented only by environmental rRNA gene sequences, so the status of this group needs to be tested and consolidated by further studies of gene sequences and descriptions of organismal properties.[7] The branches between these main evolutionary clades of Archaea are the

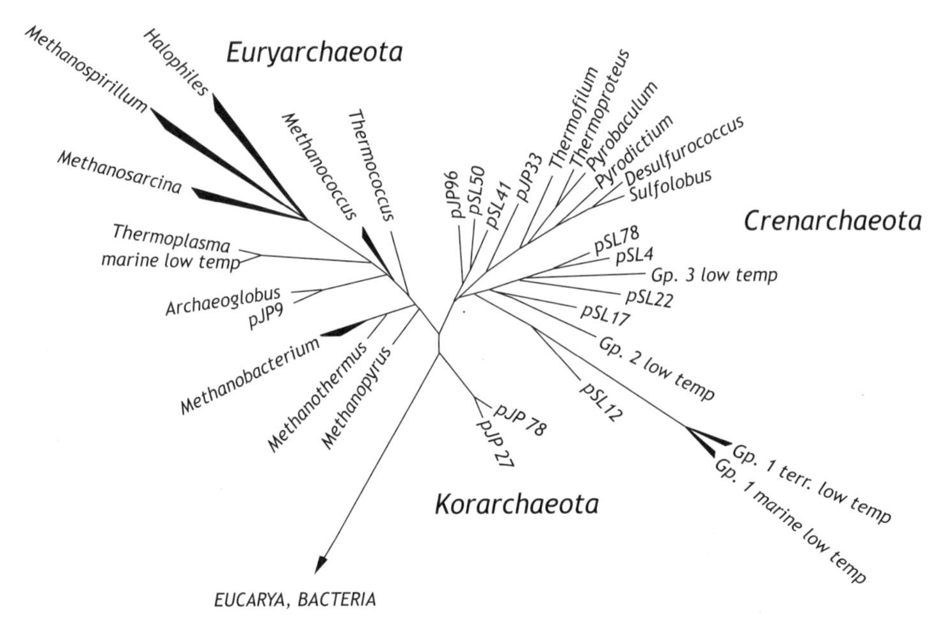

Figure 2.4. Diagrammatic representation of the phylogeny of Archaea. Wedges indicate that several representative sequences fall within the indicated depth of branching. Names correspond to organisms or groups of organisms, or environmental clones.

deepest within any of the three domains. The great depth of separation of Euryarchaeota and Crenarchaeota also is indicated by many biochemical properties and genomic features. For instance, even DNA is packaged differently in these two kinds of organisms. Whereas euryarchaeotes use histones to package chromatin, much as do eucaryotes, crenarchaeal genomes evidently lack histone genes.[22] The mode of packaging DNA by the latter organisms is not known.

There are cultured representatives of most of the main lineages of Euryarchaeota. Molecular analyses of environmental sequences have revealed no major new groups that diverge deeply in the euryarchaeal tree. In contrast, most of the known extent of crenarchaeal rRNA diversity is known only from environmental sequences. All cultured crenarchaea are thermophilic and often are obtained from geothermal environments. The properties of these organisms did much to popularize the notion of archaea as being exclusively "extremophiles." It came as a surprise, then, when abundant, phylogenetically diverse crenarchaeal rRNA gene sequences were discovered in more moderate habitats ranging from shallow and deep marine waters, soils, sediments, and rice paddies to symbionts in some invertebrates.[23]

As diagrammed in figure 2.4, only one of the main relatedness groups in Crenarchaeota comprises named organisms. The other groups consist of environmental organisms represented only by sequences. These otherwise largely unknown organisms are some of the most abundant creatures on Earth. In the oceans, for instance, low-temperature crenarchaea occur at concentrations of 10^7–10^8 cells per liter throughout the water column at all latitudes, and typically constitute 10%–50% of the cells present in the sunless deep sea, the largest biome on the planet. The niche in the global ecosystem that these organisms fill is not known. Cultured crenarchaea commonly use hydrogen as an energy source, and molecular hydrogen is pervasive in the environment at very low levels.[24] Perhaps the low-temperature crenarchaea tap this ubiquitous fuel. Although low-temperature crenarchaea have so far eluded pure culture for laboratory studies, recent developments in genome science are being exploited to learn more about them. For instance, environmental DNA can be cloned as large pieces, linked together and sequenced to gain further information on the organisms identified by the rRNA sequences.[25]

Eukaryotes

Broad-scale molecular evolutionary perspective on eukaryotes has relied on a sparse collection of gene sequences that do not represent the full range of eukaryotic diversity in nature. As shown in figure 2.1, the most diverse eukaryotic rRNA sequences are derived from microbes, yet such organisms are the least known of eukaryotes and have received the least attention from molecular phylogenetic studies. More than 100,000 microbial eukaryotes, or "protists," have been described,[26] but only a few thousand have been investigated for rRNA sequence.[14] Moreover, as with the collection of bacterial rRNA sequences, the collection of eucaryal sequences is heavily biased toward only a few relatedness groups. The recent addition of environmental rRNA gene sequences to phylogenetic calculations has improved the resolution of the eukaryotic tree by providing additional diversity.[27]

A diagram that summarizes the phylogeny of the eukaryotic taxonomic kingdoms from the rRNA perspective is shown in figure 2.5. There is no convention for the taxonomic organization of sequence-based relatedness groups of eukaryotes. On the basis of various traditional or molecular classification schemes, eukaryotes have been categorized into from three to over seventy major kingdoms. Eucaryal sequences available in the databases fall into about thirty independent relatedness clusters, which can be considered the known kingdom-level relatedness groups[28] (not all shown in fig. 2.5).

From the perspective of rRNA sequences, the overall topology of the eucaryal tree is seen as a basal radiation of independent lines of descent (e.g., Diplomonads, Trichomonads), only one of which gave rise to other main lines, of which, in turn, only one culminated in the "crown radiation" of the familiar taxonomic kingdoms such as animals, plants, stramenopiles, and so forth (Fig. 2.5). The specific positions of intermediate branches in the rRNA tree are only approximate, but the successive branching order is indicated by several kinds of analyses.[27,29] This view of successive branching in the eucaryotic tree contrasts with the results of some comparisons of protein-encoding genes, with limited phylogenetic representation.[30] These latter results have been interpreted to indicate that there is no particular branching order among the main eucaryal relatedness groups; that the contemporary kingdom-level lines derived from a single expansive radiation analogous to the bacterial radiation (Fig. 2.2). Proponents of this view have argued that extensive sequence differences between basal-derived and crown-group rRNA genes do not reflect great evolutionary distances, but rather are a consequence of relatively rapid evolution in the basal lines (so-called "long branch attraction"). Some of the environmental rRNA gene sequences, however, are not rapidly evolving lineages; yet they branch more deeply in the tree than the crown radiation. These environmental sequences thus punctuate the long lines between the crown and the previously

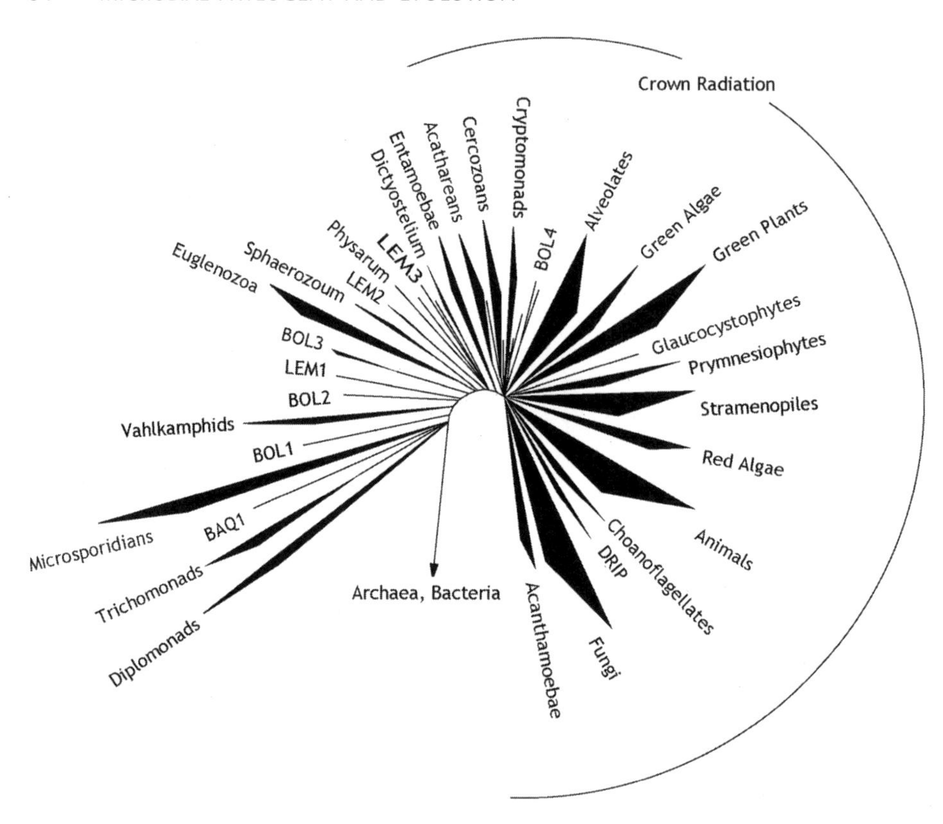

Figure 2.5. Schematic diagram of the evolution of Eucarya. The branch points of these kingdom-level groups are based on trees inferred with evolutionary distance, maximum parsimony, and maximum liklihood, and representative sequences. The areas of the wedges reflect nonlinearly the relative numbers of small-subunit rRNA sequences of these groups in GenBank. Groups named LEM, BOL, and BAQ are represented only by environmental ribosomal RNA gene clones.

identified basal divergences. The occurrence of deeply divergent eukaryotic lines with slow substitution rates (short lines) indicates that the high rates (long lines) previously ascribed to the basal divergences in rRNA trees are not the norm. The accuracy with which the kingdom-level lines can be resolved will improve as the sequence collection available for analysis grows. Phylogenetic trees based on a single gene, SSU rRNA in this case, of course cannot reflect the genealogies of all the genes that specify organisms because of the potential influence of lateral transfer. Genes with phylogenies that are not congruent with the rRNA tree possibly have undergone lateral transfer in their evolution.

Large-Scale Pattern in the Tree of Life

A calculated three-domain tree, such as shown in figure 2.1, based on a specific method and the particular suite of sequences, cannot capture the uncertainties of calculations or the idiosyncrasies of the sequences. Nonetheless, from the collection of domain-level studies discussed above, a discrete but low-resolution, large-scale pattern emerges, as summarized in figure 2.6. The figure represents the general patterns of change in rRNA sequences, which likely represents change in the core genetic machinery.[19] This is a molecular tree, not an organismic tree. A tree that would track the evolution of organisms would be far more complex because of losses and gains of genes and lateral transfers between divergent lines of descent.

The general pattern of large-scale change in each of the domains is a set of unresolved radiations, punctuations in evolution that resulted in biological diversification. The earliest line segments in all the domains have no surviving branches. This period of evolution presumably took place before acquisition of sufficient cellular sophistication to propagate a vertical line of descent, independent of some nuturing replicative focus, the source of the specificity of the domains.[10] Only at the evolutionary stage of genetically independent cells, Woese's "Darwinian Threshold,"[10] could modern genetic diversification of lineages occur. The attainment of this stage of evolution possibly is reflected in the unresolved radiations at the base of each of the domains (fig. 2.6).

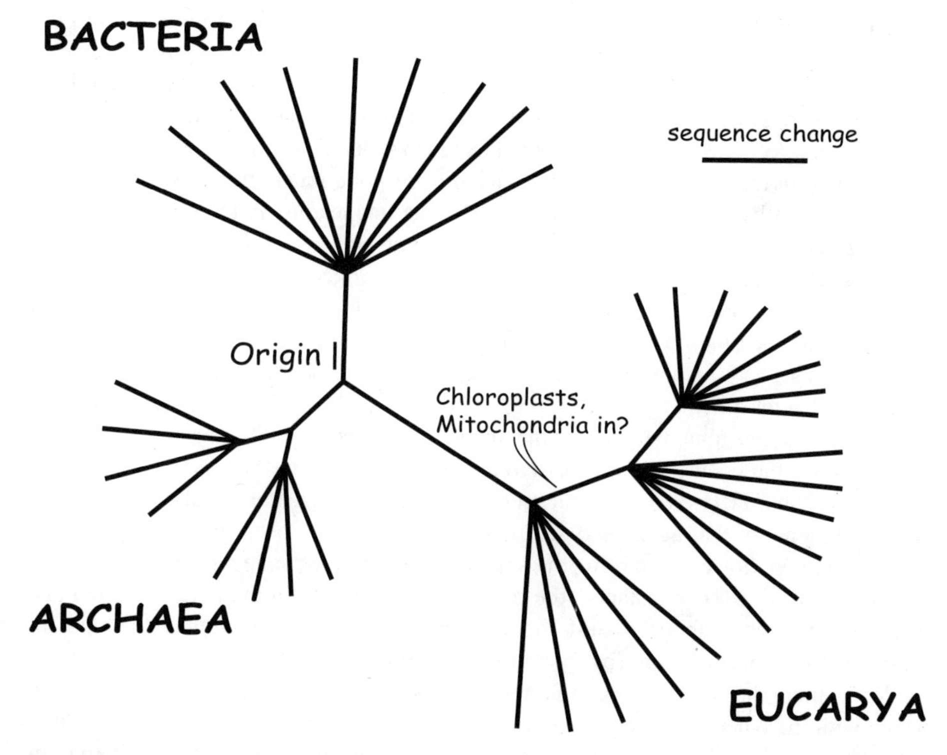

Figure 2.6. Large-scale structure of the main genetic lines of descent. The diagram models the pattern of ribosomal RNA sequence diversification, and presumably of the change in the basal genetic machinery of life. As discussed in the text, no specific branching order can be established among the approximately forty known main bacterial lineages. In contrast, eucaryotic evolution seems to have involved a succession of diversifications in a main line of descent. The scale bar corresponds to relative sequence change, not time.

The pattern of evolution in the eukaryotic line of descent seems particularly complex. The successive radiations are significant landmarks in eukaryotic history. Correlation of cellular properties or genomic sequences with rRNA trees may provide clues regarding the biological innovations that sparked these deep radiations. One noteworthy correlation may be the phylogenetic distribution of the major organelles, chloroplasts, and mitochondria. All characterized representatives of the basal lineages of eukaryotes lack mitochondria and chloroplasts, whereas organisms of more peripherally branching groups have those organelles. The clear importance of these organelles indicates that much of the modern diversity of eukaryotes was made possible by the metabolic power and light-harvesting capacity of bacteria.

Time and the Tree

Because sequences of genes change with time, it seems natural to try to infer the times of branch points in evolutionary history by the extents of sequence divergence between modern genes. Indeed, molecular phylogenetic trees often are interpreted in the context of time since the divergence of particular branches. This simple correlation between time and sequence change is not well founded, however, because different lines of descent can change at different rates. This is seen in the lengths of line segments (extents of sequence change) in the three-domain tree in figure 2.1. Thus, lines leading to modern-day archaea are systematically short compared with the lines leading to their sister group, modern eukaryotes. Moreover, the rate of change in sequences is not constant with time. This is seen in the mitochondria, which have undergone much greater sequence (and other) changes than their sister line in this tree, the line leading to the proteobacterium *Agrobacterium tumefaciens* (Fig. 2.1). Thus, a sequence-based phylogenetic tree cannot be used to date events unless the tree can be calibrated by correlating a historical occurrence with some feature in the tree. A geological and biological correlation that may estimate one time-point in the tree of life is the appearance of molecular oxygen in the rock record and the phylogenetic radiation of the only organisms that produce oxygen, cyanobacteria.

Oxygen became abundant in the atmosphere by around 2.5 billion years ago. This is evidenced by the occurrence of massive deposits of Fe_2O_3, in the context of "banded iron formations" (BIFs), beds of alternating reduced (black, Fe^{2+}) and oxidized (red, Fe^{3+}) sediments of that age.[31] Oxygen production must have been going on for a long time before that, however, to bring the atmosphere to the oxic state required for precipitation of iron oxide. For instance, the oxygen concentration in the oceans alone must have accumulated to about 50 mM, the amount required for production of the ambient marine sulfate concentration (by oxidation of sulfide).

When, then, did oxygen-forming photosynthesis begin? Biologically, it presumably began with the emergence of cyanobacteria, because only cyanobacteria carry out that form of metabolism. This must have been early in evolution, at the time of the radiation of the main bacterial lines, soon after the appearance of cellular life. The earliest BIFs indicate that this could have been as early as 3.8 billion years ago.[32] If the oxygen in those ancient BIFs is of biological origin, which is by no means certain, then oxygenic photosynthesis must have been in place at that time. A biological origin of the oxygen is consistent with the carbon isotope fractionation, in the same rocks, which indicates biological processing. Because the phylogenetic line that led to chloroplasts originated at the base of the cyanobacterial radiation,[33,34] it seems likely that chloroplasts were derived as soon as cyanobacteria emerged. The branch point of a mitochondrial lineage from proteobacteria is consistent with the early appearance of that organelle, too. Thus, the phylogenetic pattern, coupled with available geological information, indicates that the modern kind of eucaryotic cell, with organelles, may have been in place by 3.8 billion years ago.

Conclusion and Prospects

The general outline of a universal phylogenetic tree is now in place. It is clear, however, that the current view of the tree incompletely portrays the breadth of biological diversity. A main reason that it is incomplete is that our understanding of microbial diversity is rudimentary. Molecular studies of environmental organisms continue to reveal major relatedness groups that were not suspected. Are there still other primary domains to be discovered? Perhaps. The methods used to hunt organisms in the environment are heavily dependent on the microbial diversity that we already know about. Are there other new bacterial divisions and eukaryotic kingdoms to be discovered? Almost certainly. The studies of microbial ecosystems so far conducted have been limited and yet have turned up remarkable novelty. The complexity of microbial ecosystems indicates that much broader diversity will be encountered.

A description of Earth's microbial diversity cannot be captured by simple descriptions of organisms. Microbial diversity is too broad, far too complex, and far too plastic to be accommodated by species counts. However, a sampling and an articulation of the extent of cellular diversity could be accomplished by sequence surveys of environmental rRNA genes. The sequences reflect the kinds of organisms that they represent, and their relative frequencies are a rough census of the microbial world. An expanded sequence representation of life's diversity would provide a basis for more accurate molecular phylogenetic reconstructions and would bring us to a closer understanding of our beginnings.

Acknowledgments I thank colleagues in my lab for comments that improved this article. My research activities are supported by the National Institutes of Health, the National Science Foundation, and the NASA Astrobiology Institute.

References

1. D.L. Swofford, G.J. Olsen, P.J. Waddell, and D.M. Hillis, "Phylogenetic inference," *in* D.M. Hillis, C. Moritz, and B.K. Mable (ed.), *Molecular Systematics*, 2nd ed. Sinauer Associates, Inc., Sunderland, MA. 1996, pp. 407–514.
2. C.R. Woese, "Bacterial Evolution," *Microbiol. Rev.* 51 (1987): 221–271.
3. E.V. Koonin, K.S. Makarova, and L. Aravind, "Horizontal Gene Transfer in Prokaryotes: Quantification and Classification," *Annu. Rev. Microbiol.* 55 (2001): 709–742.
4. C.R. Woese, "Interpreting the Universal Phylogenetic Tree," *Proc. Natl. Acad. Sci. USA* 97 (2000): 8392–8396.

5. C.R. Woese and G.E. Fox, "Phylogenetic Structure of the Prokaryotic Domain: The Primary Kingdoms," *Proc. Natl. Acad. Sci. USA* 74 (1977): 5088–5090.
6. C.R.Woese, O. Kandler, and M.L. Wheelis, "Towards a Natural System of Organisms: Proposal for the Domains Archaea, Bacteria, and Eucarya," *Proc. Natl. Acad. Sci. USA* 87 (1990): 4576–4579.
7. S.M. Barns, C.F. Delwiche, J.D. Palmer, and N.R. Pace, "Perspectives on Archaeal Diversity, Thermophily and Monophyly from Environmental rRNA Sequences," *Proc. Natl. Acad. Sci. USA* 93 (1996): 9188–9193.
8. J.P.H. Gogarten, Kibak, P. Dittrich, L. Taiz, E.J. Bowman, B.J. Bowman, M.F. Manolson, R.J. Poole, T. Date, T. Oshima, J. Konishi, K. Denda, and M. Yoshida, "Evolution of the Vacuolar H+-ATPase: Implications for the Origin of Eukaryotes," *Proc. Natl. Acad. Sci. USA* 86 (1989): 6661–6665.
9. N.K. Iwabe, K. Kuma, M. Hasegawa, S. Osawa, and T. Miyata, "Evolutionary Relationship of Archaebacteria, Eubacteria, and Eukaryotes Inferred from Phylogenetic Trees of Duplicated Genes," *Proc. Natl. Acad. Sci. USA* 86 (1989): 9355–9359.
10. C.R. Woese, "On the Evolution of Cells," *Proc. Natl. Acad. Sci. USA* 99 (2002): 8742–8747.
11. T.L. Marsh, C.I. Reich, R.B. Whitelock, and G.J. Olsen, "Transcription Factor IID in the Archaea: Sequences in the *Thermococcus celer* Genome Would Encode a Product Closely Related to the TATA-Binding Protein of Eukaryotes," *Proc. Natl. Acad. Sci. USA* 91 (1994): 4180–4184.
12. T. Rowlands, P. Baumann, and S.P. Jackson, "The TATA-binding Protein: A General Transcription Factor in Eukaryotes and Archaebacteria," *Science* 264 (1994): 1326–1329.
13. J. Sapp, *Evolution by Association: A History of Symbiosis* (New York: Oxford University Press, 1994).
14. M.L. Sogin and J.D. Silberman, "Evolution of the Protists and Protistan Parasites from the Perspective of Molecular Systematics," *Int. J. Parasitol.* 28 (1998): 11–20.
15. N.R. Pace, "A Molecular View of Microbial Diversity and the Biosphere," *Science* 276 (1997): 734–740.
16. P.B.M. Hugenholtz, B.M. Goebel, and N.R. Pace, "Impact of Culture-Independent Studies on the Emerging Phylogenetic view of Bacterial Diversity," *J. Bacteriol.* 180 (1998): 4765–4774.
17. M.A. Dojka, P. Hugenholtz, S.K. Haack, and N.R. Pace, "Microbial Diversity in a Hydrocarbon- and chlorinated-solvent-contaminated aquifer undergoing intrinsic bioremediation," *Appl. Environ. Microbiol.* 64 (1998): 3869–3877.
18. P. Hugenholtz, C. Pitulle, K.L. Hershberger, and N.R. Pace, "Novel Division Level Bacterial Diversity in a Yellowstone Hot Spring," *J. Bacteriol.* 180 (1998): 366–376.
19. J.K. Harris, S.T. Kelley, G.B. Spiegelman, and N.R. Pace, "The Genetic Core of the Universal Ancestor," *Genome Research* 3 (2003): 407–412.
20. D.M. Hillis, "Taxonomic Sampling, Phylogenetic Accuracy, and Investigator Bias," *Systematic Biology* 47 (1998): 3–8.
21. C.R.Woese, "The Universal Ancestor," *Proc. Natl. Acad. Sci. USA* 95 (1998): 6854–6859.
22. S.L. Pereira, R.A. Grayling, R. Lurz, and J.N. Reeve, "Archaeal Nucleosomes," *Proc. Natl. Acad. Sci. USA* 94 (1997): 12633–12637.
23. E.F. DeLong and N.R. Pace, "Environmental Diversity of Bacteria and Archaea," *Systematic Biology* 50 (2001): 470–478.
24. R.Y. Morita, "Is H_2 the Universal Energy Source for Long-Term Survival?" *Microbial Ecology* 38 (2000): 307–320.
25. E.F. DeLong, C. Schleper, R. Feldman, and R.V. Swanson, "Application of Genomics for Understanding the Evolution of Hyperthermophilic and Nonthermophilic Crenarchaeota," *Biological Bulletin* (*Woods Hole*) 196 (1999): 363–366.
26. D.J. Patterson and M.L. Sogin, "Eukaryote Origins and Protistan Diversity," in H. Hartman, K. Matsuno, (ed.), *The Origin and Evolution of Prokaryotic and Eukaryotic Cells.* (River Edge, NJ: World Scientific, 1993), 13–46.
27. S.C. Dawson and N.R. Pace, "Novel Kingdom-Level Eukaryotic Diversity in Anoxic Environments," *Proc. Natl. Acad. Sci. USA* 99 (2002): 8324–8329.

28. S.C. Dawson, *Evolution of the Eucarya and Archaea: Perspectives from Natural Microbial Assemblages* (Berkeley: University of California-Berkeley Press), 2000.
29. M.L. Sogin, J.H. Gunderson, H.J. Elwood, R. A. Alonso, and D.A. Peattie. "Phylogenetic Meaning of the Kingdom Concept: An Unusual Ribosomal RNA from *Giardia lamblia*," *Science* 243 (1989): 75–77.
30. H.P. Philippe, H. Lopez, K. Brinkmann, A. Budin, J. Germot, D. Laurent, M. Moreira, D. Muller, and H. Le Guyader, "Early-Branching or Fast-Evolving Eukaryotes? An Answer Based on Slowly Evolving Positions," *Proc. Royal Soc. London Ser. B Biol. Sci.* 267 (2000): 1213–1221.
31. N. Sleep, "Oxygenating the Atmosphere," *Nature* 410 (2002): 317–319.
32. S.J. Mojzsis, G. Arrhenius, K.D. McKeegan, T. M. Harrison, A.P. Nutman, and C.R.L. Friend, "Evidence for Life on Earth Before 3.800 Million Years ago," *Nature* 384 (1996): 55–59.
33. S.J., Giovannoni, S. Turner, G.J. Olsen, S. Barns, D.J.Lane, and N. R. Pace, "Evolutionary Relationships Among Cyanobacteria and Green Chloroplasts," *J. Bacteriol.* 170(1988): 3584–3592.
34. S. Turner, K.M. Pryor, V.P.W. Miao, and J.D. Palmer, "Investigating Deep Phylogenetic Relationships Among Cyanobacteria and Plastids by Small Subunit rRNA Sequence Analysis, " *J. Eukaryotic Microbiol.* 46 (1999): 327–338.
35. D.N. Frank and N.R. Pace, "Molecular-Phylogenetic Analyses of Human Gastrointestinal Microbiota," *Current Opinion in Gastroenterology* 17 (2001): 52–57.

3

Molecular Phylogeny of Bacteria Based on Comparative Sequence Analysis of Conserved Genes

WOLFGANG LUDWIG
KARL-HEINZ SCHLEIFER

Comparative sequence analysis of the small subunit ribosomal RNA (SSU rRNA) plays a central role in microbial taxonomy and identification—even today in the age of genomics. The introduction of the rRNA approach by Carl Woese allowed, for the first time in the history of microbiology, comprehensive phylogenetic studies of the living world.[1] The SSU rRNA sequence data set is the largest database currently available for a gene or gene product, and the current taxonomy of prokaryotes, as documented in the most recent edition of *Bergey's Manual of Systematic Bacteriology*, is based on the phylogenetic framework deduced from SSU rRNA data.[2] Although the advantages of this phylogenetic marker are well documented with respect to information content and the comprehensiveness of the available data set, it is generally accepted that rRNA-based conclusions can only roughly reflect evolutionary history.[3,4] Additional phylogenetic markers have also to be considered to approximate a more detailed phylogeny.

In the pregenomics era, a limited number of studies focused on other markers such as elongation factors, ATPase subunits, and RNA polymerases.[5,6] These studies had already revealed the limitations in the information content of the selected markers. Although similar overall tree topologies were apparent, marker-specific discrepancies were commonly found in detailed topologies. As discussed elsewhere, such discrepancies are to be expected, given the generally limited and differing (with respect to the documented time span of evolution) information content of the individual markers.[3] Nevertheless, the rapid rise of new genome projects led to great hopes of finding other markers that would further resolve microbial phylogeny. However, the comparison of the first fully analyzed genomes showed that the number of potential phylogenetic markers that would fulfil the criteria of universal occurrence, functional constancy, sufficient sequence conservation, and complexity was rather limited.[7] In comparison with the richness in genetic information of even the smallest prokaryotic genomes, the part that can be used for comprehensive phylogenetic studies is minimal. In this chapter, information for a selection of such potential markers was collected from genome and general sequence databases, processed with respect to alignment and conservation, and used to construct trees.

In view of the comprehensiveness of the data, the existing taxonomy, and the history of prokaryotic phylogenetics, the small SSU rRNA–based tree is used as the foundation for comparisons with other potential markers. Actually, the procedure explicitly followed here is that used by many authors regardless of whether or not they support the current rRNA-based phylogeny. We consider only those genes or molecules that meet the requirements for a universal phylogenetic marker. The majority of the analyzed data concerns protein genes. In accordance with the criterion of functional constancy, the predicted amino acid sequences were used for comparison. As already noted, complete agreement on details of tree topologies cannot be expected[4]; the comparative analyses here thus focuses on the verification of the three domains Archaea, Bacteria, and Eucarya and the phylogenetic relationships within the domain bacteria at the phylum level.

Information Content and Significance

Before discussing the similarities and differences in phylogenetic conclusions based on alternative markers, some general remarks concerning the significance of such comparisons have to be made. The phylogenetic information content of any marker molecule is generally limited.[3,4,8] One of the primary characteristics of a universal phylogenetic marker is sequence conservation. This means that only part of the primary structure is variable and, hence, informative. The maximum information content of molecules depends on the number of monomers (sequence positions) and potential character states (four nucleotides, twenty amino acids plus insertion/deletion events) per site.[4,9] However, in general, such phylogenetic markers are rather conserved, meaning that only certain positions are variable and therefore informative. Moreover, the number of permitted character states per informative site is necessarily reduced because of functional constraints and selective pressure. In most cases, the information content and resolution power of protein markers is generally considerably lower than those of the 16S and 23S rRNA molecules. There is also the problem of "plesiomorphy." Especially at highly variable positions, identical residues may be the result of multiple changes during the course of evolution, simulating an unchanged position (plesiomorphy). As a consequence, phylogenetic data analysis is hampered not only by the limitations with respect to the information content but also by a burden of noise resulting from "false" identities at plesiomorphic sites.

These problems, in combination with the shortcomings of data analysis and tree reconstruction tools, remarkably reduce the significance of local tree topologies. Attempts to circumvent such problems focus on signature analysis or rare genomic events.[10,11] Based on a simple cladistic analysis, the presence or absence of a primary or higher-order signature or an insertion or deletion is used to split the organisms into two partitions and to deduce an evolutionary succession from the resulting subsets. However, the same criteria as for phylogenetic markers have to be fulfilled; most notably, ubiquitous distribution and functional constancy of the molecule carrying the signature.

Furthermore, the procedure is only of phylogenetic relevance if the signature was introduced only once during the course of evolution. This can be checked only if comprehensive data sets are available and the "surrounding" sequence reveals an orthologous descent for the carriers of the signature. As a consequence, insertions and deletions (indels) can be considered only if they concern conserved markers. The significance of the information provided by an indel is not more than that of a meaningful single base or amino acid change.

A further consequence of the limited information content is that different markers may carry or lack information on different eras of evolution. Thus, locally different tree topologies have to be expected when comparing phylogenetic conclusions based on different alternative markers. It is also well recognized that the significance of "local" tree topologies (the relative order of neighboring branches in a tree), especially when characterized by relatively short internodes (branches), is generally low.[4] This is a consequence of effects such as shortcomings of the commonly used treeing methods, unrecognized plesiomorphic branch attractions, or imbalanced data sets. The quintessence is that a range of uncertainty[3] has to be assumed for local branching orders in trees based on the data of an individual marker and, consequently, to be taken into consideration when assessing similarities and differences between trees derived from alternative phylogenetic markers.

The orthology of the respective molecules is a crucial consideration when comparing differences in organismic phylogenies based on alternative molecular markers. The use of orthologous (direct common ancestor) genes or gene products is essential for the delineation of the monophyletic status of groups of molecules or organisms, whereas paralogous markers (indirect common ancestor, derived from duplicated genes) can be used for the relative rooting of, or within, monophyletic (sub-) trees. Thus, discrepancies of tree topologies can result from "illegitimate" inclusion of paralogous markers. Paralogous data can only be recognized as such if the respective data set contains examples that assign both copies to a given organism. Any phylogenetic conclusions may be complicated or misleading if both copies are maintained, however, one of the products changed its function, or whether one of the copies was lost during the course of evolution.

The rRNA-Based Domain and Phylum Concept

The current view of SSU rRNA–derived phylogeny is based on comparative analyses of more than 21,000 full sequences (fig. 3.1). The three-domain concept of Archaea, Bacteria, and Eucarya is clearly supported by the optimized and evaluated tree data. Given that paralogous markers are not available for the small SSU rRNA, a significant positioning of a root is not possible. The most recent edition of *Bergey's Manual of Systematic Bacteriology*[2] distinguishes 23 bacterial phyla. All these phyla contain at least a few culturable bacteria. However, sequence data obtained by applying cultivation-independent techniques indicate the presence of a number of additional groups to which the phylum status could be assigned. It is well known that phylogenetic trees are only models of the evolutionary affiliations of the organisms and may be heavily influenced and changed whenever the underlying database is extended. The definition of bacterial phyla (the major lines of descent within the bacterial domain) in the current tree version is quite similar to that shown in *Bergey's Manual*.[2] However, in some cases drawing a

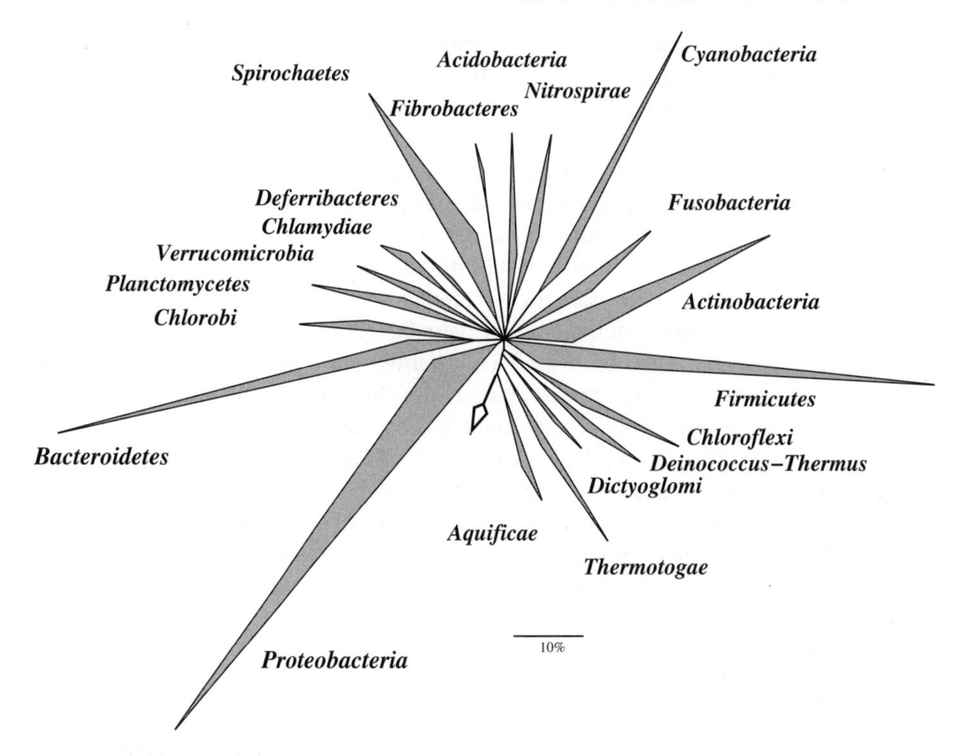

Figure 3.1. Small subunit ribosomal RNA–based phylogenetic tree visualizing the radiation of bacterial phyla. The tree was reconstructed applying the ARB-Parsimony tool[3] on a data set of about 8,000 full 16S/18S rRNA sequences. The alignment positions were selected according to the 50% conservation criterion,[3,4] with respect to all bacterial sequences included. The tree topology was evaluated and smoothed, taking into account the results obtained by using alternative treeing methods on reduced data subsets. More than 32,000 additional sequences comprising at least 800 monomers were included for tree evaluation at the phylum level.

significant boundary between individual phyla, or establishing the monophyletic status of the subgroups of some phyla, has become more difficult and ambiguous (see Table 3.1). For most of the phyla, a significant relative branching order cannot be defined. A common root for the phyla Chloroflexi, Deinococcus-Thermus, and Thermomicrobia is supported by the current data set. Similarly, there is a tendency toward a common group comprising Planctomycetes, Chlamydiae, and Verrucomicrobia; however, the significance is low. The Bacteroidetes and Chlorobi represent another phylum cluster. The Firmicutes (including the Clostridia, Mollicutes, and Bacilli) represent an independent phylum like the Actinobacteria, Nitrospirae, Spirochaetes, Fibrobacteres, Bacteroidetes, Cyanobacteria, Deferribacteres, Fusobacteria, and Proteobacteria. A slightly deeper branching of the phyla Aquificae, Thermotogae, Thermodesulfobacteria, and Dictyoglomi can be seen using data sets modified according to the 50% positional conservation convention.[4] The position of these groups as well as the Planctomycetes is still under discussion.[11] Indeed, the relative branching order of the phyla changes somewhat if the conservation threshold is drastically raised.[13]

In general, removing variable positions from the data set helps to reduce potential plesiomorphic noise. However, further raising of the thresholds for selecting sequence positions also comprises the risk of losing informative characters.

The large-subunit RNA (LSU rRNA) is perhaps the most informative phylogenetic marker. Its primary structure is at least as conserved as that of the SSU rRNA, but it contains more and longer stretches of informative positions. Although the LSU rRNA gene sequences provide more information, the major drawback is currently the limited database, which comprises only about 2,000 full sequences (excluding mitochondrial sequences). Minor local differences are apparent when SSU rRNA–and LSU rRNA–derived trees are compared, but the phyla remain distinct and well-defined (fig. 3.2). A deeper branching is indicated for Aquificae, Thermotogae, and Dictyoglomi; and the Chloroflexi and the Thermus-Deinococcus phylum share a common root. Chlamydiae and Planctomycetes, and Bacteroidetes and Chlorobi, represent two phylum clusters, respectively. Local differences in LSU rRNA– and SSU rRNA–based subtrees are documented in fig. 3.3 for the species of the genus Enterococcus.

SSU and LSU rRNA genes fulfill the requirements of ideal phylogenetic markers to an extent far greater than do protein coding genes. Because of functional constraints, sequence changes may be manifested periodically rather than continuously during the

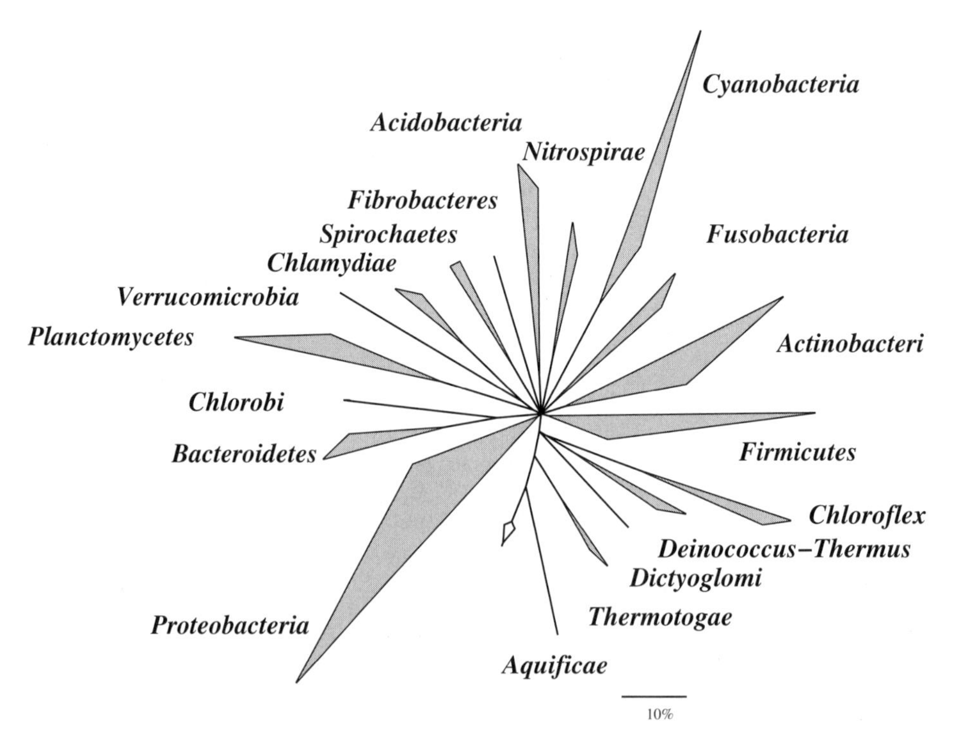

Figure 3.2. Large subunit ribosomal RNA–based phylogenetic tree visualizing the radiation of bacterial phyla. The data set comprised about 2,000 full sequences. A further 5,000 sequences comprising at least 1,000 monomers were included for tree evaluation at the phylum level. Parameters for tree reconstruction were as described for fig. 3.1.

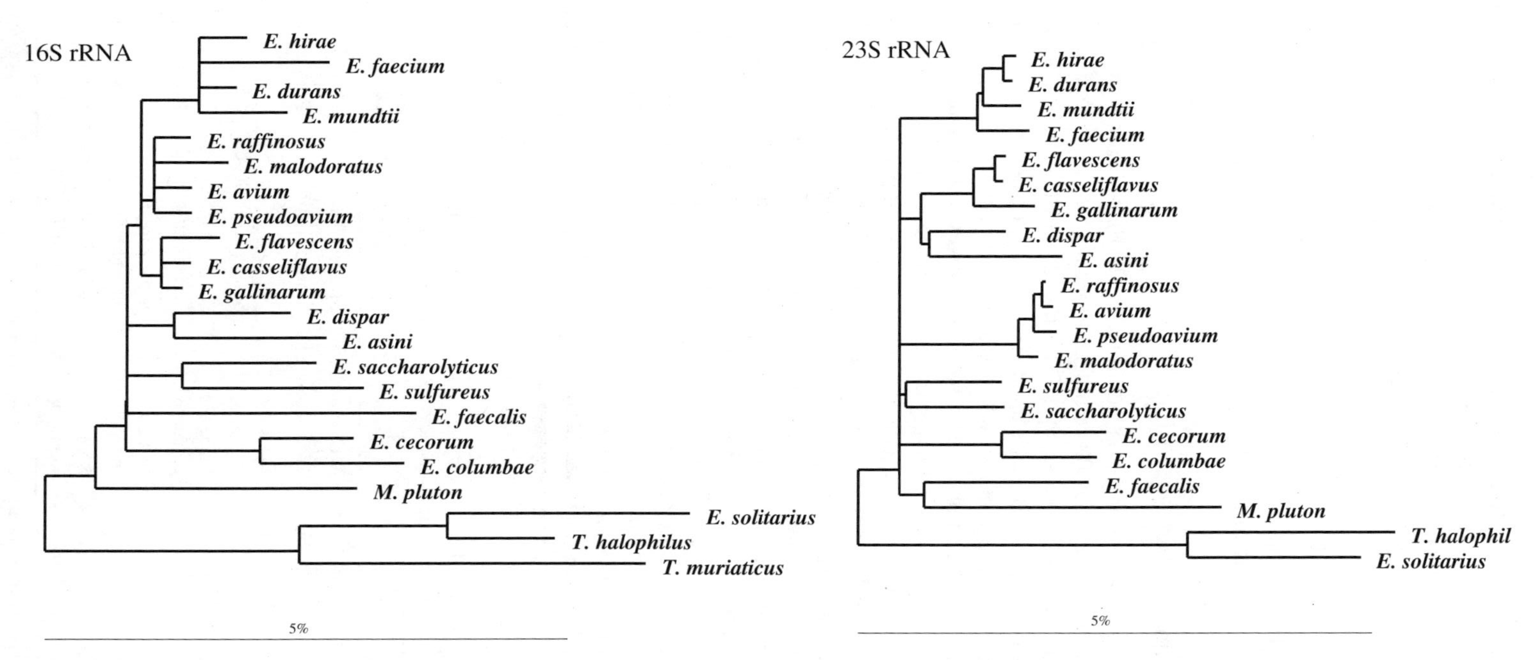

Figure 3.3. Comparative visualization of large- and small-subunit-ribosomal-RNA–based trees for the members of the genus *Enterococcus*. Parameters for tree reconstruction were as described for figure 3.1. The 50% conservation criterion was applied on all sequences from enterococci.

course of evolution. A direct correlation to a time scale cannot be postulated. Branching patterns in the periphery of a tree do not reliably reflect phylogenetic relatedness. The phylogenetic resolving power of closely related organisms (>97% similarity) is low (i.e., rRNA sequence data analyses are usually not sufficient to distinguish prokaryotes at the species level).

Alternative Marker Molecules that Support the Three-Domain Concept

Elongation Factors Tu/1 Alpha

Comparative analyses of the elongation factors (EFs) Tu (Bacteria) and 1alpha (Archaea, Eucarya) are based on more than 500 complete primary structures table 3.1. The three-domain concept is clearly supported by EF Tu–based trees (fig. 3.4). The overall sequence similarity between the Bacteria and the other two domains is only 30% in comparison with over 50% in case of the rRNA sequences. In general, the EF Tu data also support the currently defined phyla or subgroups. However, the reduced information content[3] is reflected by a lower resolving power. Furthermore, a few examples of gene duplications (*Streptomyces* sp.; see below) indicate potential paralogy problems.

As seen in the rRNA-based trees, monophyletic clusters represent the phyla Aquificae, Chlamydiae, Bacteroidetes, Chlorobi, Chloroflexi, Cyanobacteria, Thermotogae, Thermus-Deinococcus and Spirochaetes. The Proteobacteria do not appear as a monophyletic group. A common group is shown for the α-, β-, γ-, δ-proteobacteria. However, Nannocystis is separated from the Deltaproteobacteria. The Epsilonproteobacteria represent their own group. In the case of Firmicutes, the Bacilli and Mollicutes share a common root, whereas the Clostridia branch separately. The Actinobacteria represent a monophyletic lineage; however, duplicated diverged genes are documented for *Streptomyces coelicolor* and *Streptomyces* ramocissimus.[14] One of the variants appears as the deepest branch among the Bacteria. Streptomyces ramocissimus contains a third copy

Table 3.1. Molecules supporting the ribosomal RNA–based domain and phylum concepts (except aminoacyl tRNA synthetases)

Molecule	Domains	Phyla	Intraphylum	Duplicates
16S rRNA	Well defined	Well separated	Monophyletic	Present
23S rRNA	Well defined	Well separated	Monophyletic	Present
EF-Tu	Well defined	Some not monophyletic	Some paraphyletic	Present
rpoB	Well defined	Some not monophyletic	Some paraphyletic, intermixed	Not known
rpoC	Well defined	Some not monophyletic	Some paraphyletic, intermixed	Not known
Hsp60	Two domains well defined	Some not monophyletic	Some paraphyletic, intermixed	Present

The degree of support for the definition and separation of domains, phyla, and major intraphylum groups is indicated, as well as the presence of divergent multiple copies.

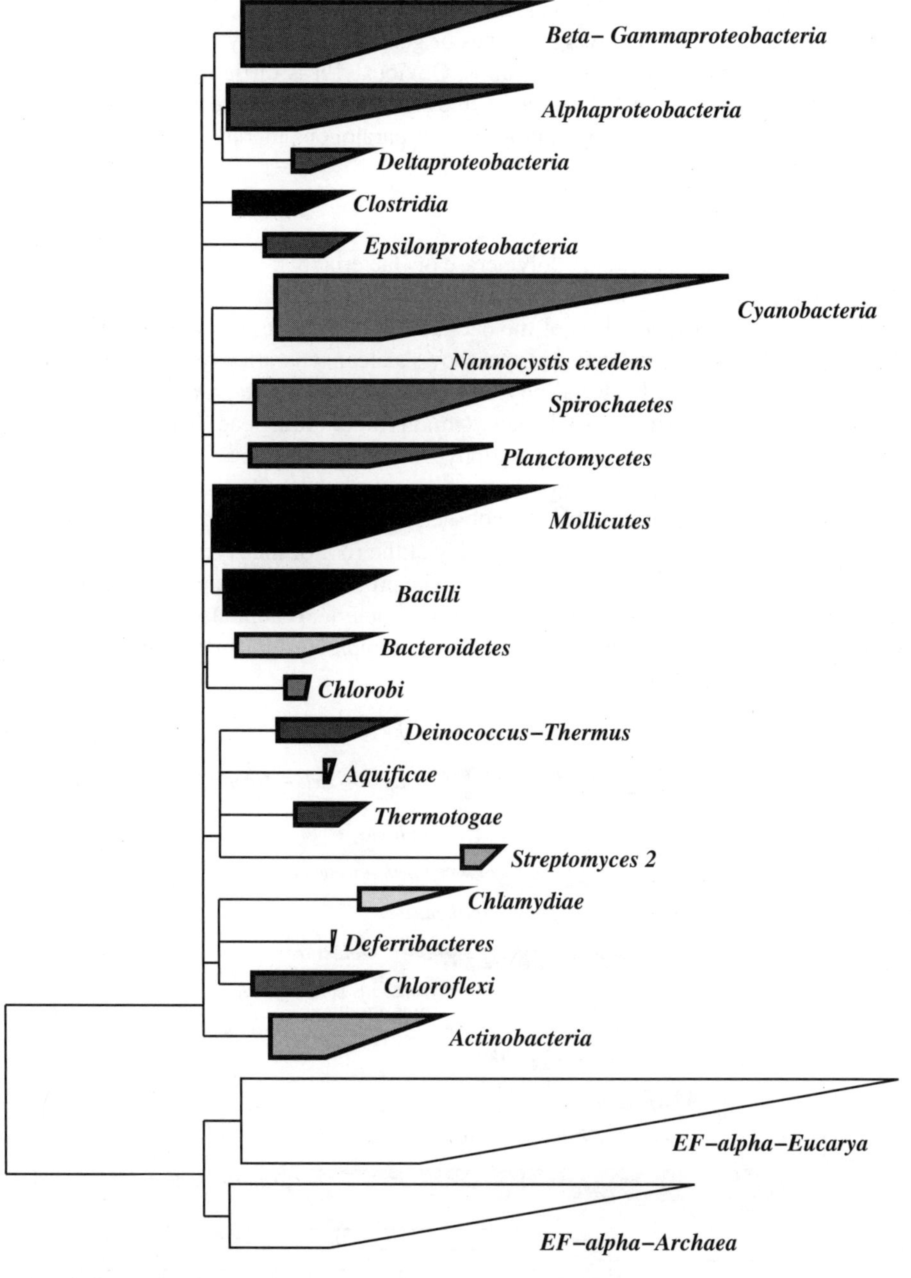

Figure 3.4. Radiation of the bacterial phyla or intraphylum groups as derived from phylogenetic analyses of elongation factor Tu amino acid sequences. Parameters for tree reconstruction were as described for Figure 3.1.

that is less diverged and that clusters among the Actinobacteria. This gene triplication demonstrates that multiple diverged genes or gene products may hamper phylogenetic analyses at various levels of relationships. Obviously, it is difficult to decide whether differences in rRNA- and EF-Tu-based trees indicate different resolution capacity or different history, or reflect the relationships of paralogous markers.

RNA Polymerase

The largest subunits of the RNA polymerase of Bacteria (β, β'), Crenarchaeota (A', A"), Euryarchaeota (B', B"), and Eucarya are highly conserved and ubiquitous. The comparative sequence data analysis of the bacterial beta subunit of RNA polymerase and their archaeal and eucaryal counterparts shows a clear separation of the three domains (fig. 3.5). Many of the bacterial phyla are represented in the data set. The Actinobacteria, Cyanobacteria, Spirochaetes, Chlamydiae, Aquificae, Thermotogae, and Thermus-Deinococcus appear as monophyletic groups. Although a significant relative branching order of the major groups is not seen, the deeper branching of the Cyanobacteria is rather unusual and is not seen in the trees based on other markers analyzed thus far. The Mollicutes branch deeply at the root of the Bacteria and cluster independently from the Bacilli and Clostridia. The proteobacterial subclasses are grouped together, with the exception of the Epsilonproteobacteria. A similar situation was found in the case of the beta' subunit (fig. 3.6). The common cluster of Proteobacteria is dis-

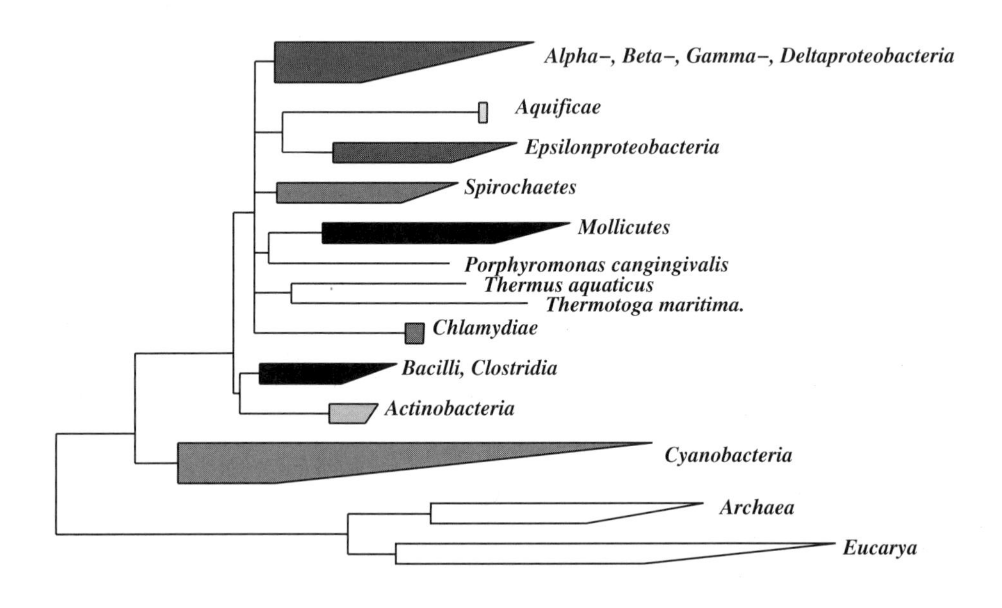

Figure 3.5. Radiation of the bacterial phyla or intraphylum groups as derived from phylogenetic analyses of bacterial DNA–directed RNA polymerase β' subunit amino acid sequences, including the homologous counterparts of the Eucarya and Archaea. Parameters for tree reconstruction were as described for figure 3.1.

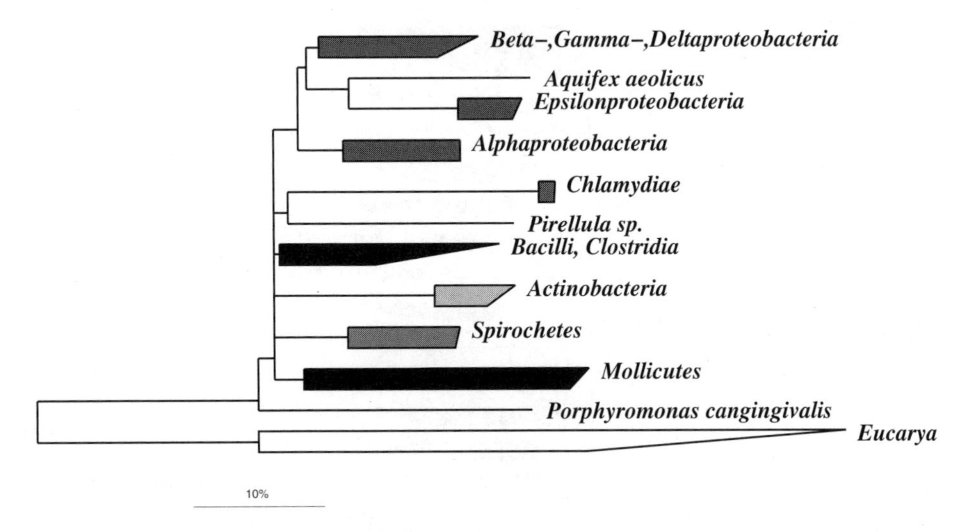

Figure 3.6. Radiation of the bacterial phyla or intraphylum groups as derived from phylogenetic analyses of bacterial DNA-directed RNA polymerase beta subunit amino acid sequences including the homologous counterparts of the Eucarya and Archaea. Parameters for tree reconstruction were as described for figure 3.1.

turbed by the inclusion of Aquifex. In both trees a moderate relationship of the latter to Campylobacter is indicated.

Hsp60

The domains of the Archaea and Bacteria are well defined and separated when Hsp60 protein sequences (Bacteria: GroEL; Archaea Tf-55) are phylogenetically analyzed (fig. 3.7). Most of the bacterial phyla represented in the data set appear as monophyletic clusters such as the Actinobacteria, Bacteroidetes, Cyanobacteria, Thermus-Deinococcus, Thermotogae, and Spirochaetes. A relative branching order of the phyla is only partially resolved, and with low significance. Monophyletic status of neither Proteobacteria nor Firmicutes could be observed. Although the Beta- and Gammaproteobacteria are unified in a common group, the representatives of the Alphaproteobacteria are split into three lineages. One is associated with the Beta-, Gammaproteobacteria cluster (e.g., rhizobia), whereas the second group (e.g., Rickettsia and relatives) is slightly separated from the majority of bacterial lines. The third group is represented by Holospora. The Epsilonproteobacteria are an additional separate lineage. A similar situation is seen in the case of the Firmicutes. Whereas the Bacilli and Clostridia represent a common cluster in the radiation of the bacterial phyla, the Mollicutes represent a somewhat deeper branch. This situation may reflect differences in information content and resolving power; however, gene duplication or lateral transfer effects may be of importance, as indicated by the multiple variants of Chlamydophila genes. One of the copies clusters among the bacterial phyla, whereas the additional versions of Hsp60-like molecules represent deeper branches in the tree.

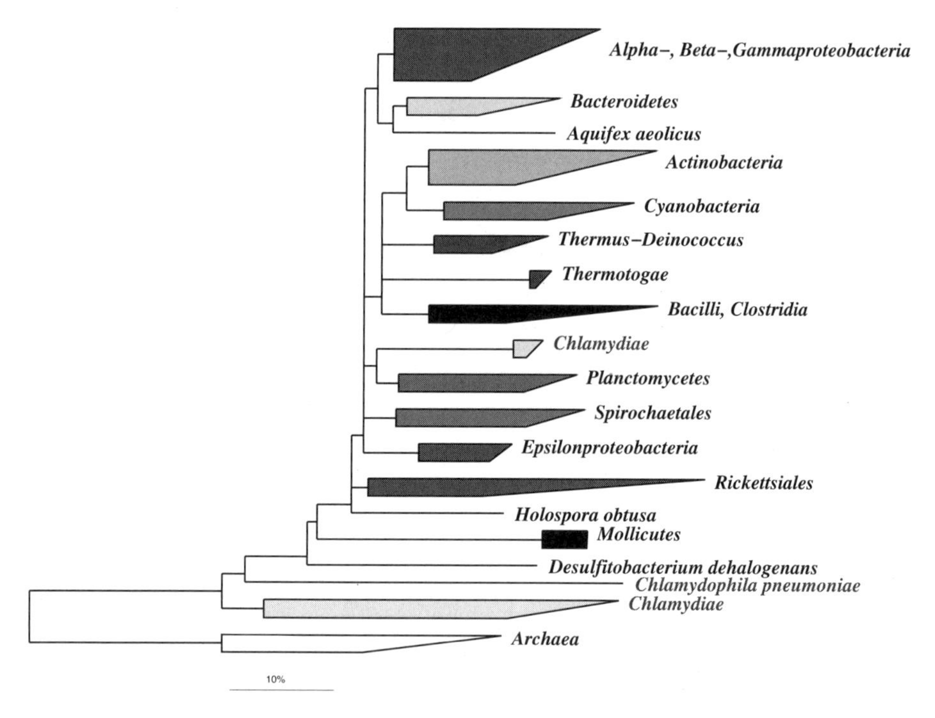

Figure 3.7. Radiation of the bacterial phyla or intraphylum groups as derived from phylogenetic analyses of Hsp60 heat shock protein amino acid sequences. Parameters for tree reconstruction were as described for figure 3.1.

Aminoacyl–tRNA Synthetase Markers that Support the rRNA Three-Domain Concept

The three-domain concept is clearly supported by tree topologies based on aspartyl, leucyl, tryptophanyl, and tyrosyl tRNA synthetases (table 3.2). An example is shown in figure 3.8 for aspartyl–tRNA synthetase. The phyla concept among the Bacteria domain is only partly supported by the respective analyses of these proteins.

Aspartyl tRNA Synthetase

The bacterial phyla Spirochetes, Deinococcus-Thermus, Actinobacteria, and Chlamydiae are well defined by comparative analyses of aspartyl–tRNA protein sequences. However, a relative branching order is not significantly supported. Among the Proteobacteria the Beta- and Gammaproteobacteria represent a common group and are clearly separate from the individually branching Epsilon- and Alphaproteobacteria. Again, the Mollicutes occupy rather deep branches within the Bacteria separated from Bacilli and Clostridia.

Table 3.2. Aminoacyl tRNA synthetases supporting the ribosomal RNA–based domain concept

Molecule	Domains	Phyla	Intraphylum	Duplicates
AspRS	Well defined	Some not monophyletic	Some paraphyletic, intermixed	Not known
LeuRS	Well defined	Some not monophyletic	Some paraphyletic, intermixed	Not known
TrpRS	Well defined	Partly defined	Some paraphyletic, intermixed	Present
TyrRS	Well defined	Partly defined	Some paraphyletic, intermixed	Present

Leucyl–tRNA Synthetase

The Proteobacteria cluster together, but a potential monophyletic structure is disturbed by the positioning of Aquifex. The phyla Chlamydia and Actinobacteria are well defined, but the Firmicutes are split up in three separate clusters (Bacilli, Mollicutes, and Clostridia).

Tryptophanyl–tRNA Synthetase

Among the Bacteria the phylum concept is poorly supported for the Chlamydiae and *Actinobacteria*. The proteobacterial subclasses are defined; however, they do not form a monophyletic group. The Mollicutes represent their own lineage, whereas representa-

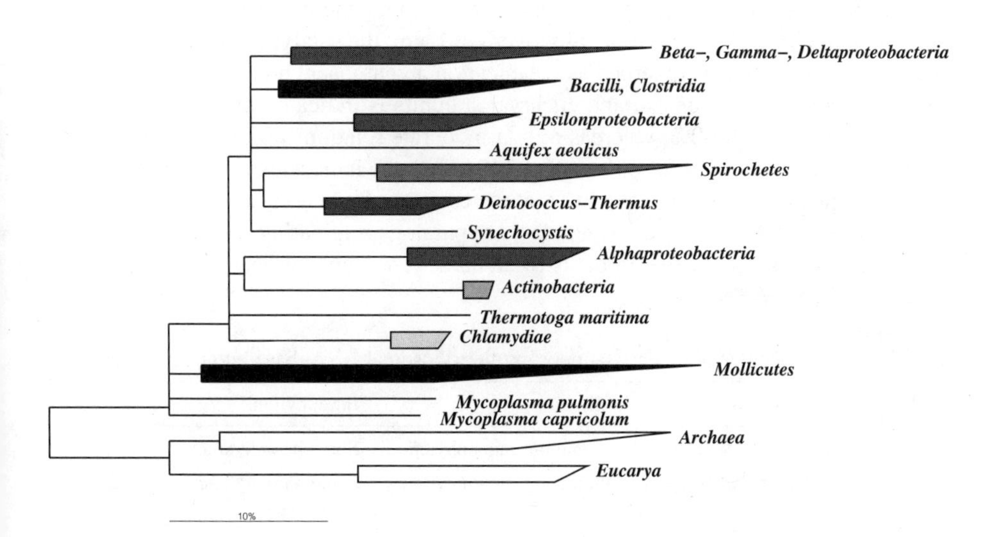

Figure 3.8. Radiation of the bacterial phyla or intraphylum groups as derived from phylogenetic analyses of aspartyl–tRNA synthetase amino acid sequences. Parameters for tree reconstruction were as described for figure 3.1.

tives of the Bacilli and Clostridia occur in three distant clusters. This situation may result from the analysis of diverged duplicated genes, as indicated by the positioning of *Bacillus halodurans* and *Deinococcus radiodurans*.

Tyrosyl–tRNA Synthetase

The tyrosyl–tRNA synthetase data globally support the three-domain concept. However, thus far two major clusters seem to exist among the Bacteria. Each of them exclusively contains part of rRNA-defined phyla or subgroups such as Epsilonproteobacteria, Spirochetes, Chlamydia, and Actinobacteria. However, members of mainly the Gammaproteobacteria and Firmicutes are dispersed in both bacterial main clusters. Again, there are examples of duplicated genes resulting in dispersed positioning of the respective organisms: *Bacillus subtilis, Clostridium acetobutylicum*, and *Vibrio cholerae*.

Phylogenetic Markers that Do Not Support the Three-Domain Concept

ATPase

A reasonable data set of sequences is available for the different subunits of the proton translocating ATPase (table 3.3). In general, the domains are clearly separated in the combined tree of alpha and beta subunits of F_1F_0 and vacuolar-type ATPases. The placing of some Bacteria among the Archaea and vice versa may likely be the result of gene duplication events. Historically, the catalytic beta subunit of the F_1F_0-type ATPase was among the first proteins tested as an alternative phylogenetic marker, and the results supported the SSU rRNA bacterial phylogeny.[15] Later, the catalytic subunit of the vacuolar type of ATPase was regarded as the eucaryal and archaeal homolog. A closer relationship between the archaeal and eucaryal subunits is indicated by overall sequence identity of more than 60%, whereas only a moderate relationship to the bacterial version is indicated by identities of less than 24%. With the increasing database of ATPase subunit sequences, it became evident that the catalytical and noncatalytical subunits of both types of ATPase (F_1F_0 and vacuolar) are paralogs originating from a common ancestor, as descendants of the products of an early gene duplication.[3,16]

Table 3.3. Molecules not clearly supporting the ribosomal RNA–based domain concept

Molecule	Domains	Phyla	Intraphylum	Duplicates
ATPase	Partly defined	Some not monophyletic	Some paraphyletic	Present
DNA Gyrase	Partly intermixed	Some not monophyletic	Some paraphyletic, intermixed	Present
Hsp70	Partly intermixed	Some not monophyletic	Some paraphyletic, intermixed	Present
RecA	Only Bacteria	Some not monophyletic	Some paraphyletic, intermixed	Present

All of this taken together prompted Woese and others to position the root of the three domains on the Bacteria branch. It was generally assumed that the vacuolar type of ATPase is a characteristic component of the members of the latter two domains, whereas the F_1F_0-type was regarded as unique for the Bacteria domain. Later, it was shown that some representatives of the Bacteria domain contained the genes and proteins for both the F_1F_0 as well as the vacuolar type of ATPase.[17] A similar situation was shown for an archaeon.[18] These findings indicate that not only the alpha and beta subunits originate from an early gene duplication but also their F_1F_0 and vacuolar-type versions from independent duplication events. In the case of *Enterococcus hirae*, it could be shown that the F_1F_0-type ATPase is responsible for proton translocation, whereas the vacuolar type works as a sodium pump.[19] The F_1F_0-type ATPase is always present in *Enterococcus* species, whereas the vacuolar type is to be found only in a few of these species[17] argue that changing the function of one of the ATPases also changes the selection pressure. Apparently, the additional vacuolar type is no longer essential in certain bacteria and may be changed at a higher rate or completely lost in the course of evolution. As a consequence, the phylogenetic positioning of such copies among the essential variants may be problematic or misleading.

The presence of additional vacuolar ATPase genes other than the F_1F_0 ATPase could be shown for representatives of different bacterial phyla. In the case of the *Firmicutes* V-ATPase, genes could be found in strains of *Clostridium, Enterococcus, Streptococcus*, and *Eubacterium* species.[17] In contrast, F_1F_0-type ATPase genes were demonstrated for the archaeon *Methanosarcina* in addition to the common vacuolar type genes.[18] However, there are bacterial taxa (members of the phyla Deinococcus-Thermus and probably also Spirochaetes and Chlamydiae) that seem to carry only the vacuolar version. This may indicate that the V-ATPase became the essential version in the time of their diversification. Because "ancient" genes or gene products are compared, it is difficult to recognize whether multiple genes originated from duplications or from early lateral gene transfer events.

Within the F_1F_0-type ATPase alpha subunit–based tree, the Actinobacteria and Cyanobacteria represent monophyletic phyla (fig. 3.9). In the case of the Proteobacteria, the Gammaprobacteria, Betaproteobacteria, and Alpha- Deltaproteobacteria form individual groups, and the Epsilonproteobacteria are placed separately. Rickettsia keeps its own line separated from the other Alphaproteobacteria. The Bacilli and Mollicutes are not unified in a Firmicutes phylum but form individual clusters. It has to be mentioned that, recently, further gene duplications were detected; namely, two diverged copies of the alpha-like subunit in the genomes of Gemmata and Pirellula (members of the Planctomycetes),[20] as well as for a representative of the Deltaproteobacteria, respectively (F.O. Gloeckner, pers. comm.).

The beta subunit data of the F_1F_0-type ATPase support the distinct phylum status of Actinobacteria, Thermotogae, Aquificae, Bacteroidetes, Chlorobi, and Cyanobacteria (fig. 3.10). The Gamma- and Betaproteobacteria share a common root, whereas the Alpha-, Epsilon-, and Deltaproteobacteria represent their own lineages. However, the situation is complicated by Stigmatella, which is separately positioned from the other Deltaproteobacteria closer to the Alphaproteobacteria branch. A similar situation is seen with the Firmicutes. The Bacilli, Mollicutes, and Clostridia appear as independent clusters, and Peptococcus as well as Acetobacterium represent separate lineages. In all cases of multiple genes described for the alpha subunit, the respective pendant was also found for the beta subunit.[20]

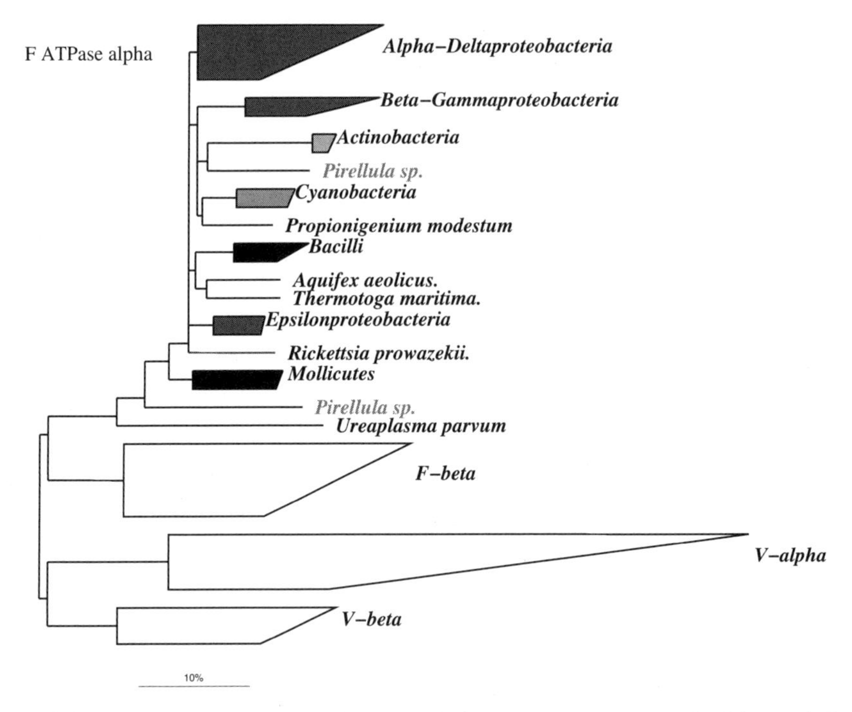

Figure 3.9. Radiation of the bacterial phyla or intraphylum groups as derived from phylogenetic analyses of F_1F_0-type ATPase alpha subunit amino acid sequences including paralogous reference subunits. Parameters for tree reconstruction were as described for Figure 3.1.

DNA Gyrase Subunit A

The bacterial phyla Spirochetes, Firmicutes, Chlamydiae, Actinobacteria, Cyanobacteria, and *Deinococcus*-Thermus are well defined, as are the proteobacterial subclasses of Gamma-, and Betaproteobacteria and the individual subclasses Alpha- and Epsilonproteobacteria (fig. 3.11). However, a common ancestry of the Proteobacteria is not supported by the gyrase subunit A data. Most notably, the representatives of the Archaea are not clearly separated from the Bacteria. The Archaeoglobus and Thermoplasma lineage cannot be significantly placed in the tree, and the halobacteria seem to group with the Spirochetes and proteobacterial subgroups at a low similarity level.

DNA Gyrase Subunit B

The monophyletic status of the bacterial phyla Firmicutes, Actinobacteria, Spirochetes, and Deinococcus-Thermus is well supported (fig. 3.12). A common origin of Beta- and Gammaproteobacteria is indicated, as is a monophyletic cluster of the Epsilonproteobacteria. However, as for the subunit A data, a monophyletic status of the proteobacterial cannot be inferred from the gyrase B data. Moreover, the representatives of the Archaea are not clearly separated from the Bacteria. With low significance, they branch off in the neighborhood of Thermotoga, Aquifex, and the Actinobacteria. Recently, a gene duplication was found for Gemmata and Pirellula, representatives of the Planctomycetes.[20]

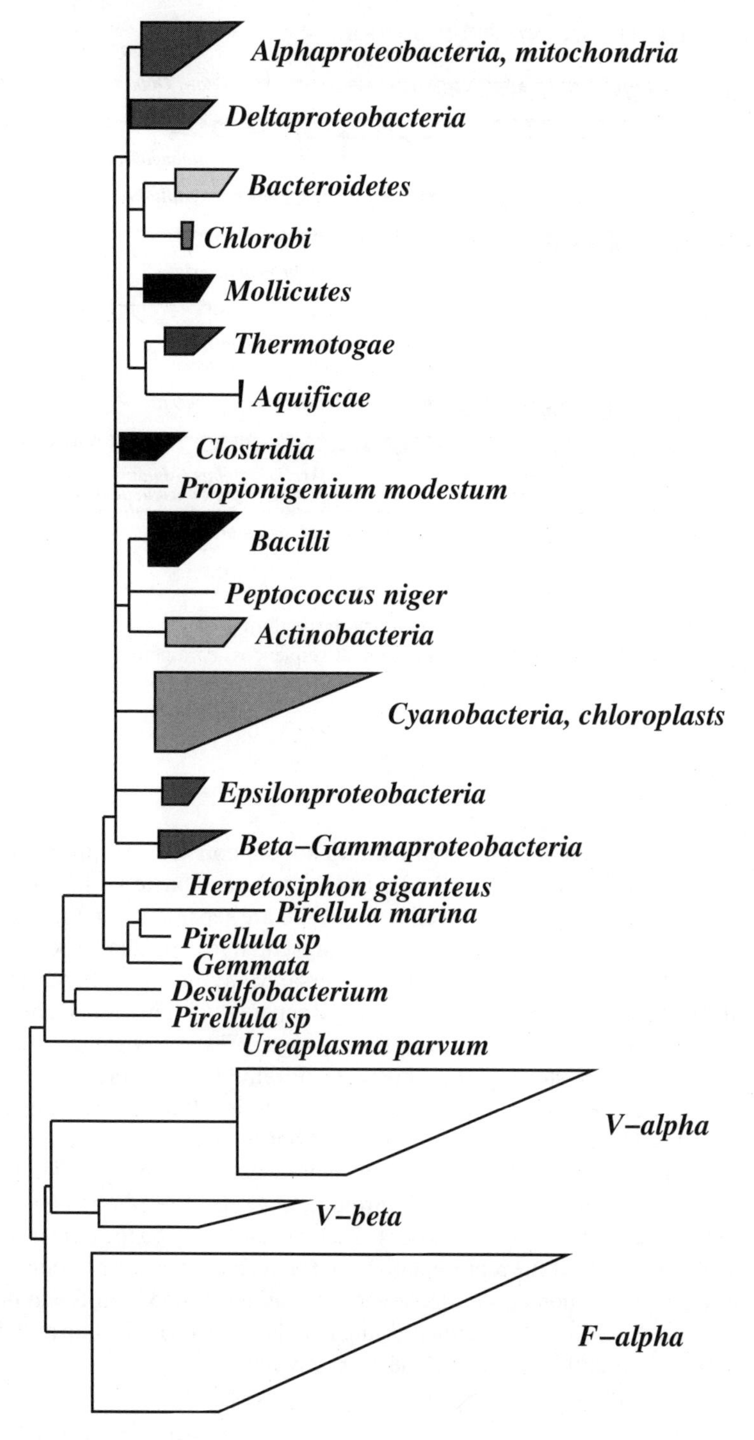

Figure 3.10. Radiation of the bacterial phyla or intraphylum groups as derived from phylogenetic analyses of F_1F_0-type ATPase beta subunit amino acid sequences including paralogous reference subunits. Parameters for tree reconstruction were as described for Figure 3.1.

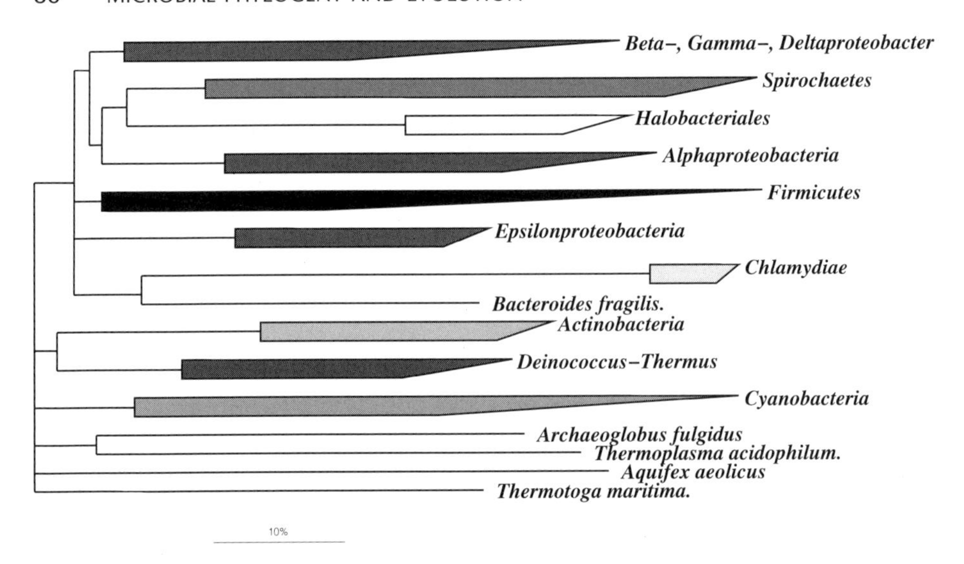

Figure 3.11. Radiation of the bacterial phyla or intraphylum groups as derived from phylogenetic analyses of DNA gyrase subunit A amino acid sequences. Parameters for tree reconstruction were as described for Figure 3.1.

Hsp70

The Hsp70 (DnaK) heat shock protein data do not support a clear separation of the domains Archaea and Bacteria (fig. 3.13). The archaeal and bacterial lineages appear intermixed. However, they are still well defined and are separated at rather low levels of relationship. So far, none of the archaeal representatives appears to be closely related to a member of the Bacteria, and vice versa. Many of the bacterial phyla are supported by the Hsp70 tree, such as the Spirochetes, Chlamydia, Deinococcus-Thermus, Actinobacteria, Aquificae, and Thermotogae. The Proteobacteria subclasses are well defined and cluster together; however, Pirellula, as a representative of the Planctomycetes, seems to be related to the Proteobacteria, although the statistical significance is low. The Firmicutes cluster together but include, again at a low level of relationship, some Archaea. The currently available sequences are not sufficient to decide whether the branching pattern results from limitations of resolving power, lateral gene transfer, or gene duplications or losses. That the latter may be the case is indicated by the branching pattern of the representatives of Cyanobacteria and chloroplasts. They are not unified in a common cluster; however, two variants found in different branches are known for Synechococcus. A further example of duplicated Hsp70-like proteins was recently found for a representatives of the Planctomycetes.[20]

RecA

The RecA protein sequence–based tree contains only representatives of the Bacteria (fig. 3.14). According to the currently available data, the Thermus-Deinococcus, Actinobacteria and Chlamydia phyla appear as contiguous groups. However, a monophyletic

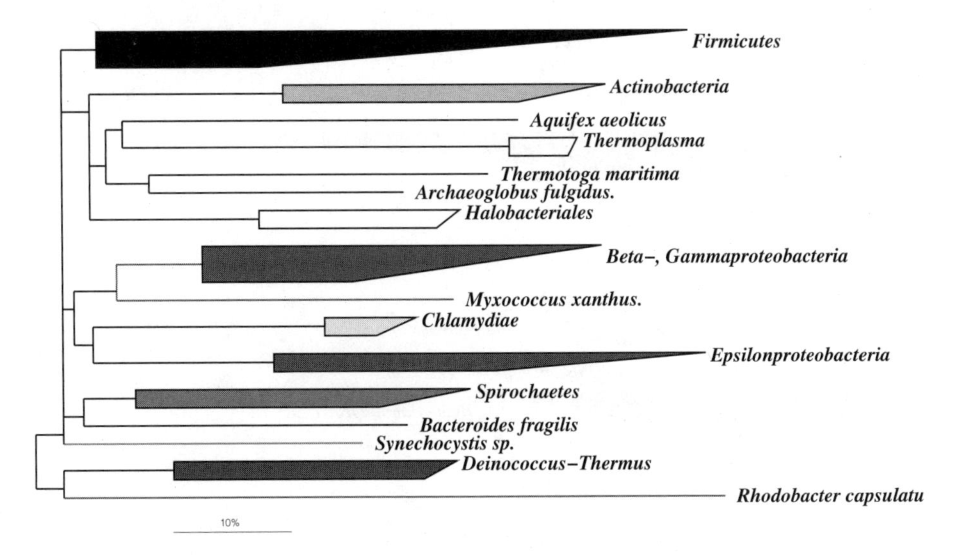

Figure 3.12. Radiation of the bacterial phyla or intraphylum groups as derived from phylogenetic analyses of DNA gyrase subunit B amino acid sequences. Parameters for tree reconstruction were as described for Figure 3.1.

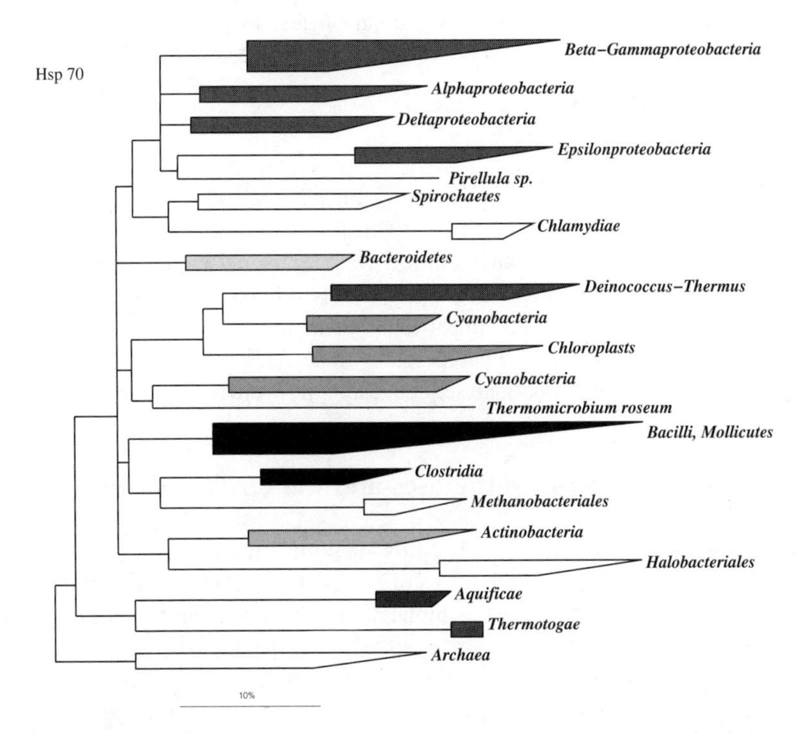

Figure 3.13. Radiation of the bacterial and archaeal phyla or intraphylum groups as derived from phylogenetic analyses of Hsp70 heat shock protein amino acid sequences. Parameters for tree reconstruction were as described for Figure 3.1.

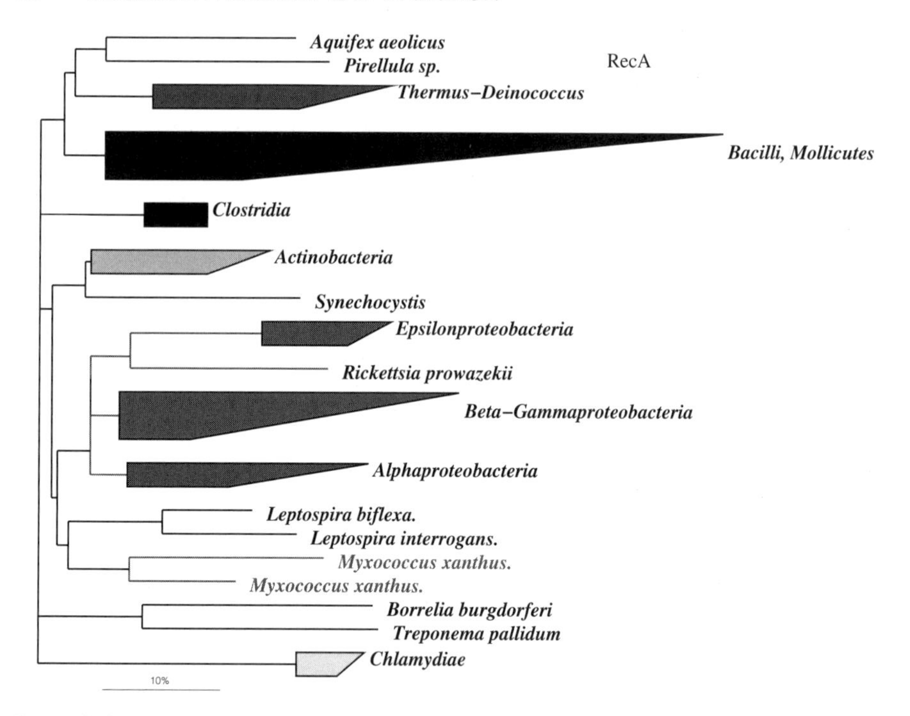

Figure 3.14. Radiation of the bacterial phyla or intraphylum groups as derived from phylogenetic analyses of RecA protein amino acid sequences. Parameters for tree reconstruction were as described for Figure 3.1.

status is not documented for the Proteobacteria, Firmicutes, and Spirochetes. In the case of Firmicutes, the subgroups of Bacilli and Clostridia are well defined and separated. The Leptospira species are somewhat remote from Treponema and Borrelia. The former disrupt the Proteobacteria cluster but are not closely related to any of them. Although the short edges forming the branching pattern of the major lineages indicate low resolving power, the pattern may also be hampered by undetected paralogy effects. Two protein sequences sharing only moderate sequence identity were reported for Myxococcus.

Aminoacyl–tRNA Synthetases and the Phyla Concept

The domain and phylum concept is only partly supported by the majority of aminoacyl–tRNA synthetases (Table 3.4). The results of comparative amino acid sequence analyses indicate a number of potential gene duplication or lateral transfer events questioning or obscuring a clonal history. However, despite these potential rearrangements, a phylogenetic signal can be seen in most cases.[21] Although phyla are often not clearly defined, many groups at or below the phylum level that are defined and supported by the other markers can be also found in the case of aminoacyl–tRNA synthetases. An example is valyl tRNA synthetase (fig. 3.15).

Table 3.4. Aminoacyl tRNA synthetases not clearly supporting the ribosomal RNA–based domain concept

Molecule	Domains	Phyla	Intraphylum	Duplicates
ArgRS	Not supported	Most not supported	Most intermixed	Not known
AspRS	Ppartly defined	Well defined	Monophyletic	Not known
CysRS	Not supported	Partly defined	Some paraphyletic. intermixed	Not known
GluRS	Well separated	Partly defined	Some paraphyletic, intermixed	Not known
HisRS	Partly intermixed	Partly defined	Some paraphyletic, intermixed	Present
IleRS	Partly intermixed	Partly defined	Some paraphyletic, intermixed	Not known
LysRS	Not supported	Partly defined	Some paraphyletic, intermixed	Not known
PheRS	Partly intermixed	Most defined	Some paraphyletic, intermixed	Not known
ProRS	Partly supported	Partly defined	Some paraphyletic, intermixed	Not known
SerRS	Partly defined	Partly defined	Some paraphyletic, intermixed	Present
ThrRS	Not supported	Partly defined	Some paraphyletic, intermixed	Present
ValRS	Partly intermixed	Partly defined	Some paraphyletic, intermixed	Not known

Alanyl–tRNA Synthetase

Although the Archaea and Eucarya appear to represent monophyletic groups, the classical three-domain concept is not supported by the alanyl tRNA synthetase sequences. The Archaea and Eucarya seem to be equivalent to bacterial phyla intermixed with those in the respective tree. The Proteobacteria, Chlamydia, Aquificae, Firmicutes, Actinobacteria, Cyanobacteria, Deinococcus-Thermus, and Spirochetes branch as distinct phyla.

Glutamyl tRNA Synthetase

At a first glance, the glutamyl–tRNA synthetase–derived tree reveals a rather confusing picture. Although Archaea and Bacteria are separated from one another, only a few phyla seem to be well defined, such as the Actinobacteria, Cyanobacteria, and Spirochaetes. Other groups such as Alpha-, Gamma-, or Epsilonproteobacteria or organisms such as Deinococcus, Aquifex and Thermotoga occur more than once in the tree, given the presence of two or more evolutionary diverged gene copies. The clusters defined by sequence similarity of the individual variants include organisms related by rRNA and other marker data.

Histidyl–tRNA Synthetase

The histidyl–tRNA synthetase data again provide examples of multiple genes exhibiting highly diverged primary structures. At a first glance, three major groups are evident. One comprises exclusively Bacteria (thus far); the second, Archaea and Bacteria; and the third comprises representatives of all three domains. Within these groups, subclusters of closer-related organisms correspond to those defined based on rRNA data. Examples in the first group are the Chlamydia, Gamma-, Beta-, Deltaproteobacteria, Actinobacteria, and Mollicutes. However, even the Bacilli are not monophyletic. Although none of the

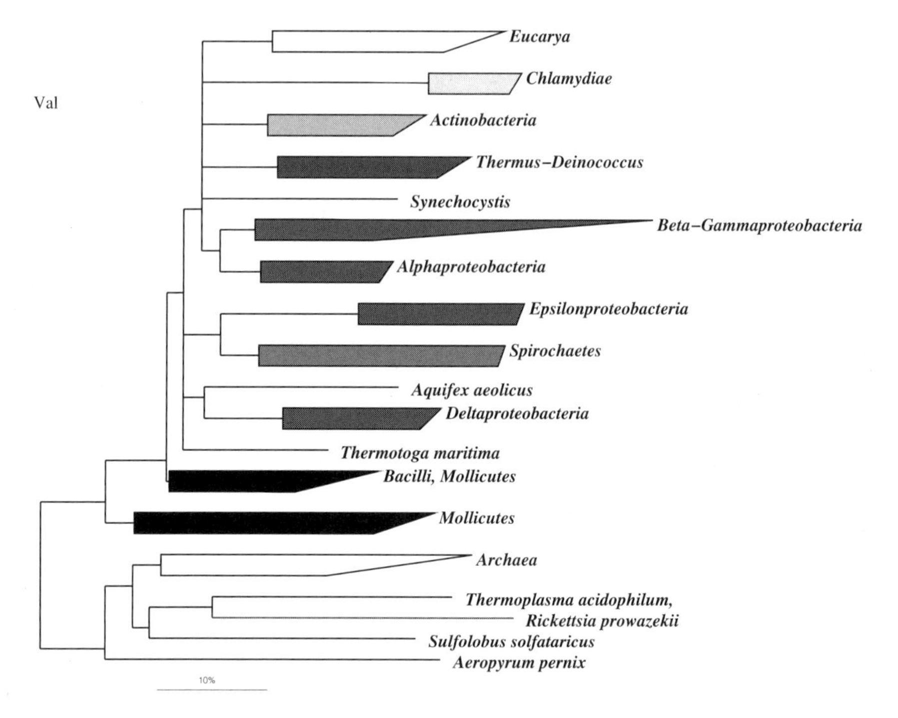

Figure 3.15. Radiation of the bacterial and phyla or intraphylum groups as derived from phylogenetic analyses of valyl tRNA synthetase amino acid sequences. Parameters for tree reconstruction were as described for Figure 3.1.

organisms is represented by multiple molecules in all three major groups, examples of duplicates are found for all possible group combinations (i.e., Thermotoga in groups one and three, Aquifex in groups one and two, and Clostridium in groups two and three.

Isoleucyl–tRNA synthetase

A division into three clusters can also be seen in the isoleucyl–tRNA-based tree. Two homogeneous clusters of Bacteria and Archaea are separated from a mixed lineage comprising monophyletic Eucarya together with bacterial subgroups and Archaea. The Proteobacteria, Aquificae, and Thermotogae, as well as the subgroups Bacilli and Mollicutes, are found in the Bacteria cluster. However, the Chlamydia, Actinobacteria, Deinococcus-Thermus, and Spirochetes branch off in the neighborhood of eucaryal representatives. Rickettsia, Staphylococcus, and Clostridium are examples for Bacteria which are positioned also here, distant from their 16S rRNA phyla.

Methionyl–tRNA Synthetase

The results of phylogenetic analyses of methionyl tRNA synthetase data only roughly support the rRNA-based domain and phylum concept. A closer relationship of Archaea and Eucarya is indicated as being separated from the majority of the bacterial groups thus far analyzed. However, Spirochetes, Chlamydia, and Clostridia are found within the Archaea–Eucarya radiation. In the subtree comprising only bacterial representatives subgroups, the represented phyla are distributed in various clusters. The separation of the different Proteobacteria subclasses is especially striking. These clusters again comprise organisms for which a relationship is documented by rRNA and other data.

Phenylalanyl–tRNA Synthetase

Both subunits (alpha, beta) of the phenylalanyl–tRNA synthetase generally support the domain status of the Bacteria separated from the Archaea and Eucarya. However, it has to be emphasized that in both trees the Spirochaetes are found among the lineages of the Archaea subtree. Within the Bacteria subtree most of the bacterial phyla such as Chlamydia, Actinobacteria, and Thermus-Deinococcus are well defined. In the alpha subunit tree, the potential monophyletic status of the proteobacteria is disturbed by the apparent affiliation of Aquifex to the Alphaproteobacteria. Similarly, the unity of the Firmicutes is abandoned by their clustering with Thermotoga (among the Firmicutes). In the beta subunit tree, the Mollicutes are separated from Bacilli and Clostridia, as are the Gamma-Betaproteobacteria from the Alpha-, as well as from the Epsilonproteobacteria.

Seryl–tRNA Synthetase

Seryl–tRNA synthetase–based tree topologies generally support only with low significance the three-domain concept. Seryl rRNA trees place representatives of the archaeal halophiles within the radiation of bacterial groups, whereas the archaeal methanogens apparently contain a highly diverged version of the protein. The phyla Spirochetes, Deinococcus-Thermus, Actinobacteria, and Chlamydia as well as the subgroups Beta-Gammaproteobacteria, Alphaproteobacteria, Epsilonproteobacteria, Mollicutes, and Bacilli-Clostridia, are supported. Clostridium acetobutylicum contains two versions of the protein, one indicating a position among the Bacilli–Clostridia group, the second separately branching in the neighborhood of the Chlamydiae.

Arginyl–tRNA Synthetase

Arginyl–tRNA synthetase–based tree topology does not generally support the three-domain concept. Deinococcus is found among Archaea, and Eucarya group among bacterial clusters. The bacterial phyla or subgroups are distributed over distantly related groups. Although Gamma- and Betaproteobacteria group together, the Alphaproteobacteria are disrupted and the Epsilonproteobacteria are clearly separated. A similar situation can be seen with respect to the positioning of subgroups of the Firmicutes or Actinobacteria.

Cysteinyl–tRNA Synthetase

The cysteinyl–tRNA synthetase data generally do not support a clear-cut separation of the domains. However, the phylum structure is partly maintained, at least in the case of the Proteobacteria, Spirochetes, and Chlamydia. As in the Arginyl–tRNA synthetase–

based tree, subgroups of the Firmicutes and other phyla are also dissected. However, again, the well-defined clusters comprise organisms related to one another in the trees inferred from rRNA and other marker data.

Lysyl–tRNA Synthetase

The three domains are not clearly separated in trees based on lysyl–tRNA synthetase data. However, the Eucarya represent a monophyletic group, and the bacterial phyla such as Actinobacteria, Chlamydia, Deinococcus-Thermus, and Firmicutes are well defined. Even the Gamma- and Betaproteobacteria represent a monophyletic group together with the Alphaproteobacteria, whereas the Epsilonproteobacteria branch more distantly.

Prolyl–tRNA Synthetase

A bipartition of a Bacteria cluster and a second group comprising Archaea and Bacteria is indicated by the prolyl–tRNA synthetase data. Some phyla or subgroups combine organisms that are significantly related, according to the rRNA data, such as the Beta- and Gammaproteobacteria, Alphaproteobacteria, Bacilli, Actinobacteria, and Chlamydia within the first cluster and the Mollicutes in the second cluster. No duplicated molecules are known thus far. However, representatives of the Clostridia, Spirochetes, and Actinobacteria are distributed among the two primary clusters.

Threonyl–tRNA Synthetase

The comparative sequence data analysis of the threonyl–tRNA synthetase indicates the existence of two types of the protein rather than a clear separation of Archaea and Bacteria. One of the two major clusters comprises the majority of the bacterial representatives, the Eucarya and some Archaea, whereas the majority of Archaea and some Bacteria are unified in the second cluster. Some of the bacterial phyla such as Actinobacteria and Chlamydiae or major subgroups as Alpha-, Beta-, and Gammaproteobacteria; Epsilonproteobacteria, and a group comprising Mollicutes and Bacilli are found in the first cluster. However, this unusual picture is obscured by the fact that two diverged protein versions are present in the case of Bacillus subtilis and Clostridium acetobutylicum. In one case, Bacillus is found within the Mollicutes–Bacilli group and, in another case, in the neighborhood of Thermotoga. Clostridium, together with Thermoanaerobacter, represents a Clostrida lineage, whereas the second version of the Clostridium protein is more similar to the archaeal and bacterial sequences in the Archaea-dominated cluster.

Valyl–tRNA Synthetase

The valyl–tRNA synthetase–based trees indicate a bipartition of the Archaea on the one hand, including Rickettsia (bacterial lineage) as an exception, and a Bacteria–Eucarya cluster on the other hand. However, the Eucarya appear as equivalent to the major bacterial lineages. The bacterial phylum concept is roughly supported, as the Chlamydiae, Actinobacteria, Thermus-Deinococcus, and Spirochetes represent monophyletic branches. The Gamma-, Beta-, and Alphaproteobacteria are unified in a common group, however, separated from the Epsilon- and Deltaproteobacteria. The Mollicutes branch rather deeply and are distinctly separated from the Bacilli–Clostridia subgroup.

Discussion

Here we have drawn on a selection of conserved proteins, as alternative phylogenetic markers, to compare with rRNA-based genealogies. Our comparative analyses are generally hampered by the great imbalance of the available sequence data sets for protein genes when compared to the comprehensive small subunit rRNA sequence database. Comprehensiveness with respect to both numbers of different sequences as well as to the spectrum of phylogenetic diversity is needed to recognize variable and conserved sequence positions as such and to estimate the significance of the separation or unification of the molecules or organisms in the respective trees. As mentioned earlier, the comparisons and evaluations focused on the verification of the rRNA defined domains, phyla, and major subgroups.

Three clearly defined and separated domains are supported by genes for several highly conserved proteins: the elongation factor Tu (1 alpha); heat shock protein Hsp60; and RNA polymerase subunits aspartyl–, leucyl–, phenylalanyl–, tyrosyl–, and tryptophanyl–tRNA synthetases. If we assume that the F_1F_0- and vacuolar-type ATPases represent paralogous markers, the former cannot be used to evaluate the three-domain concept because orthologous eucaryal and archaeal sequence data are missing or represented only by duplicates of different or uncertain function (e.g., Methanosarcina). Using the vacuolar data and excluding those "bacterial" sequences that are duplicates, a picture emerges of moderately related Archaea and those Bacteria representatives (members of Thermus-Deinococcus, Chlamydiae, and Spirochaetes) that exclusively contain V-ATPases (according to genome data) separated from the Eucarya (fig. 3.16). Although the Chlamydiae and Spirochetes are separated from the Archaea, the Thermus-Deinococcus phylum deeply branches among the archaeal lines. However, this intermingling is caused by a small number of positions with a low positional variability (with respect to the number of different character states). The currently available dataset is still too small to estimate whether these are generally conserved positions or whether the few evolutionarily allowed character states are rapidly changing.

The three domain concept is only partly supported by glutamyl–, methionyl–, alanyl–, histidyl–, isoleucyl–, and seryl–tRNA synthetases. This may be because of missing data for one of the domains, heterogeneity of one or more of them, or low resolution, and it does not necessarily indicate lateral gene transfer. The concept is not supported by DNA gyrase, heat shock protein Hsp70, arginyl–, cysteinyl–, lysyl–, prolyl–, and threonyl–tRNA synthetases. The concept cannot be checked using RecA protein data, as there are currently no homologues known for Archaea and Eucarya.

The rRNA-based bacterial phylum concept is at least partly supported by all alternative markers. However, in most cases, subclasses of Proteobacteria as well as the Firmicutes do not appear as monophyletic phyla. Whereas the Beta- and Gammaproteobacteria are almost always unified in a common group, in many cases including the Alphaproteobacteria, the Delta- and especially the Epsilonproteobacteria more often represent their own lineages.

A tendency to split off the Epsilonproteobacteria from the other subgroups is also seen with increasing 16S and 23S rRNA sequence databases (W. Ludwig, unpublished

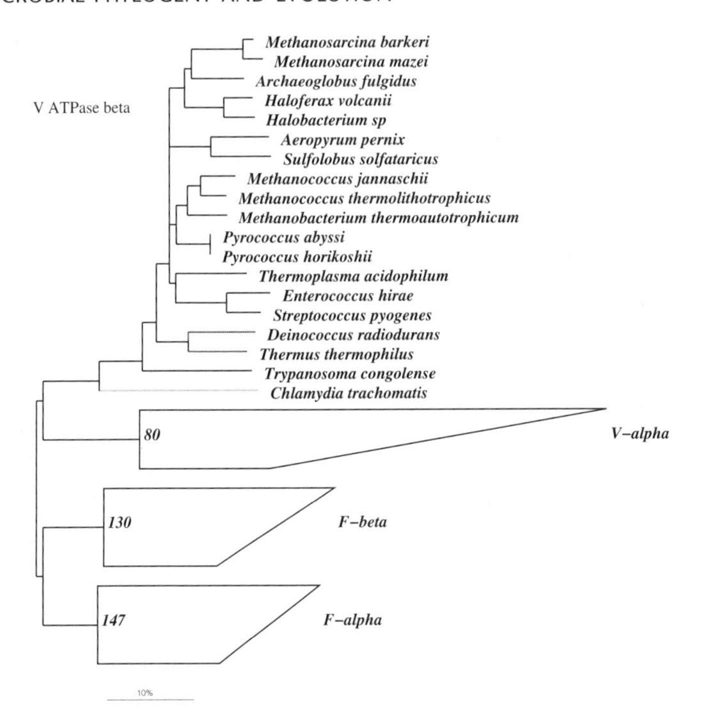

Figure 3.16. Radiation of representatives of the three domains derived by comparative analysis of vacuolar-type ATPase alpha subunit sequence data. Parameters for tree reconstruction were as described for Figure 3.1.

data). A further splitting of the major subgroups of the phyla as defined in rRNA-based trees was found for the Alpha- and Deltaproteobacteria in the case of elongation factor Tu. The Alphaproteobacteria are not monophyletic within the heat shock protein Hsp60–based tree. In rRNA-based trees, the Mollicutes, Bacilli, and Clostridia share a common root. However, they are clearly separated from one another. The only alternative markers supporting the monophyletic status of the Firmicutes are alanyl–, phenylalanyl–, and lysyl–tRNA synthetases and DNA gyrase. In the case of heat shock protein Hsp70, they are unified in a group together with some Archaea. The Mollicutes appear as monophyletic together with the Bacilli only in the case of elongation factor Tu. A common group of Bacilli and Clostridia is seen when RNA polymerase, heat shock protein Hsp60, and valyl–, seryl–, and phenylalanyl–tRNA data are subjected to phylogenetic analyses. The Bacilli and Clostridia are dispersed according to tryptophanyl–tRNA synthetase data. In the case of tyrosyl–tRNA synthetase, a dissection of the Gammaproteobacteria as well as the subgroups of Bacilli and Clostridia can be observed. The Deltaproteobacteria and Clostridia are split within F_1F_0-type ATPase beta subunit trees.

The Alphaproteobacteria, Bacilli, and Clostridia do not form monophyletic groups according to isoleucyl tRNA synthetase data. The Cyanobacteria are dissected in the case of heat shock protein Hsp70. The Clostridia, Spirochetes, and Actinobacteria appear dispersed in prolyl–tRNA synthetase–based trees, as do the Alphaproteobacteria in the context of valyl–tRNA synthetase. An unusual scattered branching of the Proteobacteria subclasses is seen with arginyl–and methionyl–tRNA synthetase data.

There are also other notable incongruities with respect to the rRNA–based view. For example, there are the rather unusual positions of Aquifex and Thermotoga in phenylalanyl–tRNA synthase–derived trees and the positioning of Stigmatella apart from the other Deltaproteobacteria on the Alphaproteobacteria branch according to F_1F_0-type ATPase beta subunit data. Multiple genes or proteins are so far known for elongation factor Tu, heat shock protein Hsp60, tryptophanyl, tyrosyl–tRNA synthetase, ATPase, glutamyl, histidyl, seryl–tRNA synthetase, DNA gyrase subunit B, heat shock protein Hsp70, threonyl–tRNA synthetase, and the RecA protein. It is generally difficult to decide whether these paralogous genes resulted from duplications or from horizontal gene transfer.

Conclusions

The currently accepted view of bacterial phylogeny and taxonomy with regard to species and higher ranks is supported by comparative analysis of more than 30,000 small subunit rRNA full and partial sequence data.[2] A sound critical evaluation or supplementation of these comprehensive data by the comparative study of alternative phylogenetic marker molecules remains problematic today, at the dawn of the new era of genome sequencing. The major reason for these problems is the limited number of currently available sequence data. Of course, phylogenetic trees, constructed on the basis of the currently available data, are dynamic structures that may change their local topologies as new data become available. Thus, many of the inconsistencies and discrepancies may be resolved in the near future.

There are several other problems when comparing sequence data of phylogenetic marker molecules. One problem, apparently underestimated in the past, is multiple evolutionary diverged versions of homologous markers. Many examples of duplicates with different degrees of sequence divergence (phylum to genus level) were found in the databases of phylogenetic markers included in this study. The resulting problems that hamper sound phylogenetic analysis are diverse (e.g., to recognize paralogous markers as such, to define the orthologous versions, to recognize the copies sharing the same function and selection pressure, and to differentiate clonal duplication and lateral gene transfer).

Commonly, testing for xenologous origin of potential markers is based on comparative analysis of codon usage or oligonucleotide profiles. Given that "ancient" events of duplication or lateral gene transfer have to be recognized as such, it is highly likely that potential peculiarities (codon usage, oligonucleotide frequency profiles) have been adapted to the "host genome" during the course of time. Furthermore, the assumption that in case of duplicates a relaxed selection pressure acting on nonessential versions should be expressed by a higher rate of change in comparison with that of the essential component is not necessarily correct. The function of the respective product may still require sequence conservation, whereas the loss of the gene may occur at a higher frequency.

Concurrent with the many ongoing genome projects and the findings of a much higher fraction of xenologous genes than initially expected, nowadays a tendency can be seen to question any clonality among microbes.[22] It is under discussion whether frequent lateral gene transfer and recombination may have completely obscured phylogenetic signals by disrupting vertical heredity. For closely related organisms, it was postulated that genomic similarity results from frequent exchange of genes rather than from common ancestry.[23] With respect to lateral gene transfer and recombination, ironically, the highly conserved genes commonly used for phylogenetic investigations should be preferred targets.[22] At least in microbial communities of organisms living in close proximity, the primary structures of these genes should be rapidly equalized. Obviously, that is not the case. Clonality could still be abandoned if these genes are frequently transferred, replacing the residing homolog without effective recombination. If this occurred frequently, more incongruities would have to be expected when comparing the marker genes. Even if recombination events concern short fragments of conserved genes,[24–28] this should not strongly effect the overall picture. A significant detailed resolution of tree topologies cannot be expected in any case, as discussed above.

Gogarten et al. listed phylogenetic incongruities for almost all markers used in this study and postulated horizontal gene transfer as the causative factor.[21] However, as discussed above, the divergent multiple copies found in almost all cases may actually indicate gene duplications. As a consequence, one cannot simply assume lateral gene transfer as the only possibility. In any case, it is presently impossible to unambiguously determine whether the discrepancies observed are substantial or result from "illegitimate" comparison of paralogous markers. In this context, it has to be considered that multiple rRNA genes are common among bacteria; however, a survey of multiple SSU and LSU rRNA sequences from representatives of the different phyla showed that the range of divergence usually does not exceed 1.5%–2%.[29] This is within the normal "noise" of rRNA-based phylogenetic trees.[3] A few exceptions are known, such as *Thermobispora bispora* and *Haloarcula marismortui* for which 6.4% or 5% sequence divergence of SSU rRNA genes have been reported.[30,31]

Because of limitations of the information content and shortcomings of data analysis, local topologies in trees based on any marker are of low significance.[3] Still, the overall tree topologies derived from the primary structure data of markers compared in this study correspond at least to the rRNA-defined subgroups of bacterial phyla. Many of bacterial phyla are well defined in the major part of the trees based on the protein markers, although a detailed significant relative branching order could not be unambiguously determined in most cases or, if determined, often differs from the rRNA-based picture. However, as discussed here and elsewhere,[4] it is important to emphasize that detailed agreement cannot be expected when comparing alternative markers.

Despite all these analytic problems, there seems to exist an organismal genealogical trace that goes back in time to the universal ancestor.[21,32] The bacterial phyla and subgroups are even supported by those aminoacyl tRNA synthetases that had undergone major reshuffling during their history. It indicates that a record of a collective history has persisted despite or irrespective of whether horizontal gene transfer or gene duplication and loss occurred.[32,33] There is convincing support of the idea that there is high conservation (and hence clonality) of that part of a cell's components that is involved in maintaining and processing genetic information and basic energy acquisition. Lateral gene transfer or gene duplications seem to be more restricted in the case of intrinsic compounds such as rRNA and elongation factors and more relaxed for modular components such as the aminoacyl tRNA synthetases.

Though the domain and phylum concept are supported by a number of the protein markers, it is evident that no individual marker molecule can correctly reflect the evolutionary history of the organisms. Although there are some discrepancies concerning both detailed and global tree topologies, the overall rRNA-based picture can only be really challenged if potential paralogy and gene transfer issues can be fixed for the respective alternative marker. In the current state of knowledge, the rRNAs remain the most informative molecules for phylogenetic analyses. Moreover, in consideration of the large data set, it is certainly justified to found bacterial taxonomy at the higher ranks upon the results of comparative rRNA analysis. Progress in comparative sequence analyses of most of the other molecules studied here will help us to better understand the phylogenetic meaning of gene or protein sequences with respect to organismic genealogy, to fix weak spots in the rRNA-based framework, and to thus improve and extend our understanding of bacterial evolution.

References

1. G.E. Fox, K.R. Pechman, and C.R. Woese. 1977. "Comparative Cataloging of 16S Ribosomal Ribonucleic Acid: Molecular Approach to Procaryotic Systematics," *International Journal of Systematic Bacteriology* 27 (1977): 44–57; 44.
2. G.M. Garrity, *Bergey's Manual of Systematic Bacteriology*, 2nd edition (New York: Springer, 2001).
3. W. Ludwig, O. Strunk, S. Klugbauer, N. Klugbauer, M. Weizenegger, J. Neumaier, M. Bachleitner, and K.H. Schleifer, "Bacterial Phylogeny Based on Comparative Sequence Analysis," *Electrophoresis* 19 (1998): 554–568; 554.
4. W. Ludwig, and H.P. Klenk, "Overview: A Phylogenetic Backbone and Taxonomic Framework for Procaryotic Systematics," in George M. Garrity, ed., *Bergey's Manual of Systematic Bacteriology*, 2nd edition (New York: Springer, 2001), 49–66; 49.
5. W. Ludwig, J. Neumaier, N. Klugbauer, E. Brockmann, C. Roller, S. Jilg, K. Reetz, I. Schachtner, A. Ludvigsen, M. Bachleitner, U. Fischer, and K.-H. Schleifer, "Phylogenetic Relationships of Bacteria Based on Comparative Sequence Analysis of Elongation Factor Tu and ATP-Synthase ß-Subunit Genes" *Antonie van Leeuwenhoek* 64 (1993): 285–305; 285.
6. H.P. Klenk, P. Palm, and W. Zillig, "DNA-Dependant RNA Polymerases as Phylogenetic Marker Molecules," *Systematic and Applied Microbiology* 16 (1994): 638–647; 638.
7. M.A. Huynen and P. Bork, " Measuring Genome Evolution," *Proceedings of the National Academy of Sciences USA* 95 (1998): 5849–5856; 5849.
8. R.S. Gupta and E. Griffiths, "Critical Issues in Bacterial Phylogeny," *Theoretical Population Biology* 61 (2002): 423–434; 423.
9. W. Ludwig and K.H. Schleifer,"Phylogeny of Bacteria Beyond the 16S rRNA Standard," *ASM News* 65 (1999): 752–757; 752.
10. A. Rokas and P.W.H. Holland, "Rare Genomic Changes as a Tool for Phylogenetics," *Trends in Ecology and Evolution* 15 (2000): 454–459; 454.
11. C. Brochier and H. Philippe, "A Non-Hyperthermophilic Ancestor for Bacteria," *Nature* 417 (2002): 244; 244.
12. Ludgiw, W. and 31 others, (2004) ARB: a software environment for sequence data. Nucleic Acids Res. 32, 1363–1371.
13. H. Brinkmann and H. Philippe, "Archaea Sister Group of Bacteria? Indications from Tree Reconstruction Artifacts in Ancient Phylogenies," *Molecular Biology and Evolution* 16 (1999): 817–825; 825.
14. G.P. van Wezel, E. Takano, E. Vijgenboom, L. Bosch, and M.J. Bibb, "The *tuf3* Gene of *Streptomyces coelicolor* A3(2) Encodes an Inessential Elongation Factor Tu that is Apparently Subject to Positive Stringent Control," *Microbiology* 141 (1995): 2519–2528; 2519.
15. R. Amann, W. Ludwig, and K.H. Schleifer, "β-Subunit of ATP-Synthase: A Useful Marker for Studying the Phylogenetic Relationship of Bacteria," *Journal of General Microbiology* 134 (1988): 2815–2821; 2815.

16. E. Hilario and J.P. Gogarten, "The Prokaryote-to-Eukaryote Transition Reflected in the Evolution of the V/F/A-ATPase Catalytic and Proteolipid Subunits," *Journal of Molecular Evolution* 46 (1998): 703–715; 703.
17. Judith Neumaier, "Gene der katalytischen Untereinheit der V-Typ- und der F1F0-ATPase bei Bakterien: Vorkommen und vergleichende Sequenzanalyse, " Ph.D. Thesis (Technische Universität München, 1996).
18. M. Sumi, M. Yohda, Y. Koga, and M. Yoshida, "F_1F_0-ATPase Genes from an Archaebacterium, *Methanosarcina barkeri*," *Biochemical and Biophysical Research Communications* 241 (1997): 427–433; 427
19. Y. Kakinuma, K. Igarashi, K. Konishi, and I. Yamato, "Primary Structure of the Alpha-Subunit of Vacuolar-Type Na(+)-ATPase in *Enterococcus hirae*. Amplification of a 1000-bp Fragment by Polymerase Chain Reaction." *FEBS Letters* 292 (1991): 64–68; 64.
20. F.O. Glöckner, M. Kube, M. Bauer, H. Teeling, T. Lombardot, W. Ludwig, D. Gade, A. Beck, K. Borzym, K. Heitmann, R. Rabus, H. Schlesner, R. Amann, and R. Reinhardt, "Complete genome sequence of the marine planctomycete Pirellula sp. strain 1," *Proceedings of the National Academy of Sciences USA* 100 (2003): 8298–8303; 8298.
21. C.R. Woese, "Interpreting the Universal Phylogenetic Tree," *Proceedings of the National Academy of Sciences USA* 97 (2000): 8392–8396; 8392.
22. J.P. Gogarten, W.F. Doolittle, and J.G. Lawrence, "Prokaryotic Evolution in Light of Gene Transfer" *Molecular Biology and Evolution* 19 (2002): 2226–2238; 2226.
23. E.J. Feil, E.C. Holmes, D.E. Bessen, M.S. Chan, N.P. J. Day, M.C. Enright, R. Goldstein, D.W. Hood, A. Kalia, C.E. Moore, J.J. Zhou, and B.G. Spratt, "Recombination within Natural Populations of Pathogenic Bacteria: Short-Term Empirical Estimates and Long-Term Phylogenetic Consequences," *Proceedings of the National Academy of Sciences USA* 98 (2001): 182–187; 182.
24. P.H. Sneath "Evidence from *Aeromonas* for Genetic Crossing-Over in Ribosomal Sequences," *International Journal of Systematic Bacteriology* 43 (1993): 626–629; 626.
25. N.H.E. Smith, C. Holmes, G.M. Donovan, G.A. Carpenter, and B.G. Spratt, "Networks and Groups within the Genus *Neisseria*: Analysis of *argF*, *recA*, *rho*, and 16S rRNA Sequences from Human *Neisseria* Species," *Molecular Biology and Evolution* 16 (1999): 773–783; 773.
26. K. Ueda, T. Seki, T. Kudo, T. Yshida, and M. Katoka, "Two Distinct Mechanisms Cause Heterogeneity of 16S rRNA," *Journal of Bacteriology* 181 (1999): 78–82; 78.
27. Y.Z. Wang and Z. Thang "Comparative Sequence Analyses Reveal Frequent Occurrence of Short Segments Containing an Abnormally High Number of Non-Random Base Variations in Bacterial rRNA genes," *Microbiology* 146 (2000): 2845–2854; 2845.
28. M.W. Parker, "Case of Localized Recombination in 23S rRNA Genes from Divergent *Bradyrhizobium* Lineages Associated with Neotropical Legumes," *Applied and Environmental Microbiology* 67 (2001): 2076–2082; 2076.
29. M. Engel, "Untersuchungen zur Sequenzheterogenität multipler rRNS-Operone bei Vertretern verschiedener Entwicklungslinien der *Bacteria*. "Ph.D. Thesis (Technische Universität München, Germany, 1999)
30. Y.Z. Wang, Z. Thang, and N. Ramanan, "The Actinomycete *Thermobispora bispora* Contains Two Distinct Types of Transcriptionally Active 16S rRNA Genes," *Journal of Bacteriology* 179 (1997): 3270–3276; 3270.
31. S. Mylvaganam and P.P. Dennis, "Sequence Heterogeneity between the Two Genes Encoding 16S rRNA from the Halophilic Archaebacterium *Haloarcula marismortui*," *Genetics* 130 (1992): 399–410; 399.
32. C.R. Woese, G.J. Olsen, M. Ibba, and D. Soell "Aminoacyl-tRNA Synthetases, the Genetic Code, and the Evolutionary Process," *Microbiology and Molecular Biology Reviews* (2000): 202–236; 202.
33. C.R. Woese, "On the Evolution of Cells," *Proceedings of the National Academy of Sciences USA* 99 (2002): 8742–8747; 8742.

4

Evolving Biological Organization

CARL R. WOESE

Seeing the Present

The science of biology enters the twenty-first century in turmoil, in a state of conceptual disarray, although at first glance this is far from apparent. When has biology ever been in a more powerful position to study living systems? The sequencing juggernaut has still to reach full steam, and it is constantly spewing forth all manner of powerful new approaches to biological systems, many of which were previously unimaginable: a revolutionized medicine that reaches beyond diagnosis and cure of disease into defining states of the organism in general; revolutionary agricultural technology built on genomic understanding and manipulation of animals and plants; the age-old foundation of biology, taxonomy, made rock solid, greatly extended, and become far more useful in its new genomic setting; a microbial ecology that is finally able to contribute to our understanding of the biosphere; and the list goes on.

All this is an expression of the power inherent in the methodology of molecular biology, especially the sequencing of genomes. Methodology is one thing, however, and understanding and direction another. The fact is that the understanding of biology emerging from the mass of data that flows from the genome sequencing machines brings into question the classical concepts of organism, lineage, and evolution at the same time it gainsays the molecular perspective that spawned the enterprise. The fact is that the molecular perspective, which so successfully guided and shaped twentieth-century biology, has effectively run its course (as all paradigms do) and no longer provides a focus, a vision of the biology of the future, with the result that biology is wandering willy-nilly into that future. This is a prescription for revolution—conceptual revolution. One can be confident that a new paradigm will soon emerge to guide biology in this new century. One can also be confident that the new paradigm will be as powerful as its predecessor and will inspire the same sense of awe and beauty as did the molecular paradigm in its heyday. One can be equally assured that the new paradigm will be revolutionary and basically different from its predecessor.

Yes, we could continue to follow the molecular path, and many would, but where does it take you? Molecular biology has ceased to be a genuine paradigm, and it is now only a body of (very powerful) technique. The only road it knows is the road of technological adventurism. Our understanding of biology still has a far way to go. We are not yet—and probably never will be—at a stage at which all that remains to do in biology is fill in the details, becoming bioengineers. The time has come to shift biology's focus from trying to understand organisms solely by dissecting them into their parts to trying to understand the fundamental nature of biological organization, of biological form.

This realization is the starting line for the proverbial journey of 1,000 miles, and in the vast collection of facts accumulated in the twentieth-century (principally through genomic sequencing) are those that will allow us to take the first few steps on that journey. All we have to do is recognize what lies before us. One aspect of biological organization that genomics is bringing within scientific reach is the age-old problem of "whence the cell, the basic unit of life?" What follows is my attempt to begin painting a picture of the process.

Conceptualizing Cells

We should all take very seriously an assessment of biology made by the physicist David Bohm over 30 years ago (and universally ignored):

> It does seem odd . . . that just when physics is . . . moving away from mechanism, biology and psychology are moving closer to it. If the trend continues . . . scientists will be regarding living and intelligent beings as mechanical, while they suppose that inanimate matter is too complex and subtle to fit into the limited categories of mechanism.[1]

The organism is not a machine! Machines are not made of parts that continually turn over and renew; the cell is. A machine is stable because its parts are strongly built and function reliably. The cell is stable for an entirely different reason: It is homeostatic. Perturbed, the cell automatically seeks to reconstitute its inherent pattern. Homeostasis and homeorhesis are basic to all living things, but not to machines.

If not a machine, then what is the cell? A common childhood experience seems to suggest, if not an answer, then a useful metaphor. A child playing in a woodland stream finds endless delight in poking with a stick at some eddy in the flowing current, causing it to disperse, and then watch in fascination as it reforms—again and again. A cell, or any living system, is of this nature: a pattern in a flux. Life exists in energy gradients of various kinds, and one might look at life as being the eddys that form in such energy flows. I will grant that the metaphor leaves much to be desired—organisms as we know them are more complex, stable, and internally coherent than an eddy in a turbulent stream, and patterns in flow cannot be dissected into parts—yet in my experience, patterns in flow provide a conceptually richer, more inspiring metaphor when imagining organisms, ecologies, and evolution than does the machine. In addition I am willing to bet that the turbulent flow metaphor will prove very useful when we reach the point at which the problem of life's origin can be conceptualized in the framework of complex dynamic systems.

I would stress one insight in particular that stems from this analogy: that the organism and its evolution are one. An organism's being cannot be separated from its becoming; the two are but different facets of the same germ. It follows, at least I feel it does, that the homeostatic and homeorhetic properties inherent in living systems inhere as well in the evolutionary processes that produces them. Natural processes, even chaotic ones, have certain definite characteristics that tend to be maintained when the process is perturbed.

Exorcising the Past

We are not quite ready to take on the grand problem of cellular evolution. Jacob Marley's ghost remains to be exosied, as it were, for us to become aware of the chain of scientific, and even cultural, prejudices that fetter our thinking. These prejudices are not solely the residue of the molecular era just ending; they go back well into biology's past, including culturally established perceptions such as the enormous, unbridgeable gap dividing the animate world from the inanimate. Even that cornerstone of our worldview has now to be reshaped from self-evident, unquestioned truth into an open question, a puzzle.

When one has worked one's entire career within the framework of a powerful paradigm, it is almost impossible to look at that paradigm as anything but the proper, if not the only possible, perspective one can have on (in this case) biology. Yet despite its great accomplishments, molecular biology is far from the "perfect paradigm" most biologists take it to be. This child of reductionist materialism has nearly driven the biology out of biology. Molecular biology's reductionism is fundamentalist, unwavering, and procrustean. It strips the organism from its environment, shears it of its history (evolution), and shreds it into parts. A sense of the whole, of the whole cell, of the whole multicellular organism, of the biosphere, of the emergent quality of biological organization, all have been lost or sidelined.

Our thinking is fettered by classical evolutionary notions as well. The deepest and most subtle of these is the concept of variation and selection. How we view the evolution of cellular design or organization is heavily colored by how we view variation and selection. From Darwin's day onward, evolutionists have debated the nature of the concept, and particularly whether evolutionary change is gradual, saltatory, or of some other nature.[2,3] However, another aspect of the concept concerns us here more. In the terms I prefer, it is the nature of the phase (or propensity) space in which evolution operates. Looked at one way, variation and selection are all there is to evolution: The evolutionary phase space is wide open, and all manner of things are possible. From this "anything goes" perspective, a given biological form (pattern) has no meaning outside of itself, and the route by which it arises is one out of an enormous number of possible paths, which makes that evolution completely idiosyncratic and, thus, uninteresting (molecular biology holds this position: the molecular biologist sees evolution as merely a series of meaningless historical accidents).

The alternative viewpoint is that the evolutionary propensity space is highly constrained, being more like a mountainous terrain than a wide open prairie: Only certain paths are possible, and they lead to particular (a relatively small set of) outcomes. Generic biological form preexists in the same sense that form in the inanimate world does. It is not the case that "anything goes" in the world of biological evolution. In other words, biological form (pattern) is important: It has meaning beyond itself; a deeper, more general significance.[4,5] Understanding of biology lies, then, in understanding the evolution and nature of biological form (pattern). Explaining biological form by variation and selection hand-waving argumentation is far from sufficient: The motor does not explain where the car goes.

Classical biology has also saddled us with the phylogenetic tree, an image the biologist invests with a deep and totally unwarranted significance. The tree is no more than a representational device, but to the biologist it is some God-given truth. Thus, for example, we agonize over how the tree can accommodate horizontal gene transfer events, when it should simply be a matter of when (and to what extent) the evolution course can be usefully represented by a tree diagram: Evolution defines the tree, not the reverse. Tree imagery has locked the biologist into a restricted way of looking at ancestors. It is the tree image, almost certainly, that caused us to turn Darwin's conjecture that all organisms might have descended from a simple primordial form into doctrine: the doctrine of common descent. As we shall discuss below, it is also the tree image that has caused biologists (incorrectly) to take the archaea and eukaryotes to be sister lineages. Much of current "discussion/debate" about the evolutionary course is couched in the shallow but colorful and cathected rhetoric of "shaking," "rerooting," "uprooting," or "chopping down" the universal phylogenetic tree.[6,7]

The received wisdom concerning the origin of the eukaryotic cell is a very big link in the chain that weighs biology down. The notion of symbiosis played a significant role in the thinking of nineteenth-century biologists. It was only natural that the biologists of that time interpret the various inclusions in the eukaryotic cell in such terms. (Given the primitive state of nineteenth-century microscopy, almost any preconceived notion could be realized at the "other end of the microscope barrel." Let me recommend to you Jan Sapp's book on the history of symbiosis, an excellent overview of the topic.[8]) Indeed, many of these early biologists saw the eukaryotic cell as totally chimerical, a collage of symbionts; every discernible cytological feature represented one or another kind of bacterium or other (sub)microscopic life form.[8] This notion has carried through into modern times and dominates thinking about the origin of the eukaryotic cell today.

The New is Old

In modern dress these old notions are stale and stilted. They no longer inspire; they are counterproductive. Look at what our unquestioned, uncritical importation of nineteenth-century notions of cellular origins into the twentieth-century molecular and biochemical context has done. It has focused discussion of cellular origins almost exclusively on the origin of the eukaryotic cell. (In the nineteenth century, the evolution of eukaryotic cells was the only game in town, which, of course, is no longer true.) This focus in turn has made the evolution of the eukaryotic cell appear qualitatively different from those of the other two cellular types. It also accords the eukaryotic cell no intrinsic character: That cell is merely some admixture of features from bacteria, archaea, and perhaps other types of cells (no longer extant). Finally, it takes for granted, and therefore asserts, that a given cell design can be formed by the melding of two or more other (mature) cellular designs.

In my opinion, a total rethinking of the problem of cellular evolution is called for, and I would like to offer two tenets that I feel should apply:

(1) All of the extant major cell designs evolved in a similar and related fashion; all reflect the same underlying evolutionary dynamic, although each clearly is highly unique.
(2) A (mature) cell design is evolutionarily homeostatic and homeorhetic: If evolutionarily perturbed, that design tends to reestablish the same general design type, but under no circumstances can it reform as a basically new type of design.

Note that the second tenet denies that the essential eukaryotic cell design is the product of cellular endosymbiotic interactions. The design of the eukaryotic cell is *sui generis*, not an amalgam of other cell designs.

These two principles invite the obvious counterpoint: What then about the endosymbioses that have given rise to the eukaryotic organelles? The answer is that these interactions are of a fundamentally different nature than those that biologists past and present invoke to generate the eukaryotic cell in the first instance. The eukaryotic cell, in essence, is eukaryotic regardless of whether it carries chloroplasts or mitochondria. (Granted, without one or the other organellar type, the cell cannot live under the usual conditions; but it can, nevertheless, survive, and it remains organizationally eukaryotic.) In the organellar evolutions, the initial contacts between endosymbiont and host are of a superficial nature—small molecules are interchanged—which disrupts the fundamental organization of neither participant.[9] Only gradually then does the organization of the endosymbiont (but not that of the host) erode. None of this can be said of the hypothetical symbioses and the like proposed for the origin of the eukaryotic cell. In these cases, the central and defining components of the eukaryotic cell are taken to arise from the postulated symbiotic interactions—interactions that would significantly perturb preexisting cell designs. (Keep in mind that my argument is taken to hold only in the case of fairly advanced cell designs, but these are the kind that biologists would invoke in their eukaryotic cellular origin scenarios.)

The Concept of "Prokaryotes"

"Oh, how the ghost of you clings" run the lyrics of an old show tune ("These Foolish Things: by Jack Strachey, Harry Link, Holt Marvell, and Eric Mashwitz). For the biologist, the name of that tune is "The Prokaryote–Eukaryote Dichotomy." Born of the microscopic observations (and their interpretation) of Edouard Chatton in the 1920's, "prokaryote–eukaryote" became inviolable biological doctrine by the middle of the last century: All life on Earth divides into two primary structural types and two corresponding phylogenetic categories, the small, featureless "prokaryotes" and the larger, more interesting eukaryotic cells, which display a collection of intracellular structures, starting with the nuclear membrane and the organelles.[10,11]

The factual basis for the dichotomy lies, as stated, in a cyto-structural distinction: The eukaryotic cell shows characteristic intracellular features; prokaryotic cells show no characteristic features—which makes "prokaryote" a catch-all category right from the start. From this structural distinction, Chatton made a sweeping and bold extrapolation: What holds in the cyto-structural sense also holds in the phylogenetic sense, and thus, two primary cell types demand two primary phylogenetic categories.[10] So strong was the belief that bacteria comprise a monophyletic unit that later attempts to define the general characteristics of the "prokaryote" were perfunctory) being in effect confined to the properties of microbiology's icon, *Escherichia coli.*[12]

The structure–phylogeny equivalence in the prokaryote–eukaryote dichotomy was broken in 1977 with the discovery of a third primary phylogenetic category, then called the "archaebacteria."[13,14] Two cellular architectures had demanded two primary phylogenetic categories. Why, then by the same token, did not three primary phylogenetic categories demand three basic cellular organizations? The logic passed unnoticed. Proponents of the prokaryote–eukaryote dichotomy begrudgingly conceded the issue of the number of primary phylogenetic categories, but they were not about to throw away their battered but beloved dichotomy: Its structural aspect remained and still remains dogma.[15] What biologists are unknowingly and cavalierly doing here is providing an answer to one of biology's most fundamental questions; namely, what is the number of basic cellular designs (organizations) that exist on Earth.

Without doubt, the eukaryotic cell is unique and special. It is more complex and intricately organized than either of the two prokaryotic cell types. It is the eukaryotic cell alone that has given rise to the rich and complex world of multicellularity, and with this world to that intriguing evolutionary puzzle we call "embryonic development."

At the same time, however, it cannot be taken for granted that the phylogenetically distinct archaea and bacteria are alike in their cellular organizations. In our complacency, we have never really examined their organization. The fact that profound differences exist (i.e., the "canonical pattern") between archaeal and eubacterial homologs of core cellular componentry implies something, as does the fact that fundamental differences in biochemistry (e.g., in their coenzyme complements) separate them and the fact that they differ fundamentally in chromosome organization and in the chromosome replication mechanism.[16–19] Then there is the fact that on the order of 300 genes exist that are characteristic of and unique to the archaea.[20] We need to understand the nature of the three types of cellular organization that inhabit this planet. This is no time to glibly assume an answer to a basic biological question (thereby effectively burying it).

How the Cell Assembled Itself

Finally, we are ready to take on the evolution of the cell. Without relying on the prejudices of the past, then, what can be taken as reasonable starting points for developing a concept of the evolution of cellular organization? There are three such points. The first is horizontal gene transfer (HGT), which, as we shall see, is the essence of cellular evolution. The second is the nature of the cellular translation apparatus, for the evolutions of translation and cellular organization are part and parcel of one another. The third is the fact that the cell and its evolution are complex dynamic systems/processes. Eventually, our understanding of cellular evolution (and of evolution in general) will be in terms of complex dynamics systems.

Evolution's Horizontal Dimension

The phenomenon of HGT has long been recognized. However, it is only recently, with the advent of microbial genomics, that its scope, significance, and grandeur have come to be fully appreciated. The question that concerns us here is the evolutionary dynamic of HGT: What is (are) the main factor or factors determining the quality of HGT, its frequency, type, and phylogenetic scope? A number of factors, such as defenses against alien DNA entry, destruction of alien DNA, and so forth (none of which is perfect), and evolutionary ecological situations that influence what DNAs are present in a potential recipient's environment, all tend to become insignificant (blurred out) in the larger evolutionary picture.[21,22] This leaves prominent only the innate selective constraints imposed by the potential recipient cell on which alien genes become (stably) incorporated.

A number of biologists have invoked HGT (glibly, I feel) in their attempts to "reroot," "uproot," or "chop down" the universal phylogenetic tree that has been inferred through rRNA sequence comparisons.[23–25] It is clear that these excesses of interpretation result principally from a failure to take into account sufficiently (or to understand) the dynamics of HGT. Our discussion of HGT here, then, will serve two purposes: The more immediate purpose of addressing what I have called "excesses of interpretation," and the deeper and more important purpose being an understanding of HGT's critical role in the evolutionary dynamic.

Much of HGT is "inconsequential," having few if any phenotypic consequences. Without selection to sustain them, alien genetic elements have but transient residence in organismal populations (on the evolutionary time scale).[21,22,26] A lesser fraction of alien genetic elements do have phenotypic significance, but in the main these too are transient: their significance is "local," having to do, for example, with adaptation to specialized niches. They occur in the population only as "patches" and persist only so long as the niche they inhabit persists.[26] Antibiotic resistance genes are perfect examples here. Finally, there are the relatively rare, more permanent, and significant horizontal transfers, those that introduce novelty that alters permanently the fabric of the cell or affects the organismal genealogical trace by displacing endogenous functions with alien equivalents. The latter, the genealogy altering and eroding transfers, are at the heart of the controversy over the universal phylogenetic tree. The former, however, that which introduces genuine novelty, is the HGT critical to the evolution of cellular organization.

Perhaps the best system in which to study the dynamic of HGT is the set of aminoacyl–tRNA synthetases. Most of the HGT that affects these enzymes is effectively neutral; an indigenous version of a given synthetase being displaced (or sometimes supplemented) by some alien equivalent. Evolutionarily trivial or not, neutral HGT of this type is particularly informative: Selective factors, which could confuse our interpretations, are taken out of consideration. By the same token, it helps that the aminoacyl–tRNA synthetases comprise a set of functionally related proteins, with the set as a whole providing a powerful framework within which to interpret the phylogenetic idiosyncrasies of the individual synthetases. Finally, the fact that the aminoacyl–tRNA synthetases are the only major components of the translation system to be appreciably affected by HGT is going to provide a key insight into the puzzle of cellular evolution.

An extensive analysis of the phylogenies of aminoacyl–tRNA synthetase sequences has come to the following major conclusions[16]: roughly two-thirds of these enzymes yield phylogenetic trees in accordance with the basal "canonical" branching pattern, characteristic of the universal tree inferred from rRNA sequences; however, in each case the canonical phylogenetic pattern is, to one extent or another, varied, eroded (presumably) by HGT; and, as would then be predicted, all of these variations in the universal pattern are idiosyncratic—those seen for one aminoacyl–tRNA synthetase type do not carry over to another. These facts force the conclusion that there must exist an organismal genealogical record (carried in the common histories, a particular subset of the organism's genes) that extends back to the universal common ancestor stage, and although it has been altered in different ways and to different extents by HGT, that ancestral pattern has not been totally erased thereby.[16] Thus, the phylogenetic tree based on rRNA sequence analyses truly depicts the organismal evolutionary course (with one caveat; see below).

Why the aminoacyl–tRNA synthetases alone among the components of the translation apparatus are strongly affected by horizontal gene displacement is easy to understand[27]: only the synthetases are modular in their function. They are loosely integrated into the cellular fabric; they interact minimally with others of the cell's componentry; and their interactions are each confined to a small subset of the tRNAs, which themselves are small molecules, simple, uniform, and universal in structure. Without external structural constraints, the exact sizes and shapes of the aminoacyl–tRNA synthetases matter little. None of this can be said of the other translational componentry, all of which is highly and idiosyncratically interconnected structurally (and dynamically). Put otherwise, the cellular design specifications for the aminoacyl–tRNA synthetases are less stringent than those for others of the translational componentry, and so, aminoacyl–tRNA synthetases could more easily be mimicked and, therefore, horizontally displaced by alien equivalents—even by taxonomically distant, structurally dissimilar equivalents.[16] (A perfect example here is the archaeal Class I lysyl–tRNA synthetase, which also occurs scattered among the bacteria [which characteristically employ the structurally completely unrelated Class II lysyl-tRNA synthetases].)[16]

Thus, it is the degree to which (and the ways in which) the various componentry is integrated into a cell, the cell design, that determines what is and what is not horizontally transferred. It is precisely that design that changes (and changes drastically) during the evolution of the cell, however. Early on, cell designs were necessarily very simple; cellular organization at these stages almost certainly had to be quite loose, and thus tolerant of change. HGT at these early stages almost certainly was a very different animal from the "domesticated" version of it that we encounter today (through genomics): rampant, pervasive, playing havoc with the simple primitive cell designs. Indeed, HGT early on must have been the dominant force in evolution, the prime mover of the process. How else could the huge amount of novelty required to bring forth the basic cellular functions have arisen, except through forging in the fires of different craftsmen (cell designs), with the occasional constructive marriage of disparate parts? Only now are we beginning to pick around the edges of this problem. A true understanding of cellular evolution remains a speck on the scientific horizon.

As you can easily see, the above scenario has profound consequences vis-á-vis organismal histories and genealogical traces. Pervasive HGT would have the capacity to obliterate an organismal and genealogical trace; any and all "indigenous" elements in the cell could be displaced sooner or later by alien equivalents—be they orthologs, paralogs, or analogs. Primitive cells should not be looked at in terms of lineages in the modern sense, but as cell lines, transient at least in the sense that their genealogical traces are ephemeral. Moreover, in an era of rampant HGT, organismal evolution would be basically collective.[28,29] It is the community of organisms that evolves, not the various individual organismal types. HGT-driven evolution is the tide that lifts all boats. In this communal era, all evolving cellular entities contribute to a common field of genes, and all in turn benefit in one way or another from that field. The phylogenetic tree image fails to usefully depict the evolutionary course at this stage. At what stage does the tree image become useful? We shall see.

Evolving Translation

Evolutionary limitations imposed by a primitive translation mechanism. One cannot look at the cellular translation apparatus without being overwhelmed by its complexity, by the number of parts and their possible interactions. It is even more daunting to contemplate the evolution of such a mechanism. In a very real sense the evolution of translation is the evolution of the cell: Translation is the heart of the evolving cell design. Cellular evolution requires entire suites of novel proteins never before seen on Earth, and it is the performance characteristics of the primitive translation apparatus that determine what general types of proteins can and cannot evolve.

A translation apparatus today must do two main things: accurately match codons with corresponding amino acids across an entire message RNA (perhaps thousands of nucleotides in length) and maintain the correct reading frame throughout the process. It seems impossible that a simple primitive translation mechanism could perform with the requisite precision to accurately produce a large (modern) protein.[30] (The point here is not only common sense but can be inferred from the fact that the structure of the genetic code appears to have been optimized to reduce the phenotypic consequences of codon recognition error.[31]) Primitive cells, then, would comprise only small proteins, which, of course, has broad implications as to the nature of the evolving cells. In almost all cases the primitive version of a particular function would be less sophisticated and precise than its modern counterpart.[30]

When it comes to the replication of modern genomes (genomes comprising, say, 1,500–2,000 genes), the accuracy of the process has to be exceedingly high.[32] Add to this the fact that the enzymatic complex that replicates the modern genome contains large proteins critical to its function, and it becomes clear that a primitive genome replication apparatus, comprising small proteins only, could not function as precisely as does a modern one. As a consequence, primitive genomes had necessarily to be smaller than modern ones.[30] There undoubtedly were many other consequences of primitive translation's lack of precision, which shaped primitive cell designs in general ways, affecting their general complexity, intricacy, integration, and so on.

A name has been given to cells that have primitive translational capacities. The name, "progenote," signifies that the genotype–phenotype link has yet to complete its evolution.[30] Given that evolution early on also appears to be communal (see above), it is useful to refer to this communal stage in evolution as the progenote era, using the term to denote the community as a whole.

How translation might have begun. If we knew how modern translation worked, we would be on far safer grounds in conjecturing how it began. Nevertheless, the overall characteristics of the mechanism and the general things we know about its workings do offer some obvious clues as to the origin of the process.

Several decades ago it was taken for granted that the translation mechanism was defined by the proteins that interact with the ribosome; namely, the ribosomal proteins and the elongation factors. Of course there was the tRNA molecule, on the back of which the amino acid entered the mechanism, but from the start tRNAs were never viewed as defining of the translation process itself. tRNA was merely an "adaptor," something that matched the amino acid to its corresponding codon or codons. tRNAs were processed by the translation mechanism (just as were mRNAs); they were not a part of the mechanism.

Slowly, protein by protein, this picture crumbled. Ribosome reconstitution experiments, which held out great hopes initially for defining which proteins performed which translation functions,[33] ultimately showed only that omitting individual proteins from the reconstitution reaction caused the ribosome either not to reassemble correctly or to function suboptimally. At best, the ribosomal proteins appeared to facilitate the translation function, not define those functions. At this point, however, biologists still pinned their hopes on two proteins to define key steps in translation: the enzyme "peptidyl transferase," which supposedly transferred the growing peptide chain from one tRNA to the next, and the protienaceous elongation factor EF-G, which effected "translocation," the movement of the ribosome from one codon to the next. Peptidyl transferase, however, always eluded detection, not to mention isolation. It finally disappeared for good when the 2.4 Å-resolution structure of the 50S ribosomal subunit showed that the area surrounding the peptidyl transfer site was totally devoid of protein.[34,35] The elongation factor EF-G was (earlier) rendered dispensable when "factor-free" translation was discovered (an in vitro system that translated perfectly well without any elongation factors or without their energy source, guanosine triphosphate (GTP).[36] (All that was required was to remove ribosomal protein S12 from the small subunit ribosomal particle.[36]) Could it be that all the critical steps in translation (except tRNA charging) were defined by RNA? The paradigm had totally shifted: "the ribosome is a ribozyme."[37]

The idea that translation is RNA based is really not all that surprising.[38] Transfer RNA had been there all along, and if it had not been looked at from the start merely as an "adaptor," someone might have suggested an active role for the molecule in defining translation.[39] Even the father of the Adaptor Hypotheses, Francis Crick, initially felt tRNA too large to be his postulated "adaptor."[40] You can more or less see it coming: tRNA will be perceived as the defining core of the translation process. It is not a far jump, then, to see a tRNA-like molecule as being the core of the first translation apparatus.

The Darwinian Threshold: The Road from There to Here

Cellular evolution basically starts from scratch. This is not to say that biological organization did not exist at the beginning of cellular evolution, but merely that all the proteinaceous structures that are the staples of our concept of cells did not initially exist,[37] which makes the evolution that spawned them very different in character from the later evolution, concerned only with maturing, embellishing, and otherwise refining the designs of cells. Starting from scratch also means starting simple in all respects. As demanded by the nature of the progenote translation mechanism at the very least, the cell's protein components are initially smaller in size, in number (of kinds), and in sophistication (specificity) than their modern counterparts. The organization of the cell, its design, had to be comparably simple: design specifications for proteins and pathways were crude by modern standards; the design itself was loosely defined, flowing; and the cellular organization that existed was largely horizontal, with the componentry being loosely coupled, minimally integrated into the whole.[30] Inexact parts specifications and loose coupling implies that individual proteins would tend to be compatible with a variety of cell designs, so that various members of a gene family (what we now view as "paralogs") could equally well fulfill a given specification. Individual genes, then, tended to be cosmopolitan, with some of them being more characteristic of particular environments than of particular organisms.

The progenote model sees organisms as genetically communal and the community as evolving as a whole, not the individual cell lines therein. There is a definite lack of individuality (of the kind we associate with organisms today) in the progenote. Would a progenote genetic community be communal in other ways as well? Probably. Nature has few new tricks; it reuses old ones in different guises. Bacteria naturally reside in communities, in ecosystems. It is hard to find a bacterial niche that does not comprise hundreds or thousands of different species, all interacting in intricate, delicate ways, to make a fascinatingly complex and stable whole. As it is now, so it probably was in the progenote era—only more so. One can picture an intensely interacting community of progenotes in great variety, trading in metabolites, proteins, and of course, genes (or larger genetic units). The small genomes of the day would make for a minimal metabolic capability in individual cell lines, perhaps insufficient for any one of them to survive on its own. In the context of the collective "genome" of the community, however, all flourish. I picture the progenote as a grouping of metabolic specialists that cannot survive without one another.

The real mystery, however, is how this incredibly simple, unsophisticated, imprecise communal progenote—cells with only ephemeral genealogical traces—evolved to become the complex, precise, integrated, individualized modern cells, which have stable organismal genealogical records. This shift from a primitive genetic free-for-all to modern organisms must by all accounts have been one of the most profound happenings in the whole of evolutionary history. Although we do not yet understand it, the transition needs to be appropriately marked and named. "Darwinian threshold" (or "Darwinian Transition") seems appropriate: crossing that threshold means entering a new stage, where organismal lineages and genealogies have meaning, where evolutionary descent is largely vertical, and where the evolutionary course can begin to be described by tree representation.

The most important, if not the only, thing that can be said right now about the progression from the pre-Darwinian progenote to cells typical of the Darwin era (i.e., modern cells), is that in the process the cell design becomes more integrated. Connectivity, coupling (among componentry) is the key to the nature of that transition. The cell is a complex dynamic system. Complex dynamic systems characteristically undergo saltations at "critical points." Drastic changes in the system result. An increase in the connectivity of a system is one factor that can bring it to such a critical point. Does the Darwinian Threshold, then, denote a critical point in the evolutionary process? I say it does. We can be confident in any case that in the full evolutionary course, from an abiotic earth to modern cells and organisms, evolutionary saltations must have occurred. The transition from the nondescript, horizontally intermeshed, and simple progenote to the complex individual cell lineages (with stable genealogical traces and vertical descent) that we know surely has the feel of a saltation.

Crossing the Darwinian Threshold does not mean that HGT ceases, merely that it changes in character (because the cell design has drastically changed). On crossing the threshold, the evolutionary emphasis shifts from establishing the basic mechanisms of the cell to refining them, and in the process developing additional layers of (more sophisticated) control. This is a period of refinement: integration, control, and specificity are of its essence. This is the period in which organismal and molecular idiosyncrasy blossom, along with molecular specialization. These factors in turn restrict the ways in which a cell design can be changed and reduce the likelihood that an alien homolog can sufficiently mimic its indigenous counterpart to successfully displace it.

When a Darwinian Threshold is crossed and further refinement of cellular design follows, HGT undergoes taxonomic "regression": its phylogenetic scope diminishes. Initially after the Darwinian Transition, HGT's scope tends to center on the individual domains, and then, with further refinement in cellular design, that scope shifts progressively down the taxonomic ladder, settling ultimately into the kind of HGT we know today. When the scope of HGT (phylogenetic distance between donor and recipient) becomes restricted enough, the nature of HGT comes to resemble the kind of variation associated with vertical descent (i.e., variations on existing themes). This is saying no more than that the breadth and scope, the general character, of HGT (inversely) mirror the complexity of organisms.

The Universal Phylogenetic Tree

The Darwinian Transition makes clear the nature of the universal phylogenetic tree. This is no ordinary tree, at least in terms of its root and the initial bifurcation, which seems to define a common ancestor for the archaeal and eukaryotic lineages. These two nodes cannot be interpreted in the classical way. The root does not represent some bottleneck in the evolutionary process, a single organism or species that gave rise to a rich variety of descendant forms—a concept drilled into our minds by the classical (X-shaped) Pirie diagram[41] and by the Doctrine of Common Descent. There has never been any such singularity in the course of cellular evolution. From its beginnings, cellular evolution has progressed on a broad front, as an enormous surging wave. There was never a universal common ancestor. The Doctrine of Common Descent has deceived us. But we are deceived, too, if we think Darwin is responsible for this doctrine. Yes, Darwin did conjecture that "all the organic beings which have ever lived on this earth may be descended from some one primordial form," but in the same discussion, he said: "this inference . . . is chiefly grounded on analogy and it is immaterial whether or not it be accepted. No doubt it is possible . . . that at the first commencement of life many different forms were evolved."[42] Thus, it is those who followed him that made doctrine of Darwin's conjecture. For Darwin, the idea was interesting, but irrelevant and "immaterial" to the Theory of Evolution.

If not some ancestor of all ancestors, what, then, does the root of the universal phylogenetic tree represent? Tree diagrams cannot usefully depict the evolutionary course in the pre-Darwinian, progenote, era. Only when the Darwinian era comes fully into its own can they do so. However, between these two stages an intermediate phase exists, in which a tree diagram represents the evolutionary course only partially. The root of the universal tree denotes the beginning of this transition phase. The tree's root denotes the point at which the first of the cell designs crosses its Darwinian Threshold. The topology of the tree demands that the first be the bacterial cell design. In shifting into a predominantly vertical mode of descent with variation, bacterial evolution has become amenable to tree representation.

At that point, though, both the archaeal and eukaryotic designs remain in the pre-Darwin, progenote, condition: still heavily immersed in the universal HGT field, still in the throes of shaping major features of their respective designs; and so, their evolutions cannot be represented in tree form. In other words, the node in the conventional phylogenetic tree that denotes a common ancestor of the archaea and eukaryotes does not actually exist. The two cell designs are not specifically related; it is just that the tree representation has made them "sisters by default."

Although there is no direct way of inferring whether it was the archaea or the eukaryotic cell design that next crossed its Darwinian Threshold, a simple argument indicates it to have been the archaeal[43]: of the basic cellular componentry common to the archaea and eukaryotes, the archaeal versions almost always are the simpler; they have fewer or less ornate subunits. Three good examples of here are the proteasome, the transcription mechanism (especially transcription initiation), and histones.[44–46] For me, the most reasonable explanation of this is that the archaea crossed their Darwinian Threshold before the eukaryotes did, thereby "locking in" more primitive, simpler versions of various functions.

The Wheat and the Chaff

It must be recognized that evolving the cell is not about evolving genomes per se, not about evolving this or that specific detail (e.g., particular metabolic pathways or particular trans-membrane pumps). Evolving the cell is about evolving its design, its organization—something not generally apparent—certainly not when the problem is viewed from a one-dimensional genomics perspective. The genes in a genome, even large sections of the genome, can come and go without affecting the basic cell design. Most of the genes are concerned with the life style of the cell and have no bearing on the cell's basic organization.

Cell designs obviously do change, and change drastically when cellular organization itself is coming into being, but I would maintain that even early on, the specific makeup of the genome has very little to do with the evolution of cell design. As was true later, only a particular subset of genes define that design, whereas the remaining genes were concerned with maintenance and specific life styles. The core of design-defining genes tends to become firmly and intricately woven into the cellular design fabric and, thus, refractory to HGT. When one speaks of an organismal genealogy, one is concerned primarily with a small cadre of genes; not with the majority of genes in the genome, many (or all) of which can come and go (on an evolutionary time scale). Phylogenetic tree inference (where the major branching is concerned) should be based primarily, if not exclusively, on the genetic core that defines the organismal character: its organization.

There is little new in this, generally speaking. Darwin understood that the genealogy of an organism is represented only by a special subset of characters:

> It might have been thought . . . that those parts of the structure which determined the habits of life, and the general place of each being in the economy of nature, would be of very high importance in classification. Nothing can be more false. . . . It may even be given as

> a general rule, that the less any part of the organisation is concerned with special habits, the more important it becomes for classification.[42]

The Nature of Evolutionary Transitions

Biologists have always known in one way or another that evolution is not a smooth progression.[47] At rare junctures, major (qualitative) changes occur. The pieces in the game seem to change, as does the arena in which the game is played: evolution shifts to a new venue and occurs on a different (higher) level. Obvious examples here are the emergence of eukaryotic multicellularity, the advent of language, and the evolution of translation. Certainly the Darwinian Transition qualifies as such an evolutionary saltation as well. As alluded to above, the entire origin of life must have been a long series of evolutionary saltations, emerging biological organization moving in succession from one level of organization to a higher one. I do not know whether at base these saltations have a common evolutionary dynamic. However, the evolution of language shows us one way a saltation can be brought about.

Language is an example of a preexisting existing biological form (in this case a metazoan primate) gaining the capacity to map what it is and what it does into a new (qualitatively different) medium. The mapping in effect brings into existence a new evolutionary propensity space, pregnant with evolutionary potential.[48] As a result, *Homo sapiens* is completely unlike its otherwise very close primate relatives; *Homo sapiens* now evolves on a different evolutionary plane than do the others. A major evolutionary transitions seems, then, to occur when a vast new evolutionary propensity space come into being.

The advent of translation, the capacity to map nucleic acid sequence into amino acid sequence, similarly creates a representational phase space. However, this dynamic (the creation of a representational phase space) does not seem to apply to such junctures as the advent of multicellularity or to the Darwinian Transition.

Backtracking

It is time to make all of the above into a coherent whole, an emerging picture of cellular evolution. As mentioned, the evolution that wrenched the cellular designs into being must be unique, totally unlike that which occurred after the Darwinian Transitions, when evolution merely refined and brought to fruition preexisting cellular designs. The amount of novelty that had to be generated early on was staggering; how it was generated was unimaginable (in any detail). Something as unique as the proteinaceous cells has to have arisen at a particular, clearly definable stage in evolution. What was that stage?

The obvious point at which modern cell designs may have begun is with the onset of translation. It marks the beginnings of proteins as we know them, the transformation of an "RNA world" into a proteinaceous RNA world (or, as some like to say, an RNP world), and as stated, it creates an enormous new evolutionary propensity space. There can be no doubt about this being a critical stage in evolutionary history, yet there may also have been a later stage, between the onset of translation and the Darwin Transitions, where the evolution of modern cell designs began. We shall explore both possibilities.

Transforming an RNA World

"RNA world"[49] has different meanings to different scientists. I see it as the "era of nucleic acid life," a time when (RNA) programmed protein synthesis did not yet exist and nucleic acid, the most evolutionarily versatile and malleable polymer then extant, defined and dominated biological organization.[50] This is not to say that polypeptides were altogether absent at this stage.[38] They existed to the extent that the primitive (bio)chemical mechanisms of the day were able to generate them. (Such peptides, however, would be relatively simple and, most important, basically immutable in sequence. In other words, in an era of nucleic acid life, it is the nucleic acids that evolve to accommodate the peptides, not the reverse.) A source of the monomers and energy to make biopolymers is also required, which calls for a "metabolic network" that produces a variety of organic compounds, such as (catalytic) cofactors, and is biochemically rich and complex enough to be self-sustaining—which right now is little more than a deus ex machina.

Just as cells today are the fundamental units of biological organization, one can expect analogous higher-order architectures, designed around nucleic acid componentry, to fill that role in the RNA world. For want of a better term let us call them "supramolecular aggregates" (SMAs), simply to distinguish them from the proteinaceous cells we know. Within these aggregates, nucleic acids presumably performed roles analogous to many of those performed by proteins today, and as life today exists in varied profusion, so a profusion of these hypothetical SMAs populated the biosphere of the RNA world.

When translation entered the picture, it produced proteinaceous representations of preexisting RNA sequences.[51] Initially, most of these translations must have had no relevance in terms of existing SMAs, but a small fraction of them likely did—especially those that resembled nontranslationally produced polypeptides already incorporated into the various SMAs—and it may not have been all that difficult to find peptides that facilitated some of the reactions in the sustaining biochemical network. Translationally produced peptides would have meaning in either of two contexts: in terms of the SMA of which the translated RNA was already a functional part, or in terms of other SMAs (and other entities) for which the translated RNA per se had no significance. Here, then, with the second possibility, was something unique in the (RNA) world; namely, nucleic acids whose primary value lay in their protein coding capacity. Herein, too, began the need for commerce in (protein) coding nucleic acids.[51]

An essential requirement for any community that trades in goods is a lingua franca, and in the ancient RNA world, where nucleic acids are becoming valuable for their proteinaceous representations, commerce in these aboriginal genes required translation systems that shared a common genetic code. (No matter how the genetic code we know came into being, its order has remained constant and universal because it is evolution's lingua franca.)

It is obvious that in an RNA world into which translation had been introduced, a variety of SMA types would start to become proteinized. It is conceivable that modern cell designs had their origins here, in effect beginning to build on a variety of different SMA designs, RNA platforms. Whether this is the actual scenario or not, what is important is that there be many such SMAs that begin to proteinize. Evolution at this stage proceeded on a broad front: exploration of the new propensity space began from numerous starting points, not just one (there was no bottleneck). Multiple beginnings allow far more of the propensity space to be explored and realized than otherwise. When the capacity for horizontal exchange among the various SMAs is added to this mix, the resulting parallel integrated exploration increases still further the capacity to sample the propensity space. As far as I am concerned, such a parallel, "hand-in-hand" wandering of the space is the only way in which the amount and degree of novelty needed to bring forth proteinaceous cell designs could have been generated. (While the discussion is in the area of exploring evolutionary phase spaces, I would point out that imprecise translation may not have been solely an impediment to evolutionary progress; it might have been a distinct advantage in exploring protein phase space. Today, evolution explores the space of possible protein sequences in effect by mapping one point in the nucleic acid space to a corresponding point in the protein space [ignoring codon degeneracy]. That corresponds to shooting a rifle at a target. The chances of hitting the target would be greatly increased, however, if a shotgun had been used. The "shotgun" effect is precisely what imprecise translation provides: If a nucleic acid sequence is translated imprecisely, the resulting protein products can comprise a set of related sequences, all of which are (different) approximate translations of the gene, and few, if any, exact translations. If one of this set of approximate translations turns out to have selective value, then the evolving cell is poised to mutationally exploit the protein space vicinity surrounding that sequence. In this way a gene sequence optimal for the function in question can be relatively rapidly discovered. This "shotgun" procedure effectively renormalizes the phase space, from a large collection of points to a smaller collection of "locales" of related points.)

The explosion of an RNA world into an RNP world, this dance of creation and destruction, may have been violent enough initially that even the existing primitive cellular designs were unstable. This may have been merely a stage, in effect, of "subcellular" invention, of "semi-autonomous subcellular entities that somehow group to give 'loose' (ill-defined) cellular forms."[28] Conceivably little in the way of overall cell designs existed at such a stage, and were they to, they would be only ephemeral. Perhaps an RNP world antedated the era we are trying to envision, when stable modern cell designs first began their evolutions. This would indicate that between the onset of translation and the Darwinian Transitions, another evolutionary saltation occurred.

The scenario being painted rests on there initially being many starting points for cell designs. Is there any evidence to support the idea that many, not just three, evolving cell designs contributed to the pre-Darwinian universal gene pool? Although far from a proof, there are facts that readily lend themselves to such an interpretation. For example, in some of the aminoacyl–tRNA synthetases phylogenetic trees, a number of lineages can be seen that arise deeply in the general area of the archaea.[16] In other words, the synthetases in question are of the general archaeal genre. Today these lineages are represented by one or a scattering of bacterial taxa, which obviously have acquired the genes in question though HGT. However, the original donors of these genes were in no way bacterial, and as the phylogenetic branching orders show, neither were they strictly archaeal nor eukaryotic. These donors are most easily pictured as extinct cell designs still evolving in the pre-Darwinian mode subsequent to the bacterial lineage's crossing its Darwinian Threshold. In some cases the branching pattern indicates that the donors existed in the pre-Darwinian mode even subsequent to archaea crossing their Darwinian Threshold (i.e., the lineages in question appear to arise deeply on the eukaryotic branch, well below the radiation of the known eukaryotic lineages). These phylogenetic anomalies, then, might represent molecular "fossil" traces of long-gone cell designs.

Final Remarks

Cellular evolution is not about evolving genomes; it is about evolving cellular design. A given cellular design is consistent with many configurations of the underlying genome, and only a small fraction of the genes in that genome reflect and define the cellular design in any case. It is this small cadre of genes that requires identification and our attention. Much of the controversy in which genomics is enmeshed today is the result of not clearly distinguishing between the evolution of cellular organization and that of the genome.

Our customary evolutionary metaphors do not help in understanding the evolution of cellular organization. The amount and degree of novelty required to build cellular designs is beyond our ken; how it is generated is a mystery. The nature of primitive simple cellular designs and their transformation into the complex cellular organizations of today are beyond our purview as well. Yet these are the things we must come to know if we are to understand biological organization, and knowing them will consolidate biology and integrate it into the fabric of science as never before. There will come a new and deeper understanding of the evolutionary process.

The following summarizes what we know (or what I think I know) about the evolution of cellular organization. First, cellular evolution had to have begun at a special, and so definable, nexus in the evolutionary course, a stage of enormous evolutionary potential. That nexus may have been the advent of translation. The capacity to map nucleic sequence into peptide sequence generated a never-before-encountered representational phase space that unfolds into an endless world of possibilities. Alternatively, the starting point for modern cell designs may have come about at a somewhat later stage, after the chaos of the evolving RNP world had given rise to a new type of higher-level nucleic acid/proteinaceous organization. It is too early to consider this possibility seriously. What is important, however, is that when modern cellular designs began to evolve, they did so by building on preexisting "platforms," which were basically RNA-like or RNP-like entities.

Second, nature shows translation to be essentially an RNA machine, a mechanism that originated in some ancient RNA world. Thus, cellular evolution too may have begun in an RNA setting, the era of nucleic acid life.

Third, primitive cells did not begin as proteinaceous aggregates. They started by gradually "proteinizing" the preexisting nucleic acid-based biological organizations that populated and defined the RNA world. Importantly, cellular evolution began from many, not one or a few, distinct starting platforms.

Fourth, a primitive peptide representation of a given RNA sequence can have significance both for the particular nucleic acid–based organization of which it is a part or for other such organizations of which it is not a part. Here, then, begins a commerce in protein-coding RNAs (i.e., HGT, which inevitably leads to the establishment of a universal genetic code [the "lingua franca" of genetic commerce] including a universal decoding mechanism).

Fifth, multiple starting points for primitive cellular designs ensure a far broader sampling of the possibilities inherent in the representational phase space, and the sharing of inventions (brought about through HGT) allows further areas of the phase space, otherwise not accessible, to be explored. Herein lays the rich source of the novelty required to evolve proteinaceous cellular organization.

Sixth, the general types of proteins that can evolve early on are limited by the performance characteristics of the primitive translation apparatus. Therefore, an evolutionary premium for improving translation function exists. In a sense, the evolution of translation leads the wave of cellular evolution.

Seventh, interestingly, an error-prone primitive translation apparatus may have proved advantageous in exploring protein phase space.

Eighth, cellular evolution begins with simple components, and relatively few of them: small proteins, small genomes, and sparse coupling were among the componentry. In other words, primitive cellular organization has to have been loose (highly modular), largely horizontal, and fluid. Component specifications are relatively ill-defined (by modern standards). There is little in the way of biological specificity and individuality (idiosyncrasy) in such primitive forms.

Ninth, at this early stage, evolution was communal. It was the organismal community, defined by the sharing of genes that evolved, not individual cell lines. This was also a time, with cell designs simple, fluid, and modular, that all of an organism's genes could be displaced by alien equivalents. Proteins tended to be generic and cosmopolitan; only later would they become idiosyncratic and contextual. Any organismal genealogical records that existed at this time were ephemeral.

Tenth, cellular evolution is a continuing thrust into greater complexity, which inevitably brings with it idiosyncrasy, specificity, and hierarchical organization. Complexity also means greater connectivity among cellular parts. Eventually, a degree of connectivity, integration, is attained that the evolving cell reaches a critical point, where the basic nature of the cell design changes rapidly and radically. The cell becomes less an ill-defined collection of parts and more an entity in its own right, having a more distinctive overall character.

Eleventh is the "Darwinian Threshold." It separates a pre-Darwinian (progenote) world, in which organismal genealogies are fluid and ephemeral, and evolution is primarily a collective affair, from the Darwinian world we know, in which individual lineages and vertical evolution dominate, and speciation occurs.

Finally, extant life is descended not from one, but from three distinct primordial cellular organizations (designs). However, these three designs have developed and have matured in a communal fashion, along with many other such designs that along the way became extinct, with most perhaps never reaching a Darwinian Threshold.

References

1. D. Bohm, "Some Remarks on the Notion of Order," in C. H. Waddington, ed., *Towards a Theoretical Biology: 2. Sketches*. (Edinburgh: Edinburgh Press 1969), 18–40.
2. W. Bateson, *Problems of Genetics* (New Haven, CT: Yale University Press, 1913), 88, 90.
3. T.N. Eldredge and S.J. Gould, "Punctuated Equilibria: An Alternative to Phyletic Gradualism," in T.J.M. Schopf, ed., *Models in Paleontology* (San Francisco: Freeman, Cooper, 1972), 82–115.
4. R. Solé and B. Goodwin, *Signs of Life: How Complexity Pervades Biology* (New York: Basic Books, 2000).
5. B.C. Goodwin, "The Life of Form. Emergent Patterns of Morphological Transformation," 2000 *C R Acad Sci III* 323 (2000): 15–21.
6. E. Pennisi, "Genome Data Shake Tree of Life," *Science* 280 (1998): 672–674.
7. E. Pennisi, "Is it Time to Uproot the Tree of Life?" *Science* 284 (1999): 1305–1307.

8. J. Sapp, *Evolution by Association* (New York: Oxford University Press, 1994).
9. C.G. Kurland and S.G. Andersson, "Origin and Evolution of the Mitochondrial Proteome," *Microbiol. Mol. Biol. Rev.* 64 (2000): 786–820.
10. E. Chatton, "Pansporella Perplexa. Reflexions sur la Biologie et la Phylogénie des Protozoaires," *Ann. Des Sc. Nat. Zool. 10e serie*, VII, 1925 1–84; 76. Edouard Chatton, *Titres et travaux scientifiques (1906—1937) de Edouard Chatton* (Sottano, Italy: Sette, 1938).
11. R.Y. Stanier and C.B. van Niel, "The Concept of a Bacterium," *Archiv fur Mikrobiologie* 42 (1962): 17–35.
12. R.G.E. Murray, "Fine Structure and Taxonomy of Bacteria." In G.C. Ainsworth and P.H.A. Sneath, eds. *Microbial Classification. Soc. Gen. Microbiol. Symp.* 12 (1962): 119–144.
13. G.E. Fox, L.J. Magrum, W.E. Balch, R.S. Wolfe, and C.R.Woese, "Classification of Methanogenic Bacteria by 16S Ribosomal RNA Characterization," *Proc. Natl. Acad. Sci. USA* 74 (1997): 4537–4541.
14. C.R. Woese and G.E. Fox, "The Phylogenetic Structure of the Prokaryotic Domain: The Primary Kingdoms," *Proc. of the Natl. Acad. of Sci. USA* 74 (1977): 5088–5090.
15. Franklin M. Harold, *The Way of the Cell* (Oxford: Oxford University Press, 2001).
16. C.R. Woese, J.G., Olsen, M. Ibba, and D. Soll, "Aminoacyl–tRNA Synthetases, the Genetic Code, and the Evolutionary Process," *Microbiol. Mol. Biol. Rev.* 64 (2000): 202–36.
17. C.R. Woese, "Bacterial Evolution," *Microbiol. Revs.* 51 (1987): 221–271.
18. R.S. Wolfe, "Biochemistry of Methanogenesis," *Biochem. Soc. Symp.* 58 (1992): 41–49.
19. G.J. Olsen and C.R. Woese, "Lessons from an Archaeal Genome: What Are We Learning from *Methanococcus jannaschii*?" *Trends Genet.* 12 (1996): 377–379.
20. D.E. Graham, R. Overbeek, G.J. Olsen, and C.R. Woese, "An Archaeal Genomic Signature," *Proc. Natl. Acad. Sci. USA* 97 (2000): 3304–3308.
21. H. Ochman, J.G. Lawrence, and E.A. Groisman, "Lateral Gene Transfer and the Nature of Bacterial Innovation," *Nature* 405 (2000): 299–304.
22. J.G. Lawrence and H. Ochman, "Reconciling the Many Faces of Lateral Gene Transfer," *Trends Microbiol.* 10 (2002): 1–4.
23. W.F. Doolittle, "Lateral Genomics," *Trends Cell Biol.* (1999): M5–M8.
24. W.F. Doolittle, "Phylogenetic Classification and the Universal Tree," *Science* 284 (1999): 2124–2129.
25. J.P. Gogarten, W.F. Dolittle, J.G. Lawrence, "Prokaryotic Evolution in Light of Gene Transfer," *Mol. Biol. Evol.* 12 (2002): 2226–2238.
26. O.G. Berg and C.G. Kurland, "Evolution of Microbial Genomes: Sequence Acquisition and Loss," *Mol. Biol. Evol.* 19 (2002): 2265–2276.
27. C.R. Woese, "Interpreting the Universal Phylogenetic Tree," *Proc. Natl. Acad. Sci. USA* 97 (2000): 8392–8396.
28. C.R. Woese, "Archaebacteria and Cellular Origins. An Overview," *Zbl. Bakt. Hyg. I. Abt. Orig.* C3 (1982) 1–17.
29. C. Woese, "The Universal Ancestor," *Proc Natl Acad Sci USA* 95 (1998):6854–6859.
30. C.R. Woese and G.E. Fox, "The Concept of Cellular Evolution," *J. Mol. Evol.* 10 (1977): 1–6.
31. S.J. Freeland and L.D. Hurst, "The Genetic Code is One in a Million," *J. Mol. Evol.* 47 (1998): 238–248.
32. J.W. Drake, "A Constant Rate of Spontaneous Mutation in DNA-based Microbes," *Proc. Natl. Acad. Sci. USA* 88 (1991): 7160–7164.
33. K. Hosokawa, R.K. Fujimura, and M. Nomura, "Reconstitution of Functionally Active Ribosomes from Inactive Subparticles and Proteins," *Proc. Natl. Acad. Sci. USA* 55 (1966): 198–204.
34. P. Nissen, J. Hansen, P.B. Moore, and T.A. Steitz, "The Structural Basis of Ribosome Activity in Peptide Bond Synthesis," *Science* 289 (2000): 920–930.
35. N. Ban, P. Nissen, J. Hansen, P.B. Moore, and T.A. Steitz, "The Complete Atomic Structure of the Large Ribosomal Subunit at 2.4 Å Resolution," *Science* 289 (2000): 905–920.
36. L.P. Gavrilova, V.E. Koteliansky, A.S. Spirin, "Ribosomal Protein S12 and 'Non-Enzymatic' Translocation," *FEBS Lett.* 45 (1974): 324–328.
37. T. Cech, "The Ribosome is a Ribozyme," *Science* 289 (2000): 878.

38. C.R. Woese, *The Genetic Code: The Molecular Basis for Genetic Expression* (New York, London: Harper & Row, 1967).
39. C.R. Woese, "Translation: In Retrospect and Prospect," *RNA* 7 (2001): 1055–1067.
40. F.H.C. Crick, "The Biological Replication of Macromolecules," *Symp. Soc. Exp. Biol.* 12 (1958): 138–163.
41. N.W. Pirie, "The Meaninglessness of the Terms 'Life' and 'Living'," in J. Needham and D. Green, eds., *Perspectives in Biochemistry* (Cambridge: Cambridge University Press 1938), 11–22.
42. C. Darwin, *On the Origin of Species 1859*, a facsimile of the first edition (Cambridge, MA: Harvard University Press, 1964).
43. O. Kandler, "Cell Wall Biochemistry in Archaea and Its Phylogenentic Implications," *J. Biol. Phys.* 20 (1994): 165–169.
44. J.L. Bouzat, L.K. McNeil, H.M. Robertson, L.F. Solter, J.E. Nixon, J.E. Beeve, H.R. Gaskins, G. Olsen, S. Subramaniam, M.L. Sogin, and H.A. Lewin, "Phylogenomic Analysis of the alpha Proteasome Gene Family from Early-Diverging Eukaryotes," *J Mol. Evol.* 51 (2000): 532–543.
45. S.D. Bell and P. Jackson, "Transcription and Translation in Archaea Trends in Microbiology," 6 (1998): 222–228.
46. F. Marc, K. Sandman, R. Lurz, and J.N. Reeve, "Archaeal Histone Tetramerization Determines DNA Affinity and the Direction of DNA Supercoiling," *J. Bio. Chem.* 277 (2002): 30879–30886.
47. E. Szathmary and J. Maynard Smith, "From Repicators to Reproducers: the First Major Transitions Leading to Life," *J. Theor. Biol.* 187 (1997): 555–571.
48. Marcello Barbieri, *The Organic Codes: An Introduction to Semantic Biology* (Cambridge: Cambridge University Press, 2003)
49. W. Gilbert, "The RNA World," *Nature* 319 (1986): 618.
50. C.R. Woese, "The Emergence of Genetic Organization," in C. Ponnamperuma, ed., *Exobiology* (Amsterdam: North-Holland Publishing Co., 1972), 301–341.
51. C.R. Woese, "On the Evolution of Cells," *Proc. Natl. Acad. Sci. USA* 99 (2002): 8742–8747.

5

If the Tree of Life Fell, Would We Recognize the Sound?

W. FORD DOOLITTLE

Taxonomy and Phylogeny

The only figure in Darwin's *Origin of Species* is a tree. The accompanying text focuses mostly on divergence within populations, but Darwin thought that the tree metaphor would hold much more deeply; indeed, that the pattern of relationships among all living things was one of successive branchings, or lineage splittings. He wrote that, "the affinities of all the beings of the same class have sometimes been represented by a great tree. I believe this simile largely speaks the truth. The green and budding twigs may represent existing species, and those produced during each former year may represent the long succession of extinct species."[1]

Furthermore, the hierarchical pattern produced by such lineage splittings was seen by Darwin as causally related to the hierarchical classificatory schemes of preevolutionary biologists like Linnaeus: "the limbs divided into great branches, and these into lesser and lesser branches, were themselves once, when the tree was small, budding twigs, and this connection of the former and present buds by ramifying branches may well represent the classification of all extinct and living species in groups subordinate to groups."

In other words, the process of phylogeny could be seen to explain the pattern of taxonomy. In the Darwinian perspective, the goals of systematists and evolutionists converge, and the only truly natural classification scheme for all life would be a universal (phylogenetic) Tree of Life.

Molecular Phylogeny and the Hegemony of SSU rRNA

It was surely clear to those who followed Darwin that the practices already in use by comparative morphologists and physiologists to construct hierarchical taxonomies were also those needed to reconstruct the universal phylogenetic tree. That they might be successfully applied to animals and plants, both living and extinct, seemed obvious. In the first half of the twentieth century, several courageous microbiologists attempted to extend such practices "downward" to their own chosen objects of study, developing comprehensive evolutionary schemes for prokaryotes and simple eukaryotes based on cell size, shape, and biochemistry. By the midpoint of that century, however, Stanier and van Niel, the field's frontrunners, called such speculations to a halt.[2,3] Microbes, and in particular bacteria, provided nothing useful in terms of a fossil record, and too little in the way of analyzable morphology—and perhaps too much in the way of biochemical versatility—to be grafted onto the base of the tree of life in any believable way. We must make taxonomies for many reasons, but we must not claim that these are "natural" or phylogenetic.

Arguably, Stanier and van Niel threw in the towel just a little too soon. Less than a decade later we had a new discipline, molecular phylogenetics, which could in principle bypass morphology and fossils. Indeed, as Zuckerkandl and Pauling pointed out in a 1965 article that is one of molecular phylogeny's founding documents, molecular and organismal approaches have the potential for mutual confirmation.[4] Specifically, Zuckerkandl and Pauling asserted that, "the topology of branching of molecular phylogenetic trees should in principle be definable in terms of molecular information alone. It will be determined to what extent the phylogenetic tree, as derived from molecular data in complete independence of the results of organismal biology, coincides with the tree constructed on the basis of organismal biology. If the two phylogenetic trees are mostly in agreement with respect to the topology of branching, the best available single proof of the reality of macroevolution will be furnished."

Where both kinds of data can be used (among mammals, let's say), molecular and organismal information can confirm each other. In addition, Zuckerkandl and Pauling, like most molecular biologists today, felt that molecular phylogenies are in a sense more directly "true." Genotype is, after all, the cause of phenotype, not just a correlate. On the morc practical side, molecular sequence approaches offered a solution for microbes, where "organismal biology" had let us down.

Implicit in the molecular sequence method in those early days was the notion that all proteins (and thus all genes) should tell the same story: That the only criteria by which we might choose one over another were practical—how widely distributed, conservative, and easily sequenced was the molecule. Early universal treeing efforts based on proteins soon gave way to those based on small-subunit ribosomal RNA (SSU rRNA). Woese's choice of this molecule and its encoding gene was a wise one.[5] SSU rRNA is ubiquitous and suitably conservative, and substantial sequence information from diverse species could be obtained (in the form of "oligonucleotide catalogs" of partial sequences) more economically than was possible for any protein before the invention and widespread use of DNA cloning and sequencing. An expanding cadre of Woese's students and associates as well as many converts to these methods soon made SSU rRNA the gold standard for phylogeny.[6] Not only could all living species be united into a single tripartite Tree of Life but thousands of different isolates of hundreds of prokaryotic species could be identified and taxonomically micropigeonholed by sequencing this molecule or its gene.

The three-domain SSU rRNA tree is now hegemonic: it is reproduced in all the textbooks, is the central organizing scheme of all modern biologists, and is often seen on TV. Its construction and elaboration are clearly the crowning achievement of the science of molecular phylogenetics that Zuckerkandl and Pauling fathered, and the salvation of environmental microbiology as well as the road map for the future of comparative microbial genomics. The 2003 award to Woese of the Crafoord Prize was very richly deserved.

Dissent

However, there are problems. The burgeoning field of comparative microbial genomics reveals that it matters very much what gene we pick when we reconstruct prokaryotic phylogeny. Different genes give different results, because genes have often been transferred across species, phylum, or domain lines. The puzzles that such lateral gene transfer (LGT) poses for phylogeneticists are as much philosophical (what does the Tree of Life really represent?) as they are practical (how can we best construct it?). Indeed, Woese himself now asks: "What does it mean . . . to speak of an organismal genealogy when nearly all of the genes in the cell—genes that give it its general character—do not share a common history?"[7]

Evidence for lateral gene transfer is of several sorts, but it seems convenient here to distinguish methods that rely on measures of phylogenetic relationship between genes and those that look at differences in gene content among sequenced genomes. The former include analyses of differences in base composition and codon usage between the putatively transferred gene and those surrounding it, discordant BLAST (sequence similarity search) scores, and incongruity of gene phylogenies, however reconstructed. Such analyses often identify different sets of genes,[8] and each has flaws. Base composition and codon usage can vary between genes in a genome for reasons other than differing origins, BLAST scores are poor measures of phylogenetic affinity, and many gene phylogenies suffer from one or another artifact or insufficient phylogenetic signal. Indeed, molecular phylogeneticists are getting very good at deconstructing each other's trees by methodological arguments alone.

Gene content comparisons provide a different kind of evidence that LGT has occurred, and they are often, in my view, more compelling in the logic of their conclusions. I will focus on them here. One can often be quite certain that fully sequenced genome X has gene A, while genome Y does not, and, using parsimony methods spelled out in the next section, base inferences about transfer on such presence or absence information. The most stunning and unexpected examples of this sort of pattern—which we can call "patchy distribution"—have emerged as genomes of different "strains" of a single bacterial "species" have become available.

Patchy Distributions of Genes within Species

Most close comparative studies focus on pathogens and have been undertaken to identify specific genes involved in virulence or resistance. Pathogenicity-related genes are indeed found, but so are many more. Three sequenced strains of *Escherichia coli*, K12 (the harmless lab workhorse), O157:H7 (a killer, causing bloody diarrhea), and CFT073 (a urinary tract invader) share only 2,996 genes, 39.2% of the total number of different genes found in all three.[9] K12 has 778 and 1,099 genes not found in O157:H7 and CFT073, respectively, O157H7 has 1,550 and 1,860 genes missing from K12 and CFT073, and CFT073 has 1,827 and 1,816 genes absent from K12 and O157:H7. Some of these strain-specific genes encode toxins or other obvious virulence determinants, and some derive from phages or other mobile genetic elements, but a good portion has honest, everyday functions.

These results are typical in intraspecific genomic studies of pathogens. Strains within the same species (whether defined phenotypically or by the possession of SSU rRNA sequences differing by less than 3%) can differ by up to 25% in the genes they contain. Such differences can be assessed by direct sequencing, subtractive hybridization, or microarray analysis and have lead to general acceptance of the "species genome" concept of Lan and Reeves.[10] A species genome consists of two components: a "core" of genes present in all (or almost all) strains, which would be expected to be highly similar in sequence and produce the same topology in phylogenetic reconstructions, and an "auxiliary set," comprising all genes that are found only in some strains, thus exhibiting patchy distribution. Often there will be more genes in the latter component than the former.

That such variation is characteristic of all prokaryotes, not just those hotly engaged in mortal combat with the immune systems and pharmacopoeias of their hosts, is not yet firmly established. However, our own subtractive hybridization experiments with close relatives of the sequenced hyperthermophilic bacterium *Thermotoga maritima*[11] indicate that, at least for this free-living extremeophilic species (with no known eukaryotic hosts, prey, or enemies), intraspecific genomic variability is also the rule. Strains we have assessed so far differ by up to 20% in gene content. The strain-specific genes encode a full range of interesting activities, such as ATPases, rhamnose catabolism, and teichoic acid synthesis. Subtractive hybridization recovers only fragments of genes, but when these fragments are used to select lambda clones of unsequenced strains for sequencing, these larger inserts are seen often to carry entire gene clusters (often operons), whose status as LGTs is supported not only by a patchy strain-specific distribution pattern but by phylogenetic trees showing alien origin.

Why Patchy Distributions Imply LGT

Patchy distribution patterns can be interpreted in two ways: as evidence for LGT or as evidence for differential loss from a common ancestor. In figure 5.1A, the presence of a gene G in taxon A that is absent from taxon B could reflect either loss of G from lineage B since divergence from their shared extinct ancestor, X, or addition (by de novo invention or LGT) to lineage A. If gene G is known from distant taxa (i.e., has significant best BLAST hits to genes outside the A/B lineages), we can eliminate the de novo invention explanation and need only decide between loss and LGT.

We still cannot make that decision, however, without knowledge about the state of X (whether or not it had G). By examining an outgroup taxon, C, we can make an informed guess about X and thus about loss or gain of G, using parsimony (figure 5.1B).

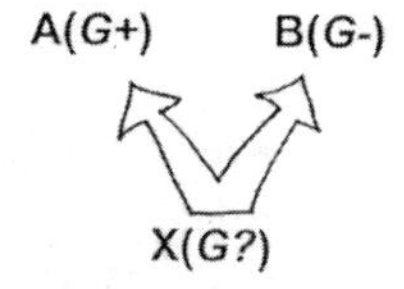

1B

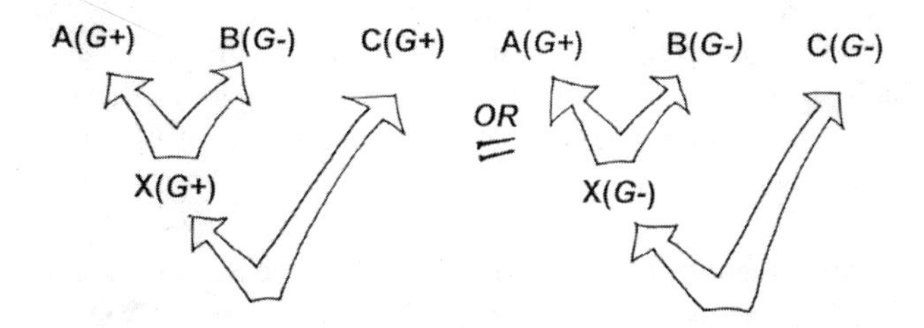

1C

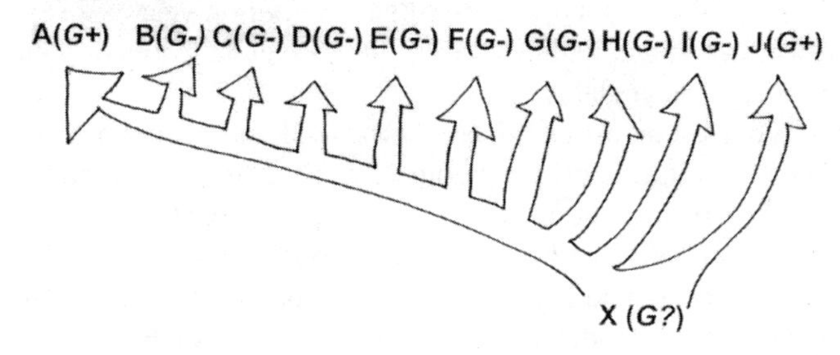

Figure 5.1. Use of reasoning by parsimony to decide between loss and gain (by LGT or other processes) of genes in genomes. (A) Either taxon A has gained gene G or taxon B has lost it. If loss and gain are equally likely as events, we cannot tell which is the best explanation here without some way of guessing whether extinct ancestor X had G. (B) Knowledge of the state of taxon C will allow us to make this guess and to decide whether A has gained or B has lost. If C has G, then one event of loss in the line leading to B from a G-containing X would explain the pattern and is more parsimonious (requires fewer steps) than the minimum of two events that would be required if X did not have G (e.g., one gain in the line leading to A from X and another somewhere in the ancestry of C). (C) The gain of G by A is the better explanation than the loss of it in the eight lineages B-I diverging from X before A (eight steps), even though we would also here require another event of gain to explain G's presence in J (so two steps overall). If the two genes G are clearly homologous, then gain by invention in lineage A (or J) and transfer to J (or A) becomes the best overall explanation, barring other knowledge.

If C has G, then G's presence in X and loss from B requires one event and is to be preferred as an explanation to the alternative—independent gains in A and C (two events). If C lacks G, then a single addition to A is a better explanation than two independent losses (in lines B and C).

Of course the assumption, implicit here, that loss and LGT events are overall equally frequent as processes, is just an assumption. By assigning different penalties to one versus the other, we can completely alter the outcomes of our analyses. However, there are some overriding considerations that should guide us in our efforts to make sense of patchy gene distributions and that indicate that equal frequency is not an absurd first assumption.

First, in many cases, the patchy pattern will look like that in figure 5.1C. Here, an explanation for G's presence in A involving only losses requires that G was retained at each intermediate branching on the path from X to A, and then lost many times independently in each of the lines diverging from it, while the gain explanation still needs only a single event. No matter what the relative probabilities of individual losses and gains, gain will sometimes be the clearly best explanation.

Second, although we have no global measures of relative rates of gene loss versus gene gain (LGT plus invention) throughout the 4 billion years of prokaryote history, if they were not in fact very nearly equal over most of that time, genomes would now be enormous or would have disappeared. Life's first efforts at genome construction no doubt produced tiny assemblages of few genes, but even if we restrict ourselves to cells that are modern in form and function (let's say the last 2 billion years), this argument has considerable force: Individual lineages can experience great losses or gains, but a global steady state seems unavoidable.

Third, gain, of course, comprises two processes: LGT and "de novo invention," which must include both recruitment of noncoding sequences (to give "orphan genes") and the functional reassignment of duplicates of preexisting genes (generating paralogs). If there were no de novo invention, the steady-state model would seem to demand that on average 50% of the genes found in any strain A but not strain B were present in X, and 50% were added to A after its divergence from B. There is, of course, de novo invention, but the 50% rule should still generally apply to patchily distributed genes with obvious homologs outside the A/B clade, as invented genes should be lineage specific.

Finally, each individual gene G whose patchy distribution is attributed to differential loss must be attributed to the genome of ancestor X. Although loss is no more onerous than gain as an explanation of any single gene's presence in one strain but not another, its application in all such cases leads to a chain of postulated ancestors of ever-increasing genome size.

Patchiness at Depth

Interstrain, intraspecies comparisons provide many of the most compelling cases of LGT. Presence or absence of particular genes can usually be unambiguously scored, and often genes that have been recently introduced will bear additional evidence of alien origin: distinctive base composition or codon use, insertion near or in tRNA-encoding genes, or inclusion within some more complex plasmid, prophage, conjugative transposon, or integron-like genetic element. Nevertheless, there are many instances of genes and suites of genes patchily distributed over much greater phylogenetic distances that can also only be explained parsimoniously by LGT. For instance, Yan Boucher finds that all haloarchaea contain genes for a type I isopentenyl diphophate isomerase, as do many bacteria, but no other archaea (Boucher and Doolittle, in preparation). It is more parsimonious to see this as a transfer from bacteria into an ancestor of the haloarchaea than as multiple independent losses from all other archaea. Similarly, the discovery by DeLong and collaborators[12] of a deep clade of proteobacteria bearing homologs of rhodopsin genes—light-sensitive proteins first extensively characterized among and thought to be confined to haloarchaea—bespeaks LGT, although here one could argue about the direction of transfer.

In my lab we have recently looked (bioinformatically) at the distribution among "phyla" or "divisions" of bacteria of many of the major metabolic pathways or complex physiologies that have traditionally been used to define such prokaryotic groups, and that are in most cases under multigene control—such fundamental features as respiration, photosynthesis, and nitrogen fixation. These traits cannot in general be hung from the branches of the SSU rRNA tree in such a way as to require only a single invention. Were we to redraw the tree so as to group all phyla showing one feature (photosynthesis, say) into a single clade, such a tree would not similarly simplify the evolutionary history of other traits. Again, the absurdity of invoking a succession of ever more gene-rich ancestors rules out differential loss as a general explanation for patchy distribution. The attribution of all complex traits (or individual genes) that are found in at least one bacterium and one archaean to the last universal common ancestor (LUCA) would require that this cell employ almost all modern biochemical reactions (methanogenesis and photosynthesis are, not all notable exceptions) and have a genome larger than that of any modern prokaryote. I call this the "genome of Eden" and consider it an unlikely model for LUCA.

Some have sought to avoid this unpleasant consequence (the giant LUCA) by asserting that the last common ancestor was not a single cell, but a population of heterogeneous primitive entities whose different genomes have contributed different genes to the modern global prokaryotic complement. However, there is no way in which different genes that were present in different genomes in the past can be brought together into common genomes now—and even less be patchily distributed among them—that is not formally equivalent to LGT. Indeed, I too think that "LUCA was a population," but I argue that this in fact means that there actually was no LUCA. Instead, we can imagine life's gradual transition from a population of "monogenic" RNAs in the "RNA world," to a population of multigenic genomes in some precellular early DNA/protein world, to a population of real cells ancestral to the modern prokaryotic world, with gene (or informational sequence exchange) playing a vital role at each stage. To be sure, we can infer that any gene family represented in modern genomes ultimately derives from a single ancestral gene in a single ancestral genome in a single ancient cell, but the ancestors of different gene families trace to different genomes in different cells at different times in the past, just as most genes in the human population have different coalescents at different times (and in different individuals) in our human (or primate) ancestry.

Issues Around the Idea of the "Core"

I suspect that most comparative microbial genomicists now accept a model for the typical prokaryotic genome that looks something like figure 5.2. The "shell" comprises all those genes in the genome that are so frequently exchanged as to be patchily distributed even among other strains of the same "species." Each individual genome's shell will, of course,

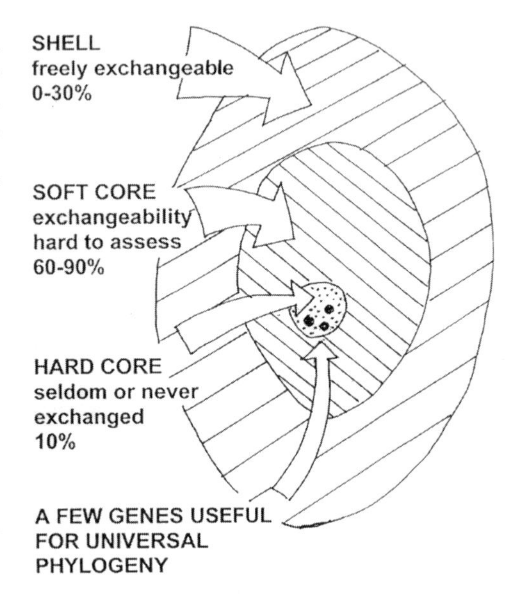

Figure 5.2. The current consensus. A typical prokaryote genome has a shell (up to 30% of its genes) made up of genes that are quickly gained or lost and are patchily distributed among strains of its "species." It also has a "hard" core of genes shared with (common to) all prokaryotes, although this may be very small and include only a few genes useful for phylogeny. For some of us, "few" includes "none." Most of a genome's genes will be in the "soft" core, comprising genes that are patchily distributed among phyla (some have them, some do not). According to the arguments presented here, these also have a history of exchange, although often this will be hard to confirm or refute, because of weak phylogenetic signal.

encompass only a fraction of the exchangeable genes available to the species in this way. I have divided the "core" itself up into a "soft" part, which would include all genes common to strains within the species (only decreasing subsets of which would be common to all species of the genus, phylum or domain), and a "hard" part. This latter comprises those (as it turns out surprisingly few) genes found in all prokaryotic genomes. Genes of the soft core will themselves be patchily distributed when groups more inclusive than species are the focus, and included among them will be many of fundamental physiological importance: genes for photosynthesis and respiration, for instance, as we noted above.

We might imagine that we could come up with meaningful phylogenies by ignoring the patchily distributed genes and using standard phylogenetic reconstruction methods with whatever "core" of genes is shared by the genomes whose evolutionary relationships we wish to know (the soft core for intraspecies phylogenies and the hard core for the universal tree, at the extremes). At the strains-within-a-species level, information exchange by homologous recombination means that we will in fact often get different phylogenies for different genes, however. Recent MLST (multilocus sequence typing) studies on many bacteria (mostly pathogens) produce a weblike pattern of incongruent trees, just as we might observe for different loci within the human population.[13] Indeed, Dykhuizen and Green noted some time ago[14] that the observation of such a pattern is essential for any claim that bacteria ever conform to the reigning "Biological Species Concept" championed by Ernst Mayr.

Homologous recombination falls off dramatically with sequence difference,[15] however, so trees for genes shared among different designated "species" of enterics (or similarly closely related bacteria) often do show congruence, consistent with the notion that these shared genes comprise a conserved (always present) and stable (not exchangeable) core. However, it remains unclear what happens when we dig deeper in the past, because not only does the core of genes held in common get smaller but phylogenetic signals within those fewer genes gets weaker. Much of the current debate about gene transfer and its meaning for any deep or universal phylogeny now focuses, quite properly, on the hard core: How big is it and how stable? How come it exists (if it does), and what is its value for reconstructing any universal tree?

How Big?

Any attempt to count or identify the genes shared as orthologs by any pair or larger collection of genomes must face the problems of recognizing and defining shared othologs, problems that only get worse as the genomes compared get farther and farther apart (by whatever measure of evolutionary distance). Surprisingly, though, most attempts to identify and enumerate genes shared by all prokaryotes come up with something in the range of fifty to 100 genes, at least half encoding components of the translation systems (especially ribosomal proteins). This is, of course, not enough genes to run a cell with. Potential artifacts of analysis can only partially explain the surprising result (R.L. Charlebois and W.F. Doolittle, in preparation).

BLAST sequence similarity homology searches will miss remote homologs, and some of the strategies designed to avoid double-counting of paralogs are probably too conservative, but even when BLAST cutoff requirements are loosened or the presence of genes is scored by searching for shared gene designations in annotated genomes, the number of universally shared genes does not rise much above 100.

Highly reduced genomes of intracellular symbionts and pathogens might skew the result, but when genomes of less than 1,000 or even 1,500 ORFs (open reading frames) are discounted, the number of shared genes still remains small. Indeed, if we look at the average numbers of genes shared in all possible comparisons between one, two, three or more (up to twelve) genomes taken one from each of the twelve sequence-endowed bacterial phyla, the number still soon levels off to something like 150 (R.L. Charlebois and W.F. Doolittle, in preparation).

Thus, the smallness of the core seems to have some biological meaning, and the most likely meaning is that, indeed, only 100 or so genes, central to transcription and translation, comprise the prokaryotic core. DNA replication, equally essential and foundational, is in fact differently conducted in Bacteria and Archaea, and almost everything else—catabolism of substrates, synthesis of building blocks, cell structure, and "behavior"—is under the control of exchangeable genes that have been chosen in a mix-and-match fashion from a generally available pan-prokaryotic gene pool during the assembly of bacterial and archaeal genomes.

How Stable?

The next thing we might want to know about the core, whatever its size, is whether the genes making it up have always been together (have the same phylogeny), or whether they too have experienced exchange. Such exchange would have to be of the type often called "orthologous replacement"—substitution of a resident version of gene by a functionally identical (or functionally very similar) copy from outside—and presumably would most often be driven by some selective advantage. The foreign gene might make a product that is more resistant to an antibiotic, for instance, or that functions better under environmental conditions at the edge of the species' adaptive range or more efficiently exploits a newly available resource. The literature provides many examples of such events,[16,17] even involving key components of the transcription and translation machinery.

To assess the overall extent to which this process affects core genes, it is necessary to reconstruct all their phylogenetic trees and to ask whether they are congruent. Usually, this has been done by analyzing all the unambiguously alignable genes shared by a quartet or quintet of genomes belonging to some supposedly monophyletic group (species, genus, phylum, or domain). With four taxa, there are three possible tree topologies, and with five there are fifteen: The task reduces to determining how many genes favor each of the possible phylogenies. We constructed trees for about half the 543 genes common to four sequenced euryarchaeotes,[18] and Raymond et al.[19] looked at 200 or so genes shared by five lineages of photosynthetic bacteria (in five different "phyla"). In both cases, all three (or fifteen) possible trees were favored by a substantial number of genes, as if orthologous replacement were the rule rather than the exception, at these moderate phylogenetic depths.

There is some dispute about the robustness of trees involving so few taxa, and a conundrum at the heart of such analyses. When a four-taxon study includes two reasonably close relatives (two cyanobacteria, say, and an archaean and Bacillus), then only transfers from one of the disparate lineages into one of the paired lineages that occurred since its divergence from the other will actually alter tree topology. Such events should give robust signals, but one would not expect so many of them, and we would clearly not detect many real transfers. When the four (or five) taxa are more equally distant, the net is cast much wider, but because the inner branches of the tree are short, topologies will more often disagree artifactually: The phylogenetic signal is too weak. Indeed, we have known for some time that different mitochondrial genes will give different mitochondrial phylogenies, and yet we believe this reflects noise, not conflicting signal.

In all published attempts to examine the topology of trees for the hundred or fewer genes shared by all prokaryotes, lack of robust phylogenetic signal—coupled with strong evidence for LGT for some genes—has been the major finding.[20–22] On the assumption that the core genes that do not show evidence of LGT nevertheless do have the same phylogenetic history, several investigators have used concatenated sequences of these supposedly faithful genes to make trees, and principal component analysis as a way to assess underlying congruence.[21–23] It is difficult to evaluate these results and their meaning for universal genome phylogeny when the final concatenates, in fact, involve as few as 1% of any genome's genes.

Recently, Vincent Daubin, Nancy Moran, and Howard Ochman have launched vigorous attacks on claims (especially those of Raymond et al.[19]) for extensive orthologous replacements within the core, claiming that most incongruence reflects noise, not conflicting signal.[24,25] They show that, indeed, there is little robust evidence for rampant orthologous replacement at the species-in-a-genus level, but even LGT's most ardent supporters had not claimed this. In their deeper analyses (the γ-proteobacterial core), these authors seem to operate from the assumption that if a gene's phylogeny does not significantly conflict with that of the core's concatenate, then there has been no LGT. There is little phylogenetic signal for most genes at this depth, however, and to require robust signals for transfer against a null hypothesis of vertical descent is to bias the result.

Absence of evidence is, of course, not evidence of absence. There may be a real catch-22 in assessing how much LGT as orthologous replacement afflicts the core at depth, but given that we know that orthologous replacement can happen and that rampant LGT drives genome (gene content) evolution at the strain-in-a-species level, there is no justification in retaining vertical descent as the null hypothesis and requiring stronger proof for LGT. Rather (at least with greater fairness), we might recast the notion of the existence of a stable core as a hypothesis that needs to be tested, not a truth that needs further elaboration. If the hypothesis is that there is a cadre of genes that have never been exchanged (and that thus track organismal phylogeny), and the test of it requires that there indeed be such universally shared genes that show the same phylogeny, then this hypothesis has yet to be proven.

How Come?

Still, there could be a core, and if some more sophisticated analysis shows in the future that some fifty plus mostly translational genes do track the same 4-billion-year evolutionary path, we would have to ask why they do that. The standard answer follows the lines of what Lake and colleagues call "the complexity hypothesis."[26] The translational machinery is so complex and its function so tightly scrutinized by natural selection that its individual components (ribosomal proteins and certainly ribosomal RNAs) cannot function in foreign cytoplasmic contexts, where all the coevolved molecular interactions would be differently evolved. Perhaps this is so, but the argument might also be turned on its head. Ribosomal RNAs, for instance, are so constrained by interaction with so many components that the only parts that change (coincidentally, the parts on which SSU phylogeny is based) are those that do not interact at all. The core of the machinery is generic, and indeed, functioning ribosomes can be constructed of quite heterologous components.[27] In addition, many of the translational components of the core are encoded by genes comprising clusters whose presence and structure is conserved (quite remarkably) across all the prokaryotes. One might suggest, only half facetiously, that the (still to be demonstrated) congruence of their phylogenies does not mean that they have never been transferred but, radically otherwise, that they have always been transferred together.

So What?

Current thinking about LGT and its implications for phylogeny makes up four schools, all of which do admit the evolutionary importance of the process. True conservatives, like C.G. Kurland,[28] seem to hold that, in spite of the qualitatively vital role of LGT, most genes in most genomes are pretty well behaved, with phylogenies congruent to that of SSU rRNA. For them, much of the "soft core" of figure 5.2 is actually quite firm.

Core supporters, like Herve Phillipe,[22] or Daubin, Moran and Ochman,[24,25] believe that there is a subset of (mostly "informational") genes that share the same phylogeny. Even though this may indeed be a very small subset (<10% of a genome) with a history that is very hard to reconstruct (because of weak signal), the core phylogeny should be that of the cellular lineages that have carried it over more than 3.5 billion years.

"Ship of Theseus" phylogeneticists, like Gary Olsen[29] accept that lateral transfer may be so frequent that no two genes in the same genome have remained together for all time. (Note, this does not require that the frequency of interspecies exchange be any greater than one LGT per gene per 10^{10-12} generations.) Still, they would argue that because few genes are exchanged in any one event, and most exchanges are over short "phylogenetic distances," various standard and new (gene-content-based) phylogenetic methods have a real chance to reconstruct the organismal Tree of Life. I name this view after the philosophical puzzle of the "Ship of Theseus," whose planks were replaced, one by one, as they rotted. The conundrum is this: after all the planks are renewed, is it still the same ship? If not, how many planks must be replaced before its identity is lost? What if all the planks, not yet fully rotten, are used one by one to build an isomorphic ship, further along the shore—which is Theseus' ship then? Philosophers, I believe, have not solved this problem, so Olsen's option,[29] to call such an ephemeral structure the Tree of Life, is perfectly defensible, although the belief that various methods now used do in fact recover its structure still needs support in theory stronger that the complexity hypothesis, and in evidence stronger than the concatenated trees we have so far seen.

The fourth group, which includes Jeff Lawrence, Peter Gogarten, and myself[30] (and many of our students) are what Ragan and Charlebois[31] have called "enthusiastic lateralists." We claim that there are no data to contradict the possibility that every gene we find in any genome today has experienced at least one between-species LGT in the 3–4 billion years since life began. To be sure, the major phyla or divisions of Bacteria and Archaea (and these domains themselves) are recreated in phylogenies of many genes, but if these phyla are themselves maintained cohesively by preferential within-phylum LGT (just as animal species are maintained cohesively by virtue of sharing genes through recombination within but not between species), then there is no evidence for deep phylogenetic signal, and the fact that most deep bacterial phylogenies resemble "stars" or "bushes" reflects this lack of signal (not some early adaptive radiation). We propose this new model not because we believe that all (or even most) of the data support it unambiguously, but because we believe that no data so far can distinguish between it and the traditional view of the meaning of the universal SSU rRNA–based Tree of Life. We would argue that any "Ship of Theseus" phylogeny is not unambiguously constructable, and is so far from the original conceptual understanding of the Tree of Life as to require a radical reworking of this understanding, not some subtle terminological negotiation.

Tree of Organisms, Web of Genomes

There probably really is (or was) a Tree of Life, if we mean by that a bifurcating branching pattern tracking every speciation event (for sexual organisms) or organismal replication (for asexual ones). If we had a continuous videotape of all such events in the last 4 billion years, we could reconstruct this tree (various heterologous cell fusions and spe-

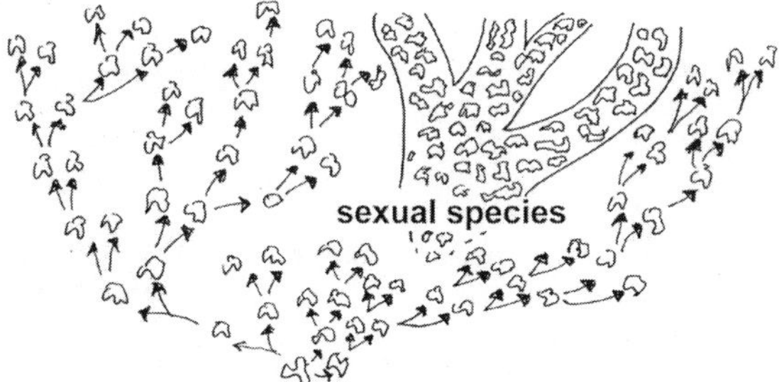

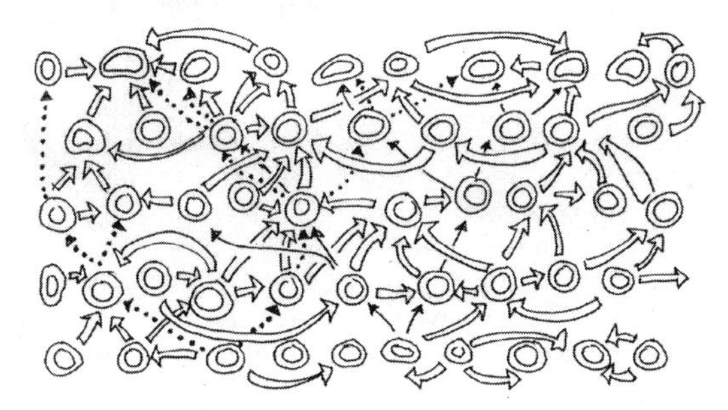

Figure 5.3. The Tree of Organisms and the Web of Genomes. (A) Almost everyone believes that life is fundamentally a bifurcating tree, although where sex is obligatory, it is populations that bifurcate, not lineages of replicating cells. This pattern is the Tree of Life, and it is unique. (B) Here we argue that because of LGT, there is no unique treelike pattern by which we can relate genomes, only a complex web. Genes within genomes (black and dotted lines) can be said to evolve in a treelike way, but without further unproven assumptions, we have no way of mapping the web of (B) to the tree of (A).

cies' hybridizations aside) in an unambiguous fashion. At least at levels higher than species, this reconstruction should yield a unique pattern of relationships between organisms (fig. 5.3A). We do not have such a video, however; all we have are genome sequences, and LGT means that there is no unique pattern of relationships between genomes (fig. 5.3B). Perhaps there is a plurality (most favored) pattern, or one tracked by several genes that we consider important, but this has yet to be proven. In any case, there is no compelling reason why this plurality pattern needs to correspond, by any simple mapping, to the tree of speciations and cell divisions. We cannot infer a unique tree of organisms from the pattern of relationships among genomes without making further assumptions about evolutionary processes that are just that: still-unproven assumptions. We have, for several decades, thought that our job was to uncover the structure of a Tree of Life, whose reality we need not question. But really, what we have been doing is testing Darwin's hypothesis that a tree is the appropriate representation of life's history, back to the beginning. Like any hypothesis, it could be false.

Acknowledgments All my current students and postdocs have helped in the elaboration of this view, even those who do not share it, so I must thank Yan Boucher (and Rebecca Case), Ellen Boudreau, Joel Dacks, David Walsh, Camilla Nesbø, Thane Papke, Christophe Douady, Uri Gophna, and Maureen O'Malley for their ideas and the Canadian Institutes for Health Research, Genome Atlantic, the Canadian Institute for Advanced Research, and the Killam Foundation for supporting us in various ways.

References

1. C. Darwin, *The Origin of Species by Means of Natural Selection* (New York: Avenel Books, 1859).
2. C.B. van Niel, "The Classification and Natural Relationships of Bacteria," *Cold Spring Harbor Symposium on Quantitative Biology* 11 (1946): 285–301.
3. R.Y. Stanier and C.B. van Niel, "The Concept of a Bacterium," *Archiv fur Mikrobiologie* 42 (1962): 17–35.
4. E. Zuckerkandl and L. Pauling, "Evolutionary Divergence and Convergence in Proteins," in V. Bryson and H.J. Vogel, eds., *Evolving Genes and Proteins* (New York: Academic Press, 1965), 97–166.
5. C.R. Woese, "Bacterial Evolution," *Microbiology Reviews* 51 (1987): 221–271.
6. N.R. Pace, "A Molecular View of Microbial Diversity and the Biosphere," *Science* 276 (1997): 734–740.
7. C.R. Woese, "On the Evolution of Cells," *Proceedings of the National Academy of Sciences USA* 99 (2002): 8742–8747.
8. M.A. Ragan, "On Surrogate Methods of Detecting Lateral Gene Transfer," *FEMS Microbiology Letters* 201 (2001): 187–191.
9. R.A. Welch, V. Burland, G. Plunkett 3rd, P. Redford, P. Roesch, D. Rasko, E.L. Buckles, S.R. Liou, A. Boutin, J. Hackett, D. Stroud, G.F. Mayhew, D.J. Rose, S. Zhou, D.C. Schwartz, N.T. Perna, H.L. Mobley, M.S. Donnenberg, and F.R. Blattner, "Extensive Mosaic Structure Revealed by the Complete Genome Sequence of Uropathogenic *Escherichia coli*," *Proceedings of the National Academy of Sciences USA* 99 (2002): 17020–17024.
10. R. Lan and P.R. Reeves, "Intraspecies Variation in Bacterial Genomes: The Need for a Species Genome Concept," *Trends in Microbiology* 8 (2000): 396–401.
11. C.L. Nesbø, K.E. Nelson, and W.F. Doolittle, "Suppressive Subtractive Hybridization Detects Extensive Genomic Diversity in *Thermotoga maritima*," *Journal of Bacteriology* 184 (2002): 4475–4488.
12. O. Beja, E.V. Koonin, L. Aravind, L.T. Taylor, H. Seitz, J.L. Stein, D.C. Bensen, R.A. Feldman, R.V. Swanson, and E.F. DeLong, "Comparative Genomic Analysis of Archaeal Genotypic Variants in a Single Population and in Two Oceanic Provinces," *Applied and Environmental Microbiology* 68 (2002): 335–345.
13. E.J. Feil and B.G. Spratt, "Recombination and Population Structures of Bacterial Pathogens," *Annual Review of Microbiology* 55 (2001): 561–590.
14. D.E. Dykhuizen and L. Green, "Recombination in *Escherichia coli* and the Definition of Biological Species," *Journal of Bacteriology* 173 (1991): 7257–7268.
15. J. Majewski and F.M. Cohan, "DNA Sequence Similarity Requirements for Interspecific Recombination in Bacillus," *Genetics* 153 (1999): 1525–1533.
16. Y. Boucher and W.F. Doolittle, "The Role of Lateral Gene Transfer in the Evolution of Isoprenoid Biosynthesis Pathways," *Molecular Microbiology* 37 (2000): 703–716.
17. C. Brochier, H. Phillippe, and D. Moreira, "The Evolutionary History of Ribosomal Protein RpS14," *Trends in Genetics* 16 (2000): 529–533.
18. C. Nesbø, Y. Boucher, and W.F. Doolittle, "Defining the Core of Nontransferable Prokaryotic Genes: The Euryarchaeal Core," *Journal of Molecular Evolution* 53 (2001): 340–350.
19. J. Raymond, O. Zhaxybayeva, J.P. Gogarten, S.Y. Gerdes, and R.E. Blankenship, "Whole-genome Analysis of Photosynthetic Prokaryotes," *Science* 298 (2002): 1616–1620.
20. S.A. Teichman and G. Mitchison, "Is There a Phylogenetic Signal in Prokaryote Proteins?" *Journal of Molecular Evolution* 49 (1999): 98–107.
21. J.R. Brown and W.F. Doolittle, "Archaea and the Prokaryote to Eukaryote Transition," *Microbiology and Molecular Biology Reviews* 61 (1997): 456–502.
22. C. Brochier, E. Bapteste, D. Moreira, and H. Philippe, "Eubacterial Phylogeny Based on Translational Apparatus Proteins," *Trends in Genetics* 18 (2002): 1–5.
23. O. Matte-Tailliez, C. Brochier, P. Forterre, and H. Philippe, "Archaeal Phylogeny Based on Ribosomal Proteins," *Molecular Biology and Evolution* 19 (2002): 631–639.
24. V. Daubin, N.A. Moran, and H. Ochman, "Phylogenetics and the Cohesion of Bacterial Genomes," *Science* 301 (2003): 829–832.

25. E. Lerat, V. Daubin, and N.A. Moran, "From Gene Trees to Organismal Phylogeny in Prokaryotes: The Case of the Gamma-Proteobacteria," *PloS Biology* 1 (2003): E19.
26. R.C. Jain, M.C. Rivera, and J.A. Lake, "Horizontal Gene Transfer Among Genomes: The Complexity Hypothesis," *Proceedings of the National Academy of Sciences USA* 96 (1999): 3801–3806.
27. T. Asai, D. Zaporojets, C. Squires, and C.L. Squires, "An *Escherichia coli* Strain with all Chromosomal rRNA Operons Inactivated: Complete Exchange of rRNA Genes between Bacteria," *Proceedings of the National Academy of Sciences USA* 96 (1999): 1971–1976.
28. C.G. Kurland, B. Canback, and O.G. Berg, "Horizontal Gene Transfer: A Critical View," *Proceedings of the National Academy of Sciences USA* 100 (2003): 9658–9662.
29. G.J. Olsen, "The History of Life," *Nature Genetics* 28 (2001): 197–198.
30. J.P. Gogarten, W.F. Doolittle, and J.G. Lawrence, "Prokaryotic Evolution in Light of Gene Transfer," *Molecular Biology and Evolution* 19 (2002): 2226–2238.
31. R.L. Charlebois, R.G. Beiko, and M.A. Ragan, "Microbial Phylogenomics: Branching Out," *Nature* 421 (2003): 217.

6

Woe Is the Tree of Life

WILLIAM MARTIN

In an ideal scientific world, where everything is simple and straightforward, the analysis of genome sequences would have fully uncovered the basic backbone of life's history by now. As it stands, however, the only thing that is certain about microbial genome sequencing is that it can be done. Many biologists imagined that a golden age of molecular evolution would emerge from genomics—an era of genome phylogenies in which the position of all organisms great and small was fully resolved in a unified tree of evolutionary history. In a perfect world, the genes of all genomes would be related by one and the same bifurcating process, and combining all these genes into one grand alignment would produce the ultimate tree, biology's key to the past, a genome-enabled-time machine. That tree would have resolved all the branches and issues about which evolutionary biologists and systematists had ever quibbled. It would have put all organisms with a sequenced genome in their proper place in the larger scheme of things and would have allowed biologists to go about the enjoyable business of mapping out the evolution of morphological and biochemical characters among those lineages.

But genomes have not uncovered a perfect world. They have uncovered abundant evidence for lateral gene transfer (LGT) among prokaryotes, and they have uncovered abundant evidence for chimaerism in eukaryotes.[1–3] Eukaryotes possess a mixture of genes, some of which clearly reflect a eubacterial ancestry, and some of which clearly reflect an archaebacterial ancestry. Various eubacteria and archaebacteria also possess mixtures of genes that they have acquired and passed on both to their progeny and to various casual acquaintances from distant prokaryotic taxa via horizontal gene transfer. All genomes studied to date also contain many genes that lack easily identifiable homologues among other lineages; these might be lineage-specific gene inventions, fast-evolving genes that have simply lost the trace of their origin, or both.

Putting specific numbers on the amount of LGT that has occurred in the evolution of individual prokarytic genomes is no easy matter. Case studies indicate that the fraction of horizontally transferred genes in genomes is substantial, with estimates reaching up to 30% in some cases, or even more.[1–7] At the same time, phylogeneticists are warning that many claims for horizontal gene transfer may largely reflect our inability to properly reconstruct the evolution of genes, our poor sampling of critical lineages, or both.[1–10] Notwithstanding the difficulties of properly quantifying it, LGT does exist.

Although LGT does not dash all hopes of piecing together life's early history, it does make the puzzle of cell evolution much more difficult to reconstruct from the standpoint of genomes, but that is not completely bad news. It is good news for biologists, because it makes it all the more important to look for independent evidence for life's history; for example, in the geological record in the form of fossils or isotopic evidence. Among the eukaryotes, it also prompts the search to identify major events in cell and biochemical evolution that can help to define assemblages of related cell lineages (and their ancestor cell lineages).

What is Wrong with the Ribosomal RNA Tree?

Those interested in microbial evolution would like to link up various kinds of mutually consistent evidence for early cell history in such a way as to present a general schematic view of life's microbial history. Many biologists think that trees of small subunit ribosomal RNA (rRNA), which are sometimes called the tree of life (sometimes even called the "Tree of Life," capitalized as if it warrants religious reverence) already do provide a general schematic view of life's unicellular history, but the rRNA tree, in whatever form, is a result produced by one or the other tree-building algorithm as programmed into a computer, and as such, it is just as error prone as every other gene tree. Some scientists, however, see that matter differently. Woese, for example, states, "The universal phylogenetic tree based on rRNA is a valid representation of organismal genealogy. But it is unlike any other phylogenetic tree" (p. 8395).[11] That view implies that there is something special about the rRNA tree that lends it some kind of infallability, as suggested eleswhere: "The universal phylogenetic tree in one sense brought classical evolution to culmination" (p. 8742).[12] However, some biologists are unconvinced that the rRNA tree culminates classical evolution. Until methods of phylogenetic inference and our understanding of gene evolution become perfect, which is very unlikely ever to occur, there is no cause to regard the rRNA tree as either "valid" or "unlike any other tree." The rRNA tree depicts, within its limits of resolution, the evolution of only one gene, and it therefore cannot depict the evolution of whole genomes, with genes coming and going via LGT and cell lineages merging via endosymbiosis.

In particular, because the rRNA tree is a strictly bifurcating one, there is no way that it can depict the evolution of eukaryotes in the overall scheme of things properly. This is because the evolution of eukaryotes entails the symbiotic origins of mitochondria and of plastids. In endosymbiosis, two distinct branches in the tree of cell lineages unite into a fundamentally novel, bipartite, and internally compartmentalized cell lineage. Endosymbiosis involves the origin of novel taxa at higher level via the combinatorial union of cell lineages. Many biologists still have problems with this fact, because it is extremely non-Darwinian.

Nothing could possibly be less Darwinian than creating novel taxa at higher levels by combining two different and highly divergent cells into one. Darwin discovered the most important principles of evolution as it occurs in organisms visible to the naked eye, natural variation among offspring to generate diversity and natural selection to prune among variants and to shape that diversity over time, but during the endosymbiotic origins of chloroplasts and mitochondria, one full-fledged (unicellular) organism got inside the cytoplasm of another one, thereby giving rise to a fundamentally new kind of organism that previously did not exist. The product of endosymbiosis is a viable bitartite cell that gives rise to progeny and that some time after the merger returns to a standard fundamental mode of Darwinian evolution via natural variation and natural selection. The compartmentation of two divergent and initially autonomously replicating genomes that arises at the onset of endosymbiosis (the genome of the symbiont and the genome of the host) gives rise to a new mechanism of natural variation that does not exist in organisms that lack an endosymbiont: Gene transfers from the chromosomes of the endosymbiont to those of the host result in the amalgamation of two genetic systems into a new, highly chimeric, and compartmentalized genetic entity.

Because mitochondria and chloroplasts arose from free-living eubacteria, they each brought along a full genome's worth of prokaryotic gene diversity to their host. As long as there was more than one endosymbiont per cell during the initial stages of these symbioses, occasional lysis of an endosymbiont would have released a full eubacterial genome's worth of genes into the cytosol of their host, ready for (illegitimate) recombination with the host's chromosomes, thereby providing a rich and virtually inexhaustible source of genetic starting material that nonsymbiogenic organisms do not possess. Furthermore, before the invention of a protein import apparatus for mitochondria and chloroplasts, respectively, gene transfers from organelles to their host simply increased the genetic variability of the host without deterring from the genetic autonomy of the organelle. In this way, endosymbiosis (more specifically, endosymbiotic gene transfer[1]) affords a specific and powerful mechanism of natural variation that spawns unique opportunities for the invention of biological novelty among unicellular organisms.

Endosymbiosis is rare, but very real, in eukaryotic evolution. The origin of chloroplasts from cyanobacteria more than 1.2 billion years ago founded the lineages that contain all plants and algae. It also appears that the origin of mitochondria founded the eukaryotic lineage that we know today, because all major lineages of eukaryotes have now been found to contain a mitochondrion of some sort.[10] Those two events—the origin of mitochondria and that of chloroplasts—are, from the standpoint of endosymbiotic theory, the mechanistic basis for the differences that distinguish eukaryotes from prokaryotes.

It is a very unfortunate circumstance for evolutionary biology that Darwin did not discover the principle of endosymbiosis (Constantin Mereschkowsky did[13]). Had Darwin discovered it, endosymbiosis would have a normal status in evolutionary thinking, which it does not. As Woese put it: "Because endosymbiosis has given rise to the chloroplast and mitochondrion, what else could it have done in the more remote past? Biologists have long toyed with an endosymbiotic (or cellular fusion) origin for the eukaryotic nucleus, and even for the entire eukaryotic cell" (p. 8742).[12] Biologists have also long toyed with the idea that the rRNA tree is a valid representation of organismal genealogy and is unlike any other tree. Endosymbiosis is a thorn in the side of systematists who wish to put all organisms onto their proper branch in a unified, bifurcating tree (of life), because on a time line or on a tree, endosymbiosis is bifurcation backwards, and endosymbiosis is real.

LGT (which Darwin also did not discover, because he had no mechanisms for natural variation—he just observed that natural variation exists) poses much the same sort of a problem as endosymbiosis, but differs in the details. In endosymbiosis, two branches of a bifurcating tree unite outright into a new, more complex lineage. In LGT, little pieces of one lineage hop across lineage boundaries and take up residence in a new lineage. In contrast to endosymbiosis, which is unspeakably rare (it apparently happened only twice involving prokaryotes in the last 4 billion years), LGT is very common among prokaryotes (it happens all the time).

The point here is that the two processes that foul up things for the tree of life the most are non-Darwinian processes: endosymbiosis and LGT. They are non-Darwinian for the simple reason that Darwin did not discover them, probably because technology was not ripe and because he was not so interested in microbes. Darwin did not envisage lineages of highly distinct cells entering into a situation where one cell lived stably within another highly disparate one, thereby giving rise to novel taxa at higher levels via lineage combination (endosymbiosis). Nor did he envisage a mode of evolution in which natural variation includes the donation of natural variation from one lineage to a different, distantly related one (LGT). Nonetheless, today we have a widely (but not universally) accepted classification scheme for unicellular organisms, which are placed into three domains as defined by their ribosomal RNA sequences. Note that because of the way that prokaryotic taxa are currently defined under rRNA systematics, any gene linked to an α-proteobacterial rRNA sequence is, by definition, an α-proteobacterial gene. Therfore, until we have sequenced all α-proteobacterial genomes, we will not know which genes are α-proteobacterial. This is important when it comes to identifying genes of α-proteobacterial origin in eukaryotic genomes.

By no means do I question the fundamental utility of rRNA sequences as a taxonomic character, but the dogma of the rRNA tree has reached a point at which its proponents either no longer see a need to put species names among the eukaryotic lineages,[11,12] lest their position in the tree be known to be incorrect on the basis of other data, with the microsporidia being the prime example,[10] or deny any biological relevance of LGT among prokaryotes[14] in the face of evidence that tells us otherwise.[1]

The classification scheme of cells that is manifest in the rRNA tree will be unaffected by this chapter, but biologists who take endosymbiosis in cell evolution seriously cannot work within the framework of the (strictly bifurcating) rRNA tree. Thus, it seems to me that there is a schisma abrew in cell evolution, with the rRNA tree and proponents of its infallibility on the one side and other forms of evidence, proponents of LGT, or proponents of a symbiogenetic origin of eukaryotes on the other. The former camp is well organized behind a unified view (be it right or be it wrong, still a view) and is arguing that we already have the answers to microbial evolution. The latter camp is not organized into castes of recognized leadership and followers, meaning that (if we are lucky) concepts and their merits, not position or power, will determine the outcome of the battle as to what ideas might or might not be worthwhile entertaining as a working hypothesis for the purpose of further scientific endeavour.

What Sorts of Evidence Are There Besides the rRNA Tree?

The geochemical evidence for life's early history is all too often overlooked by gene tree specialists, and the rRNA tree does not, by itself, yield any direct insight into biochemical evolution, but there is abundant evidence in the geochemical record for early evolution.[15] The Earth is 4.5 billion years old, and the ocean had condensed by ~4.4 Ga (billion years before present).[16] Life arose on Earth by ~3.8 Ga, because carbon isotope data provide evidence for biological CO_2-fixation in sedimentary rocks of that age.[17–19] Recent criticism has been launched at some of the carbon isotope data from sediments at 3.8 Ga,[20] but microprobe studies of those materials are still accepted as indicating biological CO_2 fixation at 3.8 Ga.[15,19] Microbial communities at hydrothermal vents existed by at least 3.2 Ga.[21]

By about 2.7 Ga, prokaryotic communities were beginning to look very similar to many modern prokaryotic communities in terms of carbon and sulfur cycles.[17,21–24] This includes the isotopic trace of methanogenesis and methanotrophy by 2.7 Ga, indicating that both methanogens (archaebacteria) and methanotrophs (α-proteobacteria, but possibly also including other groups) were present at that time.[24,25] Stromatolites, which are preserved microbial mats deposited by photosynthetic prokaryotes, were present as early as 3.5 Ga and have a more or less continuous record to the present.[26]

Such evidence indicates that most of the biochemical pathways that drive modern prokaryotic carbon, sulfur, and nitrogen cycles were in place by as early as 3.5 Ga, and by 2.7 Ga at the latest.[17,24] Accordingly, it seems reasonable to assume that major lineages of eubacteria and archaebacteria were present and well diversified by that time.

The Earth's early atmosphere contained either no O_2 at all or only very minor trace amounts. Today's O_2 stems from oxygenic photosynthesis in cyanobacteria and plastids.[27] The time at which O_2 production in the oceans began—that is, the time of the origin of oxygenic photosynthesis—is uncertain. The abundance of ultralight organic carbon bearing the isotope signature of methanotrophy (an oxygen-dependent pathway in eubacteria) and the study of microbial communities strongly indicates that oxygen was available at least as early as 2.7 Ga,[28] consistent with evidence from cyanobacterial biomarkers at 2.7 Ga.[29] Evidence from carbon cycles indicates that global oxygen production has been constant within an order of magnitude over the last 3.5 Ga.[17] Overall, it seems safe to surmise that oxygen production in the oceans (and hence the origin of cyanobacteria) occurred at least by 2.7 and possibly as early as 3.5 Ga. It also seems safe to assume that as soon as oxygen was available in the oceans, prokaryotes immediately discovered ways to use its power as an electron acceptor.

Various lines of geochemical evidence indicate that oxygen did not start accumulating in the atmosphere until about 2 billion years ago, at which point atmospheric O_2 rose sharply from <1% of present atmospheric levels (PAL) to about 15% PAL during a small window of time from 2.2 to 2.1 Ga.[17,30–33] The sulfur isotope record and carbon deposition rates indicate that a second sharp rise in atmospheric O_2 approaching present levels occurred around ~0.6 Ga,[34] but during the time from 2.2 to ~0.6 Ga, when atmospheric oxygen levels were about 15% PAL, deep ocean water was, according to newer findings, still anoxic and furthermore highly sulfidic.[35,36] That is, it contained no oxygen and high levels of sulfide as $HS^–/H_2S$. The evidence for this stems from stable sulfur isotope studies that reveal high activities of marine biological sulfate reduction, which produces sulfide, during that time.[36]

Taken together, that evidence indicates that cyanobacteria existed by at least 2.7 Ga; that there was little oxygen in the atmosphere or ocean before 2.2 Ga; that between 2.1 Ga and 0.6 Ga there was roughly 15% PAL O_2 in the atmosphere but none in deep ocean water, which was furthermore rich in sulfide; and that at ~0.6 Ga O_2 levels in the atmosphere and deep ocean water approached present levels. That means that the eukaryotic lineage, which arose well before 0.6 Ga, underwent the brunt of its diversification in a largely anoxic and sulfidic world.

Eukaryotes Are Younger than Prokaryotes

There is no consensus among biologists concerning the position of the eukaryotes in the overall scheme of cell evolution.[1] Current opinions on the origin and position of eukaryotes span a broad spectrum including the views that eukaryotes arose first in evolution and that prokaryotes descend from them,[37] that eukaryotes arose contemporaneously with eubacteria and archaebacteria and hence represent a primary line of descent of equal age and rank as the prokaryotes,[12] that eukaryotes arose through a symbiotic event entailing an endosymbiotic origin of the nucleus,[38–41] that eukaryotes arose without endosymbiosis,[42] and that eukaryotes arose through a symbiotic event entailing a simultaneous endosymbiotic origin of the flagellum and the nucleus,[43] in addition to many other models, which have been reviewed and summarized elsewhere.[44]

Under the view of autotrophic origins of life, the heterotrophic lifestyle of microbes had to arise later than the autotrophic lifestyle, because without preexisting autotrophs to produce ample reduced organic compounds, heterotrophs cannot survive.[45–50] All eukaryotes are ancestrally heterotrophs, they gain their energy through the oxidative breakdown of reduced carbon compounds (e.g., carbohydrates) that they obtain from the environment. Thus, under any scheme of cell evolution embracing autotrophic origins, eukaryotes have to postdate prokaryotes in origin,[47,50] but postdate by how much? If prokaryotes arose by at least 3.5 Ga, then what does the geological record say about the age of eukaryotes?

By about 1.5 Ga, acritarchs, fossil unicellular organisms that are almost certainly eukaryotes[51] and probably algae by virtue of an easily preserved cell wall, become reasonably abundant. By 1.2 Ga, very well preserved multicellular red algae appear.[52] Evidence of this type is accepted by palaeontologists[15] as indicating that eukaryotes are at least 1.5 billion years old and that the diversification of the red algal lineage (which is not the most ancient lineage of algae) into multicellular forms occurred at least 1.2 billion years ago.

There have been reports of more ancient remains claimed to be eukaryotes, but they are often viewed with skepticism.[42,47] For example, the filamentous fossil *Grypania* occurs at 2.1 Ga,[53] but it could just as easily be a filamentous prokaryote as a filamentous eukaryote, because the cellular structure of the material is not preserved. This is in contrast to *Bangiomorpha* at 1.2 Ga,[52] the large-celled, truly multicellular structure of which is strikingly preserved. Steranes were recently found in 2.7-Ga sediments, and it was claimed that these biomarkers provide evidence for the existence of eukaryotes at that time,[54] but several groups of prokaryotes including methanotrophic α-proteobacteria,[55] myxobacteria,[56] and cyanobacteria,[57] which make the same kinds of compounds (e.g., cholesterol), claimed to be eukaryote-specific, such that the sterane evidence appears to document biochemically diverse prokaryotes, rather than the existence of eukaryotes.

Thus, there is convincing evidence that eukaryotes (probably algae) were in existence by 1.5 Ga and that multicellular red algae existed by 1.2 Ga. Because the origin of algae entails the origin of plastids from cyanobacteria, and because the host that acquired plastids possessed mitochondria, the origin of mitochondria should be sought well before 1.2 Ga and somewhere before 1.5 Ga.

Origins of Mitochondria and Eukaryotes in Anaerobic Times

The evidence summarized above indicates that deep ocean water was anoxic and sulfidic up until about 0.6 Ga and that the origin of mitochondria dates back to at least 1.5 Ga. Therefore, mitochondria must have arisen in a global setting in which marine oxygen levels were extremely low and sulfide levels were high. Furthermore, the first ~1 billion years (at least) of eukaryote diversification occurred in a marine environment marked by low oxygen, widespread anoxia, and high sulfide. It is, therefore, not surprising that many eukaryotes still thrive today in anaerobic environments, some of which, such as marine sediments, are also sulfide rich.[58,59] On the basis of their ATP-synthesizing pathways, modern anaerobic eukaryotes can be divided into three unnatural groups: those that possess anaerobically functioning mitochondria,[60] the so-called type II eukaryotes that synthesize ATP in hydrogenosomes, and the so-called type I eukaryotes that possess neither typical mitochondria nor hydrogenosomes and that synthesize all of their ATP in the cytosol.[61–63]

It was once thought that parasitic eukaryotes such as the microsporidians[64] or type I eukaryotes such as the diplomonad *Giardia lamblia*,[65] which gain ATP without the help of mitochondria or hydrogenosomes, might be the most ancient among contemporary groups and that they might have never possessed a mitochondrion at all, but starting about 1995, numerous studies revealed eukaryotes that lack mitochondria to have possessed a mitochondrion in their evolutionary past,[66–68] or even to still possess a long-overlooked, highly reduced remnant mitochondrion with no apparent function in ATP synthesis called a mitosome.[69,70] Accordingly, it seems that mitochondria are as ancient as eukaryotes themselves and that the loss of mitochondria has occurred many times independently in various eukaryotic lineages.[10,63,66–70] Hydrogenosomes (the double-membrane bounded, H_2-producing, and ATP-producing organelles of various anaerobic eukaryotes[10,44,58,61,71,72]) figure prominently in understanding early eukaryotic history. Hydrogenosomes are specifically suited to eukaryotic life in anaerobic environments, and they harbor many O_2-sensitive enzymes such as pyruvate:ferredoxin oxidoreductase, [Fe]-hydrogenase, and pyruvate-formate lyase.[10,71–75] Hydrogenosomes occur in at least four highly disparate groups of eukaryotes—trichomonads, ciliates, amoeboflagellates, and chytridiomycete fungi[58]—and are now known to be anaerobic forms of mitochondria.[10,44,58,66] The evolutionary significance of hydrogenosomes is evident: They bridge the gap between ATP synthesis in aerobic and anaerobic eukaryotes, because they contain enzymes common both to mitochondria and to cytosolic ATP synthesis in type I eukaryotes.[62,63] Hydrogenosomes forge a biochemical link between the largely anaerobic ancient phases of eukaryotic history and the more recent past (the last 600 million years), during which time aerobic niches have become more widespread and anaerobic environments (e.g., sediments) have become more restricted. A model that specifically accounts for the common origin of mitochondria and hydrogenosomes from a single (facultatively anaerobic) eubacterial ancestor, that specifically predicts no eukaryote to be primarily amitochondriate, that specifically accounts for the origin of heterotrophy in eukaryotes, and that specifically accounts for anaerobic mitochondria has been presented elsewhere, and in sufficient detail as to account for an endogenous origin of the nucleus subsequent to the origin of mitochondria.[47,60,63] Alternative models for the origins of eukaryotes mentioned above do not directly account for hydrogenosomes and are designed to account for other things.

Eukaryote Phylogeny: A Tree Turned Upside Down

Traditional views of eukaryote phylogeny are based in the classical rRNA tree, which depicts various anaerobic and amitochondriate eukaryotes branching deeply and the animals, fungi, and plants emerging as the latest lineages of eukaryotic evolution, but that view is now outdated. Newer investigations of many genes (rather than just a single gene) are uncovering evidence for the existence of a relatively small number of major eukaryotic lineages. These include well-recognized groups such as animals, fungi, and plants with primary plastids, but also including new and surprising groups, sometimes with unfamiliar names such as excavates, amoebozoa, opisthokonts, chromalveolates, and the like.[42,66,67,76–81]

Those are exciting developments, but perhaps more important in the overall scheme of things than the sorting out of "who belongs where" in terms of groupings is the position of the root in the eukaryotic tree; that is, the question of which lineages of eukaryotes might be the oldest. Because of the way that phylogeny algorithms work, the rRNA tree seems to have consistently produced a severe artefact with regard to the placement of the root. This is mainly because when eukaryote rRNA sequences are linked up to prokaryote rRNA sequences in the same tree, the outgroup (prokaryote) branch will tend to fall among the longest eukaryote branches, regardless of whether those long-branched sequences are the most ancient or whether they are simply the most different.[66,67]

New evidence from the study of a particular gene fusion involving dihydrofolate reductase and thymidylate kinase that is found only among some eukaryotes has strongly suggested that the root in the eukaryotic tree lies on or very near the branch that separates animals and fungi from all other eukaryotes.[78] This rooting is highly compatible with the new handful of perhaps six eukaryotic "supergroups" that are currently emerging from multigene phylogenies.[79]

Osmotrophy and Prototrophy: Fungi Have Ancient Traits

From the standpoint of energy metabolism, and beyond the strength of the gene fusion data itself, the rooting of Stechmann and Cavalier-Smith,[78] or "opisthokont root," is very attractive, because it would implicate the fungi as one of the most ancient eukaryotic lineages ("opisthokonts" is a term coined by Cavalier-Smith to designate the group comprising animals and fungi on the basis of locomotion in unicellular stages). Compared with prokaryotes, eukaryotes have only a miniscule diversity of core energy metabolic pathways for sustained ATP-synthesis, but on the basis of available data, it seems that fungi have the broadest energy metabolic (physiological) diversity of any eukaryotic group. The fungi encompass many species with typical aerobic mitochondria, species with anaerobic mitochondria that can perform nitrite respiration,[82] species with hydrogenosomes,[58,75,77,83,84] species that can perform a hitherto unique feat among eukaryotes called ammonia fermentation,[85] groups with extremely reduced mitochondria,[70] and groups that perform methylotrophy; that is, they can live from methanol as their sole carbon and energy source,[86] something no other eukaryotes to the author's knowledge can.

Furthermore, the fungi as a group are osmotrophs, not phagotrophs. They take up their nourishment with the help of membrane-localized importers, just like phagotrophs do, but they do not phagocytose large particles as food vacuoles. The digestion enzymes that phagocytotic eukaryotes excrete into food vacuoles, fungi excrete into their environment. The importers that phagocytotic eukaryotes use to import digest from food vacuoles reside on the plasma membrane in fungi. It is conceivable that the fungi as a group could have diverged from the main stem of eukaryotic evolution before proper phagocytosis had evolved. That notion is not likely to become popular, because most biologists still tend to lean firmly toward the view that phagocytosis was a prerequisite for the origin of mitochondria, a view that is, however, founded more in tradition than in evidence. Examples of prokaryotic endosymbionts that live within prokaryotic hosts incapable of proper phagocytosis are known.[87] By analogy, the origin of mitochondria need not have absolutely demanded phagocytosis of its host.

The view that osmotrophy had to precede phagotrophy in eukaryotic evolution is compelling because without importers, food vacuoles are useless. That all fungi are osmotrophs and that none are phagotrophs could mean that their common ancestor was either primitively or secondarily nonphagotrophic. This leads to the subtle question of how eukaryotes became osmotrophs in the first place. Osmotrophy requires substrate importers at the plasma membrane and a cytosolic carbon metabolism suited to the heterotrophic lifestyle (ATP synthesis through the oxidation of reduced organic compounds). In yeast, heterotrophy entails eubacterial importers in the plasma membrane, eubacterial carbon metabolism in the cytosol, and a eubacterial organelle.[88,89] This observation at its level of resolution is generally compatible with the predictions that stem directly from three current models for the origin of eukaryotes: first, the host that acquired the mitochondrion was a member of the actinobacteria (a group of gram-positive eubacteria that includes actinomycetes) that had become a phagotrophic eukaryote.[42] Second, eukaryotes arose through symbiosis in which a methanogen became the nucleus in a δ-proteobacterial host.[40] Third, the host that acquired the mitochondrion was an autotrophic archaebacterium that acquired, through endosymbiotic gene transfer, the prexisting heterotrophic lifestyle of its α-proteobacterial symbiont.[63] Notwithstanding LGT,[1] discriminating genome analyses are needed to see whether eukaryotic genes, particularly those involved in osmotrophy, share more similarity with their homologs distributed among actinobacteria,[42] δ-proteobacteria,[40] or a-proteobacteria.[63]

The classical endosymbiont hypothesis, as formulated by both Doolittle and Cavalier-Smith, has always assumed that the host cell that acquired the mitochondrion was phagotrophic.[42,90–92] Protistan phagotrophs are predators. They hunt down or entrap and then engulf food particles or (more specifically) microbial cells. Many euglenids provide an excellent example of phagotrophy (see videos at http://bio.rutgers.edu/euglena/). *Euglena gracilis*, for example, is an autotroph (it can satisfy its carbon needs from CO_2 alone), but it is not a complete autotroph, because it needs to have a few vitamins—particularly cobalamin (vitamin B_{12}) because of its unusual cobalamin-dependent ribonucleotide reductase[93]—added to its medium to survive in axenic laboratory cultures. Nonetheless, *Euglena* can synthesize all twenty amino acids. Humans have a similar but more severe problem. We cannot synthesize most B-vitamins at all, and we can only synthesize about half of the twenty amino acids (the nonessential ones). We have to obtain compounds like thiamine, cobalamine, phenylalanine, tryptophane, and so forth from the food that we eat through our predatory lifestyle.

Fungi, as a rule, do not have that problem. Fungi are, as a rule, prototrophs, although many species and strains exhibit some vitamin requirements for growth. A long list of selected ascomycetes and basidiomycetes that can synthesize all the vitamins they need to survive from reduced carbon and ammonium salts has been compiled.[94]

What came first among eukaryotes, prototrophy or phagotrophy? I think the answer is that there is just no way to become a phagotroph without being a prototroph first. To become a phagotroph, one has to have a cytoskeleton, food vacuoles, and machinery for utilizing the content of food vacuoles, but one cannot evolve those attributes (from a prokaryotic state) over some indeterminate period of time in the absence of biochemical viability (prototrophy). Once a cell has become a phagotroph, it can loose some biochemical pathways for amino acid and vitamin biosynthesis, because it can obtain those cofactors from other cells that it eats as prey (as long as the prey contains the needed nutrients; e.g., higher plants do not synthesize of need cobalamin, for which reason vegetarians have to watch out about B_{12} sources in their diet). This kind of biochemically reductive evolution has probably occurred in metazoan evolution; for example, in the lineages leading to humans. Fungi apparently never became phagotrophs, and my prediction is that they never stopped being prototrophs, either. That is, when we have good genome sampling for many microbial lineages, I predict that we will observe, by and large, that a common set of amino acid biosynthetic enzymes and cofactor biosynthetic pathways was present in the last common ancestor of all eukaryotes. (Cobalamin is a bit of a puzzle though, because not all groups of eukaryotes need it, and it seems that, most fungi do not need it at all.) However, this prediction rests on the premise that both the host that acquired the mitochondrion and the mitochondrial (hydrogenosomal) endosymbiont itself were capable of synthesizing all their cofactors and amino acids by themselves;[63] hence, there should have been at least twofold redundancy (eubacterial = symbiont and archaebacterial = host) for the brunt of such pathways in the initial mitochondrion-bearing cell, and some differential loss in the process of lineage sorting in the early phases of eukaryotic evolution is to be expected.

Regardless of how eukaryotes arose and which group ultimately turns out to be the most ancient, available evidence indicating that both the presence of a nucleus and the presence of a mitochondrial endosymbiont, which in some cases may be highly reduced, are defining features of eukaryotes.[42,76] Furthermore, available evidence indicates that mitochondria arose only once in evolution.[76,95,96] In view of the hefty number of unicellular organisms that have ever lived, the origin of mitochondria was an unspeakably rare event.

Molecular phylogeneticists have not yet looked carefully at the distribution of genes for enzymes of amino acid and vitamin biosynthetic pathways among eukaryotes, nor has much attention been given to the phylogeny of proteins involved in the phagotrophic lifestyle. One possibility is that the initial eukaryotic stem possessed a mitochondrion (hydrogenosome),[63] did not possess a nucleus;[47,97] was a heterotrophic osmotroph—not a phagotroph,[98] and was furthermore a prototroph lacking any auxotrophies whatsoever (able to live from reduced carbon and reduced nitrogen plus trace elements). These ideas are, in principle, testable with the help of genome sequence data. The rRNA tree has been slow to generate any similarly explicit predictions about the biochemical or physiological nature of early eukaryotic cells. One aspect of life-in-the-wild biology that the rRNA tree has directly addressed is the GC content and inferred growth temperature that early eukaryotes might have preferred,[99] an analysis in which eukaryotes appeared as emerging from within the archaebacteria as Lake[100] has suggested, rather than as sisters to the archaebacteria per se as the rRNA tree would have us believe.

From that standpoint of microbial physiology, the first eukaryote might have been very similar to an inferable common ancestor of contemporary fungi. Because of the bizarre diversity of life cycles, sex, and changes of generation among the various groups of fungi, which are essentially aquatic and predominantly haploid,[101,102] it is very hard to infer what sort of relationship between meiosis and mitosis might have existed in that organism. Regarding today's fungi, Raper summarizes: "Seven types of life cycles, seven distinct patterns of sexuality, and about a dozen or more basic kinds of sexual histories allow, in combination, a bewildering array of distinct sexual types" (pp. 503–504).[102] Nonetheless, the common ancestor of all eukaryotes probably had standard eukaryotic flagella, because these are present among so many groups, as well as in the zoospores of the chitridiomycetes. Curiously, Mereschkowsky[13] viewed fungi as direct descendants of prokaryotes, as having arisen independently from the other eukaryotes, not as their symbiogenic sisters, and his evolutionary inferences were based largely in physiology.

High Sulfide Up to 600 Million Years Ago

In a recent and important review, Anbar and Knoll[36] pointed out one possible consequence that the evidence for anoxic and sulfidic oceans would have had on eukaryotic diversity. Their case was that high sulfide up until ~0.6 Ga would have kept marine concentrations of certain metal ions such as iron and molybdenum very low. This, in turn (so goes the argument), could ultimately limit algal diversification by hampering prokaryotic nitrogen fixation, which requires these metals to operate. By this reasoning, one could account for low levels of eukaryotic microfossil diversity observed before ~1 Ga. Although the argument of Anbar and Knoll[36] is not fundamentally flawed, it misses an important point.

It is true that iron can limit cyanobacterial biomass in the oceans,[103] but from the standpoint of eukaryotic microbial physiology, the main consequence of anoxic and sulfidic oceans would not have been the gradual problem of dealing with low nitrogen availability (slow starvation). Rather, eukaryotes would have seen themselves faced with the immediate and (for aerobes) life-threatening problem of dealing with recurrent or permanent anoxia and sulfide (asphyxia and poisoning) on a daily basis for about a billion years, and maybe even more. The major consequences from this simple consideration are threefold.

First, in that anoxic world, anaerobic energy metabolism in mitochondrion-containing cells would have been a prerequisite for survival—an absolute must, a *conditio sine qua non*. The consequence is that ancestral eukaryotes must have possessed enzymes for sustained ATP synthesis under anoxic conditions. Accordingly, it would hardly be surprising to find a trace of that ancestral anaerobic energy metabolism in mitochondrion- or hydrogenosome-bearing cells today, particularly in such lineages as inhabit anaerobic niches. Indeed, such anaerobic biochemistry is abundant among eukaryotes.[58–63] My argument has been—and remains—that the fabric of that anaerobic biochemistry almost certainly represents a holdover from the ancestral eukaryotic state (facultatively anaerobic; in this specific case, possessing a respiratory chain and capable of anaerobic fermentations).[44,47,60,63] Importantly, the presence of anaerobic energy metabolism in ancestral mitochondria in no way excludes the presence of an additional respiratory chain. The hydrogen hypothesis is often misunderstood on this point, as it posits that the initial symbiosis between the ancestor of mitochondria and its host was mediated by anaerobic syntrophy based on the ability of the symbiont to produce molecular hydrogen under anaerobic conditions, but it (obviously) also indicates that the ancestral mitochondrion was also able to respire oxygen.[63] The inference from that premise as it regards the symbiont is that the common ancestor of mitochondria and hydrogenosomes was simply a facultatively anaerobic α-proteobacterium, one with a heterotrophic physiology perhaps similar to modern-day *Rhodobacter* or countless other photosynthetic and nonphotosynthetic representatives of the group.[60,63] It has escaped the attention of many that the facultatively anaerobic, heterotrophic physiology of members of the genus *Rhodobacter* and relatives[104] is virtually identical to that of eukaryotes in terms of excreted end products: H_2O and CO_2 in the presence of oxygen, and succinate, propionate, formate, acetate, H_2, and CO_2 in the absence of oxygen. This suggests to me that the ancestor of mitochondria possessed the genes underlying that physiology and donated the majority of them to the chromosomes of its host. Accordingly, hydrogenosomes would have preserved their ancient anaerobic biochemistry and would have secondarily lost the ability to respire in many lineages. Conversely, typical mitochondria would have preserved their ancient aerobic biochemistry (respiration), and at ~0.6 Ga, with the advent of fully aerobic environments, they would have secondarily lost fermentative pathways in many lineages. By similar reasoning, eukaryotes with primary plastids have been producing their own oxygen locally for over a billion years, and many such lineages may therefore have lost much of their (ancestrally existing) anaerobic energy metabolism early on. On the basis of the newer geochemical evidence,[34–36] the conclusion seems inescapable that eukaryotes arose and spent the brunt of their evolutionary youth in an anoxic world. Hence, the still widely held view that hydrogenosomes are merely biochemically modified mitochondria, having secondarily secondarily tacked on anaerobic enzymes to an implicitly or explicitly[105] strictly aerobic ancestral state, is inconsistent with the newer geochemical data and seems very difficult to uphold.

Second, the geochemical evidence for largely anoxic oceans before ~0.6 Ga very strongly indicates that up until that time, the anaerobic biochemistry in hydrogenosomes must have been much more widespread among eukaryotes than it is today. This would significantly help to explain why the H_2-producing fermentations in hydrogenosomes of such distantly related groups as the cytridiomycete fungi, the ciliates, and the trichomonads are so similar in overall design[52] and are performed with enzymes that were present in the common ancestor of those lineages.[10,72] It also significantly helps to explain the widespread distribution of anaerobic mitochondria, even among metazoan lineages.[60] Despite the fact that hydrogenosomes have been known for 30 years, classical endosymbiotic theory has never been able to accomodate them,[43] mainly because the theory is designed to account for aerobic mitochondria only.[105] Under the view that mitochondria had aerobic origins, the biochemical unity of hydrogenosomes from different eukaryotic lineages[62] would be altogether inexplicable: Under the view that mitochondria had anaerobic origins, hydrogenosomes are the key to the eukaryotic biochemical past.

Third, during the entire period from the time of their origins up until 0.6 Ga, eukaryotes had to deal with very high levels of sulfide, which is a potent toxin. Many contemporary marine invertebrates (metazoans) still have to deal with very high sulfide concentrations, particularly those invertebrates that live in coastal sediments. Such organisms use a mitochondrial enzyme, sulfide:quinone oxidoreductase, to oxidize sulfide to the less toxic product thiosulfate, whereby the electrons from sulfide oxidation are fed into the electron transport chain to generate chemiosmotic potential for mitochondrial ATP synthesis,[106,107] just as it occurs in many sulfide-using eubacteria today.[108,109] That biochemical trace of our sulfidic past is even preserved up into the vertebrate lineage, because chicken mitochondria can also oxidize sulfide to drive ATP synthesis.[110] The gene for mitochondrial sulfide:quinone oxidoreductase has been identified in fungi,[111] and gene phylogenies indicate a single origin for the eukaryotic enzyme, suggesting that this gene was indeed present in the respiratory chain of the ancestral mitochondrion.[112] All things considered, the ability of modern mitochondria to deal with anoxia and sulfide are most easily understood as biochemical relics from the anoxic and sulfidic beginnings of the eukaryotic lineage.

Some of the considerations presented in this chapter have been summarized elsewhere as a schematic diagram of cell lineage history, taking LGT and endosymbiosis explicitly into account and including a rough time scale.[98]

Conclusion

LGT has dashed the hopes of quick success at fully uncovering life's history with genomes, but it has opened up new ways of looking into the past with the binoculars of sequence comparisons. Geologists are telling us that the Earth's oceans were largely anoxic and highly sulfidic for much longer that was previously thought. Phylogeneticists are telling us that the classical rRNA tree has eukaryotic evolution completely upside down. From the standpoint of microbial physiology, fungi look promising as genuine early branchers, or at least they seem to have preserved many traits that can be inferred to have existed in the first eukaryotic cells. Fungi might have branched off of the eukaryotic tree before phagocytosis had evolved: Their prototrophy is either a preserved ancestral state or a derived acquisition. In the former case, fungi will tend to share the same genes for amino acid and cofactor biosynthesis as other eukaryotic groups that have retained these genes. Among the plants, a new influx of genes for such pathways from the cyanobacterial ancestor of plastids may have reinstated prototrophy in the plant lineage (assuming that the host of plastids had already become auxotrophic for some amino acids or cofactors) or may have superimposed a cyanobacterial imprint on the prototrophy of the host (assuming that the host of plastid endosymbiosis had not developed any auxotrophies).

Simple biochemical considerations can help in our efforts to understand the eukaryotic past and can help us make sense of genomes—whole genomes, not just small subunit rRNA, which has received sufficient attention already. We can be anxious to see what sorts of predictions about the day-to-day-life-in-the-wild biology the rRNA tree can generate about the biochemistry and physiology of the earliest eukaryotes. The ribosome does not evolve by itself: It needs a bunch of biochemistry to synthesize its bits and pieces, and it because it resides within a cell, it needs redox chemistry as the motor that makes that cell work. No redox chemistry, no life.[47] Perhaps the schisma abrew in cell evolution is that between the evolution of rRNA and the evolution of the rest of the genome and cell.

All organisms have a ribosome, but the number of ways that microbes can make a living in their environment (the evolutionary key to microbial biology in my view) is still untallied, although about 300 different ways that hyperthermophiles alone can make a living have been recently been reviewed.[113] Because of the evident prevalance of LGT in genomes, Ford Doolittle rightly challenged the sensibility of further pursuing the notion that a system of microbes based on the rRNA tree alone would be a natural system.[114] A few years earlier, Russ Doolittle had issued a similar challenge concerning the position of eukaryotes in the overall picture of cell evolution, taking seriously the possibility that chimaerism might be real in eukaryotic evolution, as Wolfram Zillig had also suggested.[115,116] Today, the most staunch defense of the rRNA tree turns to dialectics claiming LGT to be insignificant: "The suggestion that HGT [horizontal gene transfer] is the preferred vehicle for novel sequence evolution for modern cells is contradicted by the observations briefly surveyed here" (p. 9661).[14] Everyone is entitled to their opinion, but it is funny that the genome sequences of *Eschericia coli* K12 and *E. coli* O157 differ by 70,000 base substitutions and, furthermore, by 2,000,000 bp of differentially acquired DNA, a thirty-fold greater contribution of LGT (a non-Darwinian process of natural variation) over base substitution (a Darwinian process of natural variation) in the overall evolutionary differentness of those two genomes.[117] However, such observations will not deter proponents of rRNA systematics from believing that the ribosome reflects the evolution of the whole genome and biology of the whole cell and that the rRNA tree is just plain right, period, or as Kurland recently put it, "All in all, the available data suggest that rRNA-based phylogeny is robust and that Darwinian lineages are the essence of phylogeny" (p. 9662).[14]

The view that the ribosome embodies the essence of the cell and that the rRNA tree is not only correct but furthermore sufficent for understanding microbial evolution[14] carries a potentially dangerous corollary; namely, that that we need not—and dare not—look beyond the ribosome to understand the biochemical history of microbes. My response to that view would be a 1944 quote seldom used among biologists, but very much to the point on this occasion: "Nuts!"

Postscript

In the discussion of various lectures, I have often been asked why there are no known archaebacterial pathogens of humans, and I have never had a good answer. While writing this chapter, the (perhaps obvious) answer occurred to me: Archaebacteria require a number of fundamentally different vitamins and cofactors than eukaryotes use in their biochemistry.[47,118] When pathogens invade humans, they are looking for a meal, and as a main course, eukaryotes do not provide a complete diet for archaebacteria, except for some autotrophs in those eukaryotes that have hydrogenosomes and therefore produce H_2.[62] That brings us back to the universal tree, according to which eukaryotes and archaebacteria (rather than eukaryotes and eubacteria) should tend to use the same cofactors, which they do not. From this consideration follows a simple prediction: If archaebacterial pathogens are found, they will most likely prey on other archaebacteria.

Acknowledgements I thank the Deutsche Forschungsgemeinschaft for financial support and Herbert Gutz for pointing out Raper's review on the diversity of sex patterns in fungi.

References

1. J.R. Brown, "Ancient Horizontal Gene Transfer," *Nature Reviews Genetics* 4 (2003): 121–132.
2. D.-F. Feng, Cho, G., and Doolittle, RF., "Determining Divergence Times with a Protein Clock: Update and Reevaluation," *Proceedings of the National Academics of Science USA* 94 (1997): 13028–13033.
3. W.F. Doolittle, Boucher, Y., Nesbo C.L., Douady, C.J., Andersson, J.O., and Roger, A.J., "How Big is the Iceberg of which Organellar Genes in Nuclear Genomes are but the Tip?" *Philosophical Transactions of the Royal Society of London Series B: Biological Sciences* 358 (2003): 39–58.
4. H. Ochman, Lawrence, J.G., and Groisman, E.S., "Lateral Gene Transfer and the Nature of Bacterial Innovation," *Nature* 405 (2000): 299–304.
5. J. Eisen, "Horizontal Gene Transfer among Microbial Genomes: New Insights from Complete Genome Analysis," *Current Opinion in Genetics and Development* 10 (2000): 606–611.
6. L.A. Katz, "Lateral Gene Transfers and the Evolution of Eukaryotes: Theories and Data," *International Journal of Systematic and Evolutionary Microbiology* 52 (2002): 1893–1900.
7. U. Deppenheimer and 21 others, "The genome of *Methanosarcina mazei*: Evidence for Lateral Gene Transfer between Bacteria and Archaea," *Journal of Molecular Microbiology and Biotechnology* 4 (2002): 453–461.
8. D. Penny, McComish, B.J., Charleston, M.A., and Hendy, M.D., "Mathematical Elegance With Biochemical Realism: The Covarion Model of Molecular Evolution," *Journal of Molecular Evolution* 53 (2001): 711–723.
9. S. L. Salzberg, White, O., Peterson, J., and Eisen, J.A., "Microbial Genes in the Human Genome: Lateral Transfer or Gene Loss?" *Science* 292 (2001): 1903–1906.
10. T.M. Embley, van der Giezen, M., Horner, D.S., Dyal, P.L., and Foster, P., "Hydrogenosomes and Mitochondria: Phenotypic Variants of the Same Fundamental Organelle," *Philosophical Transactions of the Royal Society of London Series B: Biological Sciences* 358 (2003): 191–203.
11. C.R. Woese, "Interpreting the Universal Phylogenetic Tree," *Proceedings of the National Academics of Science USA* 97 (2000): 8392–8396.
12. C.R. Woese, "On the Evolution of Cells," *Proceedings of the National Academics of Science USA* 99 (2002): 8742–8747.

13. C. Mereschkowsky, "Über Natur und Ursprung der Chromatophoren im Pflanzenreiche," *Biologisches Centralblatt* 25 (1905): 593–604. [English translation in *European Journal of Phycology* 34 (1999): 287–295].
14. C.G. Kurland, Canback, B., and Berg, O.G., "Horizontal Gene Transfer: A Critical view," *Proceedings of the National Academics of Science USA* 100 (2003): 9658–9662.
15. A.H. Knoll, *Life on a Young Planet: The First Three Billion Years of Evolution of Earth* (Princeton, NJ: Princeton University Press, 2003).
16. S.A. Wilde, Valley, J.W., Peck, W.H., and Graham, C.M., "Evidence from Detrital Zircons for the Existence of Continental Crust and Oceans on the Earth 4.4 Gyr ago," *Nature* 409 (2001): 175–178.
17. E.G. Nisbet, and Sleep, N.H., "The Habitat and Nature of Early Life," *Nature* 409 (2001): 1083–1091.
18. M.T. Rosing, "^{13}C-depleted Carbon Microparticles in >3700-Ma Sea-Floor Sedimentary Rocks from West Greenland," *Science* 283 (1999): 674–676.
19. Y. Ueno, Yurimoto, H., Yoshioka, H., Komiya, T., and Maruyama, S., "Ion Microprobe Analysis of Graphite from ca. 3.8 Ga Metasediments, Isua Crustal Belt, West Greenland: Relationship between Metamorphism and Carbon Isotopic Composition," *Geochimica et Cosmochimica Acta* 66 (2002): 1257–1268.
20. A. van Zuilen, Lepland, A., and Arrhenius, G., "Reassessing the Evidence for the Earliest Traces of Life," *Nature* 418 (2002): 627–630.
21. B. Rasmussen, "Filamentous Microfossils in a 3,235-Million-Year-Old Volcanogenic Massive Sulphide Deposit," *Nature* 405 (2000): 676–679.
22. G. Nisbet and Fowler, C.M.R., "Archaean Metabolic Evolution of Microbial Mats," *Proceedings of the Royal Society of London Series B: Biological Sciences* 266 (1999): 2375–2382.
23. G. Nisbet, "The Realms of Archaean Life," *Nature* 405 (2000): 625–626.
24. N.V. Grassineau, Nisbet, E.G., Bickle, M.J., Fowler, C.M.R., Lowry, D., Mattey, D.P., Abell, P., and Martin, A., "Antiquity of the Biological Sulphur Cycle: Evidence from Sulphur and Carbon Isotopes in 2700 Million-Year-Old Rocks of the Belingwe Belt, Zimbabwe," *Proceedings of the Royal Society of London Series B: Biological Sciences* 268 (2001): 113–119.
25. W. Michaelis, Seifert, R., Nauhaus, K., Treude, T., Thiel, V., Blumenberg, M., Knittel, K., Gieseke, A., Peterknecht, K., Pape, T., Boetius, A., Amann, R., Jorgensen, B.B., Widdel, F., Peckmann, J., Pimenov, N.V., and Gulin, M.B., "Microbial Reefs in the Black Sea Fueled by Anaerobic Oxidation of Methane," *Science* 297 (2002): 1013–1015.
26. R. Walter, "Stromatolites: The Main Source of Information on the Evolution of the Early Benthos," in J.W. Schopf, ed., *Earth's Earliest Biosphere: Its Origin and Evolution* (Princeton, N.J.: Princeton University Press, 1994), 270–286.
27. F. Kasting and Seifert, J.L., "Life and the Evolution of Earth's Atmosphere," *Science* 296 (2002): 1066–1068.
28. J.M. Hayes, "Global Methanotrophy at the Archean-Proterozoic Transition," in S. Bengston, ed., *Early Life on Earth.* (New York: Columbia University Press, 1994), 220–236.
29. D.E. Canfield, "A Breath of Fresh Air," *Nature* 400 (1999): 503–504.
30. U.H. Wiechert, "Earth's Early Atmosphere," *Science* 289 (2002): 2341–2342.
31. H.D. Holland, "When Did the Earth's Atmosphere Become Oxic?" *The Geochemical News* 100 (1999): 20–22.
32. H.D. Holland and Beukes, N., "A Paleoweathering Profile from Griqualand West, South Africa: Evidence for a Dramatic Rise in Atmospheric Oxygen Between 2.2 and 1.9 BYBP," *American Journal of Science* 290-A (1990): 1–34.
33. J.A. Karhu and Holland, H.D., "Carbon Isotopes and the Rise of Atmospheric Oxygen," *Geology* 24 (1996): 867–870.
34. D.E. Canfield and Teske, A., "Late Proterozoic Rise in Atmospheric Oxygen Concentration Inferred from Phylogenetic and Sulphur-isotope Studies," *Nature* 382 (1996): 127–132.
35. D.E. Canfield, "A New Model for Proterozoic Ocean Chemistry," *Nature* 396 (1998): 450–453.
36. A.D. Anbar and Knoll, A.H., "Proterozoic Ocean Chemistry and Evolution: A Bioinorganic Bridge," *Science* 297 (2002): 1137–1142.

37. P. Forterre and Philippe, H., "Where is the Root of the Universal Tree of Life?" *BioEssays* 21 (1999): 871–879.
38. J.A. Lake and Rivera, M.C., "Was the Nucleus the First Endosymbiont?" *Proceedings of the National Academics of Science USA* 91 (1994): 2880–2881.
39. R.S. Gupta, "Protein Phylogenies and Signature Sequences: A Reappraisal of Evolutionary Relationships among Archaebacteria, Eubacteria, and Eukaryotes," *Microbiology and Molecular Biolology Reviews* 62 (1998): 1435–1491.
40. D. Moreira and Lopez-Garcia, P., "Symbiosis between Methanogenic Archaea and δ-Proteobacteria as the Origin of Eukaryotes: The Syntrophic Hypothesis," *Journal of Molecular Evolution* 47 (1998): 517–530.
41. G. Wächtershäuser, "From Pre-Cells to Eukarya—A Tale of Two Lipids," *Molecular Microbiolology* 47 (2003): 13–22.
42. T. Cavalier-Smith, "The Phagotrophic Origin of Eukaryotes and Phylogenetic Classification of Protozoa," *International Journal of Systematic and Evolutionary Microbiology* 52 (2002): 297–354.
43. L. Margulis, Dolan, M.F., and Guerrero, R., "The Chimeric Eukaryote: Origin of the Nucleus from the Karyomastigont in Amitochondriate Protists," *Proceedings of the National Academics of Science USA* 97 (2000): 6954–6959.
44. W. Martin, Hoffmeister, M., Rotte, C., and Henze, K., "An Overview of Endosymbiotic Models for the Origins of Eukaryotes, Their ATP-producing Organelles (Mitochondria and Hydrogenosomes), and Their Heterotrophic Lifestyle," *Biological Chemistry* 382 (2001): 1521–1539.
45. M.J. Russell and Hall, A.J., "The Emergence of Life from Iron Monosulphide Bubbles at a Submarine Hydrothermal Redox and pH Front," *Journal of the Geological Society of London* 154 (1997): 377–402.
46. G. Wächtershäuser, "Evolution of the First Metabolic Cycles," *Proceedings of the National Academics of Science USA* 87 (1990): 200–204.
47. W. Martin and Russell, M. J., "On the Origins of Cells: An Hypothesis for the Evolutionary Transitions from Abiotic Geochemistry to Chemoautotrophic Prokaryotes, and from Prokaryotes to Nucleated Cells," *Philosophical Transactions of the Royal Society of London Series B: Biological Sciences* 358 (2003): 59–85.
48. J.G. Peretó, Velasco, A.M., Becerra, A., and Lazcano, A., "Comparative Biochemistry of CO_2 fixation and the Evolution of Autotrophy," *International Microbiolology* 2 (1999):3–10.
49. R.S. Ronimus and Morgan, H.W., "Distribution and Phylogenies of Enzymes of the Embden-Meyerhof-Parnas Pathway from Archaea and Hyperthermophilic Bacteria Support a Gluconeogenic Origin of Metabolism," *Archaea* 1 (2003): 199–221.
50. O. Kandler, "The Early Diversification of Life," in S. Bengston, ed., *Early Life on Earth* (New York: Columbia University Press, 1994), 152–160.
51. E.J. Javaux, Knoll, A.H., and Walter, M.R., "Morphological and Ecological Complexity in Early Eukaryotic Ecosystems," *Nature* 412 (2001): 66–69.
52. N.J. Butterfield, "*Bangiomorpha pubescens* n. gen., n. sp.: Implications for the Evolution of Sex, Multicellularity, and the Mesoproterozoic/Neoproterozoic Radiation of Eukaryotes," *Paleobiology* 263 (2000): 386–404.
53. T.M. Han and Runnegar, B., "Megascopic Eukaryotic Algae from the 2.1 Billion-Year-Old Neeguanee Iron Formation, Michigan," *Science* 257 (1992): 232–235.
54. J.J. Brocks, Logan, G.A., Buick, R., and Summons, R.E., "Archean Molecular Fossils and the Early Rise of Eukaryotes," *Science* 285 (1999): 1033–1036.
55. S. Schouten, Bowman, J.P., Rijpstra, W.I., and Sinninghe-Damste, J.S., "Sterols in a Psychrophilic Methanotroph, *Methylosphaera hansonii*," *FEMS Microbiology Letters* 186 (2000): 193–195.
56. W. Kohl, Gloe, A., and Reichenbach, H., "Steroids from the Myxobacterium *Nannocystis exedens*," *Journal of General Microbiology* 129 (1983): 1629–1635.
57. T. Hai, Schneider, B., Schmidt, J., and Adam, G., "Sterols and Triterpenoids from the Cyanobacterium *Anabaena hallensis*," *Phytochemistry* 41 (1996): 1083–1084.
58. T. Fenchel and Finlay, B.J., *Ecology and Evolution in Anoxic Worlds* (Oxford: Oxford University Press, 1995).

59. M.K. Grieshaber and S. Völkel, "Animal Adaptations for Tolerance and Exploitation of Poisonous Sulfide," *Annual Review of Physiology* 60 (1998): 30–53.
60. A.G.M. Tielens, Rotte, C., van Hellemond, J., and Martin, W., "Mitochondria as We Don't Know Them," *Trends in Biochemical Sciences* 27 (2002): 564–572.
61. M. Müller, "The Hydrogenosome," *Journal General Microbiology* 139 (1993): 2879–2889.
62. M. Müller, "Energy Metabolism. Part I: Anaerobic Protozoa," in J. Marr, ed. *Molecular Medical Parasitology* (London: Academic Press, 2003): 125–139.
63. W. Martin, and Müller, M., "The Hydrogen Hypothesis for the First Eukaryote," *Nature* 392 (1998): 37–41.
64. C.R. Vossbrinck, Maddox, J., Friedman, S., Debrunner–Vossbrinck, B.A., and Woese, C.R., "Ribosomal RNA Sequence Suggests Microsporidia are Extremely Ancient Eukaryotes," *Nature* 326 (1987): 411–414.
65. M. Sogin, Gunderson, J., Elwood, H., Alonso, R., and Peattie, D., "Phylogenetic Meaning of the Kingdom Concept: An Unusal Ribosomal RNA from *Giardia lamblia*," *Science* 243 (1989): 75–77.
66. T.M. Embley, and Hirt, R.P., "Early Branching Eukaryotes?" *Current Opinion in Genetics and Development* 8 (1998): 655–661.
67. H. Philippe, Germot, A., and Moreira, D., "The New Phylogeny of Eukaryotes," *Current Opinion in Genetics and Development* 10 (2000): 596–601.
68. A.J. Roger, and Silbermann, J.D., "Mitochondria in Hiding," *Nature* 418 (2002): 827–828.
69. J. Tovar, Fischer, A., and Clark, C.G., "The Mitosome, a Novel Organelle Related to Mitochondria in the Amitochondrial Parasite *Entamoeba histolytica*," *Molecular Microbiology* 32 (1999): 1013–1021.
70. B.A. Williams, Hirt, R.P., Lucocq, J.M., and Embley, T.M., "A Mitochondrial Remnant in the Microsporidian *Trachipleistophora hominis*," *Nature* 418 (2002): 865–869.
71. G.A. Biagini, Finlay, B.J., Lloyd, D., "Evolution of the Hydrogenosome," *FEMS Microbiology Letters* 155 (1997): 133–140.
72. J.H.P. Hackstein, Akhmanova, A., Voncken, F., van Hoek, A., van Alen, T., Boxma, B., Yeo Moon-van der Staay, G. van der Staay, S., Leunissen, J., Huynen, M., Rosenberg J., and Veenhuis, M., "Hydrogenosomes: Convergent Adaptations of Mitochondria to Anaerobic Environments," *Zoology* 104 (2001): 290–302.
73. C. Rotte, Stejskal, F., Zhu, G., Keithly, J.S., and Martin, W., "Pyruvate:NADP Oxidoreductase from the Mitochondrion of *Euglena gracilis* and from the Apicomplexan *Cryptosporidium parvum*: A Biochemical Relic Linking Pyruvate Metabolism in Mitochondriate and Amitochondriate Protists," *Molecular Biology and Evolution* 18 (2001): 710–20.
74. D.S. Horner, B. Heil, T. Happe, and T.M. Embley, "Iron Hydrogenases—Ancient Enzymes in Modern Eukaryotes," *Trends in Biochemical Sciences.* 27 (2002): 148–153.
75. N. Yarlett, C.G. Orpin, E.A. Munn, N.C. Yarlett, and Greenwood, C.A., "Hydrogenosomes in the Rumen Fungus *Neocallimastix patriciarum*," *Biochemical Journal* 236 (1986): 729–739.
76. M.W. Gray, Burger, G., and Lang, B.F., "Mitochondrial Evolution," *Science* 283 (1999): 1476–1481.
77. F.D. Marvin-Sikkema, Rees, E., Kraak, M.N., Gottschal, J.C., and Prins, R.A., "Influence of Metronidazole, Carbon Monoxide, Carbon Dioxide and Methanogens on the Fermentative Metabolism of the Anaerobic Fungus *Neocallimastix* sp. strain L2," *Applied and Environmental Microbiology* 59 (1993):2678–2683.
78. A. Stechmann and Cavalier-Smith, T., "Rooting the Eukaryote Tree by Using a Derived Gene Fusion," *Science* 297 (2002): 89–91.
79. A.G.B. Simpson and Roger, A.J., "Eukaryotic Evolution: Getting to the Root of the Problem," *Current Biology* 12 (2002): R691–R693.
80. S.L. Baldauf, Roger, A.J., Wenk–Siefert I., and Doolittle, W.F., "A Kingdom-Level Phylogeny of Eukaryotes Based on Combined Protein Data," *Science* 290 (2000): 972–977.
81. E. Bapteste, Brinkmann, H., Lee, J.A., Moore, D.V., Sensen, C., Gordon, P., Durufle, L., Gaasterland, T., Lopez, P., Müller, M., and Philippe, H., "The Analysis of 100 Genes Supports the Grouping of Three Highly Divergent Amoebae: *Dictyostelium*, *Entamoeba*, and

Mastigamoeba," *Proceedings of the National Academics of Science USA* 99 (2002): 1414–1419.

82. M. Kobayashi, Matsuo, Y., Takimoto, A., Suzuki, S., Maruo, F., and Shoun, H., "Denitrification, a Novel Type of Respiratory Metabolism in Fungal Mitochondrion," *Journal of Biological Chemistry* 271 (1996): 16263–16267.
83. M. van der Giezen, Slotboom, D.J., Horner, D.S., Xue, G.P., Embley, T.M., and Kunji, E.R.S., "Conserved Properties of Hydrogenosomal and Mitochondrial ADP/ATP Carriers: A Common Origin for Both Organelles," *EMBO Journal* 21 (2002): 572–579.
84. F. Voncken, Boxma, B., Tjaden, J., Akhmanova, A., Huynen, M., Verbeek, F., Tielens, A.G.M., Haferkamp, I., Neuhaus, H.E., Vogels, G., Veenhuis, M., and Hackstein, J.H.P., "Multiple Origins of Hydrogenosomes: Functional and Phylogenetic Evidence from the ADP/ATP Carrier of the Anaerobic Chytrid *Neocallimastix* sp.," *Molecular Microbiology* 44 (2002): 1441–1454.
85. Z. Zhou, Takaya, N., Nakamura, A., Yamaguchi, M., Takeo, K., and Shoun, H., "Ammonia Fermentation, a Novel Anoxic Metabolism of Nitrate by Fungi," *Journal of Biological Chemistry* 277 (2002): 1892–1896.
86. H. Yurimoto, Sakai, Y., and Kato, N., "Methanol metabolism," in G. Gellissen, ed., Hansenula polymorpha: *Biology and Applications* (Weinheim: Wiley-VCH Verlag, 2002), 61–75.
87. C.D. von Dohlen, Kohler, S., Alsop, S.T., and McManus, W.R., "Mealybug β-proteobacterial Endosymbionts Contain γ-proteobacterial Symbionts," *Nature* 412 (2001): 433–436.
88. T. Horiike, Hamada, K., Kanaya, S., and Shinozawa, T., "Origin of Eukaryotic Cell Nuclei by Symbiosis of Archaea in Bacteria Is Revealed by Homology-Hit Analysis," *Nature Cell Biology* 3 (2001): 210–214.
89. C. Rotte and W. Martin, "Does Endosymbiosis Explain the Origin of the Nucleus?" *Nature Cell Biology* 8 (2001): 173–174.
90. W.F. Doolittle, "Revolutionary Concepts in Evolutionary Biology," *Trends in Biochemical Sciences* 5 (1980): 146–149.
91. W.F. Doolittle, "You Are What You Eat: A Gene Transfer Ratchet Could Account for Eubacterial Genes in Eukaryotic Genomes," *Trends in Genetics* 14 (1998): 307–311.
92. T. Cavalier-Smith, "The Origin of Eukaryote and Archaebacterial Cells," *Annals of the New York Academy of Science* 503 (1987): 17–54.
93. P. Reichard, "The Evolution of Ribonucleotide Reduction," *Trends in Biochemical Sciences* 22 (1997): 81–85.
94. N.J.W. Kreger-van Rij, ed., *The Yeasts. A Taxonomic Study.* 3rd ed.(Amsterdam: Elsevier, 1984), 106–962.
95. G. Burger and Lang, B.F., "Parallels in Genome Evolution in Mitochondria and Bacterial Symbionts," *IUBMB Life* 55 (2003): 205–212.
96. B.F. Lang, Gray, M.W., and Burger, G., "Mitochondrial Genome Evolution and the Origin of Eukaryotes," *Annual Review of Genetics* 33 (1999): 351–397.
97. W. Martin, "A Briefly Argued Case that Mitochondria and Plastids are Descendants of Endosymbionts, but that the Nuclear Compartment is Not," *Proceedings of the Royal Society of London Series B: Biological Sciences* 266 (1999): 1387–1395.
98. W. Martin, Rotte, C., Hoffmeister, M., Theissen, U., Gelius-Dietrich, G., Ahr, S., and Henze, K., "Early Cell Evolution, Eukaryotes, Anoxia, Sulfide, Oxygen, Fungi First (?), and a Tree of Genomes Revisited," *IUBMB Life* 55 (2003): 193–204.
99. N. Galtier, N. Tourasse, and M. Gouy, "A Nonhyperthermophilic Common Ancestor to Extant Life Forms," *Science* 283 (1999): 220–221.
100. M.C. Rivera and J.A. Lake, "Evidence that Eukaryotes and Eocyte Prokaryotes Are Immediate Relatives," *Science* 257 (1992): 74–76.
101. J.R. Raper, "On the Evolution of Fungi," in C.G. Ainsworth and A.S. Sussmann, eds., *The Fungi: An Advanced Treatise. Volume III. The Fungal Population* (New York: Academic Press, 1968), 677–693.
102. J.R. Raper, "Life Cycles, Basic Patterns of Sexuality, and Sexual Mechanisms," in C.G. Ainsworth and A.S. Sussmann, eds., *The Fungi: An Advanced Treatise. Volume II. The Fungal Organism* (New York: Academic Press, 1966), 473–511.

103. E.J. Boekema, Hifney, A., Yakushevska, A.E., Piotrowski, M., Keegstra, W., Berry, S., Michel, K.-P., Pistorius, E.K., and Kruip, J., "A Giant Chlorophyll-Protein Complex Induced by Iron Deficiency in Cyanobacteria," *Nature* 412 (2001): 745–748.
104. J.F. Imhoffand Trüpe, H.G., "The Genus *Rhodospirillum* and Related Genera," in A. Balows, Trüper, H.G., Dworkin, M., Harder, W., and Schleifer, K.-H., eds., *The Prokaryotes,* 2nd ed. (New York: Springer-Verlag, 1992), vol. 3, 2141–2159.
105. S.G.E. Andersson and Kurland, C.G., "Origins of Mitochondria and Hydrogenosomes," *Current Opinion in Microbiology* 2 (1999): 535–541.
106. J.E. Doeller, Grieshaber, M.K., and Kraus, D.W., "Chemolithoheterotrophy in a Metazoan Tissue: Thiosulfate Production Matches ATP Demand in Ciliated Mussel Gills," *Journal of Experimental Biology* 204 (2001): 3755–3764.
107. J.E. Doeller, Gaschen, B.K., Parrino, V., and Kraus, D.W., "Chemolithoheterotrophy in a Metazoan Tissue: Sulfide Supports Cellular Work in Ciliated Mussel Gills," *Journal of Experimental Biology* 202 (1999): 1953–1961.
108. M. Schütz, Shahak, Y., Padan, E., and Hauska, G., "Sulfide-Quinone Reductase from *Rhodobacter capsulatus,*" *Journal of Biological Chemistry* 272 (1997): 9890–9894.
109. C. Griesbeck, Schütz, M., Schödl, T., Bathe, S., Nausch, L., Mederer, N., Vielreicher, M., and Hauska, G., "Mechanism of Sulfide–Quinone Reductase Investigated Using Site-Directed Mutagenesis and Sulfur Analysis," *Biochemistry* 41 (2002): 11552–11565.
110. R. Yong and Searcy, D.G., "Sulfide Oxidation Coupled to ATP Synthesis in Chicken Liver Mitochondria," *Comparative Biochemistry and Physiology Part B: Biochemistry and Molecular Biology* 129 (2001): 129–137.
111. J.G. Vande Weghe and Ow, D.W., "A Fission Yeast Gene for Mitochondrial Sulfide Oxidation," *Journal of Biological Chemistry* 274 (1999): 13250–13257.
112. U. Theissen, Hoffmeister, M., Grieshaber, M.K., and Martin, W., "Single Eubacterial Origin of Eukaryotic Sulfide:Quinone Oxidoreductase, a Mitochondrial Enzyme Conserved from the Early Evolution of Eukaryotes During Anoxic and Sulfidic Times," *Molecular Biology and Evolution* 20 (2003): 1564–1574.
113. J.P. Amend, and Shock, E.L., "Energetics of Overall Metabolic Reactions of Thermophilic and Hyperthermophilic Archaea and Bacteria," *FEMS Microbiology Reviews* 25 (2001): 175–243.
114. W.F. Doolittle, "Phylogenetic Classification and the Universal Tree," *Science* 284 (1999): 2124–2128.
115. R.F. Doolittle, "Of Archae and Eo: What's in a Name?" *Proceedings of the National Academics of Science USA* 92 (1995): 2421–2423.
116. W. Zillig, "Comparative Biochemistry of Archaea and Bacteria" *Current Opinion in Genetics and Development* 1 (1991): 544–551.
117. N.T. Perna, "Genome Sequence of Enterohaemorrhagic *Escherichia coli* O157:H7," *Nature* 409 (2001): 529–533.
118. R.H. White, "Biosynthesis of the Methanogenic Cofactors," *Vitamins and Hormones* 61 (2001): 299–337.

7

The Robustness of Intermediary Metabolism

HAROLD J. MOROWITZ
DANIEL BROYLES
HOWARD LASUS

Over the first half of the twentieth century, biochemists developed the chart of intermediary metabolism that includes those reaction pathways for processing energy and the synthesizing of the molecular building blocks of biological coacervates and macromolecules. An example of a coacervate structure is a cell membrane made up of amphiphilic molecules held together in a bimolecular leaflet by other than covalent bonds. An example of a macromolecule is a polymer whose primary structure is determined entirely by covalent bonds. The distinction is not absolute because amphiphiles are structured by covalent bonds and the secondary and tertiary structures of macromolecules involve other than covalent bonding.

The chart of intermediary metabolism involves reactions with a change of covalent bonding. A chart of the type described above may be constructed in principle for every extant species. The number of such species is probably in the millions, and the number of extinct species is almost certainly larger. Autotrophic organisms construct all the molecules of their charts from one carbon and one nitrogen starting molecules. Autotrophs require only the anabolic pathways of metabolism. Heterotrophic organisms, by using molecules supplied by other species, do not require all the enzymes found in the chart of the autotrophs but may require other enzymes for catabolic and transport functions. The chart of intermediary metabolism is part of the description of the phenotype of a species.

The chart of any taxon may be divided into primary metabolism and secondary metabolism. The former consists of molecules found in all taxa, such as amino acids and ribonucleotides, whereas the later deals with species-specific molecules such as pheromones, antibiotics, and taxon-associated structural components.

The chart of primary metabolism is universal. This can be because of common descent from a universal ancestor, overwhelming translational migration of genes, or robustness of the chart as the best or unique solution of metabolism. Even if one of the first two reasons is involved, the universality implies a remarkable robustness. The universality of primary metabolism is both a biochemical generalization and a necessary principle of trophic ecology. For one species to be food for another species requires a major overlap in primary metabolism. Catabolic pathways must lead to anabolic or bioenergetic molecules.

We proceed by postulating the following hierarchy of emergences: (1) prokaryotes preceded eukaryotes, (2) protocells preceded prokaryotes, (3) metabolic networks preceded protocells, (4) geochemical processes preceded metabolic networks, and (5) the emergence of the elements governed by the Pauli exclusion principle preceded geochemical processing. Related postulates are autotrophs preceded heterotrophs, chemoautorophs preceded photoautotrophs, and reductive metabolism preceded oxidative metabolism. Steps 2 through 4 are the rise of the prokaryotes, and step 1, with the loss of independent function by the prokaryotes following endosymbiosis, represents the transformation of prokaryotes to organelles.

Following the fall of temperature to 3000 K and below during the evolution of the universe, electrons and nuclei combined following the constraints of the Pauli exclusion principle. This led to a planetary system governed by the laws of chemistry, reflected in the periodic table of the elements and covalent bonding. It is our conjecture that just as the Pauli principle made chemistry emergent from quantum mechanics, other selection principles, as yet undiscovered, will make the chart of intermediary metabolism of autotrophs emergent from chemistry, which would explain the robustness of the chart. The task is to experimentally determine the universal chart of metabolism of chemoautotrophs and then develop the principle or principles of metabolism. The universal nature of biochemistry is a key feature of biology and biogenesis.

The crucial importance of the Pauli principle from an epistemological perspective is that it provides an example of an extra dynamic principle that renders all higher levels of the hierarchy informatic as well as energetic. It provides a clue not often seen by those working in bioenergetics that thermodynamics alone is not adequate to explain the emergence of metabolic networks and higher-level phenomena. Instead of a small number of elementary particles, we are presented with a large number of distinguishable entities. These are, at one level, the building blocks of biochemical networks. The most universal of these building blocks for metabolism are carbon, hydrogen, nitrogen, oxygen, phosphorus, and sulfur.

The postulation that autotrophs preceded heterotrophs is based on the fact that experimentally, heterotrophy imposes an enormous informatic and entropic burden on the environment without providing other than random methods for generating this environment. The clue to getting around this is provided by a group of chemoautotrophs that live in a reductive world and operate by the reductive citric acid cycle. This cycle is network auto catalytic. It is the simplest network for incorporating carbon and provides the informatic core for the synthesis of all the basic biochemical building blocks. Pyruvate, oxaloacetate, and keto glutarate are termini of the pathways to amino acid synthesis, pyruvate is on the pathway to sugar synthesis, acetate is the terminus of the lipid synthesis pathway, and oxaloacetate is a terminus of the pyrimidine pathway. If one redraws the metabolic chart to correspond to this group of organisms, it consists of a set of radii coming off the reductive citric acid loop with occasional bridges connecting points on the radii. It is thus possible to index every compound on the chart indicating how many reaction steps it is away from the citric acid cycle. The lower the index number, the less costly it is to make the compound, and we may assume the earlier the compound occurred in prebiotic evolution. For amino acids the index numbers are (1) glutamic acid, aspartic acid, alanine; (2) glutamine, asparagine; (5) proline, serine; (6) threonine, valine, glycine; (7) cysteine; (8) arginine, methionine; (9) leucine; (11) lysine; (12) isoleucine; (18) phenylalanine, tyrosine; (20) tryptophan; and (24) histidine.

We will use this indexing later. First we focus on a number of generalizations that come out of a detailed study of the metabolic chart of chemoautotrophs. These generalizations will be stated with more certainty than can be justified at present, but they stand as guides in the search for rules. The key reactions at the core of metabolism must be possible without enzymes, as these reactions are necessary for the synthesis of enzymes. This may be seen in the synthesis of amino acids. Amino acids may be synthesized from keto acids, ammonia, and reductants in aqueous solution in the absence of enzymes.[1] In present-day amino acid synthesis, the amine donor is either glutamic acid or glutamine. A whole series of enzymes catalyze this function, but they share pyridoxal phosphate as a prosthetic group and proceed by the ping-pong reaction.

The generalizations are that

1. There are two possible sources of energy: photon flux and environmental oxidation-reduction couples.
2. Biological energy is stored as polyphosphates, oxidation-reduction couples such as NAD – NADH + H+, and transmembrane potentials because of charge separation and concentration differences of protons.
3. All nitrogen enters into covalently bonded biological molecules from the glutamic acid, glutamine, alpha keto glutarate, and ammonia-autocatalytic network. In autotrophs all synthetic pathways for other amino acids involve a transamination reaction with a glutamate donor. These reactions with a pyridoxal phosphate intermediate conserve chirality so that the homochirality of glutamate generates 18 amino acids of the same stereospecificity.
4. All sulfur enters into covalently bonded biological molecules from the synthesis of cysteine from O-acetyl serine and hydrogen sulfide.
5. Tetrapyrroles with metallic cores are intermediates of oxidation-reduction electron transfers.
6. Pyridoxal phosphate is a universal prosthetic group in transaminations and occurs in other enzymatic reactions including decarboxylations, racemizations, and aldol condensations.
7. Carbamoyl phosphate is the key intermediate in the synthesis of pyrimidines.
8. Pyruvate is the key intermediate in the synthesis of sugars and the alanine family of amino acids.
9. Acetyl CoA is the key intermediate in the synthesis of lipids.

This list is not complete, but it indicates the kinds of generalizations.

The index number of the amino acids allows us to group these monomers into early ($N = 1$ and 2), intermediate ($N = 3$ to 11) and late ($N = 12$ to 24). We can then go to the GenBank program and use the sequence files to determine the coded amino acid concentration in various organisms and organelles. The source for the sequence data used in this analysis was the Entrez program of the National Center for Biological Information.

Before analyzing the sequence data, we will formulate a scenario of early biogenesis based on the metabolism of the reductive chemolithoautotrophs. The first network in an environment of reductants and CO_2 is the formation of the network autocatalytic reductive citric acid cycle. In all the reaction steps, the chemicals in the cycle react with environmental molecules (CO_2, H_2, pyrophosphate, reductants, etc.), so the reactions are effectively first-order kinetically in the absence of vesicles. One set of pathways leads to amphiphile (such as fatty acid) synthesis, resulting in vesicles that capture the reaction products and make possible second-order reactions. These reactions lead to charged molecules that are trapped in the vesicle, resulting in a rise in osmotic pressure.

At this stage, the metabolizing vesicles are sensitive to blowing up osmotically, a situation that is currently prevented by a peptidoglycan cage around the vesicle that holds the membrane against the osmotic pressure. This cage is the cell wall and may consist of a single molecule formed as a closed shell. The structure consists of linear strands of a repeating dimer of N acetylglucosamine and N acetyl muramic acid. Attached to the muramic acid is a short peptide chain of D and L amino acids, which is linked to a neighboring muramic acid peptide chain by a pentaglycine cross link. The dominant amino acids are alanine and glutamic acid, which are only one step removed from the tricarboxylic acid cycle. The fact that the wall has D and L amino acids indicates that the wall may have preceded the homochirality of amino acids caused by the conservation of chirality in the enzymatic ping-pong reaction at pyridoxal phosphate, the prosthetic group for all transamination reactions. The pentaglycine link is characteristic of gram-positive bacteria, though there tends to be a direct linkage of the muramic acid peptides in gram-negative bacteria. Given the membrane to permit the transmembrane potentials and the wall to permit free migration of small molecules, yet prevent lysis resulting from osmotic pressure, a milieu is provided for the development of the full metabolic chart including the complex relationship between proteins and nucleic acids.

The multiple use of molecules and networks of molecules is a very interesting feature. In the reductive world, sugars first fulfilled a structural role in cell walls and an informatic role in nucleic acids; in an oxidative world, sugars and sugar polymers store energy. In a reductive world, the reductive tricarboxylic acid cycle is primarily an engine of synthesis in producing precursors to the major types of molecules; in an oxidative world, the cycle combines the synthesis function with an energy transduction function. Throughout, bioenergetics and bioinformatics are closely linked. Similarly, proteins and nucleic acids are structural, informatic, and catalytically functional molecules. Perhaps one should look for catalytically active carbohydrates. This interrelatedness of all molecules indicates a deeper principle within the metabolic chart.

Using the index number of amino acids enabled us to ascertain preferences for index numbers in different taxa and organelles. The clusters used were low index (alanine, glutamic acid, aspartic acid, glutamine, and asparagines) and high index (isoleucine, phenylalanine, tyrosine, tryptophan, and histidine). Protein sequences for each cell genome or organelle genome were downloaded from the Entrez program of the National Center for Biological Information. The clusters (low index, high index, others) were then counted and representatives obtained for eukaryotic genomes, bacterial genomes, mitochondria, and chloroplasts.

The sequences chosen for eukaryotes was *Homo sapiens* chromosome 6, *H. sapiens* chromosome 21, *H. sapiens* chromosome 22, *H. sapiens* chromosome X, *H. sapiens* chromosome Y, *Saccharomyces cerevisiae*, and *Caenorhabditis elegans*.

For bacteria, it was *Escherichia coli K12*, *Haemophilus influenzae*, *Chlamydia pneumoniae*, *Bacillus subtilis*, *Archeoglobis figidus*, *Methanococcus jannaschii*, *Borrelia burgdorferi*, *Methanobacterium thermoautotrophicum*, and *Cyanaphora paradoxa*.

The sequences for Mycoplasma and Rickettsiia were *Mycoplasma genetalium* and *Rickettsia prowazekii*, respectively.

For mitochondria, the sequence was *H. sapiens*, *Drosophila melanogaster*, *Ancyclostoma duodenale*, *Equus asinus*, *Gorilla gorilla*, *Tetrahymena pyriformis*, *Allomyces macrogynus*, and *Chlamdymonas reihardtii.*

For chloroplasts it was *Lotus japanicus*, *Zea mays*, *Nicotiana tabacum*, *Astasia longa*, *Aridopsis thaliana*, *Euglea gracilis*, and *Pinus thunbergii.*

Extracting from the data the number of low-index number and high-index number amino acids, we get table 7.1.

Thus, as the genome length gets smaller in the evolution of the four lower classes of organisms and organelles, there is a drift toward fewer low-index number and more high-index number amino acids. The reason is presently unclear, but something systematic seems to be happening in the transformation from prokaryote to organelle.

The next step is to review the chapters in this book by W. Ford Doolittle (chap. 5), James Lake et al. (chap. 9), Karl Heinz Schleifer and Wolfgang Ludwig (chap. 3), and William Martin (chap. 6). We find that "local tree topologies often differ depending on the molecules analyzed," "environmental factors can significantly alter horizontal transfer among prokaryotes," "complete prokaryotic genome sequences ever more numerous and accessible show that we had drastically underestimated the extent of transfer over short and long evolutionary time scales," and "proteins are retained in organelles, but most of the genes are not," respectively.

Thus, we have three categories of results about cellular information: first, the metabolic chart, which is probably about 3.5 billion years old, is enormously robust. The chart is a fundamental feature of biology in the same sense that the periodic table of elements is a fundamental feature of chemistry. Second, the proteins involved in catalyzing the steps of the chart are drifting with respect to amino acid composition. Third, the genes that encode the protein sequences undergo an enormous intertaxonomic transfer because of processes such as transduction, transfection, and transformation.

Over the last 50 years, two powerful icons of biology have become firmly enough established to embellish the walls of offices and laboratories. They are the universal chart of intermediary metabolism and the universal tree of life based on ribosomal RNA sequences.[2] The metabolic chart, because of its ubiquity across the taxa, would appear either to go back to the universal common ancestor some 3.5 to 4 billion years ago or to be the optimal solution to metabolism that arose later and became universal by lateral

Table 7.1. Distribution of amino acid index numbers in various taxa and organelles

	Low-Index Number	Percentage High-Index Number
Bacteria	27.37	19.40
Eukayotes	27.23	15.25
Mycoplasma	27.57	20.29
Rickettsia	26.45	22.28
Chloroplasts	22.55	23.32
Mitochondria	17.29	26.48

gene transfer. The phylogenetic tree presumably evolves in time by mutations in the DNA coding the ribosomal RNA.

Over the last 20 years, extensive studies on gene sequences have confirmed that the tree of life established with ribosomal RNA does not accord very closely with trees established using other genes.[3] This is now believed to be the result of the "lateral movement of genes across vast phylogenetic distances."[4]

The chart of intermediary metabolism is part of the phenotype; indeed, for prokaryotic chemolithoautotrophs it is a substantial part of the total phenotype. Thus we have a mapping between a robust phenotype and a somewhat noisy changing genotype.

This reversal of our view of the source of robustness would indicate that the fundamental laws of biology are phenotypic. The informatic embodiment is in the genome, but the emergent principles that we seek for understanding are phenotypic in character.

When we deal with the metabolic chart as phenotype, the rules are likely those dealing with networks of chemical reactions for organic molecules. Thus, as in the case of the emergent Pauli principle in chemistry, new rules may be discovered. It seems clear from a study of the chart that there are many uniformities and that general principles are anticipated. The universality of the core of the chart, primary metabolism, is a necessary feature of trophic ecology, for without it there would be a number of noninteracting pools of metabolites, and we could not speak of a single material biosphere in any meaningful sense. It is also possible that there are other phenotypic rules in the hierarchy going from atoms to complex organisms.

The "dogma of molecular biology" makes the genome the primary construct and moves from genome to proteome to metabalome to physiome to phenome. The view outlined here indicates that the primary laws relate to phenotype and that the epistemic direction is the reverse of that outlined in the dogma. This would be a major paradigm shift and would lead to more effort on the hierarchy of phenotypic laws.

References

1. H.J. Morowitz, E. Peterson, and S. Chang, "The Synthesis of Glutamic Acid in the Absence of Enzymes, Implication for Biogenesis," *Origins of Life* 25 (1995): 395–399.
2. C. Woese, "The Universal Ancestor," *Proceedings of the National Academy of Sciences USA* 95 (1998): 6854–6859.
3. W.F. Doolittle, "Evolution, Uprooting the Tree of Life," *Scientific American* 282 (2000): 90–95.
4. R. Jain, M.C. Rivera, J.E. Moore, and J.A. Lake, "Horizontal Gene Transfer in Microbial Genome Evolution," *Theoretical Population Biology* 61 (2002): 489–495.

8

Molecular Sequences and the Early History of Life

RADHEY S. GUPTA

The evolutionary history of life, spanning a period of more than 3.5 billion years (Giga annum or Ga) constitutes one of the most fascinating problems in the life sciences.[1–4] This chapter will critically examine our understanding of a number of aspects of early evolutionary history. The topics covered are critical issues in Bacterial phylogeny, lateral gene transfer (LGT) and its influence on evolutionary relationships, the relationship of Archaea to Bacteria, and the origin of eukaryotes.

Bacterial Phylogeny: Some Critical Issues

The Bacteria make up the vast majority of prokaryotes. Hence, discerning the evolutionary relationships among them constitutes a major part of understanding prokaryotic phylogeny. On the basis of branching in the 16S rRNA trees, about 25 main groups or phyla within Bacteria are recognized at this time.[5] Although Bacteria have been divided into phyla on the basis of 16S rRNA, the criteria as to what actually constitutes a phylum remain to be defined.[5,6] In the beginning, when the sequence database was limited, the main phyla could be clearly distinguished in phylogenetic trees on the basis of long, "naked" internal branches that separated them. However, the explosive increase in sequence database entries in recent years has filled most of these naked branches and, as a result, distinguishing between phyla has become increasingly difficult and imprecise.[5,6] In the absence of objective criteria for the main divisions, it is unclear at present how many phyla exist within Bacteria and how to distinguish them from their subdivisions. On the basis of 16S rRNA, the proteobacterial phylum is presently divided into five subdivisions, named α, β, γ, δ, and ϵ.[3,7] Some of these subdivisions, for example, α, β, and γ, contain from several hundreds to thousands of species, and they are much larger than most other main groups within Bacteria. The members of these subdivisions are also clearly distinguished from each other and other bacterial divisions, both in phylogenetic trees and by distinctive signature sequences.[7] Lacking objective criteria, it is unclear why these major groups have been assigned subdivision status, whereas many poorly characterized taxa consisting of only a few species are recognized as distinct divisions.

Any understanding of bacterial phylogeny requires that one can determine how main groups are related to each other and their branching order from a common ancestor. Unfortunately, phylogenetic trees based on rRNA have not been able to resolve these relationships and, thus, have yielded contradictory results.[3,5,8,9] This has led to the notion that this important problem is unsolvable, and even to the erroneous assumption that most, if not all, main groups within Bacteria may have branched off from a common ancestor at about the same time.[5,9] Over the last few years we have developed a means for understanding bacterial phylogeny.

Signature Approach for Determining Bacterial Phylogeny

Our new approach is grounded on conserved inserts and deletions (referred to as indels or signature sequences) in protein sequences for deducing phylogeny. On the basis of the presence or absence of shared conserved indels, different species can be divided into distinct groups, and their specific evolutionary relationships can be revealed. The rationale for our approach is that when a conserved indel of defined length and sequence is present in the same position in a given gene or protein from all members from one or more groups of bacteria, but not in the other groups, the simplest and most parsimonious explanation is that the indel was introduced only once, in a common ancestor of the group of species that possess this characteristic. All evolutionary useful signatures need be flanked on either side by conserved regions to ensure their reliability.

The signatures that we have identified are of two main kinds. Many of them are group specific: they are uniquely present in protein homologs from particular phyla or subdivisions (referred to as groups) of Bacteria. One example of a group-specific signature is provided in figure 8.1, where a conserved insert of 18–21 aa is present in the DNA polymerase I (Pol I) from various cyanobacteria, but not in any other groups of bacteria. Cyanobacteria-specific signatures are also present in many other proteins including DNA helicase II, ADP-glucose pryophosphorylase, FtsH protease, phytoene synthase, EF-Tu, Sec A, ribosomal S1 protein, IMP-dehydrogenase, and the major sigma factor 70 (table 8.1).[10] Similar to cyanobacteria, a large number of signatures that are distinctive for the chlamydiae have also been identified.[11] Group-specific signatures have also been identified for most other major groups within Bacteria including Proteobacteria, Aquificales, Firmicutes, Actinobacteria, Deinococcus-Thermus, Spirochetes, Cytophaga-Flavobacteria-Bacteriodetes-Green sulfur bacteria (CFBG)[4,7] (Gupta, R.S., unpublished results). These signatures provide a means for identifying different bacterial groups in clear molecular terms and for assignment of species to these groups.

A second category of signatures comprises those in which a conserved indel is commonly present in several groups of bacteria, but is absent in other groups. These signatures, which I will refer to as main line signatures, have been introduced at important

Group	Species	ID	855	(boxed insert)	919
Cyanobacteria	Synechocystis PCC 6803	16331949	GYVTTIVGRRRYFNFVTEALRQLRGKTV	TELDLVDVK MNYNDAQLL	RSAANAPIQGSSADIIKIA
	Nostoc sp. PCC 7120	17228749	---E--L------D-TNNS--K-K-SKP	EDI--SKL- NLGPY--G--	-------------------
	Trich. erythraeum	23040910	---E--L------ELESKTI-D-KD-EP	E-IN-QEL- FSQ----I-	-----S-------------
	Nostoc punctiforme	23125316	---E--L------D-TNNS--R-K-SNP	EDI--SKL- NLGPY--G--	-----------N-------
	Thermosyn. elongatus	22298882	---E--L------A-ESRE-QS----PL	DV-AD--PSKLK-SNYERG--	-A--------------C-
	Synechococcus WH 8102	23133301	---E--L----P-H-DRNG-GR-L--EP	L-I--DVAR RGGME--Q-	-A--------------V-
	Pro. marinus pastoris	23122969	---E--F--K-E-K-DKNG-GR-I--DP	Y-I--QTAR KAGME--S-	-A-----------------
	Pro. marinus MIT 9313	23131695	---E-LL----P-H-DRNG-GR-L--DP	MDI--DVAR RGGME--Q-	-A--------------V-
Proteobacteria	E. coli K12	16131704	---E-LD---L-LPDIKSSNGAR-AAAE		-A-I---M--TA-----R-
	Vib. cholerae	15640140	---E--F---LHLPEI-SRNAMR-KAAE		-A-I---M--TA-----K-
	Xyl. fastidiosa	15837705	---E-VF---L-L-SIASGNQTQ-AGAE		-A-I---M--TA-----R-
	Ral. solanacearum	17546949	---E-VF---LWLPDINGGNGPR-QAAE		-A-I---M--TA--L--LS
	Nei. meningitidis	15793464	-F-E-LF---L-LPDIRNKNTNA-AGAE		-A-I---M--TAS-L--R-
	B. melitensis	17988108	---E--F---AHYPDIRASNP-V-AFNE		-A-I------AA----RR-
	A. tumefaciens	15887465	---E--F---AHYPEIRSSNPSVKAFNE		-A-I-------A----RR-
	C. crescentus	16127694	------F--KINIPDIEAKSAAH-QFAE		-A-I------AA--VMRR-
	Ri. conorii	15893129	----NCF--KCFVPLI HDKK-KQFAE		-A-I------TNT------
	Hel. pylori	2314647	TSKA-LL--YRV-D FTGANDYVKGNYL		-EGV--IF---AS-LL-L
	Camp. jejuni	15791706	-FIV-LS--K---D FEN-KPMQIAMYE		-ESI-SIL---A--V--L-
Chlamydia, CFBG group	Chl. trachomatis	15605221	---ML--E-ILSDWESS PGA-AASG		-L-V-TR----A-EL--L-
	Chl. muridarum	15835394	---ML--E-ILSDWESS PGA-AASG		-L-V-TR----A-EL--L-
	Chlam. pneumoniae	15618522	---ML--E-IIDSWN-F PGS-AASG		-F-V-TR----A-EL--L-
	Cyto. hutchinsonii	23138023	--E--L----FLRDINSRNMTM--FAE		-N-I-------A--M--V-
	Chlorobium tepidum	21674485	-----LM-----VPDLNS-NSNI-KAAE		-VTM-T----TA-----F-
Spirochetes, Chloroflexus, Deinococcus	Bor. burgdorferi	15594893	--SE--LK----IKEINSNNYLE-SAAE		-I-I-SI----A---M---
	Tre. pallidum	15639099	----SLA-----IRTIDSRNTLE-ARAE		-M-L-TQ--S-A---V---
	Lep. interrogans	24216324	---Q-LT----PVTDINSTHKSAKEAAK		-I-I-S----T---M----
	Cfx. aurantiacus	6015002	---Q-LF----VMEDLRAS GAR-AAAE		-E-I------TA--LM-M-
	D. radiodurans	15806710	---E-LY-----VPGLSSRN-VQ-EAEE		-L-Y-M----TA---M-L-
	Thermus thermophilus	1097211	---E-LF-----VPDLNARVKSV-EAAE		-M-F-M-V--TA--LM-L-
	T. maritima	15644367	---R-LF--K-DIPQLMARD-NTQAEGE		-I-I-T----TA-----L-
Gram-positive Bacteria	Clo. acetobutylicum	15894383	--N--MN---IIPEIASSNKIVK-FGE		-L-M-T-----A--L--LS
	Fuso. nucleatum	19714234	---K-LF--K--ISGIDSKNKTIKSQAE		-M-V-TV---TA-EVL-
	Myc. tuberculosis	15608767	--TS-VL-----LPELDSSN--V-EAAE		-A-L-------A-----V-
	Str. coelicolor	21220485	--TA-LF-----LPDLNSDN--R-EAAE		-M-L------TA---V---
	Cor. glutamicum	19552567	--TE-LF-----LPEL-SDN-VA-ENAE		-A-L------TA-----V-
	Bif. longum	23336564	--TE--F------PALHSTN-VA-EAAE		-A-L-------A---M---
	Thermobifida fusca	23017878	--TA--L-----LPDLNSDN--R-EMAE		-M-L-------A-----L-
	Sta. aureus	21283362	---E-LLH----IPDI-SRNFN---FAE		-T-M-T-----A-----L-
	Oceano. iheyensis	23099618	-F-S--MK----LPDI-SRNFNM-SFAE		-T-M-T-----A-----K-
	Bac. subtilis	16079961	-----LMH----IPEL-SRNFNI-SFAE		-T-M-T-----A-----K-
	L. lactis	15674124	-F-E-MSH---KIPDINARNFNV--FAE		-T-I-S-----A---L---
	Li. monocytogenes	16803605	---E--LH----IPEIVSRNFNV--FAE		-T-M-T-----A-----K-
	Strep. pneumoniae	15899978	---E-LFK---ELPDINSRNFNI--FAE		-T-I-S-----A---L---

Figure 8.1. Sequence alignment of DNA polymerase I showing a large insert (boxed) that is specific for cyanobacteria. Dashes in the alignment indicate identity with the amino acid on the top line. Polymerase I is present in all bacterial genomes, and sequence information for only representative species is presented.

evolutionary branch points and thus are very useful in understanding the branching order and interrelationships among different groups.[4] If an indel was introduced in an ancestral lineage at a critical branch point (i.e., in the main trunk of the tree), then it is expected that all species diverging from this ancestor at later times should contain the signature, whereas all other species originating from the branches that existed before introduction of the signature should be lacking the indel.[6,12] Thus, on the basis of different main-line signatures that have been introduced in the main evolutionary trunk at different stages, the order of divergence of different groups can be established.

Two examples of main line signatures are shown. In the Hsp70 (DnaK) protein found in all bacteria (fig. 8.2), a 21–23-aa insert is present in various Proteobacteria, Chlamydiae, CFBG, Aquifex, Spirochetes, Cyanobacteria, and Deinococcus-Thermus groups of bacteria, but it is not found in any of the species from the Thermotoga, Clostridia-Fusobacteria, Actinobacteria, and Firmicute groups. This indel is also absent in various Archaea, and on the basis of the established rooting of the prokaryotic tree between Archaea and Bacteria,[13,14] this observation indicates that the groups lacking this indel are ancestral and that this indel constitutes an insert in the later branching groups. The ancestral nature of the groups lacking this indel is also supported by other lines of evidence discussed in earlier work[4,15] Another example of a main-line signature found in the RNA polymerase β-subunit (RpoB), is shown in figure 8.3. RpoB is a core compo-

Table 8.1. Sequenced bacterial genomes

Proteobacteria (γ-subdivision)
Escherichia coli K12
Escherichia coli O157:H7
Escherichia coli O157:H7 EDL933
Escherichia coli CFT073
Buchnera sp. APS
Buchnera aphidicola
Buchnera aphidicola Sg
Pasteurella mutocida
Pseudomonas aeruginosa
Pseudomonas putida KT 2400
Pseudomonas syringae
Vibrio cholerae
Vibrio parahaemolyticus
Vibrio vulnificus
Xylella fastidiosa
Xyella fastidiosa Temecula
Haemophilus influenzae
Yersinia pestsis C092
Yersina pestsis KIM
Salmonella typhimurium LT2
Salmonella typhi
Xanthomonas citri
Xanthomonas campestris
Xyellela fastidiosa
Shewanella oneidensis
Shigella flexneri 2a
Wiggelsworthia brevipalpis
Coxiella burnetii

Proteobacteria (α-subdivision)
Rickettsia prowazekii
Caulobacter crescentus
Mesorhizobium loti
Bradyrhizobium japonicum
Agrobacterium tumefaciens-Dupont
Agrobacterium tumefaciens-Cereon
Rickettsia conorii
Sinorhizobium loti
Brucella melitensis
Brucella suis
Rhodopseudomonas palustris

Proteobacteria (β-subdivision)
Neisseria meningitidis MC58
Neisseria meningitidis Z2491
Ralstonia solanacearum

Proteobacteria (δ, ϵ-subdivision)
Helicobacter pylori 26695
Helicobacter pylori J99
Campylobacter jejuni

Aquifex
Aquifex aeolicus

Chlamydia-CFBG
Chlamydia trachomatis
Chlamydia muridarum
Chlamydophila pneumoniae CWL029
Chlamydophila pneumoniae J138
Chlamydophila pneumoniae AR39
Chlorobium tepidum
Bacteroides thetaiotamicron

Spirochetes
Borrelia burgdorferi
Treponema pallidum
Leptospira interrogans

Cyanobacteria
Synechocystis sp. PCC6803
Nostoc sp. PCC7120
Thermosynechococcus elongatus

Clostridia-Thermotoga
Thermotoga maritima
Clostridium acetobutylicum
Clostridium perfringens
Clostridium tetani E88
Fusobacterium nucleatum
Thermoanaerobacter tengcongensis

Deinococcus-Thermus
Deinococcus radiodurans

Actinobacteria
Mycobacterium tuberculosis H37
Mycobacterium tuberculosis 1551
Mycobacterium leprae
Corynebacterium glutamicum
Corynebacterium efficiens
Streptomyces coelicolor
Bifidobacterium longum
Tropheryma whipplei Twist
Tropheryma whipplei TW08/27

Firmicutes
Bacillus subtilis
Bacillus halodurans
Bacillus antharics
Oceanabacillus iheyensis
Staphylococcus aureus N315
Staphylococcus aureus MW2
Staphylococcus epidermidis
Staphylococcus aureus Mu50
Streptococcus pyogenes
Streptococcus pyogenes S315
Streptococcus pyogenes S8232
Streptococcus pneumoniae R6
Streptococcus pneumoniae TIGR4
Streptococcus agalactiae 2603
Streptococcus agalactiae NEM316
Streptococcus mutans UA159
Mycoplasma genitalilum
Mycoplasma pneumoniae
Mycoplasma pulmonis
Mycoplasma penetrans
Ureaplasma urealyticus
Lactococcus lactis
Lactobacillus plantarum
Listeria innocua
Listeria monocytogenes

Group	Organism	Accession	61					123
Proteobacteria	E. coli	SP/ P04475	NPQNTLFAIKRLIG	RRF	QD	EEVQRDVSIMPFKIIAAD	NGDAWVEVKGQK	MAPPQISAEVLLKM
	X. fastidiosa	AAF85139	--K--FY-V-----	-K-	G-	A---K-LDLV-Y--TQH-	------ATADAK	L--QE---K--E--
	Pse. aeruginosa	AAG08147	------Y-V-----	---	EE	NV--K-IQMV-YS-VK--	------------	-----------K--
	Pas. multocida	15602601	--K-----------	---	--	------------E-V---	------G---E-	--------------
	Vibrio cholerae	2623605	--------------	---	E-	------IK---Y--VK--	-------A----	--A--V--------
	Ral. solanacearum	16429657	--R---Y-V-----	-K-	EE	K---K-IGL--YT-SK--	--------RDK-	-----------R--
	Nei. meningitidis	H81185	-AK--IY-A-----	HK-	E-	K-----IES---E--K-N	------KAQ-KE	LS---------R--
	Bru. melitensis	17988285	--EG----V-----	--Y	D-	PM-TK-KDLV-Y--VKG-	--------H-K-	YS-S----MI-Q--
	Mesorhizobium loti	13473981	--E--I--V-----	--Y	D-	PVTEK-KKLV-Y--VKG-	-------AG-K-	QS-S----MI-Q--
	Ri. conorii	15892156	--R--IY-V-----	-N-	I-	PM-RK-QG-V-YN-VK--	-------ADNN-	YS-S----FI-Q--
	Rh. capsulatus	1373328	--T--V--V-----	--T	T-	A--EK-KKLV-YN-VDGG	--------R-E-	FS-A-V--VI-Q--
	A. tumefaciens	SP/ P20442	--T-----V-----	--Y	E-	PT-EK-KALV--E-VKG-	------KAQDKN	YS-S----MI-Q--
	Ca. crescentus	GB/ M95799	--T-----------	-TA	S-	PV-EK-KGMV-YRSSR-R	A-----KAH-KD	YS-QEV--FI-Q--
	Myx. xanthus	1805284	--E--V--A-----	-K-	DS	P-GKKAIGVS---VASSP	-------IR-KG	YS--EV--I--M--
	Geo. sulfurreducens	ZP_00080430	--E-----------	-KY	DT	---RK-I--S----VK--	-------AR-KM	YSA-E---M--Q--
	Des. vulgaris	ZP_00130430	--ER-V--V---M-	--G	DA	P--G-WKEHS-YR-V-GA	----A---Q-RP	YSA-E---MI-G-L
	Camp. jejuni	CAB73024	--EK-IYS---IM-	LMI	NE	DAAKEAKNRL-YH -TER	--ACAI-IA-KI	YT-QE---K--M-L
	Hel. pylori	2072520	--EK-IYS---IM-	LM-	NE	DKAKEAEKRL-Y--VDRN	GA-CAI-IS-KI	YT-QE---KI-M-L
Chlamydiae/ *CFBG* Group	Chl. trachomatis	SP/ P17821	--EK--AST--F--	-K-		S--ESEIKTV-Y-VAPNS	K---VFD-EQKL	YT-EE-G-QI-M--
	Chl. muridarum	AAF39496	--EK--AST--F--	-K-		S--ESEIKTV-Y-VAPNS	K---VF--ENKL	YT-EE-G-QI-M--
	Chlam. pneumoniae	AAF38114	--EK--GS---F--	-KY		S--ASEIQTV-YTVTSGS	K---VF--D-KQ	YT-EE-G-QI-N--
	Cyt. aquatilis	AF013110	--TK-IAS---FM-	HT-		A-TTNESKRVSY-VVKGV	QQYST-DID-RL	YTAQEL--MT-Q--
	Fl. ferrugineum	AF013111	-----ITSV--FM-	-G-		N--TEEI-HWSY-VAQG-	-NTVRIDID-RL	YT-QE---M--Q--
	Fib. succinogenes	AY017382	--EK-IYS---FM-	-TA		G-CSAAEKN--Y-LVGTG	SDPVR-QIDDKQ	F---E---A--QT-
	Por. gingivalis	Q9ZAD3	--TK-IYS---FM-	ETY		DQ-S-E-ERV---VVRG-	-NTPR-DID-RL	YT-QE---MI-Q-L
	Cb.tepidum	AF130447	--K--I-S---FM-	-KY		D--PNEKKLASYDVVN-E	G-Y-K-KIGDKT	YS-QE---MI-Q--
Aquificales	C. hydrogenophilum	AY188750	D-E--IYES--F--	-KS		---KEEAKRVSY-VVPDE	K---SFD-PNAGRL	VR-EEVG-HI-K-L
	Tc. ruber	AY254212	D-E--IYES--F--	MK-		-D-KEEAIRVSYMVV-DK	K---AFD-PNAGRL	AR-D
	Hydro. marinus	AY188440	D----IYES--F--	-K-		D--KEEIKYV-Y-VV-D-	K---AFD-PNAGKI	VR-EEVG-Q--K-L
	Aqu. pyrophilus	CAA06703	D-E--VYES--F--	-K-		---KEEAKRVSY-VVPDE	K---AFDIPNAGKL	VR-EEVG-H--R-L
	Aqu. aeolicus	2983493	D-E--VYES--F--	-K-		N---KEAKRVSY-VVPDE	K---AFDIP-KL	VR-EEVG-H--R-L
Spirochetes	Tre. pallidum	3322484	--EH-IYS---F--	S--		N-LTGEAKKV-Y--VPQG	-D-VR---E-KL	YSTQE---FI-Q--
	Bor. burgdorferi	SP/ P28608	--E--IYS---FM-	---		---ASEIKMV-Y--EKGL	----R-NISNI-KQ	-S--E---AT-T--
	Lep. interrogans	AAN50903	-AV--IRSA--F--	--L		N-CESEMKHVSY-V-RSG	-EGVKF-TSAGE	FT-QE---R--M--
Cyanobacteria	Synechocystis 7942	GB/ D29668	--R--FAN---F--	--Y		D-LTDESKRV-YTVRRDP	E-NVRIVCPQLSRE	F--EEVA-MI-R-L
	Synechococcus 6803	GB/ M57518	--G--FYSV--F--	-K-		D-ITNEATEVAYSVVKDG	--NVKLDCPA-GKQ	F--EE---Q--R-L
	Anabaena variabilis	2073390	-----F--V--Y--	--Y		N-LSPESKRV-YT-RKD-	V-NIK-ARLNKE	F-AEE---M--K-L
Deinococcus-Thermus, GNS	Cfx.aurantiacus	AF130446	--E---YSV--F--	-S-		D--TEERDV----VVKGP	RN-VRIY-PQTNKE	Y--QE---M--Q-L
	D. proteolyticus	1813672	---A---EV--F--	--W		D--KDEAARS--TVKEGP	G-SVRI--D-KD	Y--E-V-----R-L
	D. radiodurans	A75557	--AA---EV--F--	--W		D--KEEAARS--TVKEGP	S-SVRI--N-KD	L--E-V-----R-L
	Thermus aquaticus	GB/ Y07826	--EG-I-E---F--	---		----EEAKRV-Y-VVPGP	D-GVR-----KL	YT-EE---MI-R-L
Thermotoga	T. maritima	3882431	--ER-IKS---KM-				T-YK-RIDDKE	YT-QE---FI-K-L
Gram-positive Bacteria	Clo. acetobutylicum	118726	--DK-IIS---KM-				TAEK-AIDDKN	YT-QE---MI-Q-L
	Clo. perfringens	SP/ P26823	--DK-IMS---HM-				T-YK-NID-KD	LS-QE---MI-Q-L
	Myc. leprae	SP/ P19993	-VDR-IRSV--HM-				S-WSI-ID-K-	YTAQE---R--M-L
	Str. coelicolor	GB/ 014499	-VDR-IRSV--HM-				T-WK-NLD-KD	FN-Q----F--Q-L
	Bif. longum	AAN24348	-VDR-ISSV--HM-				S-WT-DID-K-	WT-QE--AQILMKL
	Cor. glutamicum	Q8NLY6	-VDR-IRSV--H--				T-WS-AIDDKN	YTSQE---RT-M-L
	Bac. subtilis	GB/ U39711	-- --IMS---HM-				T-YK--IE-KD	YT-QEV--II-QHL
	M. pneumoniae	1674090	-- --IVS----M-				TSNK-T--NPD	GSTKEL--QI-SYL
	M. genitalium	1361531	-- --IVS----M-				TSNK-K-TTKE	LS-E-V--QI-SYL
	Strep. mutans	2145133	-- E-ILS--SKM-				TSEK-SANAKE	YT-QE---MI-QYL
	L. lactis	GB/ M98865	-- E-IIS--SKM-				TSEK-SAN-KE	YT-QE---MI-QNL
	Sta. aureus	GB/ X75428	-- --VQS---HM-				T-YK-DIE-KS	YT-QE---MI-QNL
Archaea	The. acidophilum	7522204	--EG-I--A--KM-				T-YKFK-FDKE	FT-Q----FI-Q-I
	Methanosarcina mazei	1169378	--D--VYS---HM-				EANYK-TLN-KD	YT-QE---MI-Q-L
	H. marismortui	232284	--DE-IQS---HM-				QD-YS--LD-EE	YT-E-V--MI-Q-I

Figure 8.2. Partial alignment of Hsp70 sequences showing two main line signatures. The large insert (box 1) is a distinctive characteristic of various gram-negative bacteria (as defined by the presence of an outer membrane), and it is not found in any gram-positive or monoderm bacteria. This insert is also not present in any Archaea. The smaller, 2-aa insert (box 2) is a distinctive characteristic of Proteobacteria.

nent of the RNA polymerase found in all bacterial genomes. The signature in this case consists of a large indel of between 90 and 133 aa that is commonly present in various proteobacteria, Aquifex, and the Chlamydia–CFBG groups, but not in any other groups of bacteria.[16,17] This indel is also not present in the RpoB homologs from Archaea, indicating that it constitutes an insert in the latter branching groups. On the basis of its specific presence only in Proteobacteria, Aquificales, and the Chlamydia–CFBG groups, this insert was likely introduced in a common ancestor of these groups after branching of the other groups (fig. 8.4).

We have described a large number of other main-line signatures that are helpful in determining Bacterial evolutionary relationships. Our analyses of these signatures, as shown here for Hsp70 and RpoB proteins, indicate that they have been introduced at

Group	Species	Accession	919	Insert	1058
	E. coli	42818	RVPNGVSGTVIDVQVFTRD	GVE(91aa) KRRKIT	QGDD LAPGVLKIVKVYLAVKR
	Pas. multocida	15603602	----S--------------	---(91aa) Q-G--I	---- -------V---------
	Pse. aeruginosa	15599466	---T-TK------------	---(103aa)-K--LQ	---- --------------I--
	Vibrio cholerae	15640355	------A------------	---(92aa) ------	---- -----------------
	X. fastidiosa	15839222	---P-MD------------	-I-(104aa)--S---	---- -------M---F-----
	Ral. solanaceum	17547753	---S-M------------E	--T(104aa)--K-L-	---E P---I-M----------
	Nei. meningitidis	15676060	-M-T-M------------E	-IQ(104aa)-KK-L-	---E -Q---Q-M---FI-I--
Proteobacteria	Rho. rubrum	ZP_00013726	-I-P--T---VE-R--N-R	---(104aa)-KVEKL	-RG-E-P-----M---FV----
	Ri. prowazekii	6652726	H--S------VE-RI-S-R	---(104aa)-VE-LQ	S--- -PQ-A--V---FI-T-H
	Bru. melitensis	17987032	-M-P-TY---VE-R--N-H	---(104aa)-VE-VQ	R--E MP---M-M---FV----
	Ca. crescentus	16124757	-L-P--A--IV--R--N-H	--D(103aa)-VD--Q	R--E -P---M-MV--FV----
	A. tumefaciens	15889250	-M-P-TF--IVE-R--N-H	---(104aa)-VE-VQ	R--E MP---M-M---FV----
	Des. desulfuricans	ZP_00129103	KV-P-IE------KL-N-R	SG-(111aa)--E-V-	E--- -P---I-MA--HI----
	Hel. pylori	15645812	YC-PSLE------K---KK	-Y-(108aa)EKLS-L	EK--I-PN--I-K--L-I-T--
	Camp. jejuni	15791842	YATASLE-V-V--KI--KK	-Y-(108aa)EKLE-L	EK--I-PS--I-L----I-T--
	Aqu. aeolicus	15606949	-C-P--E-I---------K	-TG(110aa)EKKETL	LKRRD-P---ITL---FI-N--
Aquificales	Aqu. pyrophilus	4753643	-C-P--E-I---------K	-GK(110aa)EKKETL	LKRRD-P---ITL---FI-N--
	Hydro. marinus	AY188442	---T--E-I-V-----A-K	-I-(125aa)-IKEVE	K-A- -K---NEL----I-Q--
	C. hydrogenophilum	AY188443	-C-P--E-V-V-----A-K	VG-(110aa)-EES-G	KRSE -PA--IAL----I-Q--
	Cyt. hutchinsonii	ZP-00119276	KA-PSLR-V---TKL-S-P	KKD(133aa)RE-FTL	EVG-E-PA-IVQLA---I-K--
	Cb. tepidum	AAM71403	H--A-MK-I--KTKL-S-K	KKI(110aa)EKY--N	V--E -P--IEELA---I-Q--
Chlamydiae/ CFBG Group	Por. cangingivalis	10637868	KANPSL--V--KTHL-SKA	MHS(107aa)RKKFDA	TIG-E-PN-IIQ-A--LI-K--
	Chl. trachomatis	15605036	T--P-TE-V-M--K--S-K	DRL(112aa)EAEH-K	E--AD-DH--IRQ----V-S--
	Chlam. pneumoniae	16753076	T--P-TE-V-M--K--S-K	DRL(112aa)EVEH-R	E--AD-DH--IRQ----V-S--
	Lep. interrogans	AAN50618	-M---FE-----IKR-S-E		NQ-E -PA--EEM---FV-R--
Spirochetes	Bor. burgdorferi	15594734	K--H-TE-------RI-KE		DVGN -S---EE-L---V-K--
	Tre. pallidum	15639233	---H--E-------RLR-S		E----N---SEV---LI-T--
	Nostoc sp. PCC 7120	17229086	-----EK-R-V--RL---E		---E -P--ANMV-R--V-Q--
Cyanobacteria	Synechocystis 6803	16329957	-----EK-R-V--R----E		K--E -P--ANMV-RI-V-Q--
	Thermo. elongatus	BAC08193	-----EK-R-V--R----E		---E -P--ANMV-R--V-Q--
Deinococcus-Thermus, GNS	Cfx. aurantiacus	ZP_00017399	------R-K----K--S-S		E-AE -PV--NQT-R-L-CQ--
	Thermus aquaticus	20139789	---P-EG-I-VGRLRLR-G		DPGVE-K---REV-R-FV-Q--
	D. radiodurans	15805937	--QS-QG-I-VKTVR-R-G		DEGVD-K---REM-R--V-Q--
Thermotoga	T. maritima	15643224	-L-H--E-R--R-D-YDQN		DIAE -GA----L-R--V-SRK
	Cor. glutamicum	19551731	K--H-ET-K--G-RH-S-E		DD-- -----NEMIRI-V-Q--
	Myc. leprae	418765	K--H-E--K--GIR--SHE		DD-E -PA--NEL-R--V-Q--
	Str. ceolicolor	21223036	K--H-EI-K--G-R--D-E		E--E -P---NQL-R--V-Q--
	Therm. tengcongensis	AAM25443	TM-H-SK-V-V-ILELS-E		N--E -KA--N-SIR-LV-E--
	Ocean. iheyensis	BAC12068	---H-GG-I-L--KI-N-E		D--E -P---NQLVRA-IVQ--
	Bac. subtilis	CAB11883	---H-GG-IIH--K--N-E		D--E -P---NQL-R--IVQ--
Gram -positive Bacteria	Lis. innocua	16799362	---H-GG-I-L--KI---E		A--E -P---NQL-R--IVQ--
	Clo. perfringens	18311395	---H-EA-IIV--K----E		N--- -S---NEL-RC-I-Q--
	Sta. aureus	15923532	---H-AG-I-L--K--N-E		E---T-S---NQL-R--IVQ--
	M. genitalium	12045200	K-SH-GD-I-SA-KR-SIA		N--E -ND--IEMI---VVQ--
	Strep. pneumoniae	15901784	---H-AD-V-R--KI---V		N--E -QS--NML-R--I-Q--
	L. lactis	NP_267957	---H-GG-I-H--R----E		N--E -PS--N-L-R-FI-Q--
	Halo. sp. NCR-1	AAG20693	TMRS-ED-VVDT-TLMEG-		DGSK-AK-SVRDE-
Archaea	Meth. barkeri	ZP_00079000	TMRSNET-I-DT-ILTESI		NGTRLA--KVRDE-
	Pyr. aerophilum	AAL62934	A-RR-EK-I-DK-IITESP		EGN-L---R-REL-

Figure 8.3. Partial alignment of RNA polymerase β subunit (RpoB) showing a large insert (>100 aa) that is specific for the Proteobacteria, Chlamydiae–CFBG group, and Aquificales groups, but that is not found in any other bacteria. The absence of this insert in archaeal homologs provides evidence that the groups lacking this insert are ancestral.

specific stages in bacterial evolution, as depicted in figure 8.4. On the basis of the presence or absence of these signatures, all main groups within Bacteria can be clearly distinguished, and it is also possible to logically deduce that they have branched off from a common ancestor in the order shown in figure 8.4.[6,12,17,18]

Testing the Indel Model on Bacterial Genomes

The completed bacterial genomes provide an objective means to test the reliability of the deduced branching order based on our signature sequence approach. At the end of April 2003 sequences for 100 bacterial genomes were available in the public domain (http://www.ncbi.nlm.nih.gov/genomes/MICROBES/complete.html). The main groups to which these species belong are indicated in table 8.2. The branching order of these groups as shown in figure 8.4 makes very specific predictions as to which of the different indels should be present or absent in different species. According to the indel model,

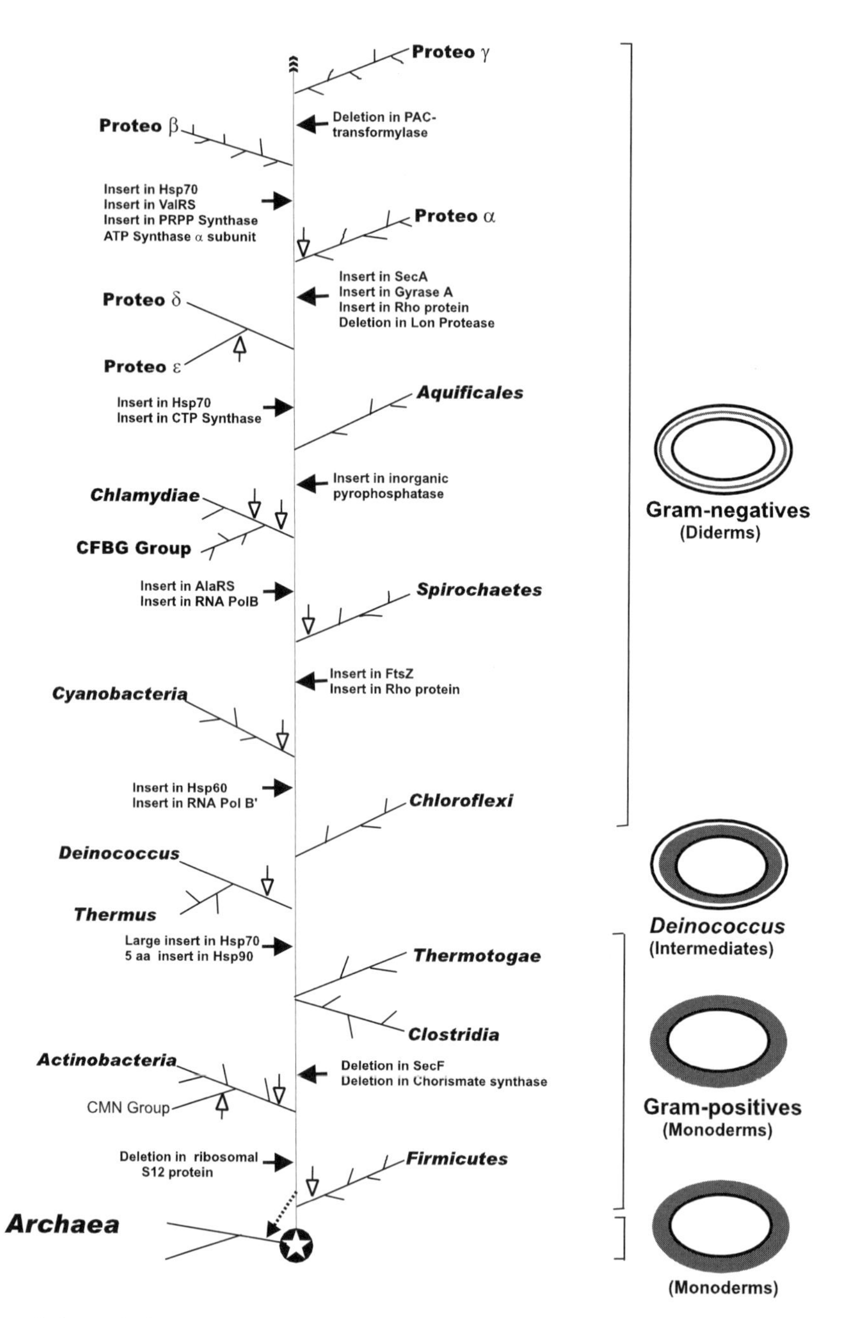

Figure 8.4. Evolutionary model based on signature sequences indicating the branching order of the main bacterial groups. The filled arrows depict the stages at which the different main-line signatures indicated in table 8.3 have been introduced. These signatures are expected to be present in bacterial groups that have diverged at a later time (i.e., those lying above the indicated insertion points), but they should be absent in the earlier branching groups. The unfilled arrows denote the positions of many group-specific signatures (not shown here). The cell structures of different groups of bacteria are indicated on the right. The dotted arrow at the bottom indicates the possible derivation of Archaea from gram-positive bacteria.

Table 8.2. Predicted versus observed distribution of indels in 100 bacterial genomes

Protein	Signature Description	No. Genomes with Protein	No. Genomes with Indels Expected/ Found	No. Genomes Lacking the Indel Expected/ Found	Exceptions Observed
Rib. S12 protein	13 aa *Firmicute* insert	100	25/25	75/75	0
Hsp70/DnaK	21–23 aa G+/G- insert	100	60/60	40/40	0
Hsp90	5 aa G+/G- insert	52	11/11	41/41	0
Chorismate Synthase	15–17 aa deletion after *Actinobacteria*	89	29/29	60/60	0[a]
SecF protein	3–4 aa deletion after *Actinobacteria*	81	15/17	56/54	2[b]
Hsp60/GroEL	1 aa insert after *Deinococcus*	98	65/66	33/32	1[c]
RNA Polymerase β^{1-} subunit	>150 aa after *Deinococcus*	100	59/59	41/41	0
FtsZ protein	1 aa insert after cyanobacteria	91	51/51	40/40	0
Rho protein	2 aa insert before spirochetes	83	56/57	27/26	1[d]
Ala-tRNA Synth.	4 aa after spirochetes	100	53/53	47/47	0
RNA Polymerase β- subunit	90–120 aa insert after spirochetes	100	53/53	47/47	0
Inorganic pyro-phosphatase	2 aa insert common to *Aquifex* and proteo.	71	45/45	26/26	0
Hsp70/DnaK	2 aa Proteo insert	100	45/45	55/55	0
CTP Synthetase	10 aa Proteo Indel	92	45/45	47/47	0
Lon protease	1 aa deletion in αβγ-proteobacteria	70	41/43	29/27	2[e]
Rho Protein	3 aa αβγ-Proteo indel	83	42/43	41/40	1[f]
DNA Gyrase A subunit	26–34 aa insert in αβγ-proteobacteria	100	42/42	58/58	0
SecA protein	7 aa αβγ-Proteo indel	100	42/42	58/58	0
HSP70/DnaK	4 aa βγ-Proteo insert	100	31/34	69/66	3[g]
ATP Synthase α-subunit	11 aa insert in βγ-proteobacteria	92	31/32	61/60	1[h]
Val-tRNA Synth.	37 aa βγ-Proteo insert	100	31/31	69/69	0
PRPP synthetase	1 aa βγ-Proteo insert	94	31/31	63/63	0
PAC-formyltransferase	2 aa γ-Proteo deletion	83	55/55	28/28	0

Abbreviations in the proteins names are: PAC, 5'- phoshoribosyl-5-aminoimidazole -4 carboxamide formytransferase; PRPP, phosphoribosyl pyrophosphate;
[a]Smaller inserts also present in this region *in D. radiodurans, A. aeolicus, Cb. tepidum*, and *C. tetani.*
[b]*T. whipplei* contains the insert that is not expected.
[c]*M. penetrans* constitutes an exception.
[d]*T. maritima* contains the insert that is not expected.
[e]*B. japonicum* and *P. putida* are exceptions.
[f]*B. thetaiotamicron* is the exception.
[g]*B. longum* and *T. whipplei* are exceptions.
[h]*Cb. tepidum* also contains this insert.

once a main-line signature has been introduced in an ancestral lineage, all species from the latter branching groups should contain the indel, whereas all species from groups that branched off before the introduction of the signature should be lacking the indel.[4] If the deduced branching order is reliable, then the observed distribution of these indels in different genomes should be close to that predicted by the model. However, if such indels could arise independently, or if the genes harboring them were subjected to frequent lateral transfer, then their presence or absence in different species would not follow the prediction of the model. Thus, the reliability of the deduced branching order can be objectively assessed by determining how closely the distribution of these signatures in different genomes follows the predictions of the model.

Results of these analyses, examining the presence or absence of different main line signatures in various bacterial genomes, are presented in table 8.3. The number of genomes in which these proteins have been found, and the number of species, we would expect to contain or lack these indels based on their postulated insertion positions, are also indicated in table 8.3. For example, for the main-line indels in RpoB and AlaRS, the model predicts that 53 of the 100 species from various groups that branched off after the insertion points of these indels (i.e., groups on the top in figure 8.4) should contain the indel, whereas all 47 species from groups that diverged before the introduction of the indels (those below the insertion point) should not possess it. Similarly, the large, 21–23-aa insert in the Hsp70 protein should be present in 60 of the 100 species branching after the Thermotoga-Clostridia clade (figure 8.4), but it should not be found in the remaining 40 species from groups branching earlier. The last few columns in table 8.3 summarize the results obtained for different indels and the number of exceptions or contradictions that were observed. The results of these studies are strikingly clear (table 8.3), as the presence or absence of various signatures in different genomes was found to be

Table 8.3. Statistical significance of cyanobacterial signatures

Protein	Signature	Presence of Indel in Cyanobacteria and Plastids	Presence of Indel in other Bacteria	χ^2 Probability
DNA Helicase II (UvrD)	6 aa. insert	10/10	0/>100	$<10^{-15}$
	7 aa insert	10/10	0/>100	$<10^{-15}$
	28 aa insert	10/10	0/>100	$<10^{-15}$
DNA Pol I	18–21 aa insert	8/8	0/>100	$<10^{-15}$
ADP-Glucose Pyrophosphorylase	14 aa insert	17/17	0/40	$<10^{-15}$
FtsH protease	3 aa insert	8/8	0/>90	$<10^{-15}$
Phytoene Synthase	11–13 aa insert	13/13	0/40	$<10^{-15}$
Elongation factor-Tu	5 aa insert	15/15	0/>100	$<10^{-15}$
Ribosomal S1 protein	2 aa and 7 aa deletions	15/15	0/>60	$<10^{-15}$
SecA protein	2 aa insert	15/15	0/>80	$<10^{-15}$
IMP dehydrogenase	2 aa deletion and 6 aa insert	9/9	0/>80	$<10^{-15}$
Major sigma factor-70	1 aa deletion	13/13	0/>80	$<10^{-15}$

Data for these signatures taken from reference 35. The χ^2 probability for the random occurrence of these indels in different bacteria was calculated using two degrees of freedom.

almost exactly as predicted by the model. In 2,079 observations examining the presence or absence of these indels in 100 genomes, only 11 exceptions or ambiguities were observed. These exceptions could be the results of LGTs or other nonspecific mechanisms. The ability of the indel model (fig. 8.4) to predict with such remarkable accuracy (>99%) the presence or absence of various indels in different genomes provides compelling evidence of its reliability and predictive power.

The relationships depicted in figure 8.4, with a few notable exceptions (to be discussed later), are generally in accordance with phylogenetic trees based on different genes and proteins.[4,8,10,16,19] Most published phylogenetic trees show groups such as Thermotoga, Deinococcus-Thermus, Cyanobacteria, and green nonsulfur bacteria to be deep branching, whereas other groups such as Proteobacteria and Chlamydiae–CFBG clade are late-branching lineages. The Spirochetes generally branch in middle in proximity of the Chlamydiae–CFBG and cyanobacterial taxa. In contrast to these groups, the branching of gram-positive bacteria (Firmicutes, Actinobacteria, Clostridia) and Aquificales is found to be highly variable in different trees.[5,7,8,16]

Despite its uncertain phylogenetic position, the deep branching of Aquifex, as seen in the rRNA trees, has become a cornerstone of our present understanding of bacterial phylogeny.[5,20] However, this view is not supported by signature sequences in various proteins. These studies instead provide strong and consistent evidence that the order Aquificales has diverged late in bacterial evolution, branching in between Chlamydiae–CFBG and the δ,ϵ-proteobacterial groups (fig. 8.4).[6,12,17] We have recently obtained evidence that this inference is not limited to Aquifex but also applies to various other species belonging to the order Aquificales (viz. *Calderobacterium hydrogenophilum*, *Hydrogenobacter marinus*, and *Thermocrinis ruber*).[21]

The branching order of different groups is also consistent with the major structural differences seen within Bacteria (fig. 8.4). Bacteria can be divided into two distinct groups, depending on whether they are bounded by one membrane (monoderms) or two different membranes separated by a periplasmic compartment (diderms).[4,22] These two groups roughly correspond to the gram-positive and gram-negative bacteria. The signature sequences support this important structural distinction and indicate that of these two groups, the monoderm or gram-positive bacteria are ancestral (fig. 8.4).The deduced branching order also places the Deinococcus-Thermus group in an intermediate position between the gram-positive and gram-negative bacteria. This placement is in accordance with the unique structural characteristics of Deinococcus species, which contain a thick peptidoglycan layer and show positive Gram staining, but they are surrounded by both inner and outer cell membranes, similar to various gram-negative bacteria.[23] The branching of cyanobacteria after the Deinococcus-Thermus group is also of interest because their cell walls are intermediate, in terms of thickness of the peptidoglycan layer and degree of cross-linking, between gram-positive and gram-negative bacteria.[24] These observations are indicative that Deinococcus-Thermus and Cyanobacteria are evolutionary intermediates in the transition from gram-positive bacteria to gram-negative bacteria. The genotype–phenotype correspondence is a central theme of biology, and the picture of bacterial phylogeny that is emerging based on signature sequences is now showing a good correspondence between these two important aspects.[6,12,17]

Signature sequences allow us to finally establish objective criteria for distinguishing the main groups within Bacteria and the major subdivisions within them. In the scheme shown in figure 8.4, all suggested main groups are required to meet two different criteria.

First, all such groups should be clearly distinguishable from the other main groups based on group-specific signature sequences. Second, their branching position should be distinct from all other identified main groups. On the basis of these criteria, the Proteobacterial phylum has been divided into four main groups (δ, ϵ-, α-, β-, and γ-), each of which is distinct from the others and also branches in a different position.[6,7,12] In contrast, a number of other groups such as Chlamydiae and the CFBG, which branch in the same position based on signature sequences, have not been assigned separate main group statuses, even though Chlamydiae are clearly distinguishable from the CFBG group by a large number of signatures.[11] We have suggested that distinct groups of species that branch in the same position should be tentatively recognized as subdivisions of a given main group or phylum. Using this criterion, the δ- and ϵ-proteobacteria, whose branching order cannot be distinguished at present,[7] are also placed in the same group. The division of Bacteria into different main groups or their subdivisions as proposed here is based strictly on genealogical considerations, and it reflects both the degrees of similarities among different groups and their hierarchal order, which is the most logical and natural way for understanding evolutionary relationships.[25]

Lateral Gene Transfer—Its Prevalence and Effects

Analyses of whole genome sequences have led to the widespread notions that LGT among prokaryotic organisms is rampant and that it poses a serious problem for discerning evolutionary relationships.[5,9,26–29] A survey of the recent literature gives the impression that LGT among prokaryotes is so pervasive that it has almost completely obliterated any phylogenetic signal resulting from vertical descent (i.e., Darwinian evolution). According to this view, LGT and homology-dependent recombination are the major mechanisms of prokaryotic evolution, and any observed similarities within a group of species is mainly a consequence of selective LGTs.[9,26] Before offering an alternative view, we first briefly examine the evidence that has led to this belief.

The occurrence of LGT has been inferred from a variety of observations. One involves the unexpected branching of species in phylogenetic trees.[30,31] The identification of LGT by this means is based on the assumption that phylogenetic relationships among the groups under consideration are well understood. The vast majority of LGTs that have been identified on this basis involves unexpected branching of an archaeal species within the Bacteria or a bacterial species within Archaea.[30,32,33] Although the Archaea and the Bacteria are presently recognized as two distinct domains, their evolutionary relationship is not well understood (see next section).[34,35] In addition to deduced LGT between Archaea and Bacteria, there have also been several reports describing the branching of individual genes or proteins in unexpected phyletic positions.[11,26,30] The number of such cases is rather limited, however, and does not support the view that LGT between different groups is rampant and occurs indiscriminately.

The major stimulus for the belief in the widespread occurrence of LGT within prokaryotes, particularly Bacteria, has come from analyses of genome sequences using a variety of indirect approaches. They include reliance on atypical base composition or codon usage, use of BLAST searches to ascertain species relatedness, and the presence or absence of genes in closely related genomes.[9,27,29,32,33,36] One of the most influential studies of this nature was by Lawerence and Ochman, who, on the basis of atypical GC content and pattern of codon usage inferred that about 17.6% of the open reading frames in *Escherihia coli* and *Salmonella* have undergone LGT, after the divergence of these species.[29] Because these species are estimated to have diverged only about 100 million years ago, the inferred high rate of LGT from this study indicated that over long evolutionary periods, LGTs can completely abrogate the evolutionary relationships among prokaryotic organisms.[29] Likewise, several authors have inferred massive LGTs between certain bacterial lineages (Thermotoga, Aquifex, Deinococcus) and Archaea, using BLAST searches to identify closest relatives.[27,32,33,36]

However, many recent studies point out the fallacies of using these approaches to infer the incidence of LGT.[31,37–40] In a detailed study examining the use of BLAST searches to identify closest relatives, Koski and Golding[39] determined that the genes that appear to be most related by BLAST are often not each others' closest relatives, phylogenetically. The extent to which this occurs depends on the availability of closest relatives in the database. Other studies provide evidence that atypical base composition or codon usage are also not reliable indicators of LGTs.[31,37,38,40] Many genes previously classified as laterally transferred using these criteria are, in fact, native.[37] In view of the incongruent results obtained using these approaches for different genes, none of these methods are reliable indicators of LGTs, and much caution needs to be exercised in interpreting the results of such studies.[31,37–41] Alarmed with the growing bandwagon of rampant LGT across different group boundaries, which began with his earlier studies, Ochman[42] has recently warned that such is not the case: "Whereas LGT has certainly been a significant factor in the rapid adaptation and speciation of many bacterial lineages, the overall stability of the genome is, in fact, what allows one to assess the role of lateral gene transfer."

There is little doubt at present that LGT constitutes an important evolutionary mechanism and that it may have affected different genes to various extents. For certain genes whose acquisitions confer selective advantage (e.g., those involved in antibiotic resistance or virulence), one expects that they would be readily transferred from one species to another. However, the extent to which LGT has affected numerous others genes involved in various essential functions remains to be determined. The main challenge before us is to determine whether such genes have also been subjects of rampant LGTs, or whether a significant number of them have undergone either minimal or no LGTs and whether they could provide a stable core of well-preserved molecular fossils, on the basis of which the early evolutionary history can be reliably deduced. I will examine this question here mainly in the context of bacterial evolution, which has been studied in most detail.

To identify LGT, an essential prerequisite is to define all the groups under consideration in unequivocal terms. Without this, it is difficult to assess or quantify LGT. Second, it is also necessary to have a reliable model as to how these groups are related to each other. In the absence of such a model, it is difficult to evaluate whether an observed relationship is natural or whether it is a consequence of LGT. Within Bacteria, a number of main groups have been identified in phylogenetic trees (e.g., Deinococcus-Thermus, Cyanobacteria, low-G+C gram-positive, high-G+C gram-positive, Spirochetes, Chlamydiae-CFBG, and Proteobacteria).[3,5,8] These groups can now be clearly distinguished on the basis of signature sequences in different proteins. Table 8.1 lists a number of signatures that have been identified for the Cyanobacteria.[10] The distribution of these signatures in cyanobacteria and other groups of bacteria is also indicated. We could now ask the question of whether any LGT for these genes has occurred between cyanobacteria and other bacteria. On the basis of a Darwinian model of evolution, these signatures were introduced in a common ancestor of the cyanobacteria at the time when this lineage evolved. The model predicts that these signatures should be present in various cyanobacterial species but not elsewhere. However, if these genes were either subjects of frequent LGTs, or if such indels could independently arise in various species, then their distribution should differ greatly from that predicted by the Darwinian model. From the observed distribution of these indels in various bacteria, it is clear that these indels are highly specific for cyanobacteria (many of them are also shared by plastid homologs that have originated from cyanobacteria via endosymbiosis),[2,43,44] but they are not found in any other bacteria, even though most of these proteins are present in all such organisms. The statistical significance of these results can be determined by means of a simple χ^2 test, with two degrees of freedom. The χ^2 probability that the observed distribution of these indels in different groups can result from random occurrence or LGTs is virtually nil ($<10^{-15}$). On the basis of other identified signatures, similar strong arguments can be made for most of the other major groups within Bacteria including Proteobacteria, Deinococcus-Thermus, Chlamydiae, Spirochetes, Aquificales, Actinobacteria, and Firmicutes.

If the major groups within Bacteria have not undergone extensive LGT, then one could inquire next whether the interrelationships between these groups have been affected by LGTs. As discussed earlier, at present there is no reliable model as to how different groups within Bacteria are related to each other or how they branched off from a common ancestor.[5,9] In the absence of a reliable model, the branching pattern of these groups in the 16S rRNA trees has been assumed as a working model.[5,8] However, there is no stable or consistent branching pattern of these groups in the rRNA trees, and there are numerous disagreements between these trees and those based on other genes and proteins.[5,8,19]

Although most of these differences have been attributed to LGTs, in the absence of any reliable model, it is difficult to determine that this is actually the case. In our work, based on signature sequences in different proteins, we have proposed a very specific model of how these groups are related and of their branching order from a common ancestor. This model, which makes very definite predictions regarding the presence or absence of indels in different species, allows one to objectively determine whether the genes in question have been affected by LGT. As discussed earlier, the excellent correspondence between the predicted and observed distribution of different main-line indels in 100 completed bacterial genomes (table 8.3) provides strong evidence that the genes containing these indels have not been affected to any significant extent by LGT. The χ^2 probability that the observed distribution of these indels could be caused by random occurrence or LGT is virtually nil ($<10^{-15}$ in all cases), indicating that these indels were introduced only once in the common ancestors of different groups (as indicated in fig. 8.4) and then passed on to other species by vertical descent.

Concerns have been privately expressed that these inferences are based on a small number of indels that may be biased toward portraying the indicated relationship. We emphasize in response that, during our extensive work on signature sequences, we have not detected any other indels that challenge the consistent picture developed here. Furthermore, in contrast to those phylogenetic trees based on a single gene or protein, all inferences drawn here are based on a large number of different and widely distributed proteins. Our inferences are not restricted to a particular group or family of genes or proteins. The proteins in question are responsible for various essential functions including transcription, translation, replication, DNA repair, protein folding, cell division, metabolic enzymes, and cell wall biosynthesis. The fact that all of these signatures yield a highly consistent picture, and that they are also in accordance with cell structural characteristics, strongly indicates that the model presented here is reliable.

The use of molecular sequences for deducing evolutionary relationships is analogous to evaluating fossil evidence to piece together the evolutionary history of extinct species. Paleontologists give greater credence to well-preserved fossils and much less to those that have been disintegrated. Similarly, molecular sequences vary greatly in terms of their degree of conservation and usefulness for evolutionary studies. During the course of evolution, although some genes and proteins have undergone extensive changes, others are less affected by such factors. Given the long evolutionary history of prokaryotes, it should not be surprising to find many examples of genes that have been affected by LGT. However, it is wrong to infer that all genes have been similarly affected. This would amount to throwing the baby with the bath. As evolutionary scientists, our focus should not be limited to finding examples of genes that have been affected by LGTs, but to identify and discover genes that are minimally affected by LGTs, on which reliable evolutionary models can be developed and validated. Such models should enable us to reliably identify the LGT events as well as to understand why they have occurred.

On the basis of this rationale, our work on signature sequences, still in its initial phases, has focused on documenting highly conserved molecular signatures that are minimally affected by LGT or gene duplication and that can be interpreted with minimal ambiguity. Our model was first developed at a time when fewer than ten bacterial genomes were available,[4] and it is now being rigorously tested for its various predictions using sequence data for >100 bacterial genome. From the outset, our model made specific predictions as to which of these indels should be present or absent in various bacteria for which no sequence information was available. At present, when sequence information for >100 bacterial genomes is available, it is gratifying to note that the predictions made by this model are borne out with such high degree of accuracy (>99.0%) in over 2,000 observations. These results strongly validate our evolutionary model and indicate that the genes and proteins on which it is based provide a stable core that is minimally affected by factors such as LGTs.

Origin of Archaea and Their Relationship to Bacteria

Unlike the evolutionary relationships within Bacteria which are beginning to be understood, the relationship between Archaea and Bacteria (and also the relationships within Archaea) have proven much more difficult to resolve.[4,28,34,35,41,45–50] That the Archaea were distinct from Bacteria (or eubacteria) was first proposed by Woese and his coworkers on the basis of pronounced differences in their 16S rRNA oligonucleotide catalogues and their discrete branching in the 16S rRNA trees.[3,46] The distinction between these two groups was also supported by a number of other characteristics including lack of muramic acid in archaeal cell walls, membrane lipids that contain ether-linked isoprenoid side chains rather than the diacyl esters found in bacteria, distinctive RNA polymerase subunits structures, differences in sensitivity profile to various toxins and antibiotics, and so forth. Subsequent studies based on duplicated pairs of gene sequences indicated that the root of the universal tree lay between Archaea and Bacteria, with eukaryotic homologs derived from the same branch as Archaea.[13,14] The inference from these studies that Archaea are the closest relatives of Eukarya has had a profound influence on their acceptance as a separate domain. However, it is now firmly established that all eukaryotic cells have received major gene contributions from both Bacteria and Archaea.[4,35,51,52] The ancestral eukaryotic cell, therefore, is not a direct descendent of the archaeal lineage. On the basis of fossil evidence, the eukaryotic organisms have also evolved much later (>2 Ga) than the prokaryotes. Therefore, if Archaea and Bacteria were the only organisms that existed for much of the early history of life, it is important for us take a closer look at them to see how they differ from each other and how such differences possibly arose.

The majority of the genes that indicate Archaea to be different from Bacteria are for the information transfer processes, such as those responsible for DNA replication, transcription, and protein synthesis.[53] Of these, the DNA replication machinery appears to be most different between the two domains. Archaea do not contain typical bacterial DNA polymerases (PolI, Polβ), helicase, or most other proteins (e.g., DnaA, SSB, and DnaG) involved in different stages of DNA replication. Although proteins that carry out analogous functions are present in Archaea, they do not show significant sequence similarity to the bacterial counterparts. In terms of transcription, the core subunits of the RNA polymerase (α, β, and β') are the same in Bacteria and Archaea, but the archaeal enzyme also contains several smaller subunits not present in bacteria.[53] Archaea also contain a variety of transcription factors not found in bacteria.[53] The translation machinery is generally quite similar between Bacteria and Archaea: All rRNAs, most r-proteins, the major elongation factors, various amino acid-charging enzymes and tRNAs, and so forth. are common to both these groups of prokaryotes. The vast majority of r-proteins in Archaea are also arranged in operons similar to that seen in Bacteria. However, Archaea differ from Bacteria in having a small number of unique r-proteins as well as many translation initiation factors.[53]

Apart from these differences and dissimilarities in their cell envelope biosynthetic enzymes,[54] Archaea and Bacteria are extensively similar.[36] Most of the metabolic pathways, which make up the vast majority of any organism's gene repertoire, are common between Archaea and Bacteria. In terms of their cell structures, *Archaea* are indistinguishable from gram-positive bacteria.[4,34] Within prokaryotes, only these two groups of organisms are bounded by a single unit lipid membrane, and they generally contain a thick sacculus of varying chemical composition. Some Archaea also show positive Gram staining, and a few of them (e.g., Thermoplasma), similar to certain gram-positive bacteria (e.g., mycoplasma), are unique in not containing any cell wall.[4] The similarity between Archaea and Bacteria extends to numerous other characteristics including their cellular size, which is much smaller (<100–1000-fold) than that of eukaryotic cells; absence of nucleus; cytoskeleton; histones; spliceosomal introns; circular organization of their genomes; organization of genes into operons; presence of 70 S ribosomes; and so forth.[47] Koonin et al.[36] have reported that about 63% of the genes in *Methanococcus janaschii* are also found in other bacteria, whereas only 5% of them are uniquely shared with Eukarya.[36] Although about one-third the total genes in this Archaea are unique (i.e., no similarity seen to any other organisms), the same is generally true for most other prokaryotic genomes.

The similarity between Archaea and gram-positive bacteria, as noted above, is not limited to their cell structures. In phylogenetic trees based on a number of different proteins, archaeal species show polyphyletic branching within gram-positive bacteria.[4,19,55] If one considers only prokaryotic homologs, then phylogenetic trees for the majority of proteins indicate that the Archaea are more closely related to gram-positive bacteria (i.e., monoderm bacteria, which include Thermotoga) than to gram-negative bacteria (R.S. Gupta, unpublished results).[4,19] Strong evidence of a closer relationship between Archaea and gram-positive bacteria, as compared with gram-negative bacteria, is also provided by several prominent signature sequences (e.g., 21–23-aa indel in Hsp70 [fig. 8.2] and 26-aa indel in GS I), which are commonly and uniquely shared by these two groups of prokaryotes.[4,15]

The question can now be asked of how Archaea and Bacteria are related to each other? Because the majority of the genes that indicate Archaea to be distinct from Bacteria are for the information transfer processes, and because these processes are of fundamental importance, it has been assumed that differences in these regards arose in the universal ancestor before separation of these two domains.[3] According to both Woese and Kandler, these two primary domains, as well as the eukaryotic cells, evolved from a precellular community that contained different types of genes that define these lineages by a process leading to the fixation of specific subsets of genes in the ancestors of these domains.[28,50,56] To account for the presence of different genes, it is postulated that the universal ancestor was not a unique organism but a loose community of precellular entities that evolved independently and also in an interdependent manner. These precellular entities did not have stable genealogy or chromosome, and they also lacked a typical cell membrane, thus allowing unrestricted LGTs among them.[28,50,56] These proposals thus postulate that all differences between Archaea and Bacteria originated at a precellular stage by non-Darwinian means, but they suggest no rationale as to how or why the observed differences between these two groups of prokaryotes arose or evolved.[28,50,56] Cavalier-Smith has suggested the possibility that the Archaea have evolved from gram-positive bacteria as an adaptation to hyperthermophily or hyperacidity,[48] but his proposal fails to explain how various differences in the information transfer genes that distinguish Archaea from Bacteria arose.

Our work offers an alternate proposal as to how Archaea and Bacteria may be related: Archaea are related to gram-positive bacteria, as seen by the striking similarities in their cell structures and by a large number of gene phylogenies.[4,34,35] An important distinctive characteristic of all Archaea is that they are resistant to a wide variety of antibiotics that are primarily produced by gram-positive bacteria.[4] Further, the majority of the genes that indicate Archaea to be distinct from Bacteria are related to either information transfer processes or to synthesis of cell wall and membrane lipids, and they provide the main cellular targets for these antibiotics.[4,35,57] These observations are of central importance for understanding the origin of Archaea.[4,34,35] If the differences that characterize Archaea evolved at a precellular stage before the formation of Bacteria, then it is difficult to understand how Archaea developed resistance to most antibiotics that are produced by gram-positive bacteria. Further, it seems too much of a coincidence that most of the genes that indicate Archaea to be distinct from Bacteria provide the main targets for these antibiotics.

To account for these observations, I have suggested that the earliest groups of prokaryotes that evolved were related to the gram-positive bacteria. The characteristics that distinguish Archaea from Bacteria, rather than evolving independently at a precellular stage, evolved from gram-positive bacteria in response to antibiotic selection pressure.[4,35] In one plausible scenario, after a certain group of gram-positive bacteria developed the ability to produce different types of antibiotics to survive in this strongly selective environment, some sensitive bacteria underwent extensive changes in genes that provided the targets for these antibiotics. The changes leading to resistance were of different kinds including mutations, insertions and deletions, nonhomologous recombinations, and replacement of the target genes with nonorthologous genes.[58,59] A prolonged and successive selection in this environment led to the eventual development of a resistant strain that had undergone extensive changes in many genes that were targets of these antibiotics, and this strain represented the common ancestor of present-day Archaea.[4,35] The evolution of Archaea in response to antibiotic selection also provides a plausible explanation for their adaptation to harsher environments such as high temperature, high salts, high acidity, and so forth,[3] which could have been a defensive strategy on their part to find niches that are "hostile" to antibiotic-producing organisms.[4,35] Thus, this proposal can logically explain the evolution of most of the distinguishing characteristics of Archaea from known groups of bacteria by normal evolutionary mechanisms without attributing such differences to the unusual properties of the universal ancestor. Because differences between Archaea and Bacteria evolved at a very early stage in prokaryotic history (fig. 8.4), the Archaea appear distinct from Bacteria in phylogenetic trees or other studies based on such characteristics.

The Origin of the Eukaryotic Cell

The origin of the eukaryotic cell and many observations central to understanding this problem have been discussed in detail in earlier work.[4,52] I will present here only a brief overview of the main hypotheses for the formation of the eukaryotic cell and examine them critically in light of the available data. It is now well established that all eukaryotic cells possess a large number of genes (representing significant portions of their genomes) that exhibit greater similarity to either Archaea or Bacteria.[4,19,45,51,52,60] Thus, the original three-domain hypothesis that Archaea and Eukarya (or Eukaryotes) are sister lineages and that the ancestral eukaryotic cell directly evolved from an archaeal ancestor is no longer valid. To account for the unique genotype and phenotype of eukaryotic cells, several hypotheses have been proposed. Some authors have suggested that the eukaryotic cells evolved first and that prokaryotic organisms originated from them by gene loss and simplification processes.[61,62] However, such hypotheses are inconsistent with the fossil and geological evidence. Other hypotheses postulate the existence of hypothetical proto-eukaryotic organisms possessing various distinctive characteristics of the eukaryotes, which later engulfed either an Archaea or both an Archaea and Bacteria to give rise to eukaryotic cells.[63,64] These hypotheses defer an understanding of the origin of the eukaryotic cell to hypothetical entities for which there is no evidence. In a recent proposal, Cavalier-Smith[48] posits that a common ancestor of eukaryotic cells and archaebacteria (termed Neomura) evolved directly from gram-positive bacteria. This proposal is unique in postulating a very late divergence (about 0.85 Ga) of both Archaea and eukaryotes, but it is not supported by the observations that bacterial genes in eukaryotes are derived from gram-negative bacteria rather than from gram-positive bacteria.[51,52]

The most widely accepted proposals for the origin of eukaryotic cells postulate it to be a chimera formed by fusion or association of distinct lineages of Bacteria and Archaea.[4,52,65–68] These proposals were put forward to account for the observations that whereas information-transfer genes of eukaryotic cells are most closely related to Archaea, their metabolic genes are primarily derived from Bacteria.[19,52,60,69] The chimeric proposals that have been suggested are of two main kinds. One key aspect that distinguishes these proposals is whether the chimeric event that led to the formation of the eukaryotic cell was the same as that which gave rise to mitochondria (mitochondria–nucleus co-origin model) or whether the two events differed from each other both in their nature and timings (nucleus first–mitochondria later models).[4,52,65–68,70] This distinction is important because mitochondria and plastids are already known to be derived from bacteria via endosymbiosis.[2,43,44]

The main impetus for the mitochondria–nucleus co-origin proposals has come from observations that a number of protist lineages that were earlier believed to lack mitochondria have now been shown to contain at least some genes or proteins that are distinctive of mitochondria.[65,71,72] This has led to the view that all eukaryotic species once harbored mitochondria and that absence of mitochondria in some of these lineages is the result of a secondary loss of the organelle. The wide acceptance of the mitochondria-early view in conjunction with the observation that several anaerobic protist lineages contain a mitochondria-related organelle, hydrogenosome (which releases gaseous hydrogen), has led to the "hydrogen hypothesis" for the formation of the eukaryotic cell.[65] According to this hypothesis, the ancestral eukaryotic cell arose as a result of symbiotic association between a hydrogen-dependent archaebacterium (a methanogen) and an α-proteobacterium, which, under anaerobic conditions, produced molecular hydrogen as a waste product. The driving force for the formation of the eukaryotic cell was dependance of the archaebacterium on molecular hydrogen produced by the bacterial symbiont.[65] This association led to the endosymbiotic capture of α-proteobacterium by the archaeal partner, leading to the formation of mitochondria as well as to various other eukaryotic cell characteristics.

Although the hydrogen hypothesis accounts for the origin of hydrogenosomes from mitochondria, it offers no explanation of how any of the main characteristics that define a eukaryotic cell (e.g., nucleus, endoplasmic reticulum [ER]) evolved. The hydrogen hypothesis is also called into question by a number of observations:[4,73] first, there exists now unequivocal evidence that all extant eukaryotic organisms have evolved from a common ancestor, thus indicating that the formation of the eukaryotic cell was a unique event occurring only once.[4] If metabolic symbiosis based on hydrogen production and use was the main driving force for the formation of the eukaryotic cell, then given the widespread association between methanogenic archaea and hydrogen-producing proteobacteria, it is difficult to envisage why this sort of syntrophy has not lead to the formation of eukaryotic-like cells on numerous independent occasions. Second, in all well-established cases of endosymbiosis (e.g., formation of mitochondria and plastids), the metabolic processes that have formed the basis of symbiosis have been retained,[2] yet eukaryotes have not retained any genes for methanogenesis, the proposed basis of their origin. Third, the α-proteobacteria, which the hydrogen hypothesis suggests as the bacterial partner in the syntrophic association, evolved much later than cyanobacteria.[17] This implies that formation of the eukaryotic cell took place in an aerobic atmosphere, but the symbiosis between an anaerobic hydrogen-producing bacterium and a strictly anaerobic methanogenic archaebacterium would have produced an anaerobic organism, which would thus be at a great selective disadvantage. Fourth, the hydrogen hypothesis provides no explanation as to why eukaryotic genes for the information transfer processes are derived from the archaeal partner.[51,52,60] Finally, molecular sequence data indicate that thermoacidophilic archaea rather than methanogens are the closest relatives of eukaryotes.[4,74]

All alternative chimeric proposals posit that the primary fusion (or endosymbiotic) event that led to the formation of eukaryotic cell was distinct and that it preceded the formation of mitochondria. According to Margulis's most recent proposal, the ancestral eukaryotic cell was formed by association of a spirochete and a Thermoplasma, in which the Thermoplasma contributed the nucleocytoplasm, whereas the motility apparatus (e.g., microtubules) was provided by the spirochete.[67] However, there is no evidence at present that any of the eukaryotic genes, including those for motility functions, have been derived from spirochetes.[4] Recently, Jenkins et al.[75] have identified a protein showing strong sequence and biochemical similarity to tubulin in Prosthecobacter, indicating for the first time a possible origin of this key cytoskeletal protein. Lake and Rivera[66] have proposed that the nucleus is an endosymbiont that arose from the engulfment of an eocyte (or Crenarchaeota) archaea by a gram-negative bacterium. However, this model ignores the fact that the nucleus is not an endosymbiont in the same sense as mitochondria and plastids, which have retained their information transfer machinery and are specifically related to their parental lineages.[2,43] Zillig[76] suggested the possibility that the ancestral eukaryotic cell was formed by primary fusion of an archaebacterium and a eubacterium. However, the nature of the proposed fusion event was not elaborated.

A detailed chimeric proposal for the nucleus-first, mitochondria-later origin has emerged from our work on some of the best-characterized protein families in eukaryotic cells.[4,52,55,70] The Hsp70 family of proteins represents one such family. Distinct homologs of Hsp70 that are encoded for by different genes are present in mitochondria, cytosol, and the ER compartments.[4,70] The mitochondrial and hydrogenosomal homologs of Hsp70 are clearly derived from α-proteobacteria, as evidenced by phylogenetic analyses and many common signature sequences.[4] In contrast, the homologs of Hsp70 that are present in the cytosol and ER compartments (referred to as nuclear-cytosolic homologs), which are also derived from bacteria, show no relationship to the mitochondrial homologs.[4,52,70] These homologs contain a large number of signature sequences that are not present in any mitochondrial, hydrogenosomal, or prokaryotic homologs. These signatures are thus uniquely eukaryotic, and they were likely introduced in the Hsp70 gene at a very early stage in the formation of the eukaryotic cell.[4,52,70] The absence of these signatures in mitochondrial and hydrogenosomal homologs provides strong evidence that nuclear-cytosolic homologs have originated independently of these organelles. Importantly, although the cytosolic and ER Hsp70 homologs are present in all eukaryotic organisms, in the earliest branching eukaryotic lineages such as Giardia, no Hsp70 gene that qualifies as a mitochondrial homolog has been detected. Morrison et al.[77] have recently identified an Hsp70 homolog in Giardia that is distinct from the nuclear-cytosolic homologs. However, this highly divergent protein shows no specific affinity to the mitochondrial homologs.

Because ER forms the nuclear envelope, an understanding of events leading to its evolution (or proteins found in this compartment) is directly relevant to the origin of the nucleus. Phylogenetic analyses of Hsp70 and Hsp90 sequences indicate that the ER and cytosolic homologs of these proteins in different eukaryotic organisms including Giardia are the results of ancient gene duplication events.[70,78] Thus, a very early event associated with the formation of the ER (and via inference the nuclear envelope) involved duplication of genes for these proteins.[70,78]

To explain these observations, and the chimeric nature of eukaryotic nuclear-cytosolic genes, we have proposed that the ancestral eukaryotic cell evolved as a result of symbiotic association between a gram-negative bacterium (related to proteobacteria or the CFBG group) and an Archaea (fig. 8.5).[4,52,73] This symbiosis developed in an aerobic environment predominated by antibiotic-producing organisms. A combination of these two selective forces (oxygen and antibiotics sensitivities) led to the association of an antibiotic-resistant and oxygen-sensitive archaea with an oxygen-tolerant (or oxygen-using) and antibiotic-sensitive bacterium, which provided mutual protection in this environment.[4,73] The association of these two groups of prokaryotes led to the surrounding of the archaea by membrane enfolds from the bacterial partner to shield it from oxygenic environment (fig. 8.5). The cell membrane of the archaea became redundant under these conditions, and it was eventually lost. At a later stage, the membrane enfolds surrounding the archaea got separated from the bacterial membrane. This led to formation of the endomembrane system (or the ER), as well as of nucleus in the cell. Because this newly formed compartment (i.e., the ER) had to communicate (i.e., import and export proteins and other molecules) with the rest of the cell, its formation was either accompanied or preceded by duplication of the genes for the Hsp70 and Hsp90

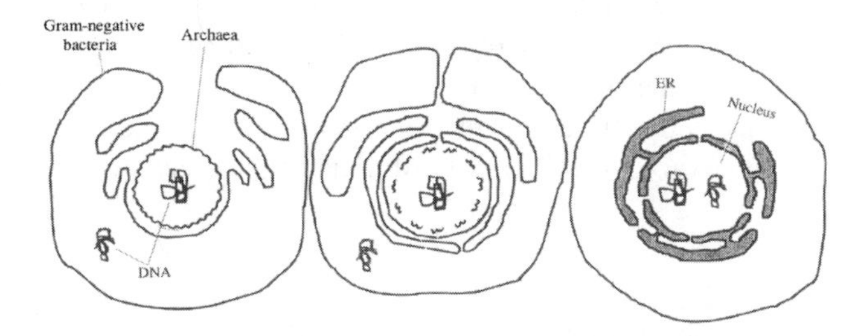

Figure 8.5. The primary fusion model for the origin of the eukaryotic cell. According to this model, the ancestral eukaryotic cell, which contained nucleus and endomembrane system, was formed before the endosymbiotic event that led to acquisition of mitochondria (not shown here). The key event leading to its formation was a long-term symbiosis between a gram-negative bacterium and an archaebacterium. This symbiosis developed in an oxygenic and antibiotic-rich environment, and its basis was sensitivity to antibiotics of the bacterial partner and the oxygen-sensitivity of the archaebacterium. As the membrane of the gram-negative bacterium surrounded the archaebacterium, its membrane (containing ether-linked lipids, shown by the wavy line) became redundant and was lost. The separation of the bacterial membrane folds surrounding the archaebacterium led to formation of the nuclear envelope and endoplasmic reticulum (ER). The resulting cell was antibiotic resistant and oxygen tolerant, and it retained the majority of the genes for the information-transfer processes, which provide main targets for antibiotics, from the archaeal partner.

chaperones, which are essential for this purpose.[70,78] The formation of this new antibiotic-resistant and oxygen-tolerant bacterium was accompanied by an assortment of genes from the two parents. During this process, most of the genes for information-transfer processes (which provide the main targets for antibiotics) were mainly retained from the archaea, whereas those for the metabolic processes were acquired from the bacterial partner.[4,73] Various eukaryotic-specific signatures were also introduced into different genes at this early stage. The transfer of all of these genes into the newly formed nuclear compartment led to integration (or primary fusion) of the original symbionts into a new type of cell, which became the prototype eukaryotic cell (fig. 8.5).[4]

Concluding Remarks

Using molecular sequence data, it is now possible to develop a reliable picture of bacterial phylogeny, where all the main groups can be clearly distinguished and their branching order can be logically inferred. This emerging picture is also consistent with the structural characteristics of prokaryotes. Our proposed model for prokaryotic evolution makes very specific predictions that are strongly corroborated by the genome sequence data. LGT, though an important evolutionary mechanism, is not a serious problem for the determination of bacterial phylogeny. The origin of the Archaea and their relationship to Bacteria remains a contentious issue. The current view is that all three domains have evolved from a precellular community containing different types of genes by an annealing process that led to stabilization of a subset of the genes in common ancestors of these domains. However, the fact that Archaea are resistant to most antibiotics produced by gram-positive bacteria, and that majority of the genes that indicate them to be distinct from Bacteria provide targets for these antibiotics, support our alternate proposal: that they could have evolved from gram-positive bacteria in response to antibiotic selection pressure. The formation of the eukaryotic cell constitutes an evolutionary discontinuity that is explained by their origin from fusion of different groups of prokaryotes. Given the unique and unusual nature of this fusion event, a clear understanding as to how the ancestral eukaryotic cell originated remains to be achieved.

References

1. J.W. Schopf, "The Evolution of the Earliest Cells," *Sci. Am.* 239 (1978): 110–120.
2. L. Margulis, *Symbiosis in Cell Evolution* (New York: W.H. Freeman, 1993).
3. C.R. Woese, "Bacterial evolution," *Microbiol. Rev.* 51 (1987): 221–271.
4. R.S. Gupta, "Protein Phylogenies and Signature Sequences: A Reappraisal of Evolutionary Relationships among Archaebacteria, Eubacteria, and Eukaryotes," *Microbiol. Mol. Biol. Rev.* 62 (1998): 1435–1491.
5. W. Ludwig and H.-P. Klenk, "Overview: A Phylogenetic Backbone and Taxonomic Framework for Prokaryotic Systematics," in D.R Boone and R.W. Castenholz, eds., 2nd ed. *Bergey's Manual of Systematic Bacteriology* (Berlin, Springer, 2001), 49–65
6. R.S. Gupta and E. Griffiths, "Critical Issues in Bacterial Phylogenies," *Theor. Popul. Biol.* 61 (2002): 423–434.
7. R.S. Gupta, "The Phylogeny of Proteobacteria: Relationships to Other Eubacterial Phyla and Eukaryotes," *FEMS Microbiol. Rev.* 24 (2000): 367–402.
8. G.J. Olsen, C.R. Woese, and R. Overbeek, "The Winds of (Evolutionary) Change: Breathing New Life into Microbiology," *J. Bacteriol.* 176 (1994): 1–6.

9. W.F. Doolittle, "Phylogenetic Classification and the Universal Tree," *Science* 284 (1999): 2124–2128.
10. R.S. Gupta, M. Pereira, C. Chandrasekera, and V. Johari, "Molecular Signatures in Protein Sequences that are Distinctive of Cyanobacteria and Plastids," *Int. J. Syst. Evol. Microbiol.* 53 (2003): 1833–1842.
11. E. Griffiths and R.S. Gupta, "Protein Signatures Distinctive of Chlamydial Species: Horizontal Transfer of Cell Wall Biosynthesis Genes *glmU* from Archaebacteria to Chlamydiae, and *murA* between Chlamydiae and *Streptomyces*," *Microbiology* 148 (2002): 2541–2549.
12. R.S. Gupta, "Phylogeny of Bacteria: Are We Now Close to Understanding it?" *ASM News* 68 (2002): 284–291.
13. N. Iwabe, K. Kuma, M. Hasegawa, S. Osawa, and T. Miyata, "Evolutionary Relationship of Archaebacteria, Eubacteria, and Eukaryotes Inferred from Phylogenetic Trees of Duplicated Genes," *Proc. Natl. Acad. Sci. USA* 86 (1989): 9355–9359.
14. S.L. Baldauf, J.D. Palmer, and W.F. Doolittle, "The Root of the Universal Tree and the Origin of Eukaryotes Based on Elongation Factor Phylogeny," *Proc. Natl. Acad. Sci. USA* 93 (1996): 7749–7754.
15. R.S. Gupta and B. Singh, "Cloning of the HSP70 Gene from *Halobacterium marismortui*: Relatedness of Archaebacterial HSP70 to its Eubacterial Homologs and a Model for the Evolution of the HSP70 Gene," *J. Bacteriol.* 174 (1992): 4594–4605.
16. H.P. Klenk et al., "RNA Polymerase of *Aquifex pyrophilus*: Implications for the Evolution of the Bacterial *rpoBC* Operon and Extremely Thermophilic Bacteria," *J. Mol. Evol* 48 (1999): 528–541.
17. R.S. Gupta, "Evolutionary Relationships among Photosynthetic Bacteria," *Photosynth. Res.* 76 (2003): 173–183.
18. R.S. Gupta, "The Branching Order and Phylogenetic Placement of Species from Completed Bacterial Genomes, Based on Conserved Indels Found in Various Proteins," *Int. Microbiol.* 4 (2001): 187–202.
19. J.R. Brown and W.F. Doolittle, "*Archaea* and the Prokaryote-to-Eukaryote Transition," *Microbiol. Rev.* 61 (1997): 456–502.
20. G. Deckert, et al. "The Complete Genome of the Hyperthermophilic Bacterium *Aquifex aeolicus*," *Nature* 392 (1998): 353–358.
21. E. Griffiths and R.S. Gupta, "Signature Sequences in Widely Distributed Proteins Provide Evidence for the Late Divergence of the Order *Aquificales*," *Int. Microbiol.* 7 (2004): 41–52.
22. R.Y. Stanier, E.A. Adelberg, and J.L. Ingraham, *The Microbial World* (Engelwood Cliffs, NJ: Prentice-Hall, 1976).
23. R.G.E. Murray, "The Family Deinococcaceae," in A. Balows, H.G. Truper, M. Dworkin, W. Harder, and K.H. Schleifer eds., *The Prokaryotes* (New York: Springer, 1992), 3732–3744.
24. E. Hoiczyk and A. Hansel, "Cyanobacterial Cell Walls: News from an Unusual Prokaryotic Envelope," *J. Bacteriol.* 182 (2000): 1191–1199.
25. Charles Darwin, *On the Origin of Species 1859 A Facsimile of the First Edition.* (Cambridge, MA: Harvard University Press, 1964).
26. J.P. Gogarten, W.F. Doolittle, and J.G. Lawrence, "Prokaryotic Evolution in Light of Gene Transfer," *Mol. Biol. Evol.* 19 (2002): 2226–2238.
27. R. Jain, M. Rivera, and J.A. Lake, "Horizontal Gene Transfer among Genomes: The Complexity Hypothesis," *Proc. Natl. Acad. Sci. USA* 96 (1999): 3801–3806.
28. C.R. Woese, "On the Evolution of Cells," *Proc. Natl. Acad. Sci. USA* 99 (2002): 8742–8747.
29. J.G Lawrence and H. Ochman, "Molecular Archaeology of the *Escherichia coli* Genome," *Proc. Natl. Acad. Sci. USA* 95 (1998): 9413–9417.
30. M.W. Smith, D.F. Feng, and R.F. Doolittle, "Evolution by Acquisition: The Case for Horizontal Gene Transfers," *Trends Biochem. Sci.* 17 (1992): 489–493.
31. M.A. Ragan, "Detection of Lateral Gene Transfer Among Microbial Genomes," *Curr. Opin. Genet. Dev.* 11 (2001): 620–626.
32. K.E. Nelson, et al., "Evidence for Lateral Gene Transfer between Archaea and Bacteria from Genome Sequence of *Thermotoga maritima*," *Nature* 399 (1999): 323–329.

33. L. Aravind, R.L. Tatusov, Y.I. Wolf, D.R. Walker, and E.V. Koonin, "Evidence for Massive Gene Exchange between Archaeal and Bacterial Hyperthermophiles," *Trends Genet.* 14 (1998): 442–444.
34. R.S. Gupta, "What are Archaebacteria: Life's Third Domain or Monoderm Prokaryotes Related to Gram-Positive Bacteria? A New Proposal for the Classification of Prokaryotic Organisms," *Mol. Microbiol.* 29 (1998): 695–708
35. R.S. Gupta, "The *Natural* Evolutionary Relationships among Prokaryotes," *Crit. Rev. Microbiol.* 26 (2000): 111–131.
36. E.V. Koonin, A.R. Mushegian, M.Y. Galperin, and D.R.Walker, "Comparison of Archaeal and Bacterial Genomes: Computer Analysis of Protein Sequences Predicts Novel Functions and Suggests a Chimeric Origin for the Archaea," *Mol. Microbiol.* 25 (1997): 619–637.
37. L.B. Koski, R.A. Morton, and G.B. Golding, "Codon Bias and Base Composition are Poor Indicators of Horizontally Transferred Genes," *Mol. Biol. Evol.* 18 (2001): 404–412.
38. J.A. Eisen, "Horizontal Gene Transfer among Microbial Genomes: New Insights from Complete Genome Analysis," *Curr. Opin. Genet. Dev.* 10 (2000): 606–611.
39. L.B. Koski and G.B. Golding, "The Closest BLAST Hit is Often not the Nearest Neighbor," *J. Mol. Evol.* 52 (2001): 540–542.
40. B. Wang, "Limitations of Compositional Approach to Identifying Horizontally Transferred Genes," *J. Mol. Evol.* 53 (2001): 244–250.
41. R.F. Doolittle, "Searching for the Common Ancestor," *Res. Microbiol.* 151 (2000): 85–89.
42. H. Ochman, "Lateral and Oblique Gene Transfer," *Curr. Opin. Genet. Dev.* 11 (2001): 616–619.
43. M.W. Gray, "The Endosymbiont Hypothesis Revisited," *Int. Rev. Cytol.* 141 (1992): 233–357.
44. C.W. Morden, C.F. Delwiche, M. Kuhsel, and J.D. Palmer, "Gene Phylogenies and the Endosymbiotic Origin of Plastids," *Biosystems* 28 (1992): 75–90.
45. R.S. Gupta, "Life's third domain (*Archaea*): An Established Fact or an Endangered Paradigm? A New Proposal for Classification of Organisms Based on Protein Sequences and Cell Structure," *Theor. Popul. Biol.* 54 (1998): 91–104.
46. C.R. Woese, O. Kandler, and M.L. Wheelis, "Towards a Natural System of Organisms: Proposal for the Domains Archaea, Bacteria, and Eucarya," *Proc. Natl. Acad. Sci. USA* 87 (1990): 4576–4579.
47. E. Mayr, "Two Empires or Three?" *Proc. Natl. Acad. Sci. USA* 95 (1998): 9720–9723.
48. T. Cavalier-Smith, "The Neomuran Origin of Archaebacteria, the Negibacterial Root of the Universal Tree and Bacterial Megaclassification," *Int. J. Syst. Evol. Microbiol.* 52 (2002): 7–76.
49. S.L. Lyons, "Thomas Kuhn is Alive and Well: the Evolutionary Relationships of Simple Life Form–A Paradigm Under Siege," *Perspect. Biol. Med.* 45 (2002): 359–376.
50. O. Kandler, "Thermophiles: The Keys to Molecular Evolution and the Origin of Life?" in J. Wiegel and W.W. Adams, eds. (Athens: Taylor and Francis, 1998), 19–31.
51. S. Ribeiro and G.B. Golding, "The Mosaic Nature of the Eukaryotic Nucleus," *Mol. Biol. Evol.* 15 (1998): 779–788.
52. R.S. Gupta and G.B. Golding, "The Origin of the Eukaryotic Cell," *Trends Biochem. Sci.* 21 (1996): 166–171.
53. G.J. Olsen and C.R. Woese, "Archaeal Genomics: An Overview," *Cell* 89 (1997): 991–994.
54. O. Kandler and H. Konig, "The Biochemistry of Archaea (Archaebacteria)," in M. Kates, D.J. Kushner, and A.T. Matheson, eds. (New York: Elsevier Science, 1993), 223–259.
55. R.S. Gupta and G.B. Golding, "Evolution of HSP70 Gene and Its Implications Regarding Relationships between Archaebacteria, Eubacteria, and Eukaryotes," *J. Mol. Evol.* 37 (1993): 573–582 .
56. C.R. Woese, "The Universal Ancestor," *Proc. Natl. Acad. Sci. USA* 95 (1998): 6854–6859.
57. J. Davies, "Inactivation of Antibiotics and the Dissemination of Resistance Genes," *Science* 264 (1994): 375–382.
58. E.V. Koonin, and M.Y. Galperin, "Prokaryotic Genomes: The Emerging Paradigm of Genome-Based Microbiology," *Curr. Opin. Genet. Dev.* 7 (1997): 757–763.

59. P. Forterre, "Displacement of Cellular Proteins by Functional Analogues from Plasmids or Viruses Could Explain Puzzling Phylogenies of Many DNA Informational Proteins," *Mol. Microbiol.* 33 (1999): 457–465.
60. M. Rivera, R. Jain, J.E. Moore, and J.A. Lake, "Genomic Evidence for Two Functionally Distinct Gene Classes," *Proc. Natl. Acad. Sci. USA* 95 (1999): 6239–6244.
61. A. Poole, D. Jeffares, and D. Penny, "Early Evolution: Prokaryotes, the New Kids on the Block," *Bioessays* 21(1999): 880–889.
62. P. Forterre and H. Philippe, "Where is the Root or the Universal Tree of Life," *Bioessays* 21 (1999): 871–879.
63. M.L. Sogin, "Early Evolution and the Origin of Eukaryotes," *Curr. Opin. Genet. Dev.* 1, (1991): 457–463.
64. H. Hartman and A. Fedorov, "The Origin of the Eukaryotic Cell: A Genomic Investigation," *Proc. Natl. Acad. Sci. USA* 99 (2002): 1420–1425.
65. W. Martin and M. Müller, "The Hydrogenosome Hypothesis for the First Eukaryote," *Nature* 392 (1998): 37–41.
66. J.A. Lake and M.C. Rivera, "Was the Nucleus the First Endosymbiont?" *Proc. Natl. Acad. Sci. USA* 91 (1994): 2880–2881.
67. L. Margulis, "Archaeal–Eubacterial Mergers in the Origin of Eukarya: Phylogenetic Classification of Life," *Proc. Natl. Acad. Sci. USA* 93 (1996): 1071–1076.
68. V.V. Emelyanov, "Mitochondrial Connection to the Origin of Eukaryotic Cell," *Eur. J. Biochem.* 270 (2003): 1599–1618.
69. G.B. Golding and R.S Gupta, "Protein-Based Phylogenies Support a Chimeric Origin for the Eukaryotic Genome," *Mol. Biol. Evol.* 12 (1995): 1–6.
70. R.S. Gupta, K. Aitken, M. Falah, and B. Singh, "Cloning of *Giardia lamblia* Heat Shock Protein HSP70 Homologs: Implications Regarding Origin of Eukaryotic Cells and of Endoplasmic Reticulum," *Proc. Natl. Acad. Sci. USA* 91 (1994): 2895–2899.
71. B.J. Soltys and R.S. Gupta, "Presence and Cellular Distribution of a 60-kDa Protein Related to Mitochondrial hsp60 in *Giardia lamblia*," *J. Parasitol.* 80 (1994): 580–590.
72. A.J. Roger, et al., "A Mitochondrial-Like Chaperonin 60 Gene in *Giardia lamblia*: Evidence that Diplomonads Once Harbored an Endosymbiont Related to the Progenitor of Mitochondria," *Proc. Natl. Acad. Sci. USA* 95 (1998): 229–234.
73. R.S.Gupta, "Origin of Eukaryotic Cells: Was Metabolic Symbiosis Based on Hydrogen the Driving Force?" *Trends Biochem. Sci.* 24 (1999): 423.
74. M.C. Rivera and J.A. Lake, "Evidence that Eukaryotes and Eocyte Prokaryotes Are Immediate Relatives," *Science* 257 (1992): 74–76.
75. C. Jenkins, et al., "Genes for the Cytoskeletal Protein Tubulin in the Bacterial Genus *Prosthecobacter*," *Proc. Natl. Acad. Sci. USA* 99 (2002): 17049–17054.
76. W. Zillig, "Comparative Biochemistry of Archaea and Bacteria," *Curr. Opin. Genet. Dev.* 1 (1991): 544–551.
77. H.G. Morrison, A.J. Roger, T.G. Nystul, F.D. Gillin, and M.L. Sogin, "*Giardia lamblia* Expresses a Proteobacterial-Like DnaK Homolog," *Mol. Biol. Evol.* 18 (2001): 530–541.
78. R.S. Gupta, "Phylogenetic Analysis of the 90 kD Heat Shock Family of Protein Sequences and an Examination of the Relationship among Animals, Plants, and Fungi Species," *Mol. Biol. Evol.* 12 (1995): 1063–1073.

9

Fulfilling Darwin's Dream

JAMES A. LAKE
JONATHAN E. MOORE
ANNE B. SIMONSON
MARIA C. RIVERA

> The time will come I believe, though I shall not live to see it, when we shall have fairly true genealogical trees of each great kingdom of nature.
>
> Charles Darwin to T. H. Huxley, 1857

Today there is enormous interest in discovering the tree of life, but as we get closer to reconstructing the universal tree, new experimental and theoretical challenges continually appear that cause us to reexamine our goals. New obstacles may initially seem insurmountable, but in reality they enrich our understanding of the evolution of life on earth.

One of the most recent evolutionary mechanism to challenge our view of genome evolution is the massive horizontal gene transfer (HGT) that has recently become so apparent.[1–7] This genetic cross-talk theoretically has the potential to erase much of the history of life that has been recorded in DNA. Indeed, some scientists think that HGT has already effectively erased the phylogenetic history contained within prokaryotic genomes.[8]

This chapter, however, is predicated on optimism. We think that the only way to decide whether the tree of life is knowable is to try our hardest to determine it. Like Darwin, we believe we shall someday have fairly accurate trees. Unlike Darwin, however, who recognized that he would not see "the genealogical trees of each great kingdom of nature," we optimistically anticipate that a fairly accurate tree of life, including life's deepest branches, will soon be available.

This chapter reviews the fits and starts of progress in this area and analyzes this progress from the unique perspective of our laboratory.

What Are the "Great Kingdoms of Nature"?

When Darwin wrote his famous quotation, he was not thinking of prokaryotic life. Rather, he probably envisioned understanding the trees of animal and plant life. In that sense, part of his dream is already a reality. We currently understand the major radiations of the bilateral animals,[9,10] and the relationships linking the major plant groups are starting to be understood.[11–15] This review, however, focuses on a goal even more ambitious than that of Darwin: understanding the radiations that occurred even before those of the plants and animals; namely, understanding the enigmatic evolution of prokaryotes and the emergence of eukaryotes.

The origin of the eukaryotes was a milestone in the evolution of life, as they are utterly different from prokaryotes in their spatial organization. For example, eukaryotes posses an extensive system of internal membranes that transverse the cytoplasm and enclose organelles, including the mitochondria, chloroplast, and nucleus. This compartmentalization has required a number of unique eukaryotic innovations. The most dramatic innovation is the nucleus, a specific compartment for storing, processing, and transcribing DNA. The nucleus and its origin are intimately connected with the origin of eukaryotes because a nucleus is present in all eukaryotes—and only in eukaryotes. The nucleus is, in fact, the defining character for which eukaryotes are named (eu = good or true, karyote = kernel, as in nucleus).

The prokaryotes, with their simple cellular organization, are generally thought to have preceded the eukaryotes. However, the root of the tree of life is uncertain and in flux because of a concern that artifacts of phylogenetic reconstruction may have led to an improper root.[16,17] Before discussing the prokaryotic groups or groups that are most closely related to the eukaryotic nucleus, we briefly survey prokaryotic diversity. The major groups of prokaryotes are the eubacteria, the halobacteria, the methanogens and their relatives, and the sulfur-metabolizing, thermophilic eocyte prokaryotes.

The eubacteria are a diverse group that includes all the photosynthetic bacteria (except for the halobacteria) as well as many nonphotosynthetic groups. Most eubacteria are mesophiles; however, the eubacteria also include extreme thermophiles, such as *Thermotoga maritima* and *Aquifex pyrophilus*, which can grow in temperatures up to 90°C and 95°C, respectively.[18–20] The lipids of eubacteria are primarily of the ester type, although Thermotoga, Aquifex, and their relatives also contain ether-linked lipids.

The halobacteria can live in highly saline environments that are not tolerated by other prokaryotes. These bacteria are carbon heterotrophs that employ an unusual photosynthesis system; namely, a light-driven proton pump based on bacteriorhodopsin. Like eubacteria, they contain the biochemical pathways for the synthesis of C_{40} and C_{50} carotenoids,[21] but like methanogens, they also contain ether lipids.

The methanogens are a phylogenetically diverse group, despite the fact that they share a common phenotype. They are strict anaerobes with the ability to chemically reduce carbon compounds to methane to provide energy. Associated with the methanogens is a phenotypically diverse group of organisms represented by such organisms as *Thermococcus celer* and *Archaeoglobus fulgidus* and, possibly, *Methanopyrus kandleri*. This last organism grows at temperatures up to 112°C. Similar to methanogens, *Methanopyrus* reduces carbon compounds to methane; however, it is not closely related to other methanogens but instead is intermediate between them and the next group of organisms.[22]

The eocytes, or crenarcheotes, include primarily thermophilic, sulfur-metabolizing organisms, many of which can grow at temperatures in excess of 100°C. The eocytes include *Sulfolobus*, *Desulforococcus*, *Thermoproteus*, and *Pyrodictium*. *Sulfolobus sulfataricus* oxidizes sulfur to H_2S, but others, such as *Acidianus infernus*, can oxidize or reduce sulfur to H_2SO_4 or to H_2S. The organisms with the highest maximum growth temperatures are *Pyrodictium occultum* (112°C), *Pyrodictium abyssum* (112°C), and *Pyrolobus fumarii* (113°C.).[23,24] The group is metabolically diverse and phylogenetically monophyletic and includes many uncultured organisms identified only through their rRNA sequences.

The First Eukaryotic Cells and the Origin of the Nucleus

Organisms do not evolve in a vacuum. Most prokaryotes live in environments teeming with life. As a consequence, their interactions with their neighbors influence their evolution. Organisms living together sometimes develop symbiotic relationships, and in a few rare cases this has led to endosymbiosis, the intracellular capture of former symbionts. Today no one doubts that mitochondria and chloroplasts are eukaryotic endosymbionts, and even the possibility that the nucleus itself is an endosymbiont is increasingly gathering support.[25]

Proposals that eukaryotic organelles were derived from prokaryotic symbionts are quite old. E.B. Wilson, in his classic book, *The Cell*, describes an "entertaining fantasy" that the mitochondrion, chloroplast, and perhaps even the nucleus might be symbionts.[26] Wilson's brief descriptions of theories for the formation of the nucleus reveal his fascination with these ideas and his hesitancy to regard them as proper science. In 1910, Merezhkowsky hypothesized that "eukaryotes resulted from a symbiotic association of a primitive non-nucleated Monera composed of amoeboplasm and a ultra-microscopic bacteria-like biococci. By ingestion of the latter (biococci) by the Monera arose a symbiotic association of the two forms, the cocci becoming chromidial granules and thus ultimately forming the nucleus."[27] Wilson continues, explaining that, "Pfeffer once considered the possibility that the cell may have been the product of a symbiosis between nucleus and cytosome, while Boveri suggested a 'symbiosis of two kinds of simple plasma-structures—Monera, if we may so call them—in such fashion that a number of smaller forms, the chromosomes, established themselves within a larger one which we now call the cytosome.'"[28] For the next 50 years, these symbiotic hypotheses, based on the light microscopic resemblance of nuclei to bacteria, remained largely forgotten.

With the advent of modern cell biology, and of rapid DNA and RNA sequencing in the early 1980s, interest in the origin of eukaryotes blossomed. Lynn Margulis's revolutionary and prophetic book, *Origin of Eukaryotic Cells*, successfully focused attention on the endosymbiotic origins of the chloroplast and mitochondria.[29] In rapid succession, chloroplasts and mitochondria were demonstrated to be endosymbionts,[25,30] but regarding the origin of the nucleus, most scientists preferred the karyogenic (or autogenous) theory for the origin of the nucleus (see Fig. 9.1).

As explained by Margulis, a mitochondrial–host symbiosis provided the biosynthetic steps required to form eukaryotic phospholipid membranes and thereby set the stage for the evolution of the nuclear membrane and the endoplasmic reticulum.[29]

KARYOGENIC HYPOTHESIS

Nucleus

mechanism unspecified

Protoeukaryote

ENDOKARYOTIC HYPOTHESIS

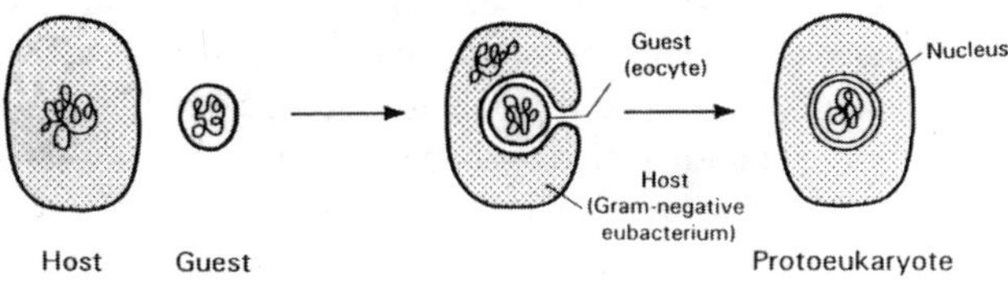

Figure 9.1. A comparison of two popular theories for the origin of the eukaryotic nucleus.

In contrast, the endokaryotic hypothesis posits that the nucleus is an endosymbiont like the chloroplast and mitochondrion. Impetus for this hypothesis initially came from phylogenetic comparisons of prokaryotic and eukaryotic ribosomes indicating that eukaryotic ribosomes were similar to those of eocyte prokaryotes; this hypothesis was reinforced by two additional observations.[31,32] First, the nucleus, like the chloroplast and mitochondrion, is surrounded by two membranes.[31,33] Second, eukaryotes contain two types of glycerolipids, one with exclusively ester linkages and a second type consiting of two variants with one ether and one ester linkages, making it unlikely that eukaryotes had come solely from eubacteria or solely from eocytes.[34,35] Taken together, these three observations, the ester and ether linkages, the double-layer nuclear membranes, and the topology of the tree linking eukaryotes to eocytes, all suggested an endosymbiotic origin for the nucleus.

Testing the Endokaryotic Hypothesis

Testing of the endokaryotic hypothesis was, ironically, delayed because the nucleus was the defining feature of eukaryotes. It is surprisingly easy to test whether the chloroplast, or the mitochondrion, is an endosymbiont, but it is difficult to test this for the nucleus. For the chloroplast, one only need show that chloroplast DNA is more closely related to that of a cyanobacterium than it is to DNA from any other organism, including the plant's own nuclear DNA. The key to this traditional test for organelles is that nuclear DNA serves as the marker for the "eukaryotic" component in four-taxon phylogenetic tests. If one is testing whether the nucleus is an organelle, however, what is the proper test? One cannot simultaneously use the nucleus and its genes as the marker for the "eukaryotic" component and as the marker for the organelle. A proper test requires first identifying two separate, independent, nuclear components and then identifying the prokaryotic antecedents of these components. This is difficult, and the appropriate phylogenetic methods are continuing to be refined and improved.

Just as the dual roles of the nucleus slowed progress in testing the endokaryotic hypothesis, several additional cell biological observations further slowed its consideration. Some of these objections were superfluous, but most were substantial and simply needed more data to resolve. These criticisms are briefly discussed below, especially in the light of new experimental findings.

The endokaryotic hypothesis has been criticized because its predictions do not match the membrane properties of the nucleus. For example, it has been suggested that the nucleus should be surrounded by three separate membranes, rather than the two that are present. Two membranes should have come from the proposed eubacterial host and one from the nuclear guest. However, the same reasoning also predicts that the mitochondrion and the chloroplast should be surrounded by three membranes rather than the two that are, in fact, present. (The mitochondrion is derived from an α-proteobacterium with two membranes and the chloroplast from a gram-negative cyanobacterium with two membranes, and the host for both has one.) Thus, that criticism is moot, but the reason that endosymbionts are surrounded by only two membranes remains an unsolved problem.

A related criticism of the endokaryotic hypothesis is that the proposed host for a nuclear endosymbiont is an eubacterium, whereas the mitochondrion and chloroplast are eubacterial guests, so that there is no direct precedent for a eubacterial host. In fact, there now are precedents. Recently, von Dohlen et al.[36] found that mealybugs contained an endosymbiont, shown by rRNA sequencing to be derived from a β-proteobacterium. Much like a series of Russian nesting Matryoshka dolls, each endosymbiont contains its own endosymbiont, a γ-proteobacterium. Hence, eubacteria can indeed host endosymbionts.

Until recently, it was also thought that the nucleus differed from the mitochondria and chloroplast because nuclear ribosomes could not actively synthesize proteins. However, recent experiments have demonstrated extensive protein synthesis within eukaryotic nuclei. Iborra et al. estimate that nuclear translation accounts for about 10%–15% of protein synthesis in eukaryotes.[37] Furthermore, Hentze points out that ribosomal subunits are assembled in the nucleus, that the translation initiation and elongation factors reside in the nucleus, and that even aminoacyl-tRNA synthetases function in the nucleus.[38] He also reviews the work of Iborra and colleagues and concludes that they "have mounted a case of unprecedented strength in support of nuclear translation." Thus, strong evidence indicates that the nucleus functions in all three of the fundamental processes of life: replication, transcription, and translation.

The availability of complete and partial genomes from diverse organisms has motivated several phylogenetic analyses that support the endokaryotic hypothesis. Notably, Gupta et al., in a pioneering paper, concluded from analyses of *Giardia* heat shock protein HSP70 that the eukaryotic nucleus resulted from an endosymbiotic event between a gram-negative eubacterial host and an eocyte prokaryote, based on two paralogous HSP70 genes found in eukaryotes.[39] One paralog is present in the cytoplasm, and the second, based on a characteristic N-terminal signal sequence and on a C-terminal endoplasmic retention sequence, is present in the endoplasmic reticulum. A comprehensive review of these experiments has been provided elsewhere.[40] A second paper, strongly supporting the endokaryotic hypothesis, was based on the whole-genome-analysis method of Horiike et al.[41] After removing mitochondrially related open reading frames from the genomes being studied, the authors analyzed yeast orthologs and compared these with those from six Archaea and nine Bacteria at several levels of discrimination.

Their paper is unique in using several levels of analyses to improve the sensitivity and reliability of these analyses. The authors conclude that their analyses strongly support the endokaryotic theory.

Finally, any hypothesis for the origin of the nucleus needs both to be metabolically reasonable and to provide a selective advantage for both partners of the initial symbiosis. A number of authors, most notably Martin and Muller,[42] have carefully considered the selective advantages of complementary host and guest metabolisms and have established plausibility arguments for the initial steps of endosymbiotic relationships. A thoughtful early paper of Searcy[43] summarized potential roles of sulphurous compounds in symbioses. Martin and Muller's classic paper proposed an explicit hypothesis for the roles of H_2, CO_2, and CH_4 in the origin of mitochondria: the hydrogen hypothesis. This hypothesis couples the metabolism of H_2, CO_2, and CH_4 so that the metabolic waste product of one symbiotic partner matches the metabolic needs of the other, and vice versa, thereby deriving a model for the origin of the mitochondrion and the hydrogenosome, a mitochondrion-like organelle. In their model, the waste product of the host, H_2, is used as the nourishment for the guest, a methanogen. However, these authors also point out that it is possible to recycle H_2 several ways, including using elemental sulfur, S^o. In addition, Moreira and Lopez-Garcia[44] proposed a synthropic theory for the origin of the nucleus that involves the coming together of many cells that together act as the host.

Using the hydrogen hypothesis as a guide, we have outlined a detailed model in which the first nucleated cell is derived from a eubacterium and an eocyte.[45] Purple sulfur bacteria and eocytes are naturally complementary endosymbiotic partners, with the metabolic waste products of each matching the metabolic needs of the other. That model, shown in figure 9.2, is briefly summarized as follows: The sulfur bacteria (the host species) are unicellular photoautotrophs capable of growing anaerobically in the light using CO_2 as a carbon source, so that ATP is provided by cyclic photophosphorylation and reducing power is provided by H_2S, which is oxidized anaerobically to S^o and to H_2SO_4. Anaerobic eocyte prokaryotes typically generate energy by the reduction of sulfur with hydrogen, forming H_2S, and many can be grown on a range of carbon sources. Through endosymbiotic association, based on hydrogen and sulfur recycling, the endosymbiont guest would provide the host with a reliable supply of H_2S, and the host would provide the endosymbiont a reliable supply of S^o and H_2. This recycling would thus decouple ATP generation by photophosphorylation from H_2S availability, thereby nearly eliminating the need for sulfur and requiring only sufficient H_2 to replace that lost by diffusion. Once a relationship was established between host and endosymbiont, if they were removed from a concentrated source of sulfur, both would be irreversibly dependent on each other. This metabolically driven model indicates that a symbiotic relationship between an eubacterium and an eocyte could have great selective advantages and allow the endosymbiotic complex to range beyond its initial sulfur-rich habitat.

Thus, experiments conducted during the last decade are laying the groundwork for understanding the evolution of eukaryotes. It is indeed possible that one of the greatest events in biological history, the origin of the eukaryotic nucleus, occurred when two disparate prokaryotes joined together to form an organism that would be less dependent on its physical environment.

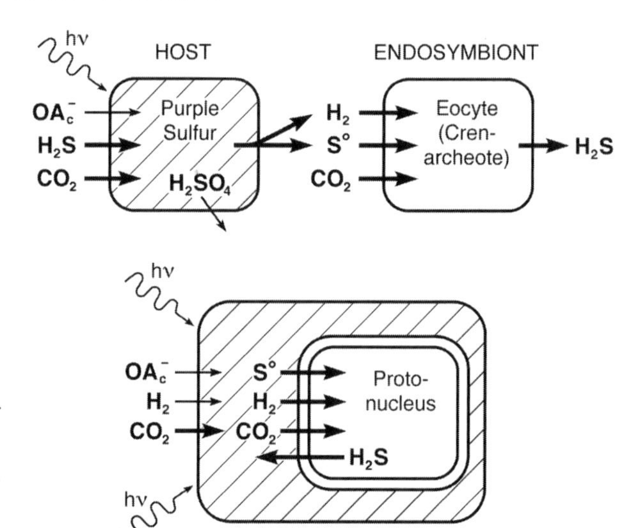

Figure 9.2. A metabolically driven model for the origin of the nucleus utilizing the selective principles first proposed in the hydrogen hypothesis.

How Are Eukaryotes Related to the Prokaryotes?

Phylogenetic analyses of the tree of life still provide the principal information for understanding the origin of eukaryotes. The two most intensively studied prokaryotic genes spanning the tree of life, 16/18S ribosomal RNA (rRNA) and protein synthesis factor EF-Tu (EF-1 alpha in eukaryotes), together with whole-genome analyses, remain the primary sources of information for understanding the origin of eukaryotes.

The two trees shown in figure 9.3 are almost always obtained when the phylogenetic tree of life is reconstructed from 16/18S rRNA or EF-1 alpha genes. In the tree at the left, the archael tree [46]eukaryotes are the sister group of (i.e., most closely related to)

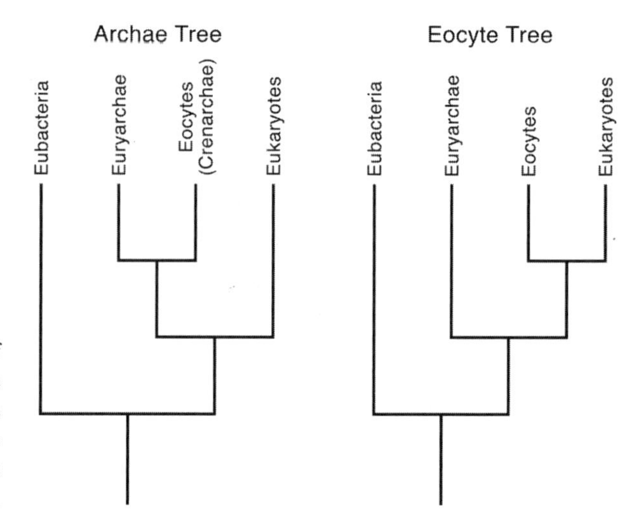

Figure 9.3. A comparison of the Archael and Eocyte theories illustrating the differing phylogenetic positions of eukaryotes in the respective trees.

the last common ancestor (cenancestor) of the euryarchaea + eocytes (the archaea). In the tree at the right, the eocyte tree, eukaryotes are the sister group of the eocytes. These two trees, although appearing similar, make greatly different predictions about the origin of the eukaryotes.

Many initial studies using 16/18S sequences supported the archael tree.[46–51] However, some early studies supported the eocyte tree,[52,53] and recently, some very prominent 16/18S studies have also tended to support the eocyte tree rather than the archael one.[54]

A different pattern was observed when EF-1 alpha genes were used to reconstruct the tree of life. Early studies supported both trees, but in contrast to 16S/18S studies, almost all studies since the early 1990s have supported the eocyte tree.[55–61]

These changes may have resulted from recent improvements in phylogenetic reconstruction methods, as reconstructing the tree of life is inextricably bound up in details of phylogenetic reconstruction that can lead to incorrect trees. A brief explanation of these effects follows: Artifacts in alignment and tree reconstruction that can produce incorrect trees are known as long-branch attraction. The name refers to the observation that such attraction occurs when taxa are evolving at different rates. The effect of this artifact is to cluster rapidly evolving taxa (corresponding to long branches in the tree) with other rapidly evolving taxa in an evolutionary tree, and to cluster slowly evolving taxa (short branches) with other slowly evolving ones in the tree, even though the taxa may not be phylogenetically related. Starting in the late 1980s, some phylogeneticists began to conclude that long-branch attraction between the long branches of the rapidly evolving eubacteria and rapidly evolving eukaryotes was causing these divergent groups to be placed together in the archaeal tree, and that this tree is simply an artifact.[53]

Recently, Tourasse and Gouy reevaluated the RNA-based tree of life and concluded that the archael tree is an artifact of long-branch attraction.[51–54] Although Gouy was initially an extremely influential supporter of the 16/18S Archael tree,[48] this recent reappraisal, using improved distance methods, now rejects the archael tree in favor of the eocyte tree. The authors conclude, "All these results suggest that obtaining monophyletic Archaea may be an artifact due to underestimation of branch lengths."[54] Furthermore, they explain that, "this may be a result of the long-branch mutual attraction phenomenon. That is, mutual attraction between the long bacterial and eukaryotic branches force together the two shorter archaeal branches, resulting in a misleading archaeal monophyly." The authors observed this effect with sequences from both small-subunit (SSU) and large-subunit (LSU) rRNAs. Figure 9.4 shows their branch lengths and illustrates why they think this happened. They note that the branches leading to the eukaryotes and eubacteria are the two longest in the tree. The long branches are obvious in both their SSU and LSU trees. Note, however, that when distances are corrected for multiple substitutions using more accurate algorithms, the topology switches so that the short-branch eocytes are connected with the long-branch eukaryotes, and the short-branch euryarchaea are connected with the long-branched eubacteria. Also note that the branches leading from the central nodes to the eukaryotes and to the eubacteria are longer for the corrected data than for the uncorrected data. This is precisely because most of the multiple nucleotide changes, which normally are missed when uncorrected distances are used, are found in the longest branches. However, these multiple changes only fully appear when the distances are properly corrected. This classic paper serves as a model documenting the difficulties in calculating phylogenetic trees that encompass the entire span of diversity of life on Earth.

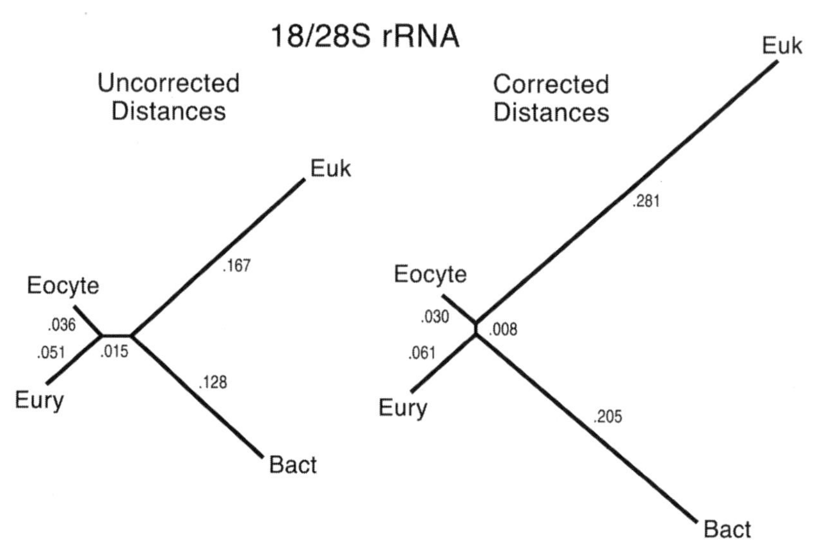

Figure 9.4. Evidence that the Archael tree is an artifact of long branch attraction based on the analysis of ribosomal RNA sequences, data from Tourasse and Gouy[54].

The algorithm on which the archael tree is based has a history of producing trees that are demonstrably incorrect because of long-branch attraction.[46] Two prominent examples using their algorithm are known to have resulted from long-branch artifacts. Specifically, the first tree to maintain that an amitochondriate eukaryote branched from the base of the eukaryotic lineage was the analysis of the position of the *V. necatrix* sequence.[62] That finding sparked the archaeozoa proposal—that the earliest eukaryotes lacked mitochondria—and has subsequently been compellingly rejected.[63–70] Subsequently, their algorithm produced a second, highly visible, incorrect tree. This time the tree indicated that the bilateral animals had evolved multiple times.[71] Although the data and experimental concepts were visionary, the phylogenetic analysis itself was demonstrably biased by long-branch attraction artifacts that marred an otherwise pioneering work. Immediately following its publication, that analysis provoked an uproar of protest because it was inconsistent with other data on the morphology of the bilateral animals, and the concept was subsequently abandoned.[72–74] Given a history of faulty reconstructions, it is plausible, and even likely, that long-branch attraction has biased the archael tree toward an incorrect long-branched topology. This is especially probable because the divergence times for the deep branches of the combined prokaryotic and eukaryotic trees of life are far older than the deepest branches of the eukaryotic tree. Hence, the prokaryotic/eukaryotic tree is much more prone to long-branch attraction.

From this perspective, it is better to think of the part of the eukaryotic genome involved in protein synthesis as having evolved from a, possibly sulfur-metabolizing, eocyte prokaryote, rather than from a vague ancestor that is intermediate between eubacteria and the last common ancestor of the archaea.

The Horizontal Gene Transfer Revolution and the Tree of Life

The Horizontal Gene Transfer Revolution

The possibility of analyzing complete genomes has awakened interest in prokaryotic genome evolution and is profoundly changing our understanding of genome evolution. Before the first genomes were sequenced, there was nearly unanimous scientific agreement that prokaryotic genomes were evolving clonally, or approximately so. In other words, as generation after generation of bacteria divided, each bacterium would contain the DNA it inherited from its parent, except that occasionally a single DNA nucleotide might have mutated, causing a minor change in the daughter genome. Thus, it was thought that the family tree derived from any one gene would look like the family tree from any other gene. Diploid eukaryotic cells with two copies of each gene per cell slightly complicated this picture, but they too were thought to be evolving as a clonal tree. Thus, everyone felt comfortable that reliable family trees could be calculated from gene trees. In particular, ribosomal RNA genes were favored, as rRNA was easy to sequence, and it was assumed that trees calculated from rRNA would probably be the same as those calculated from any other genes.

Because so much attention was focused on the approximately clonal evolution of rRNA in the pregenomic era, only a few genes were noticed that indicated horizontal gene transfer (HGT). Early on, results began to indicate a chimeric origin for the nuclear-coded eukaryotic genes. In the mid-1980s, as previously discussed, it was noted that the evolution of lipids was inconsistent with the evolution of other molecules in eukaryotes.[35] Zillig et al.,[75] using trees based on DNA-dependent RNA polymerase genes, found additional evidence for chimeric eukaryotic origins, and Sogin,[76] noting major differences between rRNA and protein gene trees, proposed a novel type of chimeric eukaryotic origin. Golding and Gupta,[77] using heat shock protein (hsp70) gene sequences, interpreted their results as being consistent with earlier chimeric proposals.[77] Simultaneously in this pregenomic era, a few studies on isolated genes were also suspiciously indicating HGT.[3,78–81]

Once complete genomes were available, the pace of discovery accelerated, as highlighted in analyses of complete, or nearly complete, genome studies from the laboratories of R. Doolittle, F. Doolittle, Golding, and ourselves.[82–85] These and even more recent studies of the evolution of life, based on complete genomes, which are described below, have revealed the flaws in the old, clonal view. Scientific opinion has now shifted, and most now favor a significant role for HGT in genome evolution.

HGT Has Profoundly Affected Our Understanding of Genome Evolution

Three amazing new findings, based on analyses of whole genomes, have engendered appreciation for this new paradigm. First, HGT is now generally recognized to be rampant among genomes (rampant at least on a geological timescale). Second, not all genes are equally likely to be horizontally transferred. Informational genes (involved in transcription, translation, and related processes) are rarely transferred, whereas operational genes (involved in amino acid biosynthesis and numerous other housekeeping activities) are readily transferred. Third, biological and physical factors appear to have altered horizontal gene transfer. These include intracellular structural constraints between proteins (the "complexity hypothesis"), interactions between organisms, and interactions with the physical environment. These three findings are described below.

Evidence for Extensive HGT

As early as 1996, the complete sequence of the methanogen *Methanococcus janaschii* revealed that its genome consisted of certain groups of genes much more similar to eukaryotes than bacteria, whereas other groups of genes were much more closely related to their bacterial homologs.[86] Koonin et al.[87] confirmed that the *M. jannaschii* genes for translation, transcription, replication, and protein secretion were more similar to eukaryotes than to bacteria. The authors interpreted this as methanogens being derived from eukaryotes and eubacteria. Using rigorous phylogenetic methods, our lab discovered the presence of two super-classes of genes in prokaryotes that had different relationships to eukaryotic genes. In studies of the *Escherichia coli*, *Syneccocystis PCC6803* (a cyanobacterium), *M. jannaschii*, and *Saccharomyces cerevisiae* genomes, the *M. jannaschii* informational genes, consisting of gene products responsible for such processes as translation and transcription, were found to be most closely related to those found in eukaryotes.[85,86,88–90] The operational genes of the eukaryote, responsible for the day-to-day operation of the cell (housekeeping genes), however, were most closely related to their counterparts found in *E. coli* and *Syneccocystis*.[85] This provided definitive evidence that the 16S rRNA tree does not reflect the evolution of all the genes in a genome and also provided evidence that the eukaryote is a chimera of eubacteria and a member of the methanogens or eocyte part of the tree (an eocyte genome was not available then). A stylized illustration of these results is shown in figure 9.5.

Further evidence for extensive HGT came from the observation that another methanogen, *Methanobacterium thermoautotrophicum*, contains several regions that have an approximately 10% lower G+C content than the G+C content of the whole genome on average.[91] Open reading frames (ORFs) in these regions exhibit a codon usage pattern

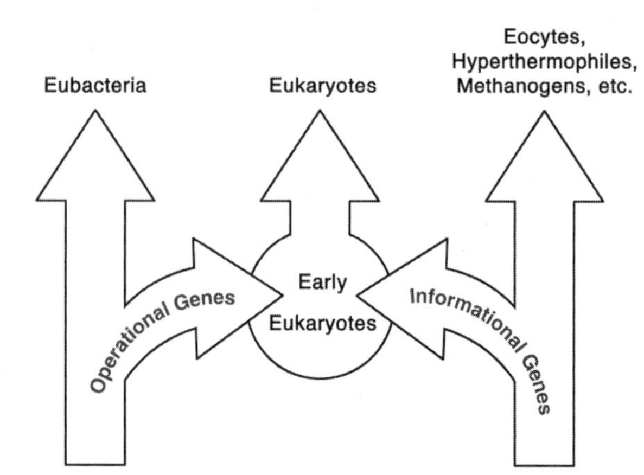

Figure 9.5. A stylized illustration of the flow of eubacterial-operational genes and methanogen- or eocyte-informational genes into the protoeukaryote.

atypical of *M. thermoautotrophicum*, indicating that the DNA sequences may have been acquired by HGT.[91] By counting the presence or absence of ORFs between *M. thermoautotrophicum* and *M. jannaschii*, 36 new acquisitions in *M. jannaschii* seem to be absent in *M. thermoautotrophicum*, providing more evidence for HGT.[91]

Additional evidence for HGT came from a thermophilic relative of the methanogens, *Archaeoglobus fulgidus*.[92] ORFs in the functional categories of translation, transcription, replication, and some essential biosynthetic pathways are very similar to those of *M. jannaschii*, but these two genomes differ in many of their housekeeping functions, such as environmental sensing, transport, and energy metabolism. The tryptophan biosynthesis pathway in *A. fulgidus* seems very closely related to the eubacterium *B. subtilis*, even though these two are separated by large distances on the 16S tree. These observations indicate that the extent of gene exchange that has occurred in the methanogens and their relatives is tremendous.

Among the extreme thermophiles, some of which live in temperatures in excess of the boiling temperature of water, HGT is equally prevalent.[93] Lecompte et al.[94] compared the three proteomes of the high-temperature methanogen relatives *Pyrococcus abyssi*, *Pyrococcus furiosus*, and *Pyrococcus horikoshii*. In this analysis, the ORFs encoding translation proteins, transcription proteins, and some others gave fairly consistent distances among the three species, indicative of a clonal evolutionary pattern. Most other ORFs, such as the housekeeping genes, gave a wide distribution of distances. The existence of a distribution was interpreted as evidence of HGT.[94] In addition, *P. furiosus* is capable of transporting and metabolizing maltose and maltodextrin, properties that are absent in *P. horikoshii*. Of two maltose/maltodextrin import systems in *P. furiosus*, one has greatest similarity to the transport system of *E. coli*, a finding best explained as a lateral transfer of the system from *E. coli* to *P. furiosus*.[95, 96] Comparison between *P. furiosus* and *P. abyssi* has revealed linkage between restriction-modification genes. These genes are responsible for protecting the cell from invasion by foreign DNA, such as viruses, while protecting the genome by DNA modifications, such as methylation. Through codon bias analysis, some restriction–modifications systems in the *Pyrococcus* genomes seem to have been acquired by horizontal transfer.[97]

HGT is also widely prevalent in the eubacteria. It has been demonstrated in *A. aeolicus*, where little consistency was seen among trees reconstructed from a number of operational genes.[98] For *E. coli*, comparative analyses of *E. coli* ORFs showed that 675 *E. coli* ORFs have greatest similarity to *Synechocystis*, 231 ORFs to *M. jannaschii*, and 254 ORFs to the eukaryote *S. cerevisiae*.[90] Using skewed base composition and codon usage as a measure of an alien gene, Ochman and colleagues[99] argued that 755 of 4,288 *E. coli* ORFs have been horizontally acquired in 234 lateral transfer events since *E. coli's* divergence from *Salmonella* approximately 100 million years ago.

The *B. subtilis* genome also harbors a number of foreign genes, as evidenced by many prophage-like regions encompassing approximately 15% of the genome.[100] Another species of the same genus, *Bacillus halodurans,* an alkaliphilic, similar to its close relative *B. subtilis*, possesses regions with a G+C content similar to that of some viruses.[101] As a consequence of this similarity, those DNA sequences were proposed to have been obtained by lateral transfer.[101] The genome of *Clostridium acetobutylicum* contains genes missing in *B. subtilis*. These genes have a number of different phylogenetic relationships. For example, 49 genes reveal an immediate relationship between *C. acetobutylicum* and eukaryotes, and another 195 are most closely related to several noneubacterial extremophiles.[102]

The cyanobacterium *Synechocystis PCC6803* is another bacterium whose genome supports extensive HGT among prokaryotes. The genome of *Synechocystis* contains a number of insertion sequences, termed IS elements. The DNA around the IS elements displays features of *E. coli* DNA, indicative of horizontal genetic acquisitions.[103] It is also interesting to note that some genes from *Synechocystis* have representatives only in eukaryotes, such as humans. Because the cyanobacteria are thought to be the endosymbionts that gave rise to the chloroplast, *Synechocystis* has many genes that are most closely related to plants.[88]

Although HGT Is Rampant, It Is Not Random: The Complexity Hypothesis

In a subsequent phylogenetic analysis, our lab examined the frequency of lateral transfer of operational genes among six prokaryotic proteomes, *E. coli*, *Synechocystis PCC6803*, *B. subtilis*, *A. aeolicus*, *M. jannaschii*, and *A. fulgidus*.[86,88,90,92,98,100] We performed three tree topology tests on orthologs of the six prokaryotes to measure the extent of HGT. All the tests significantly showed that operational genes have been continually laterally transferred among prokaryotes since the last common ancestor of life. Moreover, to explain why operational genes undergo HGT more frequently than informational genes, we proposed the complexity hypothesis, which posits that informational genes are not as likely to undergo horizontal transfer because they are members of large complexes. Operational genes, however, are generally not parts of large complexes and, thus, are more readily transferred.[104]

HGT Accelerates Genome Innovation and Evolution

It is becoming clear that HGT has had a major effect on the evolution of life on Earth. It is a major agent, perhaps *the* major agent, responsible for spreading genetic diversity by moving genes across species boundaries.[105] By rapidly introducing newly evolved genes into existing genomes, HGT circumvents the slow step of *ab initio* gene creation and thereby accelerates genome innovation, but HGT can only affect organisms that readily exchange genes (exchange communities). An analysis of approximately twenty thousand genes contained in eight free-living prokaryotic genomes indicated that HGT preferentially occurs among organisms that have similar environmental and genomic factors in common. These include genome size, genome G/C composition, carbon utilization, and oxygen tolerance.[105] On the basis of the numbers of prokaryotic species within exchange groups, that study estimated that HGT has accelerated the introduction of new genes into species by a factor of ten thousand. Indeed, HGT may be responsible for a remarkable increase in genome innovation that greatly exceeds anything that could have been accomplished by clonal evolution alone.

HGT Greatly Complicates Reconstructing the Universal Tree of Life

W. Ford Doolittle has recently reviewed the state of "Phylogenetic Classification and the Universal Tree" and asks, "What if phylogenetic classification is just let go?"[8] It is a bold question, and an appealing one given what is being learned about HGT. Furthermore, the difficulties run deeper than just HGT. Those include additional obstacles that must be overcome to accurately reconstruct microbial phylogenies. Although Doolittle's observations are for the most part on the mark, our lab's outlook is much more optimistic. We think these are just temporary obstacles that can, and will, be overcome. Our working hypotheses are that future fundamental improvements in molecular phylogeny reconstruction techniques and other new theoretical advances will make it possible to hurdle these barriers in the near future.

It would be disingenuous to pretend that the difficulties are not sizable. They are not only enormous, but there are many of them. However, they are not necessarily insurmountable. In fact, the only way to discover whether they could destroy Darwin's dream is to assume that they are not insurmountable and to use every effort to solve them. Some of barriers to reconstructing the tree of life are described below.

The rRNA Tree of Life Wanes

Consider what has happened to the once-ebullient field of ribosomal RNA phylogenies. For years, phylogenies based on ribosomal RNAs had been the holy grail of microbial phylogenetics. To be sure, ribosomal RNA-based phylogenies have been responsible for many successes including both the demonstrations that the mitochondrion and chloroplast are endosymbionts and the new animal phylogeny.[9,10,25,30,106,107] However, prokaryotic phylogenies are another story. One only has to read the latest *Bergey's Manual* or to read the chapter by Schleifer and Ludwig in this volume to realize that the tree of prokaryotic life is fuzzy and unresolved.[108] So much so, in fact, that rRNA-based trees, although capable of identifying to which phylum a prokaryote belongs, are incapable of determining how the phyla relate to each other.

Another seldom-mentioned barrier to reconstructing trees accurately is the finite length of genomes. Given our current fascination with completely sequencing ever-longer genomes, it is hard to appreciate that genomes are of finite length. However, unfortunately, they are, and this length limits the feasible time resolution within trees. Thus, there has never been a possibility that we could reconstruct perfect trees, so perhaps the question we should be asking is, How much information can one obtain?

While asking whether we should just let go of phylogenies, Doolittle[8] has pointed out the specific challenges to classification that HGT presents as follows: "If, however, different genes give different trees, and there is no fair way to suppress this disagreement, then a species (or phylum) can 'belong' to many genera (or kingdoms) at the same time: There really can be no universal phylogenetic tree of organisms based on such a reduction to genes." In other words, Doolittle suggests that the gene mixing resulting from HGT is so extensive that it precludes one from reconstructing the tree of life.

Although he may be correct, we subscribe to an alternative view. Yes, HGT is extensive, and yes it imposes limits to phylogenetic reconstruction, but we view the difficulties of HGT as just another temporary setback on the search for Darwin's goal.

Phylogeny and Classification are Inextricably Linked?

Doolittle implicitly assumes that classification must be phylogenetic, and most scientists—but not all—implicitly concur. Hence, it is useful to consider the alternatives. Ernst

Mayr, for example, has made a strong and passionate case that classifications can include a number of factors such as rates of evolution and overall similarity.[109] As an articulate counter to this, E.O. Wiley has argued that phylogenetics is the only basis for classification, and most evolutionists support this view.[110] In phylogenetic classification, the only allowable groups are those that include all descendents of, and only descendents of, the most recent common ancestor of the group. It has been shown that any other rules for grouping organisms must produce nonphylogenetic classifications. These acceptable phylogenetic groups are named monophyletic groups because they contain a complete clade. In contrast, groups which either do not contain all the descendents, or which contain members not derived from the most recent common ancestor are paraphyletic groups. For example, assuming for the moment that informational genes are useful for following cell lineages, if the archaea (euryarchaea + eocytes) are monophyletic then archaea would be the acceptable phylogenetic classification. On the other hand if eocytes and eukaryotes are monophyletic, then the only proper group would be the karyota.[35] The beauty of phylogenetic classifications is that the seemingly arbitrary nature of naming the branches of the tree of life, becomes instead a question of phylogeny, which is testable. Obviously classification will also be complicated by endosymbiotic events, but first we need to know the history of events.

How Does HGT Affect our Ability to Know the Tree of Life?

To determine phylogenies accurately, one needs to know how extensive HGT has been and its overall affect on phylogenetic reconstructions. If no HGT, or very little, has occurred, then current methods of analysis will allow one to reconstruct the clonal tree of life. At the other extreme, if all genes undergo HGT once a year, then all gene trees will be undecipherable. In between these extremes lies a continuum of results. In practice, one hopes to find a common thread of similarity between all the gene trees within a set of genomes, to deconvolute the phylogeny of the genomes from the corrupting HGT. In any such deconvolution one's success will necessarily depend on the extent of the corrupting process. If HGT is sufficiently large, then it will effectively preclude phylogenetic reconstruction. Hence, predictions about its severity depend on obtaining accurate estimates of the extent of HGT.

An example serves to give one an idea of how HGT affects phylogenies. Jain et al. studied six complete prokaryotic genomes from diverse prokaryotes with the goal of determining whether, and how, HGT was changing with time.[104] The genomes were from four eubacteria and two euryarchaea; *Aquifex aeolicus* (an eubacterial extremophile), *B. subtilis*, *E. coli*, *Synechocystis 6801* (a cyanobacterium), *M. jannaschii* (a methanogen), and *Archaeoglobus fulgidus* (a methanogen relative), respectively. They identified 312 sets of orthologous genes of known function that were present in all six genomes, aligned them, and calculated phylogenetic trees. These genes included 203 operational (higher-HGT) and 109 information (lower-HGT) genes. Plotted on each internal branch of the reference tree in figure 9.6 are two numbers representing the local deviations of informational and operational gene trees from the reference EF-1 alpha topology. For example, the average percentage of support for the clade of *Escherichia* + *Synechocystis* is better for informational genes (40%) than for operational genes (31%). Because there are three alternative topologies possible for each internal branch, the 31%

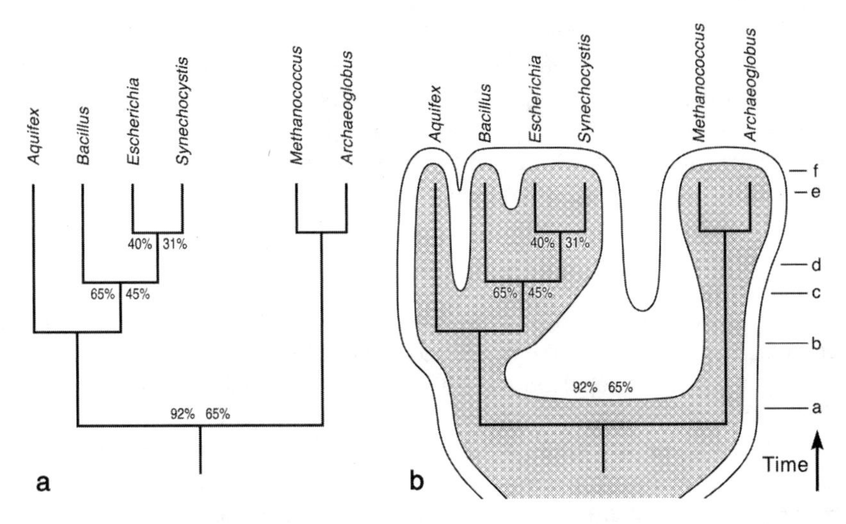

Figure 9.6. A comparison of the inferred amount of horizontal transfer of operational and informational genes in different regions of a reference (zero HGT) tree. Details are provided in the text. In panel b, this figure is interpreted as representing platypatric speciation, and should be compared with figure 9.7.

figure is not statistically different from random (33 1/3%), indicating that HGT between *E. coli* and *Synechocystis* is extensive for operational genes. In contrast, the 40% figure observed for informational genes across this branch indicates that a phylogenetic signal exists for informational genes, although it is small.

As one moves down the tree, the percentage of trees that match the reference tree increases, indicating that there is less mixing by HGT between the groups at the higher taxonomic levels. For example, the next internal branch of the tree shows 65% agreement for informational genes and 45% agreement for operational genes. Note, however, that the informational genes consistently have higher percentage scores; that is, fewer deviations from the reference trees.

Our schematic interpretation of these results is shown in figure 6b. The ability of informational genes to resolve groups is shown by two contours. The informational genes are indicated by the shaded contours, and the operational genes are indicated by the outermost contour. The vertical axis represents a very approximate measure of time, such that present day is located at the top and the most distant times at the bottom. The contour lines are drawn in such a way that a horizontal line drawn across the figure at a given time approximately indicates what can and can not be resolved. For example, at present there has been sufficient HGT that the average operational gene tree can resolve *Aquifex* from the remainder of the eubacteria, but it probably can not resolve *E. coli*, *B. subtilis*, and *Synechocystis* from each other. In contrast, the average informational gene can separate both *Aquifex* and *Bacillus* from the remainder of the eubacteria. Obviously the informational genes that are least subject to HGT can do considerably better. Notice that the separation between the eubacteria and the euryarchaea is routine for both informational and operational genes. However, if we were to go back in time, say to time b, then the average informational genes could perform that separation but the average operational gene could not. Furthermore if an observer had been present at time a, neither gene type could separate the groups. Thus, we see can start to see the nature of the confounding effect of HGT. At a sufficiently late time, one can observe the separation of groups as distinct entities, but as long as recently divergent groups are still members of the same exchange group[105] (i.e., still rapidly conducting HGT, then it will be extremely difficult to pick apart their phylogenetic relationships).

Platypatric Speciation

The mechanisms of animal speciation have been extensively studied.[111] Speciation can be determined by geography (allopatric speciation), ecological niches within the normal range of the parental population (sympatric speciation), and additional mechanisms (e.g., founder effect, parapatric speciation, heterochrony, etc.). For example, figure 9.7 illustrates a hypothetic species evolving, perhaps through a combination of allopatric and sympatric speciation. In this figure, the range of the original parental species is shown as a shaded area. As the group evolves in time and space, ultimately the two terminal groups are unable to exchange genes.

Similarly, HGT can cause the evolution of new species that is based on the ability of genes to be horizontally transferred (platypatric speciation; see the Glossary for a complete definition). As for animals, the platypatric speciation shown in figure 9.6b can be the result of interactions between environmental, genetic, ecological, or other factors with HGT.[105] Also note that the two contour levels in figure 9.6b represent the two groups

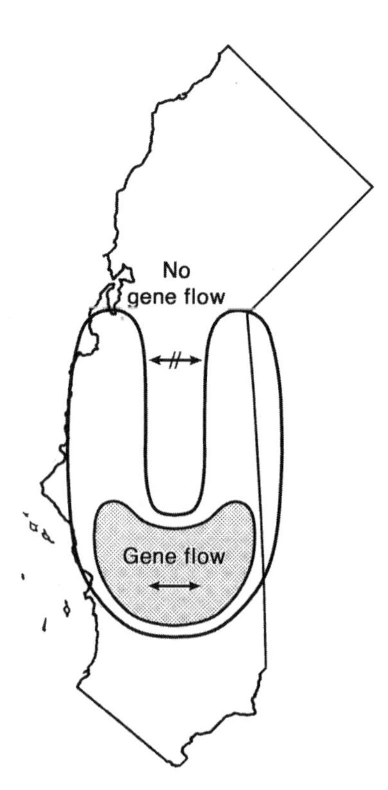

Figure 9.7. A representative example of allopatric and sympatric speciation, to be compared with the platypatric (HGT) speciation shown in figure 9.6.

of genes that could potentially exchange information by HGT: the informational (shaded) and operational (not shaded) genes. It is becoming apparent that HGT may be a significant factor in prokaryotic species evolution, mimicking speciation effects in animal evolution.

The challenge, then, is to understand phylogenetic relationships in sufficient detail that we can relate the tree of life to the evolution of life on earth. Then we can truly probe major questions about the evolution of life on our planet: What were environmental conditions like early in the evolution of life on Earth? Did life start in a hot environment? What was the role of HGT during these times? Did carbon heterotrophy evolve before autotrophy, or vice versa?

HGT should not be viewed as a negative participant in evolution. It is becoming clear that it has played a major role in the evolution of prokaryotic life on earth by spreading genes across species boundaries, thereby accelerating genome innovation and evolution to an extent that greatly exceeds anything that could have been accomplished by clonal evolution alone. It is noteworthy that William Thomson (later Lord Kelvin) criticized Darwin's theory because it required a longer time to evolve life than his calculations dated the age of the earth, at least according to his preradioactivity estimations. The discovery of radioactivity dramatically lengthened the estimates of the age of the earth and removed this criticism of Darwin's work. Now a remarkable new mechanism for change, HGT, appears to have again effectively extended the time available for prokaryotic evolution.

Glossary of Specialized Terms Related to Horizontal Transfer

Given the enormous interest in, and recognition of the importance of, horizontal gene transfer there is a pressing need to specifically and accurately refer to individual steps in the acquisition and spread of genes by horizontal transfer. The following list defines common terms, as well as several new ones, used throughout this chapter. We hope that this glossary will improve the precision of scientific communication within this emerging scientific discipline.

Acquisition. The capture of genes from exogenous sources by an organism. Acquired genes are not necessarily in a form in which the genes may be clonally passed on to descendants. May be used in combinations with horizontal gene transfer as in "Horizontal Gene Acquisition" or "Horizontal Acquisition."

Clonal Inheritance. The generational transfer of genetic information from parents to descendants.

Exchange Groups. A group of organisms that may actively share genetic information via horizontal transfer. The individuals within a group are not necessarily within physical proximity, as exchange groups may have global distributions. In addition, physically proximate organisms are not necessarily members of the same exchange group.

Gene Dissemination. The spread of a gene within an exchange group through a combination of horizontal transfer, incorporations, subsequent selection, and perhaps other methods. Dissemination is to horizontal gene transfer what fixation is to clonal evolution.

Genome Innovation. The creation of a novel gene that may subsequently be acquired, incorporated, and shared by members of an exchange group.

Horizontal Gene Transfer. The acquisition and incorporation of complete genes from exogenous sources into the host genome in a form in which they may be clonally inherited. Also called lateral gene transfer.

Horizontal Transfer. The acquisition and incorporation of genetic information from exogenous sources in a form in which it will be clonally inherited. The information may consist of parts of genes, regulatory regions, or noncoding DNA and may be as small as a single nucleotide. Also called lateral transfer.

Incorporation. The incorporation of genes acquired from exogenous sources into the host genome in a form in which they may be clonally passed on to descendants. May be used in combinations with horizontal gene transfer as in "Horizontal Gene Incorporation" or "Horizontal Incorporation."

Platypatric speciation. Evolution of new species based on the ability of genes to be horizontally transferred. Platypatric speciation can result from the effect of geographical, genetic, environmental, or other factors on horizontal gene transfer. Because different genes have different probabilities of horizontal transfer, at any given time some genes may define species, wheareas other more readily transferred genes may still freely exchange among higher taxonomic groups. Platypatric speciation may occur in concert with clonal speciation mechanisms.

References

1. S. Karlin, J. Mrazek, and A.M. Campbell, "Compositional Biases of Bacterial Genomes and evolutionary implications," *Journal of Bacteriology* **179** (1997): 3899–3913.
2. J.G. Lawrence and H. Ochman, "Molecular Archaeology of the *Escherichia coli* Genome," *Proceedings of the National Academy of Sciences USA* **95** (1998): 9413–9417.
3. J.P. Gogarten, W.F. Doolittle, and J.G. Lawrence, "Prokaryotic Evolution in Light of Gene Transfer," *Molecular Biology and Evolution*, **19** (2002): 2226–2238.
4. W.F. Doolittle, "Lateral Genomics," *Trends in Genetics* **15** (1999): M5–M8.
5. A.M. Campbell, "Lateral Gene Transfer in Prokaryotes," *Theoretical Population Biology*, **57** (2000): 1–77.
6. H. Ochman, "Lateral and Oblique Gene Transfer," *Current Opinion in Genetics and Development* **11** (2001): 616–619.
7. E.V. Koonin, K.S. Makarova, and L. Aravind, "Horizontal Gene Transfer in Prokaryotes: Quantification and Classification," *Annual Review of Microbiology* **55** (2001): 709–742.
8. W.F. Doolittle, "Phylogenetic Classification and the Universal Tree," *Science* **284** (1999): 2124–2128.
9. A.M.A. Aguinaldo et al., "Evidence for a Clade of Nematodes, Arthropods and other Moulting Animals," *Nature* **387** (1997): 489–493.
10. K.M. Halanych et al., "Evidence from 18s Ribosomal DNA that the Lophophorates Are Protostome Animals Inarticulate," *Science* **267** (1995): 1641–1643.
11. P.S. Soltis et al., "The Phylogeny of Land Plants Inferred from 18S rDNA Sequences: Pushing the Limits of rDNA Signal?" *Molecular Biology and Evolution* **16** (1999): 1774–1784.
12. D.L. Nickrent et al., "Multigene Phylogeny of Land Plants with Special Reference to Bryophytes and the Earliest Land Plants," *Molecular Biology and Evolution* **17** (2000): 1885–1895.
13. K.G. Karol et al., "The Closest Living Relatives of Land Plants," *Science* **294** (2001): 2351–2353.
14. K.M. Pryer et al., "Horsetails and Ferns are a Monophyletic Group and the Closest Living Relatives to Seed Plants," *Nature* **409** (2001): 618–622.

15. K.M. Pryer et al., "Deciding among Green Plants for Whole Genome Studies," *Trends in Plant Science* **7** (2002): 550–554.
16. H. Philippe and P. Forterre, "The Rooting of the Universal Tree of Life is Not Reliable," *Journal of Molecular Evolution* **49** (1999): 509–523.
17. D. Penny and A. Poole, "The Nature of the Last Universal Common Ancestor," *Current Opinion in Genetics and Development* **9** (1999): 672–677.
18. R. Huber et al., "*Aquifex-Pyrophilus* Gen-Nov Sp-Nov Represents a Novel Group of Marine Hyperthermophilic Hydrogen-Oxidizing Bacteria," *Systematic and Applied Microbiology* **15** (1992): 340–351.
19. R. Huber, T.A. Langworthy, H. Konig, M. Thomm, C.R. Woese, U.B. Sleytr, and K.O. Stetter, "*Thermotoga-maritima* sp-nov. Represents a New Genus of Unique Extremely Thermophilic Eubacteria Growing up to 90-degrees-C," *Archives of Microbiology* **144** (1986): 324–333.
20. R. Huber and K.O. Stetter, "Aquificales", in *Encyclopedia of Life Sciences*, (London: Nature, 1999).
21. T.W. Goodwin, *The Biochemistry of the Carotenoids*, Vol. 1. (London: Chapman and Hall, 1980).
22. M.C. Rivera and J.A. Lake, "The Phylogeny of *Methanopyrus kandleri*," *International Journal of Systematic Bacteriology* **46** (1996): 348–351.
23. K.O. Stetter, "Hyperthermophilic Procaryotes," *Microbiology Reviews* **18** (1996): 149–158.
24. K.O. Stetter et al., "Hyperthermophilic Microorganisms," *Microbiology Reviews*. **75** (1990): 117–124.
25. M.W. Gray, "Evolution of Organellar Genomes," *Current Opinion in Genetics and Development* **9** (1999): 678–687.
26. E.B. Wilson, *The Cell in Development and Heredity* (New York: The Macmillan Company, 1925).
27. C. Mereschkowsky, "Theorie der zwei Plasmaarten als Grundlage der Symbiogenesis, einer neuen Lehre von der Entstehung der Organismen. Biologisches Centralblatt," **30** (1910): 277–303.
28. T.H. Boveri, "Ergebnisse uber die Konstitution der chormatischen Kernsubstanz" 1904 cited in E. B. Wilson, *The Cell in Development and Heredity* (New York: Macmillan, 1925).
29. L. Margulis, *Origin of the Eukaryotic Cells* (New Haven: Yale University Press1970), 252.
30. Z. Schwarz and H. Kossel, "Primary Structure of 16s Rdna from Zea-Mays Chloroplast Is Homologous to *Escherichia-Coli* 16s Ribosomal-RNA," *Nature* **283** (1980): 739–742.
31. J.A. Lake et al., "Mapping Evolution with Ribosome Structure—Intra-Lineage Constancy and Inter-Lineage Variation,"*Proceedings of the National Academy of Sciences USA* **79** (1982): 5948–5952.
32. J.A. Lake, "Ribosome Evolution: The Structural Bases of Protein Synthesis in Archaebacteria, Eubacteria, and Eukaryotes," *Progress in Nucleic Acid Research* **30** (1983): 63–194.
33. G.E. Palade and P. Siekevitz, "Liver Microsomes. An Integrated Morphological and Biochemical Study," *Journal of Biophysical and Biochemical Cytology* **2** (1956): 171–200.
34. L. Stryer, *Biochemistry* (San Francisco: W.H. Freeman and Company, 1974).
35. J.A. Lake, "Origin of the Eukaryotic Nucleus Determined by Rate-Invariant Analysis of Ribosomal RNA Sequences," *Nature* **331** (1988): 184–186.
36. C.D. von Dohlen et al., "Mealybug beta-proteobacterial Endosymbionts Contain Gamma-Proteobacterial Symbionts," *Nature* **412** (2001): 433–436.
37. F.J. Iborra, D.A. Jackson, and P.R. Cook, "Coupled Transcription and Translation within Nuclei of Mammalian Cells," *Science* **293** (2001): 1139–1142.
38. M.W. Hentze, "Protein Synthesis—Believe It or Not—Translation in the Nucleus," *Science* **293** (2001): 1058–1059.
39. R.S. Gupta et al., "Cloning of Giardia-Lamblia Heat-Shock Protein Hsp70 Homologs—Implications Regarding Origin of Eukaryotic Cells and of Endoplasmic-Reticulum," *Proceedings of the National Academy of Sciences USA* **91**(1994): 2895–2899.
40. R.S. Gupta, "Protein Phylogenies and Signature Sequences: A Reappraisal of Evolutionary Relationships among Archaebacteria, Eubacteria, and Eukaryotes," *Microbiology and Molecular Biology Reviews* **62** (1998): 1435–1491.
41. T. Horiike et al., "Origin of Eukaryotic Cell Nuclei by Symbiosis of Archaea in Bacteria is Revealed by Homology-Hit Analysis," *Nature Cell Biology* **3** (2001): 210–214.

42. W. Martin, and M. Muller, "The Hydrogen Hypothesis for the First Eukaryote," *Nature* **392** (1998): 37–41.
43. D. Searcy, "Origins of Mitochondria and Chloroplasts from Sulfur-Based Symbioses," In, H.H. and K. Matsuno, eds., *The Origin and Evolution of the Cell* (Singapore: World Scientific Press, 1992), 47–48.
44. D. Moreira, and P. Lopez-Garcia, "Symbiosis between Methanogenic Archaea and Delta-Proteobacteria as the Origin of Eukaryotes: The Syntrophic Hypothesis," *Journal of Molecular Evolution* **47** (1998): 517–530.
45. J.A. Lake, R. Jain, J. Moore, and M. Rivera, "Genome Evolution and the Impact of the Physical Environment," in A. Lister and L. Rothschild, eds., *Evolution Planet Earth* (London: Academic Press, 2002), 121–130.
46. G.J. Olsen et al., "Sequence of the 16s Ribosomal-RNA Gene from the Thermoacidophilic Archaebacterium *Sulfolobus-Solfataricus* and Its Evolutionary Implications," *Journal of Molecular Evolution,* **22** (1985): 301–307.
47. C.R. Woese, and G.J. Olsen, "Archaebacterial Phylogeny—Perspectives on the Urkingdoms," *Systematic and Applied Microbiology* **7** (1986): 161–177.
48. M. Gouy and W.H. Li, "Phylogenetic Analysis Based on Ribosomal-RNA Sequences Supports the Archaebacterial Rather than the Eocyte Tree," *Nature* **339** (1989): 145–147.
49. N. R. Pace, G.J. Olsen, and C.R. Woese, "Ribosomal-RNA Phylogeny and the Primary Lines of Evolutionary Descent," *Cell* 45 (1986):325–326.
50. C.R. Woese, L.J. Magrum, and G.E. Fox, "Archaebacteria," *Journal of Molecular Evolution* **11** (1978): 245–252.
51. N. Galtier, N. Tourasse, and M. Gouy, "A Nonhyperthermophilic Common Ancestor to Extant Life Forms," *Science* **283** (1999): 220–221.
52. J.A. Lake, "Prokaryotes and Archaebacteria Are Not Monophyletic Rate Invariant Analysis of Ribosomal RNA Genes Indicates that Eukaryotes and Eocytes Form a Monophyletic Taxon," *Cold Spring Harbor Symposia on Quantitative Biology*, **52** (1987): 839–846.
53. J. Volters and V.A. Erdmann, "The Structure and Evolution of Archaebacterial Ribosomal-RNAs," *Canadian Journal of Microbiology* **35** (1989): 43–51.
54. N.J. Tourasse and M. Gouy, "Accounting for Evolutionary Rate Variation among Sequence Sites Consistently Changes Universal Phylogenies Deduced from rRNA and Protein-Coding GeneS," *Molecular Phylogenetics and Evolution* **13** (1999): 159–168.
55. J.A. Lake, "Recontructing Evolutionary Trees from DNA and Protein Sequences—Paralinear Distances," *Proceedings of the National Academy of Sciences USA* **91** (1994): 1455–1459.
56. R. Creti et al., "Nucleotide-Sequence of a DNA Region Comprising the Gene for Elongation Factor-1-Alpha (Ef-1-Alpha) from the Ultrathermophilic Archaeote Pyrococcus-Woesei—Phylogenetic Implications," *Journal of Molecular Evolution* **33** (1991): 332–342.
57. B. Cousineau et al., "The Sequence of the Gene Encoding Elongation Factor-Tu from Chlamydia-Trachomatis Compared with Those of Other Organisms," *Gene* **120** (1992): 33–41.
58. S.L. Baldauf, J.D. Palmer, and W.F. Doolittle, "The Root of the Universal Tree and the Origin of Eukaryotes Based on Elongation Factor Phylogeny," *Proceedings of the National Academy of Sciences USA* **93** (1996): 749–7754.
59. M. Hasegawa, T. Hashimoto, and J. Adachi, "Origin and Evolution of Eukaryotes as Inferred from Protein Sequence Data," in H. Hartman and K. Matsuno, eds., *The Origin and Evolution of Prokaryotic and Eukaryotic Cells* (Singapore: World Science, 1993).
60. T. Hashimoto, and M. Hasegawa, "Origin and Early Evolution of Eukaryotes Inferred from the Amino Acid Sequences of Translation Elongation Factors 1 alpha/Tu and 2/G," *Advances in Biophysics*, **32** (1996):73–120.
61. B. Runnegar, "Proterozoic Eukaryotes: Evidence from Biology and Geology," in S. Bengston, ed., *Early Life on Earth* (Cambridge: Cambridge University Press, 1993).
62. C.R.Vossbrinck, et al., "Ribosomal-RNA Sequence Suggests Microsporidia Are Extremely Ancient Eukaryotes, *Nature* **326** (1987): 411–414.
63. A.J. Roger, "Reconstructing Early Events in Eukaryotic Evolution," *American Naturalist* **154** (Suppl) (1999): S146–S163.
64. A.J. Roger et al., "An Evaluation of Elongation Factor 1alpha as a Phylogenetic Marker for Eukaryotes," *Molecular Biology and Evolution* **16** (1999): 218–233.

65. A.J Roger et al., "A mitochondrial-like chaperonin 60 gene in Giardia lamblia: Evidence that diplomonads once harbored an endosymbiont related to the progenitor of mitochondria," *Proceedings of the National Academy of Sciences USA* **95** (1998): 229–234.
66. A.J. Roger, C.G. Clark, and W.F. Doolittle, "A Possible Mitochondrial Gene in the Early-Branching Amitochondriate Protist *Trichomonas vaginalis*," *Proceedings of the National Academy of Sciences USA* **93** (1996): 14618–14622.
67. C.G. Clark and A.J. Roger, "Direct Evidence for Secondary Loss of Mitochondria in *Entamoeba histolytica*," *Proceedings of the National Academy of Sciences USA* **92** (1995): 6518–6521.
68. E.T.N. Bui, P.J. Bradley, and P.J. Johnson, "A Common Evolutionary Origin for Mitochondria and Hydrogenosomes," *Proceedings of the National Academy of Sciences USA* **93** (1996): 9651–9656.
69. A. Germot, H. Philippe, and H. Le Guyader, "Presence of a Mitochondrial-Type 70-kDa Heat Shock Protein in *Trichomonas vaginalis* Suggests a Very Early Mitochondrial Endosymbiosis in Eukaryotes," *Proceedings of the National Academy of Sciences USA* **93** (1996): 14614–14617.
70. P.J. Bradley C.J. Lahti, E. Plumper, and P. Johnson, "Targeting and translocation of proteins into the hydrogenosome of the protist *Trichomonas*: similarities with mitochondrial protein import," *The EMBO Journal*, **16** (12) (June 1997): 3484–3493.
71. K.G. Field et al., "Molecular Phylogeny of the Animal Kingdom," *Science* **239** (4841) (1988): 748–753.
72. C. Nielsen, "Phylogeny and Molecular-Data," *Science* **243** (1989): 548–548.
73. W.F. Walker, "Phylogeny and Molecular-Data—Response," *Science* **243** (1989): 548–549.
74. H.R. Bode and R.E. Steele, "Phylogeny and Molecular-Data—Response," *Science* **243** (1989): 549–550.
75. W. Zillig, P. Palm, and H.P. Klenk, "A Model of the Early Evolution of Organisms: The Arisal of the Three Domains of Life from the Common Ancestor," in H. Hartman and K. Matsuno, eds., *The Origin and Evolution of the Cell* (Singapore: World Scientific, 1992), 47–78.
76. M.L. Sogin, "Early Evolution and the Origin of Eukaryotes," *Current Opinion in Genetics and Development* **1** (1991): 457–463.
77. G.B. Golding and R.S. Gupta, "Protein-Based Phylogenies Support a Chimeric Origin for the Eukaryotic Genome," *Molecular Biology and Evolution* **12** (1995): 1–6.
78. E.Hilario and J.P. Gogarten, "Horizontal Transfer of Atpase Genes—The Tree of Life Becomes a Net of Life," *Biosystems* **31** (1993): 111–119.
79. M.W. Smith, D.F. Feng, and R.F. Doolittle, "Evolution by Acquisition: The Case for Horizontal Gene Transfers," *Trends in Biochemical Sciences* **17** (1992): 489–493.
80. M. Syvanen, "Horizontal Gene-Transfer—Evidence and Possible Consequences," *Annual Review of Genetics* **28** (1994): 237–261.
81. R.F. Doolittle, et al., "A Naturally-Occurring Horizontal Gene-Transfer from a Eukaryote to a Prokaryote," *Journal of Molecular Evolution* **31** (1990): 383–388.
82. R.F. Doolittle and J. Handy, "Evolutionary Anomalies among the Aminoacyl-tRNA Synthetases," *Current Opinion in Genetics and Development* **8** (1998): 630–636.
83. J.R. Brown and W.F. Doolittle, "Gene Descent, Duplication, and Horizontal Transfer in the Evolution of Glutamyl- and Glutaminyl-tRNA Synthetases," *Journal of Molecular Evolution* **49** (4) (1999): 485–495.
84. S. Ribeiro and G.B. Golding, "The Mosaic Nature of the Eukaryotic Nucleus," *Molecular Biology and Evolution* **15** (1998): 779–788.
85. M.C. Rivera et al., "Genomic evidence for two functionally distinct gene classes," *Proceedings of the National Academy of Sciences USA* **95** (1998): 6239–6244.
86. C.J. Bult et al., "Complete Genome Sequence of the Methanogenic Archaeon, *Methanococcus jannaschii*," *Science* **273** (1996): 1058–1073.
87. E.V. Koonin et al., "Comparison of Archaeal and Bacterial Genomes: Computer Analysis of Protein Sequences Predicts Novel Functions and Suggests a Chimeric Origin for the Archaea," *Molecular Microbiology* **25** (1997): 619–637.
88. T. Kaneko et al., "Sequence Analysis of the Genome of the Unicellular Cyanobacterium

Synechocystis sp. Strain PCC6803. II. Sequence Determination of the Entire Genome and Assignment of Potential Protein-Coding Regions," *DNA Research* **3** (1996): 109–136.

89. A. Goffeau et al., "Life with 6000 Genes," *Science* **274** (1996): 546–557.
90. F.R. Blattner et al., "The Complete Genome Sequence of *Escherichia coli* K-12," *Science* **277** (1997): 1453–1462.
91. D.R.Smith et al., "Complete Genome Sequence of *Methanobacterium hermoautotrophicum* Delta H: Functional Analysis and Comparative Genomics," *Journal of Bacteriology* **179** (1997): 7135–7155.
92. H.P. Klenk et al., "The Complete Genome Sequence of the Hyperthermophilic, Sulphate-Reducing Archaeon *Archaeoglobus fulgidus,*" *Nature* **390** (1997): 364–370.
93. K.S. Makarova et al., "Comparative Genomics of the Archaea (Euryarchaeota): Evolution of Conserved Protein Families, the Stable Core, and the Variable Shell," *Genome Research* **9** (1999): 608–628.
94. O. Lecompte et al., "Genome Evolution at the Genus Level: Comparison of Three Complete Genomes of Hyperthermophilic Archaea," *Genome Research* **11** (2001): 981–993.
95. J. DiRuggiero et al., "Evidence of Recent Lateral Gene Transfer among Hyperthermophilic Archaea," *Molecular Microbiology,* **38** (2000): 684–693.
96. D.L. Maeder et al., "Divergence of the Hyperthermophilic Archaea *Pyrococcus Furiosus* and *P. horikoshii* Inferred from Complete Genomic Sequences," *Genetics* **152** (1999): 1299–1305.
97. A. Chinen, I. Uchiyama, and I. Kobayashi, "Comparison between *Pyrococcus horikoshii* and *Pyrococcus abyssi* Genome Sequences Reveals Linkage of Restriction-Modification Genes with Large Genome Polymorphisms," *Gene* **259** (2000): 109–121.
98. G. Deckert et al., "The Complete Genome of the Hyperthermophilic Bacterium *Aquifex aeolicus,*" *Nature* **392** (1998): 353–358.
99. J.G. Lawrence, "Molecular Archaeology of the *Escherichia coli* genome," *Proceedings of the National Academy of Sciences USA* **95** (1998): 9413–9417.
100. F. Kunst et al., "The Complete Genome Sequence of the Gram-Positive Bacterium *Bacillus subtilis,*" *Nature* **390** (1997): 249–256.
101. H. Takami et al., "Complete Genome Sequence of the Alkaliphilic Bacterium *Bacillus halodurans* and Genomic Sequence Comparison with *Bacillus subtilis,*" *Nucleic Acids Research* **28** (2000): 4317–4331.
102. J. Nolling et al., "Genome Sequence and Comparative Analysis of the Solvent-Producing Bacterium *Clostridium acetobutylicum,*" *Journal of Bacteriology* **183** (2001): 4823–4838.
103. C. Cassier-Chauvat, M. Poncelet, and F. Chauvat, "Three Insertion Sequences from the Cyanobacterium Synechocystis PCC6803 Support the Occurrence of Horizontal DNA transfer among bacteria," *Gene* **195** (1997): 257–266.
104. R. Jain, M.C. Rivera, and J.A. Lake, "Horizontal Gene Transfer among Genomes: The Complexity Hypothesis," *Proceedings of the National Academy of Sciences USA* **96** (1999): 3801–3806.
105. R. Jain, M.C. Rivera, J.E. Moore, and J.A. Lake, "Horizontal Gene Transfer Accelerates Genome Innovation and Evolution," *Molecular Biology and Evolution* **20** (10) (2003): 1398–1602.
106. A. Adoutte et al., "Animal Evolution—The End of the Intermediate Taxa?" *Trends in Genetics* **15** (1999): 104–108.
107. A. Adoutte et al., "The New Animal Phylogeny: Reliability and Implications," *Proceedings of the National Academy of Sciences USA* **97** (2000): 4453–4456.
108. D. Boone and R.W. Castenholz, "The Archea and the Deep Branching and Phototrophic Bacteria," in D. Boone and R.W. Castenholz, eds., *Bergey's Manual of Systematic Bacteriology* (New York: Springer, 2001).
109. E. Mayr, *Principles of Systematic Zoology* (New York: McGraw-Hill, 1969).
110. E.O. Wiley, *Phylogenetics: The Theory and Practice of Phylogenetic Systematics* (New York: Wiley, 1981).
111. E. Mayr, *Animal Species and Evolution* (Cambridge, MA: Belknap Press of Harvard University Press, 1963).

10

Paradigm Lost

C.G. KURLAND

In a refreshingly candid note, M. Syvanen summarizes attempts to detect facile gene transfers to bacteria from transgenic plants that had been modified with antibiotic resistance markers. In this chapter, the ease of transfer of such markers from DNA to bacteria in pure culture is contrasted with the repeated failure to detect the expected transfer from plants in the field. Syvanen concludes his comments by saying, "If horizontal transfer of genes from plants to bacteria is found to occur naturally, perhaps the most significant outcome would be the reshaping of our evolutionary paradigms."[1] Here, I will attempt to explain why it would be very difficult to identify such transfers in nature. Furthermore, I will suggest that even though it may be easy to demonstrate horizontal transfer under other well-defined circumstances, its persistence at low frequencies as well as its limited range within natural populations do not necessitate a reshaping of our evolutionary paradigms.

In addition, I suggest that the purported conflict between HGT and Darwinian evolution in modern organisms is a misunderstanding, very like the one that characterized an earlier unproductive dispute concerning the effect of neutral molecular evolution.[2] Though often more elaborate than neutral mutations, gene transfers, like neutral sequences, are simply another expression of the play of mutations on genomes. Unless mutant sequences, alien or domestic, improve the fitness of genomes, their ultimate fate is extinction.[3] Furthermore, as will be discussed below, modern cells erect selective barriers to HGT in the form of structural and functional constraints on the compatibility of novel sequences within highly tuned physiological systems.[4–7] For these reasons it is a priori unlikely that Global HGT could be a formidable disruptive force for the phylogeny of modern organisms. Most important, hard data indicate that the evolution of contemporary organisms is unambiguously Darwinian.

Finally, I will suggest that current views of the origins of eukaryotic genomes as fusion products of archaeal and bacterial genomes have been based on the uncritical application of inadequate phylogenetic methods. More robust data indicate that eukaryotes like archaea and bacteria are independent lineages that diverged from a common ancient ancestor.[8–13]

Horizontalism

What is nowadays called horizontal gene transfer (HGT) has been under study for roughly half a century. Indeed, the discoveries of the different modes of gene transfer in bacteria represent the very origins of molecular genetics. The reinvention of gene transfer as the "essence of the phylogenetic process" was a remarkable social event.[14] More than once during a recent EMBO workshop participants were reminded that "Lateral gene transfer is the only new thing in evolution since Darwin." So much for the population biology of the twentieth century.

The obsession with gene transfer during the previous decade arose in a context that was unusually amenable to hyperbole. It was early days for genomics, and that meant that sequence data was accumulating at unmanageable rates, extravagant promises were being made to patrons of the sciences, and presumptuous editorial policy was in the ascendance at the most prestigious journals. Lead articles in *Science*, *Nature*, and elsewhere advertised nothing less than a paradigm shift for evolution.[14–16] In one such screed W.F. Doolittle foretold the demise of the Darwinian view and dubbed its successor, HGT "the essence of the phylogenetic process."[14] Indeed, the very language used points to an unusual social context. One can only guess the nature of the post-Kuhnian trauma that would move a working scientist to announce that he or she had discovered a new paradigm.

The most remarkable aspect of the horizontal view of evolution is the virtual absence of supporting evidence. A recent *Science* paper illustrates this gap. This exemplary paper[17] describes an attempt to reconstruct a coherent phylogeny from a subset of coding sequences selected from five cyanobacterial genomes. In fact, Raymond et al.[17]are unable to construct a unique tree that resolves all the selected proteins from the five genomes. This failure is attributed to the inroads of rampant HGT, but there are no attempts to identify HGT as such in that study. This is remarkable given the fact that incoherent phylogenetic reconstructions also may result from problems as diverse as biased sequence composition, segregating paralogs, or inadequate clade selection.[18–27] Nevertheless, the bottom line of this study is the claim that phylogeny based on rRNA is unreliable.[17] An agnostic will certainly have trouble with this conclusion. However, a reassuring editorial preface reports that W.F. Doolittle was "excited" by the observations of Raymond et al.[28] This is as realistic as the horizontal paradigm gets.

More leisurely criticism of the horizontal paradigm can be found elsewhere.[7,18] Here, we need to say that HGT occurs and that it has had important evolutionary consequences, some of which will be discussed here. Nevertheless, for modern organisms, those we call archaea, bacteria, and eukaryotes, the transfer of alien sequences is limited in its frequencies, in its persistence, and in its range.[3,7,18] Like other mutations, alien transfers may be frequent, but their residence times are often negligible on an evolutionary time scale. Furthermore, persistent alien transfers are found most often in small subpopulations (patches) of microorganisms such as the antibiotic resistant patches found in bacterial populations.[3,7,18,29–33]

Estimates of the frequencies for different phylogenetic anomalies based on reconstructions with circa fifty thousand microbial coding sequences identify HGT as a minor player.[34] This catalog indicates that gene loss, gene duplication, and the segregation of paralogs, as well as the generation of open reading frame (ORF) sequences provide much more frequent challenges to genome phylogeny than does HGT. Because the calculations were not corrected for the influence of eccentric mutation rates, the maximum estimate of 15% of all such anomalies is probably an overestimate for the incidence of HGT.[34] The data are persuasive: Vertical lineages dominate the phylogeny of modern organisms.

This conclusion is strongly supported by whole-genome phylogenies reconstructed from the protein-coding sequences of completely sequenced genomes that are remarkably similar to those obtained with rRNA.[34–38] In one such study of fifty genomes, a simple distance measures is derived from the orthologous matches of all individual proteins in one genome tested against all the proteins in the other genomes.[38] These data are pooled for each genome and then used in a Neighbor-Joining algorithm to generate genome phylogeny. There is a striking coherence between such protein-based phylogeny for the fifty genomes and phylogeny based on rRNA from the same genomes. Such coherence strongly indicates that Darwinian descent is the dominant mode of genome evolution for these fifty genomes.[34]

Gogarten et al.[39] have responded to these data by presenting a model predicated on an intense coordinate transfer of both rRNA and protein domains in organisms that share a common niche. Here, coherence between rRNA and proteins is maintained by sequence transfer for both rRNA and proteins so intense that it generates a common sequence mosaic for the genomes resident in any particular environmental niche. Two sorts of observations are offered as empirical support for this model. One is the finding that two partial transfers for rRNA are known among thousands of examples of organisms for which rRNA sequences are available.[40,41] The other is that transgenic bacteria containing rRNA operons from very closely related organisms grow at rates that are circa 91% as fast as the reconstructed bacteria with native rRNA.[42] On the basis of these observations, Gogarten et al.[39] suggest that transfer of rRNA as well as proteins is facile.

In contrast, a quick calculation (see below) shows that the loss of 9% of the growth rate for a transgenic bacterium is a lethal defect that would eliminate it from a small population of "native" bacteria in a matter of days.[7] Likewise, the absence of any known example of complete replacement of rRNA operons in one organism by those of another indicates that such transfer is rare or nonexistent in nature.[7,18] Much more could be said about the absence of evidence for a mechanism to couple the transfer of rRNA and protein domains as well as the absence of evidence that communal organisms share mosaics of rRNA as well as protein domains.[7] All in all, it would seem that coordinate transfer of rRNA and protein domains is not a believable starting point for evolutionary models. On the contrary, it seems clear that rRNA-based phylogeny is quite robust.

The Progenote

The view developed here begins with the recognition that HGT is just one among many different sorts of mutational events that affect genomes.[18] We might expect that the frequencies of HGT events would tend to follow the overall intensity of mutation. However, all mutations are not automatically retained or fixed within a population of genomes. Simple mutations or complex novel sequences must be advantageous to organisms to persist under the unrelenting flux of random mutation. If novel sequences are disadvantageous or functionally neutral, they eventually will be destroyed by random mutations.[2,3,18] Accordingly, a functional sequence acquired by HGT or by sequence evolution will persist in a global population or in a subpopulation (patch) only as long as it contributes to the fitness of that population or patch.[3,18]

Woese has reasoned that the tempo of sequence evolution will be particularly high when the basic architectures of cells are emergent.[9–12] Then, the cellular machinery evolves by continually sifting mutant variants. Replication, like much else in primordial cells, will function with less coherence than in modern cells. Thus, mutation rates will be particularly high, and mutants relatively often will be selected in this early phase of evolution, when the population of genomes is referred to as a collective, the progenote. The reason we view the progenote as a collective is precisely because the correlates of intensive novel sequence acquisition, namely, intensive HGT as well as freely segregating paralogs coupled with gene loss, preclude phylogenetic delineation of genomes. Here, we might truly speak of a reticulate network of genomes rather than the treelike lineages that are most often presented as phylogeny.

In time, the cellular systems of the progenote evolve a radically slower tempo of novel sequence acquisition. Thus, global mutation rates decrease as the accuracy of evolving replication systems improves. More important, the efficiencies of individual proteins as well as the coherence of their interactions within integrated physiological systems improve with time. Accordingly, the tempo of sequence evolution throttles down and subcellular pathways as well as integrated cell systems become less accommodating recipients of novel sequences.[9–12] As this loss of tempo proceeds, the effect of the HGT mode must decline because variant structures are less likely to improve fitness. Indeed, at some point in the evolution of mutation rates, the likelihood that random mutations will improve the design and the fitness of a cell becomes exceedingly small. At this point a system will pass the Darwinian Border and will be locked into a trajectory of evolution through vertical lineages.[11,12] Here, persistent HGT will be rare unless there are exceptional conditions, as discussed below.

Regrettably, we do not have access to genome samples from the progenote; so much of this scenario is conjecture. However, it is reasonable conjecture because it is based on an eminently reasonable premise: that the highly efficient, exquisitely precise, and delicately regulated biochemical systems that are characteristic of modern cells could not have appeared initially in their present forms. They must have been rather clumsy, inefficient, and unregulated when they made their first appearances in biological systems. Accordingly, the evolution of such clumsy entities would entail replacement by mutant variants or alien transfers with progressively more efficient functions. Such variants would be retained when they confer a selective advantage on genomes; otherwise, they would be purged by mutation. Thus, this initially intense period of HGT is an expression of the relatively high probability with which mutant variants may improve the fitness of systems. As this probability to enhance the fitness shrinks, the probability of HGT as well as adaptive sequence evolution also shrinks. In other words, "selection of the fittest" eventually culminates in evolution by linear descent.

The notion of the progenote provides a coherent interpretation of global phylogenies for universal sequences such as rRNA, as well as for the ubiquity of the genetic code along with its translation system.[9,10] Indeed, recent data indicate that there are other, more mundane, coding sequences that yield global phylogeny very like that of rRNA. We will return to these in our discussion of eukaryotic origins, but before that, we need to shift our perspective away from the distinctive properties of the progenote.

Transient Patches of HGT

We can begin this shift with an observation. HGT enthusiasts seem to think of the transport of alien gene transfers between modern cells as a facilitated process, like an infection. No doubt they have infectious vehicles such as plasmids and transducing viruses in mind. Nevertheless, it is well established that viruses and plasmids are most often deleterious to their host cells in the absence of strong selection.[43–46] This means that they are not very infective. In addition, the enormous size of biological populations, particularly among microorganisms, is itself a formidable barrier to the fixation of alien sequences in global populations.

The most direct demonstration of the ways population size and random mutation influence the acquisition of alien genes is provided by the dynamics of neutral sequence diffusion. There are two instances of neutral diffusion that are particularly relevant to HGT. One situation is the diffusion of a novel neutral sequence through a large population. The other is the diffusion of a novel sequence through a large population that contains initially a functionally equivalent homologue indigenous to that population. The first case concerns the probability of global fixation by diffusion, and the other case concerns the probability of global replacement by a novel sequence.

An example of the first case would be the diffusion of a novel sequence encoding a resistance factor for some toxic agent. Here, the sequence is strictly neutral with respect to the fitness of an organism except in the presence of the toxic agent, where it is selected. In the absence of the toxin, the probability of the global fixation of the resistance factor is $1/N$ where N is the size of the population.[2] This variable is estimated to be as large as 10^{20} for some bacteria,[47] so the probability of fixation is, to say the least, very small. However, it gets worse. The more realistic description of this case would take into account the effect of destructive mutations. Here, destructive mutations that inactivate coding sequence would randomly affect the population. If the effective mutation rate (uN_e), which is determined by the mutation rate per gene (u) and the effective population size (N_e), is larger than 1, the approximate probability of global fixation is reduced to $(1/N)^{uNe}$.[3] Because for microorganisms the effective mutation rate is at least 10, inactivating mutations combined with large population size provides a virtually impenetrable barrier to the neutral diffusion of novel coding sequences, whether or not they are alien transfers.[3]

In the second case, we are interested in the probability of an alien sequence replacing a domestic gene that is otherwise required for the organism's survival. Here, the complication is that one or the other version of the gene must be retained for viability, so in this case, selection to maintain at least one copy opposes, in part, mutational attrition. Here, the probability for global fixation of the alien sequence in place of the domestic one is $1/2N$,[3,48] which, though better than for the nonreplacement case, is still close to zero probability for large populations such as those of microorganisms.

Obviously, an alien sequence that reduces the fitness of its new host will have a lower probability of fixation than a neutral sequence, and it will be purged from a population faster than a neutral sequence. However, the important point is that any sequence that is not strongly selected, including ones that modestly improve the fitness of an organism, will be excluded from global fixation in large populations.[3] In addition, global sequences that become modestly selective because of environmental variability along with neutral and counter-selected sequences will be purged from populations by random mutation.[3] This important conclusion is a departure from conventional thinking about neutral evolution, which has been focused for the most part on single site polymorphisms in small populations in the absence of destructive mutation.[2] The implication of these more general results is that without strong selection, coding sequences will be purged from populations by mutation. In short, without strong selection novel sequences in general, and HGT in particular, are neither infective nor persistent.

In addition, strong selection for a coding sequence most often is not expressed throughout the whole range of a microorganism.[3,18] Selection is more often limited to a small part of the range of an organism—a patch. In this case, a native or alien novel sequence will be retained where the selection is strong and purged by mutation from the rest of the population. Because strong selection is likely to be patchy, HGT is likely to be distributed in patches.[3,18] Antibiotic resistance is an excellent example of strong selection in patches, but there are other well-studied examples.[18,30,31,33,49,50–51]

Another source of transient patchy HGT is provided by alien sequences that, although not driven by strong selection, may be able to passively diffuse or hitchhike into small patches in which they will persist for times determined by the effective mutation rates for those patches.[2,3] Though it may not have any adaptive consequences, such HGT will also contribute to the sequence turnover of genomes.[3,18] Thus, microbial genomes as well as microbial populations are patchy in the sense that they contain many transient neutral sequences in various states of mutational decay that are in turn distributed in patches of the global population.[18,29–33]

The patchy genomes in patchy populations pattern is well documented in bacteria.[15,29–33] Up to 17% and a median of 6% of fully sequenced prokaryotic genomes have been identified on the basis of eccentric composition with this category of transient alien sequence.[15] Calculations indicate that a sequence in this fraction has a lifetime much less than 1 million years.[3] Indeed, there is a growing conviction that such transient genomic debris may be the origin of at least some of the so-called ORFan sequences found in all genomes.[52–54] Finally, the patchy distribution of transient novel sequences may account for the variability of genome size in populations of bacteria. Thus, the transient patch dynamic will support genome sizes in a quasi-steady state.[3] Here, occasional large fluctuation as in the strain O157 may generate genome size fluctuations approaching one million base pairs or more.[33] Clearly, the presence in genomes of a substantial fraction of transient, mostly neutral, novel sequence may provide phylogeneticists with some frustrating moments. Nevertheless, as Snell, Huynen, and others have shown, genome phylogeny is robust.[34–38,55] The reason that all this transient debris within the global population has little influence on genome phylogeny is that it is so heterogeneous that it cannot produce a coherent phylogenetic signal.

Syntony versus HGT

The focus in the previous section on neutral HGT highlights the ways that large population size and random mutations impede HGT. Another sort of barrier to HGT arises from the evolution of the integrated biochemical networks characteristic of modern cells.

As noted above, the selection of mutant variants in the progenote should eventually enhance not only the performance of individual components but also the coherence of their cooperative interactions within integrated molecular systems.[11,12] The vagaries of random mutation would ensure that this tendency toward an optimal harmony between components of physiological networks would be expressed in distinctly divergent ways in different cell lineages. As a consequence, the components of any physiological network from one organism may not be fully exchangeable with counterparts from another organism. Accordingly, as the networks evolve, alien transfer becomes increasingly uncommon. Woese has described the evolution from a cellular network that is amenable to HGT to one that is not as a transition through a Darwinian barrier.[11,12]

By studying transgenic cells it is possible to learn much about the physiological consequences of HGT in modern cells. It is useful in this connection to recall that the biochemical or physiological analogue of a Darwinian network has been described previously as a kinetically optimized system.[4,5] Examples of such systems are the biochemical networks responsible for any and all cellular activities ranging from stress responses to cellular differentiation. The overall performance of a cell in any given environment will depend on integrating different physiological systems so that the cellular performance as a whole is optimized. I refer to this integrative or tuning aspect of evolution as syntony. The microbial translation apparatus provides a well-studied example of syntonic evolution. A population of bacteria growing in a supportive environment is made up of cells multiplying at characteristic rates. Here, individual cells are competing with each other, with the fastest ones eventually dominating the growing population.[4,5] Maximum growth rates are demonstrably dependent on optimization of the translation system. Such optimizations involve tuning the concentrations of translation components and their binding interactions, as well their catalytic rates, so that an optimal efficiency of protein synthesis is expressed under different growth (nutrient) conditions.[4–6]

Given this highly tuned pattern of optimization, it is not surprising that mutations that alter subtle details of the kinetic performance characteristics of individual components in the translation apparatus reduce growth rates.[5] This is important, because it shows that selection, in this case for maximal growth rate, is very sensitive to small structural changes in individual components. It is worth adding that this should be completely general. Fitness will also be sensitive to small structural changes even in, for example, the chemotaxis apparatus of bacteria. Rao and Varshney provide an exceptionally useful example of how alien transfers affect the translation system of syntonic cells.[56] They have studied in vivo the functions of a ribosome release factor from *Mycobacterium tuberculosis* in the translation cycle of *Escherichia coli*. They find that, alone, the release factor from *M. tuberculosis* will not rescue *E. coli* with defective factor, but if elongation factor G and release factor from *M. tuberculosis* are introduced together, the defective *E. coli* can be rescued. Other rescue experiments with different transgenic release factors show that deletion of a five–amino acid sequence normally found at the C terminus of the factor from *Thermus thermophilus* is required for it to rescue the defective *E. coli* mutant.[57] Not surprisingly, alien factors that may rescue defective *E. coli* at one expression level will kill its new host at another level.[57] Evidently, both dissonant molecular interactions as well as inappropriate expression levels limit the compatibility of transferred sequences in transgenic cells.

Gogarten et al.[39] have reasoned that because rRNA evolves more slowly than protein-coding genes, the consequences of exchanging closely related rRNA sequences would be negligible for the growth optimizations of bacteria. Indeed, exacting genetic reconstructions of *E. coli* with rRNA from its very close relative *Salmonella typhimurium* indicate that the transgenic bacteria can grow at rates that are similar to those of the homologous reconstruction (i.e., at doubling times of 52.6 vs. 47.9 min in broth, respectively).[42] Again, the growth rate differences between the homologous and the transgenic reconstructions increase as the phylogenetic distance between the donor and recipient increases.[42] These results have been touted as evidence that rRNA transfer is facile,[39] but is that really what these experiments say?

It is no exaggeration to suggest that a 9% loss of growth rate is catastrophic and that in nature it would most certainly lead to the rapid extinction of a transgenic bacterium. Calculations show that the takeover probability in the steady state is less than 10^{-40} for a single transgenic bacterium with a 9% growth disadvantage that appears in a miniscule population of a mere one thousand native *E. coli* cells.[7] Obviously, for more realistic population sizes, the takeover probability is even closer to zero. Furthermore, one single bacterium would outgrow and displace in 10 days a population of 10^8 transgenic bacteria with a 9% slower growth rate.[7] By extension, a fraction of a 1% difference in growth rate is a definitive handicap from an evolutionary perspective. These observed growth rate defects of bacteria with transgenic ribosomes[42] go a long way to explain the rarity of rRNA transfers observed in nature.

Those few rRNA transfers detected so far are found in genomes containing both the native rRNA of the host and the alien rRNA sequences.[40,41] The rRNA operons from many organisms have been sequenced, and thousands of 16S rRNA as well as hundreds of 23S rRNA sequences have been recorded in the databases. Comparison of these figures with the two anomalous ones[40,41] probably gives a reliable measure of the frequency with which rRNA is partially transferred between different microorganisms. However, more important is the fact that there still are no examples of complete replacements of native rRNA by alien homologs.[18]

Such studies of transgenic bacteria are informative about the overall constraints on HGT. In particular, they show that it is not permissible to infer from experiments demonstrating transfer of rRNA domains between very closely related organisms that more distant alien rRNA transfer is also facile.[39,58] It seems appropriate at this point to insist that despite all the rhetoric to the contrary,[14,17,39,58] the data indicate that rRNA operons are robust phylogenetic markers.[18]

Other results from the analyses of transgenic bacteria, as well as from studies of mutant variants stabilized by strong contingent selection (e.g., antibiotic-resistant variants), point to additional barriers to HGT of protein-encoding sequences.[5,18,31–33,42–46,50,51] Thus, alien proteins are often functionally incompatible with the networks of their syntonic hosts. This means that in the absence of strong contingent selection, cells lacking the alien gene product may outgrow transgenic cells. The incompatibility of alien gene products in syntonic cells in the absence of strong selection reinforces the patchy distribution expected for HGT. Obviously, HGT limited to small patches within a microbial population has little probability of influencing global phylogeny. Conversely, the detection of an alien sequence in an isolate from a global population cannot be properly interpreted until it is determined whether or not the alien sequence is resident in a patch or globally. To my knowledge, there are no instances of phylogenetic uncertainty resulting from the identification of an alien antibiotic resistance marker in a bacterial isolate. The tendency of HGT to be distributed in patches does not preclude global fixation of alien sequences. What is precluded in modern cells is persistent global fixation of alien sequences that are not driven by strong global selective forces. Examples of the latter will be taken up below.

Eukaryotes Are Not Prokaryotes in Drag

We would not expect the fusion of a pair of distantly related syntonic cells or their genomes to produce viable chimeras. Rather, the interactions between thousands of different proteins from each cell lineage ought to produce many, perhaps hundreds, of debilitating interactions. For this reason, it seems to me that any scenario involving simple mosaics of bacterial and archaeal genomes[49] is a priori an unrealistic way to think about the origins of eukaryotes.

Beginning in the 1990s, BLAST searches were used to identify the most similar homologs in pairwise comparisons of archaeal, bacterial, and eukaryotic sequences.[59–65] The most closely related pairs in the different domains were identified then as members of orthologous lineages. The results of BLAST searches for enzymes of intermediary metabolism were considered anomalous in the sense that eukaryotic enzymes were most often identified as more closely related to bacterial homologs than to archaeal ones. Such results were interpreted as evidence for systematic HGT between archaea and bacteria, through which the first eukaryotic genomes arose as mosaics of ancestral prokaryotic sequences.[59–64] Amazingly, this is the sole basis for the interpretation that eukaryotes are genomic descendents of archaeal and bacterial mosaics.[49,66,67] In contrast, inspection of an unrooted rRNA tree simply does not support such speculation. Therefore, the best-match BLAST protocol deserves a closer scrutiny.

According to the most naive interpretations of the BLAST protocol, intensive HGT between archaeal and bacterial genomes is indicated, particularly among the thermophiles.[63,64] In contrast, the two domains are unambiguously resolved in the reconstructions of Snel, Huynen, and their colleagues, so that even thermophilic archaea and bacteria are well resolved in their reconstructions.[34,38,55] In addition, Snel et al. [34,38,55] observe that a majority of the phylogenetic anomalies in fifty fully sequenced genomes including thermophiles are not caused by HGT. To this it may be worth noting that the genomes of archaeal and bacterial thermophiles may contain sequences that are convergent in the sense that they have been selected for the same strong compositional biases. All such phylogenetic diversions are easily confused with HGT, but they are not distinguishable by the best match BLAST protocol. Accordingly, annotations for the genomes of thermophylic prokaryotes ought to be redone properly.

More recently, the BLAST protocol has been used to identify hundreds of putative bacterial transfers to the human genome.[65] Not surprisingly, it was subsequently found that most of these transfers were homologous to counterparts in invertebrate ancestors of vertebrates.[27,68] It seems that most of these human sequences are, in fact, vertically inherited though the lineages and were initially obscured by gene loss. In other words, the initial cohort was too small. Further analysis is required to determine whether the residues of forty to eighty putative gene transfers are genuine examples of HGT or whether they are caused by other phylogenetic anomalies, such as segregating paralogs.

The point of recounting these examples is that the best-match BLAST protocol is an inadequate method for reconstructing phylogeny. It does not distinguish a variety of phylogenetic anomalies from HGT events. Similarly, a recent HGT event is indistinguishable in this protocol from sequence affinity between homologs that are both descendents of an ancient common ancestor. Thus, we have every reason to be suspicious of speculations about the origins of eukaryotes that are based on results from this inadequate methodology.

Indeed, a more rigorous phylogenetic study of a large cohort of enzymes of intermediary metabolisms yields a very different interpretation of the origins of these proteins.[13] This cohort consists of the Embden-Meyerhof and of the Entner-Doudoroff pathways from eukaryotes, bacteria, and the few archaea that use recognizable versions of these glycolytic enzymes. Several of the enzymes studied had either no eukaryotic homologs at all or too few of them in public databases to be useful. A few yielded complex phylogenies that proved to be composites of two or three paralogous lineages. The reconstructions of the remaining six enzymes were relatively straightforward. Eukaryotic clades and bacterial clades are separate and clustered. In no case are the eukaryotic clades rooted in any of the canonical bacterial phyla.

The clusters representing the two domains are on separate branches that emerge from a common ancestral node very much as observed in the unrooted rRNA tree.[8] For the few glycolytic enzymes that are represented in the domain of the archaea as well as in bacteria and eukaryotes, a three-domain cluster pattern is obtained that is very like the unrooted rRNA tree. Because the clades from the archaea can serve as outgroups for clades in the other two domains, these few reconstructions with archaeal clades provide rigorous evidence that those bacterial and eukaryotic glycolytic enzymes are descendents of a common ancient ancestor.[13]

These data contradict the speculations that glycolytic as well as other enzymes of intermediary metabolism were transferred en masse to an archaeal ancestor of eukaryotes when the α-proteobacterial endosymbiont seeded the mitochondrial lineages.[66,67] The common rooting in an ancient node, taken together with their evident homology, make it difficult not to conclude that the bacterial and eukaryotic glycolytic enzymes shared the exchange mode of progenote sequence evolution.[13]

The notion that after their emergence from a common progenote ancestor the genome lineages of archaea, eukaryotes, and bacteria evolved in an essentially vertical mode is supported by other data. These include the discovery of nearly one thousand protein-encoding sequences in the genomes of archaea, eukaryotes, and bacteria that are unique to one of these domains.[71,72] Unfortunately, we cannot yet fix the ages of these signature lineages. This failure limits the value of the signature sequences as links with the progenote ancestor. Nevertheless, such domain-specific signature sequences indicate that vertical inheritance has been a common motif rather than the exception since the emergence of three modern domains from a common progenote ancestor.[8]

Mitochondria

Of course, the provocative exceptions to the theme of linear descent are provided by the organelles of eukaryotes. Indeed, the systematic transfer of coding sequences from a canonical α-proteobacterial ancestor to eukaryotic nuclear genomes provides the benchmark phylogenetic pattern for the identification of sequence transfers from a well-defined bacterial group to the genomes of eukaryotes.[13,69,72,73] It would seem that the classical theory of the endosymbiotic origin of mitochondria in a eukaryotic host[74,75] is correct, but with two provocative twists. There is agreement that the genomes of mitochondria are the descendents of an endosymbiotic α-proteobacterium.[76,77]

On the basis of the discussion in the previous section, we tentatively identify the host as a primitive eukaryote. For this reason, we might be concerned that the initial intracellular association between syntonic host and endosymbiont would generate a nonviable cellular chimera. Thus, thousands of mixed protein–protein interactions between the two syntonic proteomes are almost certain to produce more that one lethal incompatibility. However, two aspects of an endosymbiotic association would tend to mitigate such a problem. One is that the association would have been stabilized by strong selection (i.e., survival in the presence of mounting oxygen tensions). The other is that the bacterial membranes initially would provide perfectly adequate barriers to the mixing of proteins from the two syntonic cell systems. Nevertheless, extensive exchange has characterized the evolution of the proteomes of both the mitochondrion and its host on two levels.[69,72,73]

Thus, the evolution from the endosymbiont to the organelle has been accompanied by a profound reduction of the original bacterial genome. Initially, such an endosymbiont would be expected to encode perhaps as many as 1,600 proteins.[73] Genomes of modern mitochondria encode between two and 67 proteins.[76] This means that nearly all of the endosymbiont's coding sequences have been lost from the organelle's genome. Most of these seem to have simply disappeared.[69,72] In yeast there are nearly fifty α-proteobacterial genes that undoubtedly have been transferred to the nucleus.[69] Another 150 or so that also may be bacterial descendants are nuclear companions. Such nucleus-encoded descendents of the endosymbiont provide the first twist in the classical endosymbiotic scenario.

The second twist is that the largest group of nuclear genes encoding members of the mitochondrial proteome, at least two hundred in yeast, contains eukaryotic signature proteins (i.e., proteins with no alignments with archaeal or bacterial sequences).[69] The latter are evidently products of eukaryotic evolution that have transformed the symbiont into an organelle. It is likely that at least some of these sequences are derived from ancient eukaryotic gene families that originally had other functions.[69,72] The novelty here is that proteins originating in eukaryotic nuclear genomes are synthesized in the cytoplasm and then transported to the mitochondria for service in the organelle. It is difficult not to interpret this cohort of proteins as the one that has converted the endosymbiont into an organelle.

Thus, the overwhelming majority of mitochondrial proteins are synthesized in the cytoplasm of the host and then transported into the organelle. How are toxic interactions between bacterial/mitochondrial and cellular proteins avoided? Clearly, during the hundreds of million of years during which the mitochondrial proteome was evolving, there was adequate time for the selection of some mutually compatible sequences. Here it should be recalled that initially the protein networks responsible for oxidative respiration and the Krebs cycle of the endosymbiont/mitochondria were limited to the organelle. For this reason there may not have been many opportunities for such proteins to disrupt networks of the host. In addition, it has been observed in baker's yeast that mitochondrial proteins of α-proteobacterial descent tend to be synthesized on polysomes that are attached to the mitochondrial outer membrane.[78] Accordingly, while they are nascent, potentially toxic proteins, though cytoplasmic in origin, are tethered to mitochondria, out of harm's way. It remains to be seen how general this mechanism is.

Thorsness, Fox, and their colleagues have studied experimentally the exchange of coding sequences between yeast mitochondria and yeast nuclei.[79–83] Briefly, they find that it occurs at a measurable rate from organelle to nucleus but at an undetectable rate in the opposite direction. They conclude that there is at least a 10^5 bias in favor of transfer to the nucleus. On the basis of observations of peroxisome turnover in eukaryotic cells, the authors suggest that the ubiquitous phagolysosomes may consume defective mitochondria and release DNA fragments that transform the nuclear genome. Given the enormous bias of the transfer process, it can be shown that any mitochondrial coding sequence not required in the organelle genome will eventually be transferred to the nucleus.[72,84]

Could the Thorsness and Fox mechanism[79–83] provide a pathway for other sorts of HGT? Many pathogenic bacteria live in phagolysosomes, and it is conceivable that they could transfer coding sequences to nuclear genomes by this mechanism. However, there is a big gap between a required mitochondrial coding sequence that is transferred by chance to the nucleus and a bacterial one for which there is no selective requirement (see above). The first may very well be fixed in the global population, particularly if mutation purges the organelle's copy. However, the bacterial sequence has virtually no chance of survival in the cellular population unless it can meet some specific need of the host cells.

The evolutionary success of the endosymbiont and its mitochondrial descendents may be attributed to the fact that they provided a novel function that was globally selected in unidentified primitive eukaryotic hosts and their modern descendents. Thus, oxidative respiration coupled with Krebs cycle metabolism may have provided the initial function of detoxification in an environment becoming increasingly toxic to anaerobic eukaryotes.[69,72] Later, ATP production may have been nurtured with the aid of unique eukaryotic proteins that transformed the endosymbiont into an organelle.[69,72] The point is that at all times it may be assumed that selection was driving the evolution of the endosymbiont's as well as the host's exchanges of coding sequences and proteins.

Concluding Remarks

Much has been made lately about the difficulties of identifying HGT.[58] This is puzzling. The HGT associated with the evolution of mitochondria from an ancestral α-proteobacterium sticks out like a sore thumb.[13,69,72,73,77] All that was needed to identify it was a large cohort of the relevant clades and reasonable phylogenetic routines. Of course, there is a problem for those who expect to find hordes of alien genes lurking in genomes, but the problem is not a methodological one. The reality is that careful studies of more than fifty fully sequenced genomes[34–38] do not uncover hordes of alien genes. HGT is identified in these genomes as a minor fraction of the anomalies, and the overwhelming majority of sequences seem to be evolving vertically with the occasional vertical hiccup. The real problem seems to be unrealistic expectations.

These unrealistic expectations may arise from naive reliance on inadequate phylogenetic methods, repeated observations of vertical phylogenetic anomalies that may be confused with HGT, and what seems to be a total disregard for a basic fact of population genetics. Alien sequences like any novel sequences will be fixed in populations if and only if they improve the fitness of genomes.[3] That requirement is not easy to meet because in modern organisms novel sequences in general, and alien sequences in particular, tend to be neutral or deleterious to their hosts.[2,3,5,42–46,56,57,85] For these reasons, the novel sequences that persist do so in patches of global populations in which they are stabilized by local contingent selection.[3,18,32,33,43–46,85]

Indeed, the work of Ochman and others on bacterial HGT indicates that most HGT consists of transient sequences circulating through subpopulations (patches) in progressive states of mutational decay.[15,29–33] The limited range and persistence expected of HGT[3,18] does not lend credence to speculations about rampant HGT in modern cells.[14,15,16,39,58] To put it most succinctly, the data indicate that there are no icebergs "waiting to sink the ship of genome-based phylogeny," as it has been so piquantly put,[58] and it is not those who have stuck to this ship who ought to be taking to the lifeboats.

Gene families in one modern domain reveal their origins in the ancestral progenote population by their orthologous relationship to those in another modern domain.[8,10,11,13] This signature of ancient ancestry is as clearly reflected in the global phylogeny of glycolytic enzymes[13] as it is in the phylogenies of ribosomal RNA. Because of reliance on inadequate phylogenetic methods, these relationships have been systematically misrepresented. Thus, homology between coding sequences in eukaryotes with those in prokaryotes has been mistakenly identified as a signature of modern interdomain gene transfer.[18,13,7, 27,68,14,59–62,66,67] The available data, in fact, are consistent with the interpretation that homologous sequences shared by the archaea, bacteria, and eukaryotes are descendents of an ancestral progenote population that has diverged into three predominantly vertical lineages.[8–12]

Furthermore, the connection between adaptive HGT and the divergence of global populations (speciation) is probably an occasional one[3] rather than the tight relationship suggested by HGT enthusiasts.[15,16] The isolation of a patch from its global population is a prerequisite for a speciation (divergence) event. However, gene loss together with adaptive sequence evolution (not necessarily HGT) is more effective than novel sequence acquisition alone to generate the requisite isolation.[3] Indeed, sequence loss has been identified as a contributing factor in the divergence of *Yersinia pestis* from *Yersinia pseudotuberculosis*.[50]

The assertion that HGT provides a unique avenue for rapid response to environmental challenge and for the divergence of lineages[14–16,39] is not supported by observation: It is ideology. Most novel sequence evolution seems to be vertical.[34–38] In contrast, the capacity to use lactate has been paraded as an example of HGT that facilitated the divergence of *E. coli* from *S. enterica*.[15] The alternative interpretation that the loss of lactate metabolism facilitated the divergence of *S. enterica* from *E. coli* is never discussed in the horizontal literature. However, as we have seen, gene loss has been a rich source of mistaken identities in the HGT game.[27,68]

Finally, it is something of a mystery that the horizontal view, or as one of its proponents likes to refer to it, the paradigm shift, could have had such currency. No data indicate that HGT is rampant. The lists of HGT offered by its boosters are too short to qualify the phenomenon as anything more than a minor nuisance.[39] The most rigorous speculations about HGT tend to fall into the "what if" category,[14,39,58] and yet, there are these enthusiastic lead articles in *Science*, *Nature*, and other journals. What is going on?

My own take on this social phenomenon is that it is a product of the breakdown of the referee system, particularly in the offices of the most widely read journals. The interpretation that journalistic excess has promoted HGT to its current prominence is not new.[18] Indeed, one may wonder what sort of objective editorial policy would allow a sub-editor to publish a series of notes with opinions critical of the universal rRNA tree[28,86,87] without encouraging the publication of a single defense. This is a particularly relevant question in light of the demonstrably shoddy support upon which these editorial opinions rest (see above).

Acknowledgment I am grateful to Irmgard Winkler for help in preparing this manuscript. Siv Andersson, Otto Berg, and Carl Woese, each in their distinctive ways, have been sources of new information, stimulating conversation, and helpful criticism.

References

1. M. Syvanen, "In Search of Horizontal Gene Transfer," *Nature Biotech.* 17 (1999): 8833.
2. M. Kimura, *The Neutral Theory of Evolution* (Cambridge. Cambridge University Press, 1983).
3. O.G. Berg and C.G. Kurland, "Evolution of Microbial Genomes: Sequence Acquisition and Loss," *Mol. Biol. Evol.* 19 (2002): 2265–2276.
4. M. Ehrenberg and C.G. Kurland, "Cost of Accuracy Determined by a Maximal Growth Rate Constraint," *Q. Rev. Biophys.* 17 (1984): 45–82.
5. C.G. Kurland, "Translational Accuracy and the Fitness of Bacteria," *Annu. Rev. Genet.* 26 (1992): 29–50.
6. H. Dong, L. Nilsson, and C.G. Kurland, "Gratuitous Over-Expression of Genes in *Escherichia coli* Leads to Growth Inhibition and Ribosome Destruction," *J. Mol. Biol.* 260 (1996): 649–663.
7. C.G. Kurland, B. Canback, and O.G. Berg, "Horizontal Gene Transfer: A Critical Review," *Proc. Natl. Acad. Sci. USA* 100 (2003): 9658–9662.
8. C.R. Woese, "The Primary Lines of Descent and the Universal Ancestor," in D.S. Bendall, ed., *Evolution from Molecules to Men* (Cambridge: Cambridge University Press, 1983), 209–233.
9. C.R. Woese, "On the Evolution of the Genetic Code," *Proc. Natl. Acad. Sci. USA* 54 (1965): 1546–1552.
10. C.R. Woese and G.E. Fox, "The Concept of Cellular Evolution," *J. Mol. Evol.* 10 (1977): 1–6.
11. C.R. Woese, "The Universal Ancestor," *Proc. Natl. Acad. Sci. USA* 95 (1998): 6854–6859.
12. C.R. Woese, "Interpreting the Universal Phylogenetic Tree," *Proc. Natl. Acad. Sci. USA* 97 (2000): 8392–8396.
13. B. Canback, S.G.E. Andersson, and C.G. Kurland, "The Global Phylogeny of Glycolytic Enzymes,"*Proc. Natl. Acad. Sci. USA* 99 (2002): 6097–6102.
14. W.F. Doolittle, "Phylogenetic Classification and the Universal Tree," *Science* 284 (1999): 2124–2129.
15. H. Ochman, J.G. Lawrence, and E.A. Groisman, "Lateral Gene Transfer and the Nature of Bacterial Innovation," *Nature* 405 (2000): 299–304.
16. F. de la Cruz and J. Davies, "Horizontal Gene Transfer and the Origin of Species: Lessons from Bacteria," *Trends Microbiol.* 8 (2000): 128–133.
17. J. Raymond, O. Zhaxybayeva, J.P. Gogarten, S.Y. Gerdes, and R.E. Blankenship, "Whole-Genome Analysis of Photosynthetic Procaryotes," *Science* 298 (2002): 1616–1619.
18. C.G. Kurland, "Something for Everyone; Horizontal Gene Transfer in Evolution," *EMBO Rep.* 1 (2000): 92–95.

19. N. Galtier and M. Guoy, "Inferring Phylogenies from DNA Sequences of Unequal Base Compositions," *Proc. Natl. Acad. Sci. USA* 92 (1995): 11317–11321.
20. N. Galtier, N. Tourasse, and M. Gouy, "A Nonhyperthermophilic Common Ancestor to Extant Life Forms," *Science* 238 (1999): 220–221.
21. H. Philippe and P. Forterre, "The Rooting of the Universal Tree of Life Is Not Reliable," *J. Mol. Evol.* 49 (1999): 509–523.
22. Z. Yang, "A Space-Time Process Model for the Evolution of DNA Sequences," *Genetics* 139 (1995): 993–1005.
23. J. Felsenstein,"Taking Variation of Evolutionary Rates between Sites into Account in Inferring Phylogenies," *J. Mol. Evol.* 53 (2001): 447–455.
24. P.J. Lockhart, M.A. Steel, M.D. Hendy, and D. Penny, "Recovering Evolutionary Trees under a More Realistic Model of Sequence Evolution," *Mol. Biol. Evol.* 11 (1994): 605–612.
25. P. Lopez, P. Forterre, and H. Philippe, "The Root of the Tree of Life in the Light of the Covarion Model," *J. Mol. Evol.* 49 (1999): 496–508.
26. J.R. Brown, "Genomic and Phylogenetic Perspectives on the Evolution of Prokaryotes," *Systematic Biology* 50 (1999): 497–512.
27. M.J. Stanhope, A. Lupas, M.J. Italia, K.K. Koretke, C. Volker, and J.R. Brown, "Phylogenetic Analyses of Genomic and EST Sequences Do Not Support Horizontal Gene Transfers between Bacteria and Vertebrates," *Nature* 411 (2001): 940–944.
28. E. Pennisi, "Bacteria Shared Photosynthesis Genes," *Science* 298 (2002): 1538–1539.
29. J.G. Lawrence and H. Ochman, "Amelioration of Bacterial Genomes: Rates of Change and Exchange," *J. Mol. Evol.* 44 (1997): 383–397.
30. J.G. Lawrence and H. Ochman, "Molecular Archaeology of the *Escherichia coli* Genome," *Proc. Natl. Acad. Sci. USA* 95 (1998): 9413–9417.
31. S.D. Hooper and O.G. Berg, "Gene Import or Deletion—A Study of the Difference Genes in *Escherichia coli* Strains K12 and O157:H7," *J. Mol. Evol.* 54 (2002): 734–744.
32. H. Ochman and I.B. Jones, "Evolutionary Dynamics of Full Genome Content in *Escherichia coli*," *EMBO J.* 19 (2000): 6637–6643.
33. N.T. Perna, G. Plunkett III, V. Burland, et al., "Genome Sequence of Enterohaemorrhagic *Escherichia coli* O157:H7," *Nature* 409 (2000): 529–533.
34. B. Snel, P. Bork, and M. Huynen, "Genomes in Flux: The Evolution of Archaeal and Proteobacterial Gene Content," *Genome Res.* 12 (2002): 17–25.
35. S.T. Fitz-Gibbon and C.H. House, "Whole Genome-Based Phylogenetic Analysis of Free-Living Microorganisms," *Nucleic Acids Res.* 27 (1999): 4218–4222.
36. F. Tekaia, A. Lazcano, and B. Dujon, "The Genomic Tree as Revealed from Whole Proteome Comparisions," *Genome Res.* 9 (1999): 550–557.
37. J.R. Brown, C.J. Douady, M.J. Italia, W.E. Marshall, and M.J. Stanhope, "Universal Trees Based on Large Combined Protein Sequence Data Sets," *Nat. Genet.* 28 (2000): 281–285.
38. J.O. Korbel, B. Snel, M.A. Huynen, and P. Bork, "SHOT: A Web Server for the Construction of Genome Phylogenies," *Trends Genet.* 18 (2002): 158–162.
39. J.P. Gogarten, W.F. Doolittle, and J.G. Lawrence, "Prokaryotic Evolution in Light of Gene Transfer," *Mol. Biol. Evol.* 19 (2002): 2226–2238.
40. S. Mylvaganam and P.P. Dennis, "Sequence Heterogeneity between the Two Genes Encoding 16S rRNA from the Halophilic Archaebacterium *Haloarcula marismortui*," *Genetics* 130 (1992): 399–410.
41. W.H. Yap, Z. Zhang, and Y.J. Wang, "Distinct Types of rRNA Operons Exist in the Genome of the Actinomycete *Thermomonospora chromogena* and Evidence for Horizontal Transfer of an Entire rRNA Operon," *J Bacteriol.* 181 (1999): 5201–5209.
42. T. Asai, D. Zaporojets, C. Squires, and C.L. Squires, "An *Escherichia coli* Strain with All Chromosomal rRNA Operons Inactivated: Complete Exchange of rRNA Genes between Bacteria," *Proc. Natl. Acad. Sci. USA* 96 (1999): 1971–1976.
43. D.I. Andersson and B.R. Levin, "The Biological Cost of Resistance," *Curr. Op. Microbiol.* 2 (1999): 487–491.
44. C.T. Bergstrom, M. Lipsitch, and B.R. Levin, "Natural Selection, Infectious Transfer and the Existence Conditions for Bacterial Plasmids," *Genetics* 155 (2000): 1505–1519.

45. J. Björkman, I. Nagaev, O.G. Berg, D. Hughes, and D.I. Andersson, "Compensatory Mutations to Ameliorate the Costs of Antibiotic Resistance Differ with Environment," *Science* 287 (2000): 1479–1482.
46. A.M. Borman, S. Paulous, and F. Clavel, "Resistance of Human Immunodeficiency Virus Type I to Protease Inhibitors: Selection of Resistance Mutations in the Presence and the Absence of the Drug," *J. Gen. Virol.* 77 (1996): 419–426.
47. H. Ochman and A.C. Wilson, "Evolutionary History of Enteric Bacteria," in F.C. Neidhardt, ed., Escherichia coli *and* Salmonella typhimurium: *Cellular and Molecular Biology* (Washington, DC: ASM Press, 1987), 1649–1654.
48. M. Lynch, M. O'Hely, B. Walsh, and A. Force, "The Probability of Preservation of a Newly Arisen Gene Duplicate," *Genetics* 159 (2001): 1789–1804.
49. P. Lopez-Garcia and D. Moreira, "Metabolic Symbiosis at the Origin of Eukaryotes," *Trends Biotechnol.* 24 (1999): 88–93.
50. M. Achtman, K. Zurth, G. Morelli, G. Torrea, A. Gulyoule, and E. Carniel, "*Yersinia pestis*, the cause of plague, is a recently emerged clone of *Yersinia pseudotuberculosis*" *Proc. Natl. Acad. Sci. USA* 96, (1999) 14043–14048.
51. F. Taddei, M. Radman, J. Maynard-Smith, B. Touponce, P.H. Gouyon, and B. Godelle, "Role of Mutator Alleles in Adaptive Evolution," *Nature* 387 (1997): 700–702.
52. D. Fischer and D. Eisenberg, "Finding Families for Genomic ORFans," *Bioinformatics* 15 (1999): 759–762.
53. A. Mira, L. Klasson, and S.G.E. Andersson, "Microbial Genome Evolution: Sources of Variability," *Curr. Opin Microbiol.* 5 (2002): 506–512.
54. H. Amiri, W. Davids, and Siv G.E. Andersson, "Birth and Death of ORFan Genes in *Rickettsia*," *Mol. Biol. Evol.* 20 (2003): 1575–1587.
55. M.A. Huynen, B. Snel, and Bork, P, "Lateral Gene Transfer, Genome Surveys, and the Phylogeny of Prokaryotes," *Science* 286 (1999): 1443a.
56. A.R. Rao and U. Varshney, "Specific Interaction between the Ribosome Recycling Factor and the Elongation Factor G from *Mycobacterium tuberculosis* Mediates Peptidyl-tRNA Release and Ribosome Recycling in *Escherichia coli*," *EMBO J.* 20 (2001): 2977–2986.
57. M. Ohnishi, L. Janosi, M. Shuda, H. Matsumoto, T. Terawaki, and A. Kaji, "Molecular Cloning, Sequencing, Purification and Characterization of *Pseudomonas aeruginosa* Ribosome Recycling Factor," *J. Bacteriol.* 181 (1999): 1281–1291.
58. W.F. Doolittle, Y. Boucher, C.L. Nesbø, C.J. Douady, J.O. Andersson, and A.J. Roger, "How Big Is the Iceberg of Which Organellar Genes in Nuclear Genomes Are but the Tip?" *Phil. Trans. R. Soc. Lond. B* 358 (2003): 39–58.
59. D.F. Feng, G. Cho, and R.F. Doolittle, "Determining Divergence Times with a Protein Clock: Update and Reevaluation," *Proc. Natl. Acad. Sci. USA* 94 (1997): 13028–13033.
60. M.W. Smith, D.F. Feng, and R.F. Doolittle, "Evolution by Acquisition: The Case for Horizontal Gene Transfers," *Trends Biochem. Sci.* 17 (1992): 489–493.
61. R.F. Doolittle, D.F. Feng, S. Tsang, G. Cho, and E. Little, "Determining Divergence Times of the Major Kingdoms of Living Organisms with a Protein Clock," *Science* 271 (1996): 470–477.
62. M.C. Rivera, R. Jain, J.E. Moore, and J.A. Lake, "Genomic Evidence for Two Functionally Distinct Gene Classes," *Proc. Natl. Acad. Sci. USA* 95 (1998): 6239–6244.
63. L. Aravind, R.L. Tatusov, Y.I. Wolf, D.R. Walker and E.V. Koonin, "Evidence for Massive Gene Exchange between Archaeal and Bacterial Hyperthermophiles," *Trends Genet.* 14 (1998): 442–444.
64. K.E. Nelson, and 24 others, "Evidence for Lateral Gene Transfer between Archaea and Bacteria from Genome Sequence of *Thermotoga maritima*," *Nature* 399 (1999): 323–329.
65. E.S. Lander et al., "Initial Sequencing and Analysis of the Human Genome," *Nature* 409 (2001): 860–921.
66. J.P. Gogarten, L. Olendzenski, E. Hilario, C. Simon, and K.E. Holsinger, "Dating the Cenancester of Organisms," *Science* 274 (1996): 1750–1751.
67. W. Martin and M. Müller, "The Hydrogen Hypothesis for the First Eukaryote," *Nature* (London) 392 (1998): 37–41.

68. S.L. Salzburg, O. White, J. Peterson, and J.A. Eisen, "Microbial Genes in the Human Genome: Lateral Transfer or Gene Loss?" *Science* 292 (2001): 1903–1906.
69. O. Karlberg, B. Canback, C.G. Kurland, and S.G.E. Andersson, "The Dual Origin of the Yeast Mitochondrial Proteome," *Yeast* 17 (2000): 170–187.
70. D.E. Graham, R. Overbeek, G.J. Olsen, and C.R. Woese, "An Archaeal Genomic Signature," *Proc. Natl. Acad. Sci. USA* 97 (2000): 3304–3308.
71. H. Hartman and A. Fedorov, The Origin of the Eukaryotic Cell: A Genomic Investigation," *Proc. Natl. Acad. Sci. USA* 99 (2001): 1420–1425.
72. C.G. Kurland and S.G.E. Andersson, "Origin and Evolution of the Mitochondrial Proteome," *MMBR* 64 (2000): 786–820.
73. E.M.M. Macrotte, I. Xenarios, A.M. van der Bliek, and D. Eisenberg, "Localizing Proteins in the Cell from Their Phylogenetic Profiles," *Proc. Natl. Acad. Sci. USA* 97 (2000): 12115–12120.
74. L. Margulis, *Origin of Eucaryotic Cells* (New Haven, CT: Yale University Press, 1970).
75. M.W. Gray, "The Endosymbiont Hypothesis Revisited," in D.R. Wolstenholme and K.W. Jeon, eds., *Mitochondrial Genomes* (San Diego: Academic Press, 1992), 233–357.
76. M.W. Gray, G. Burger, and B.F. Lang, "Mitochondrial Evolution," *Nature* 283 (1999): 1476–1481.
77. S.G.E. Andersson, A. Zomorodipour, J.O. Andersson, T. Sicheritz-Ponten, U.C.M. Alsmark, R.M. Podowski, A.K. Näslund, A.S. Eriksson, H.H. Winkler, and C.G. Kurland, "The Genome Sequence of *Rickettsia prowazekii* and the Origin of Mitochondria," *Nature* 396 (1998): 133–140.
78. P. Marc, A. Mageot, F. Devaux, C. Blugeon, M. Corral-Debrinski, and C. Jacq, "Genome-Wide Analysis of mRNAs Targeted to Yeast Mitochondria," *EMBO Rep.* 3 (2002): 159–164.
79. T.D. Fox, L.S. Folley, J.J. Mulero, T.W. McMullin, P.E. Thorsness, L.O. Hedin, and M.C. Costanzo, "Analysis of Yeast Mitochondrial Genes," *Methods in Enzymology* 194 (1990): 149–165.
80. P.E. Thorsness, K.H. White, and T.D. Fox, "Inactivation of YME1, a Gene Coding a Member of the SEC18, CDC48 Family of Putative ATPases, Causes Increased Escape of DNA from Mitochondria in *Saccharomyces cerevisiae*," *Mol. Cell Biol.* 13 (1993): 5418–5426.
81. P.E. Thorsness and E. Weber, "Escape and Migration of Nucleic Acids between Chloroplasts, Mitochondria and the Nucleus," *Int. Rev. Cytol.* 165 (1996): 207–234.
82. P.E. Thorsness and T.D. Fox, "Escape of DNA from Mitochondria to Nucleus in *Saccharomyces cerevisiae*," *Nature* 346 (1990): 376–379.
83. P.E. Thorsness and T.D. Fox, "Nuclear Mutations in *Saccharomyces cerevisiae* That Affect the Escape of DNA from Mitochondria to the Nucleus," *Genetics* 134 (1993): 21–28.
84. O.G. Berg and C.G. Kurland, "Why Mitochondrial Genes Are Most Often Found in Nuclei," *Mol. Biol. Evol.* 17 (2000): 951–961.
85. L. E. Cowen, L.M. Kohn, and J.B. Anderson, "Divergence in Fitness and Evolution of Drug Resistance in Experimental Populations of *Candida albicans*," *J. Bacteriol.* 183 (2001): 2971–2978.
86. E. Pennisi, "Genome Data Shake Tree of Life," *Science* 284 (1998): 672–674.
87. E. Pennisi, "Is It Time to Uproot the Tree of Life?" *Science* 284 (1999): 1305–1307.

11

Contemporary Issues in Mitochondrial Origins and Evolution

MICHAEL W. GRAY

In the several decades since the modern revival of the endosymbiont theory, the mitochondrial genome has served as a key indicator of the evolutionary ancestry of the organelle in which it resides.[1–3] As the remnant genome of a eubacterial symbiont, mitochondrial DNA (mtDNA) has proven to be a rich source of phylogenetic information.[4–6] Initially, mitochondrial gene sequences argued strongly in favor of a xenogenous ("from outside") rather than an autogenous ("from within") origin for the mitochondrion, with mitochondrial genome data subsequently permitting a precise identification of the organismal group that gave rise to mitochondria.[2,3,5,6] More recently, with recognition that only a small part of the mitochondrial proteome (the collection of enzymes and structural, regulatory, and other proteins that make up the functional organelle) can be shown to have the same origin as the mitochondrial genome itself, the evolutionary perspective has shifted from one predominantly focused on mtDNA to one that includes nuclear data as well.[7–10]

In this essay, I pose a number of questions that have stimulated workers in this field over the last 30 years. Some of these questions have, for the most part, been answered: In these cases, I comment only briefly on the underlying issues and the insights that have emerged, referring the reader to published articles that offer greater depth together with the supporting evidence. The remaining questions, the ones with which we are still grappling, provide a road map for continuing investigations of mitochondrial evolution.

Question 1: What is the Evolutionary Source of Mitochondria in Eukaryotic Cells?

The simple answer, a eubacterial (specifically α-proteobacterial) symbiont, is almost certainly correct as far as the mitochondrial genome itself is concerned. Much biochemical, molecular, and other sorts of data have been marshalled in support of this conclusion.[2,3] Comparison between the most mitochondria-like eubacterial genomes, such as that of *Rickettsia prowazekii*,[11] and the most eubacteria-like mitochondrial genomes, such as that of the flagellate, *Reclinomonas americana*,[12] testify strongly to an origin of mtDNA from within the α-Proteobacteria, with the Rickettsiales, an order of obligate intracellular parasites, representing the closest extant eubacterial relatives of mitochondria.[4–6] Moreover, available information overwhelmingly points to the same origin for all of the genes in any given mtDNA (i.e., there is no evidence of an origin of the mitochondrial genome from multiple eubacterial sources).[6] Incorporation of foreign genetic information (other than mobile introns[13,14]) into the mitochondrial genome in the course of its evolution seems to have been an extremely rare event, a notable exception being angiosperm mtDNA, which has clearly received both chloroplast and nuclear DNA via interorganellar gene transfer.[15]

For the mitochondrial proteome, a rather different picture is emerging. Here, the various proteins that comprise the mitochondrion are seen to have diverse evolutionary origins.[7–10] Hence, when one poses questions about the origin of mitochondria, the answers may well be different, depending on the molecular data set (mitochondrial genome vs. mitochondrial proteome) used in the analysis.

Question 2: When Did the Mitochondrion Originate?

More specifically, did mitochondria originate at the very earliest stages of eukaryotic cell evolution, or some time later? The answer, according to classical serial endosymbiont theory, is "later": sometime after the emergence of a eukaryotic (nucleus-containing) "eating cell" capable of ingesting bacteria.[16,17] Initially, a group of putatively early-diverging eukaryotes devoid of mitochondria ("Archezoa") was offered as a modern-day example of the sort of amitochondriate cell that might originally have served as host to an endosymbiotic progenitor of mitochondria.[18,19] More recently, the Archezoa concept has been challenged (see following).

An alternative view of eukaryotic cell evolution proposes that the essence of the mitochondrion originated at the same time as the elements of the nucleus, through a fusion of two different types of prokaryotic cell (α-proteobacterium and archaeon [archaebacterium]).[20] Implicit in this alternative scenario is the assumption that there was no transitional, amitochondriate stage in the evolution of the eukaryotic cell, leading to the prediction that primitively amitochondriate eukaryotes should not exist—which prompts the next question.

Question 3: Can We Identify Any Modern Eukaryotes Whose Ancestors Diverged Away from the Main Line of Eukaryotic Evolution Before the Advent of Mitochondria (That Is, Early Diverging Eukaryotes That Are "Primitively Amitochondriate")?

This possibility seems increasingly remote, based on several lines of evidence.[21,22] First, certain protists, initially thought to have diverged early within the domain Eucarya, are now recognized as highly modified members of later-diverging, mitochondria-containing clades (e.g., microsporidia are now considered to be degenerate fungi).[23] Second, genes encoding proteins that are normally localized to and function in mitochondria, and that affiliate with the same subdivision of α-Proteobacteria from which mitochondria are descended, have been identified in the nuclear genome of many of the amitochondriate lineages originally grouped in the Archezoa.[24] Finally, several protists initially thought to lack mitochondria have now been shown to contain remnant organelles devoid of DNA but thought to be derived from mitochondria. These organelles include hydrogenosomes in a variety of protists,[25,26] the mitosome[27] (crypton[28]) in *Entamoeba histolytica*, and a small double membrane–bounded structure in the microsporidian *Trachipleistophora hominis*.[29] The fact that "mitochondrial" proteins, such as the chaperonins Hsp60 and Hsp70, are targeted to these organelles strengthens the argument that these proteins are evolutionarily related to, and perhaps derived from, mitochondrial proteins.[25]

The intermingling of mitochondria-containing and amitochondriate lineages within monophyletic assemblages is most straightforwardly explained by secondary loss of mitochondria. Indeed, reversion to an anaerobic mode of metabolism, typical of the hydrogenosome, may reflect a primitive evolutionary state in which the potential for anaerobic and aerobic metabolism was simultaneously present in the ancestral eukaryotic cell.[30] Such a thesis is an implicit, if not explicit, postulate of the hydrogen hypothesis, in which the progenitor of the mitochondrion is proposed to be a facultatively anaerobic α-proteobacterium.[20]

In my opinion, we cannot yet answer with compelling conviction the question of whether the mitochondrion originated under anaerobic[30] or aerobic[31] conditions, and whether it emerged simultaneously with or subsequent to other defining elements of the eukaryotic cell, such as the nucleus. What we can say is that fewer and fewer protist lineages remain as possible representatives of a primitively amitochondriate condition, and hence as exemplars of a premitochondriate evolutionary state in eukaryotic cell evolution.[32,33]

A few cautionary comments are in order. First, we have likely examined, let alone identified, only a very small proportion of extant protist lineages. Hence, we are currently extrapolating from a relatively small information base, albeit one that is increasingly representative of protist diversity. Second, none of the typical "mitochondrial" genes so far identified in amitochondriate protists has been found in any of the mtDNAs characterized to date. The usual assumption is that nuclear genes encoding such mitochondrial proteins (e.g., chaperonins) were transferred from the mitochondrial to the nuclear genome at an early stage in eukaryotic cell evolution. An inference of mitochondrion-to-nucleus transfer is compelling in cases in which a given mitochondrial protein is encoded by mtDNA in some organisms but by nuclear DNA in others, and in which it can be shown that the homologous proteins form a clade, regardless of which genome their genes inhabit.[34] Where only nuclear copies of a mitochondrial protein gene exist, one can invoke other evolutionary mechanisms to account for a nuclear location, such as direct α-proteobacterium-to-nucleus transfer of a gene as a result of a transient, cryptic endosymbiosis[35] or routine ingestion of food bacteria.[36] In contrast, we can point to clear and compelling examples of mitochondrial proteins (e.g., cytochrome *c*) that are exclusively encoded by nuclear genes but that almost certainly originated from the mitochondrial ancestor, based on their intimate structural and functional association with mtDNA-encoded components of the mitochondrial respiratory chain and translation system, in addition to their phylogenetic placement.

Question 4: What Was the Nature of the Proto-Mitochondrial Genome, and How Do We Go About Answering This Question?

The most gene-rich mitochondrial genomes contain at least an order of magnitude fewer genes than do the most gene-poor eubacterial genomes.[5,6] Accordingly, we infer that at an early stage in the transition from endosymbiont to organelle, massive loss/transfer of genes from the genome of the eubacterial ancestor of mitochondria must have occurred. What, then, can we say about the residual ("proto-mitochondrial") genome in terms of gene content, organization, and mode of expression?

Early work (up to the early 1990s) on animal, fungal, plant, and protist mtDNAs not only revealed a bewildering structural and organizational diversity but also highlighted the fact that none of these mitochondrial genomes looked particularly bacterial (i.e., these initially sampled mtDNAs appeared to have diverged, in some cases radically so, not only from one another but also away from an ancestral, eubacteria-like pattern, evidently by very different evolutionary pathways and mechanisms). These observations raised questions about the nature of the proto-mitochondrial genome and how one might go about inferring the ancestral pattern. The answer was that we needed much more data; specifically, a comprehensive database of complete mitochondrial genome sequences from which one might hope to make robust evolutionary deductions. Over the last decade, such a database has been created through a systematic, comparative-genomics approach focused on the mtDNA of protists (mostly unicellular eukaryotes).[37]

Sequencing protist mitochondrial genomes was (and still is) attractive for a number of reasons. Even today, the database of complete mtDNA sequences is highly skewed and nonrepresentative and still overwhelmingly dominated by animal sequences (which constitute >80% of the total number of complete mtDNA sequences in the public domain). Yet most of the evolutionary (and a preponderance of the biochemical) diversity of the eukaryotic lineage resides within protists.[38] To acquire an appropriate perspective of mitochondrial genome evolution, we need much additional information about mtDNA from the breadth and depth of the protist radiation, at least some of whose lineages likely represent early divergences within the eukaryotes. Compared with animal, fungal, and plant mtDNAs, protist mitochondrial genomes in general might be expected to more closely resemble the eubacterial, proto-mitochondrial genome to which contemporary mtDNAs trace their descent. Finally, we posit that multicellular eukaryotic groups must have evolved from unicellular eukaryotic ancestors. A comparative mitochondrial genomics approach could help to identify extant unicellular protist taxa that are specifically allied with multicellular, later-evolving lineages.

In this regard, the Organelle Genome Megasequencing Program (OGMP; http://megasun.bch.umontreal.ca/ogmpproj.html) has been particularly successful in exploring mitochondrial genome organization and diversity. A decade ago, this multi-institution Canadian consortium undertook comprehensive sequencing of protist mtDNAs; to date, the initiative has determined some three dozen complete protist mitochondrial genome sequences encompassing more than 2.5 million base pairs of annotated sequence. Although this output appears rather modest compared with bacterial and eukaryotic nuclear genome sequencing projects, the most challenging and, ultimately, rate-limiting step in this program turned out to be the culture of diverse protists and isolation of sufficient amounts of pure mtDNA for sequencing by a shotgun approach. In contrast, provision of DNA starting material is relatively trivial in the case of bacterial and nuclear genome sequencing projects.

Over the last decade, the work of the OGMP and other groups has expanded our appreciation of the structural and organizational diversity of mtDNA. Most (but not all[39]) mitochondrial genomes consist of a single molecule. Some mtDNAs are linear, but the majority map as circles (although circular-mapping genomes may not actually exist as such in vivo[40]). Characterized mitochondrial genomes range in size from a low of only 6 kbp in apicomplexa such as the malaria parasite, *Plasmodium falciparum*,[41] to more than 2400 bp—the size of a typical bacterial genome—in some flowering plants (angiosperms).[42] The large plant mtDNAs are not, however, the most gene rich: That distinction goes to the mitochondrial genomes of a relatively recently described and little investigated group of protists known as the core jakobids. The mtDNA of one of these protists, *Reclinomonas americana*, contains ~100 identified genes,[12] the largest gene complement described so far among mitochondrial genomes. In fact, *R. americana* mtDNA carries 18 protein-coding genes that at the time of its characterization[12] had not previously been found in mtDNA.

The comparative data generated by the OGMP and other groups allow us to categorize mitochondrial genomes into one of three basic organizational types: ancestral, reduced, or expanded. Ancestral mtDNAs, exemplified particularly by that of *R. americana* but also (to varying extents) by those of many other protists, tend to be compact genomes ranging in size from about 30 to 100 kbp and consisting mostly of coding sequence; they specify eubacteria-like 23S, 16S, and 5S rRNAs and a complete or almost complete set of tRNAs having conventional secondary structures; in general, they harbor few introns, they often retain eubacteria-like gene clusters, and they tend to use the standard genetic code. Reduced mtDNAs are characterized by extensive gene loss (protein-coding, tRNA, and 5S rRNA); nonconventional, often truncated rRNA and tRNA secondary structures that typically lack otherwise characteristic helical domains; split rRNA genes whose subgenic modules are frequently rearranged and interspersed with other genes at the genome level; an accelerated rate of sequence divergence in both protein-coding and rRNA genes; a highly biased pattern of codon usage, with certain codons entirely absent; and a nonstandard genetic code, the most common deviation being the use of TGA to specify tryptophan rather than termination. Reduced mitochondrial genomes, which by definition are more or less highly derived, are found particularly in animals, in chlorophycean green algae (e.g., *Chlamydomonas reinhardtii*), and in apicomplexa such as *Plasmodium*. Expanded mtDNAs mostly comprise the large mitochondrial genomes of land plants: Here, gene content approximates that of ancestral mtDNAs, and many features of the ancestral pattern are retained, including the presence of a 5S rRNA gene, conventional eubacteria-like rRNA and tRNA secondary structures, and a standard genetic code. However, the genome is greatly enlarged in size, so that it is substantially or even mostly noncoding. This expansion is attributed in large part to a pronounced increase in the content of noncoding spacer sequence as well as the appearance of a large number of repeats, introns, and intron ORFs.

In all likelihood, the proto-mitochondrial genome had an organizational pattern most closely resembling that of contemporary ancestral mtDNAs, particularly that of *R. americana*. Is it possible there are mtDNAs even more gene rich and more eubacteria-like than the *Reclinomonas* example that might exemplify an earlier evolutionary phase in the transition from a bona fide α-proteobacterial to a reduced organellar genome? So far, mitochondrial genome exploration has failed to uncover a mtDNA more ancestral than that of *Reclinomonas*. Although this does not exclude the possibility that a "transitional" mitochondrial genome may yet be discovered (there being no good way to select for such a genome among the vast assemblage of poorly sampled protists), it seems increasingly likely that *Reclinomonas* mtDNA will retain its title as the least derived of extant mtDNAs.

Question 5: Did Mitochondria Arise Only Once, or More Than Once?

Insofar as the mitochondrial genome constitutes the evolutionary essence of the mitochondrion, we may confidently assert that this organelle arose only once in evolution. Several lines of evidence support this contention.[5,6,43]

First, in any given mitochondrial genome, the genes encoding proteins (of assigned function) and rRNAs are a subset of those found in the *Reclinomonas* mitochondrial genome. If we accept that this genome is the substantially reduced product of a much-larger progenitor (eubacterial) genome, then it is highly unlikely that independent symbioses followed by independent events of mass reduction would have yielded basically the same small set of respiratory and translation genes in the resulting mtDNAs.

Second, although few traces of ancestral gene linkage remain in mtDNA, examples of clustered ribosomal protein genes have been found in both plant and protist mitochondrial genomes. These ribosomal protein gene clusters are colinear with their counterparts in eubacterial genomes, except that the mitochondrial clusters lack several of the ribosomal protein genes that are contained in the eubacterial clusters (either because these "missing" genes have been relocated elsewhere on the mitochondrial genome or were lost from it altogether). Importantly, these gene losses represent mitochondrion-specific characters: features shared by mitochondrial genomes to the exclusion of eubacterial genomes. As an example, in *Escherichia coli and R. prowazekii* (both proteobacteria), the contiguous S10–*spc* operons harbor the ribosomal protein gene assemblage –L2–S19–L22–S3–L16–L29–S17–L14–L24–L5– ("L" and "S" standing for "large subunit" and "small subunit," respectively). The same gene clustering is seen in several plant and protist mtDNAs, except that the mitochondrial clusters uniformly lack L22, L29, S17, and L24. It is highly unlikely that these missing genes reflect independent events of gene loss leading to the same pattern of gene linkage via convergent evolution. Indeed, this possibility becomes ever more remote as the mitochondrion-specific pattern continues to be reinforced by additional examples from newly sequenced plant and protist mtDNAs. The most parsimonious explanation of these data is that the mitochondrion-specific deletions were already present in the common ancestor of all mitochondrial genomes in which clustering of ribosomal protein genes is retained (i.e., these diverse mitochondrial genomes are monophyletic).

Third, in phylogenetic reconstructions based on concatenated mtDNA-encoded protein sequences, mitochondria form a statistically robust clade, to the exclusion of the

α-Proteobacteria and other eubacteria, again testifying to a monophyletic origin of the mitochondrial genome. Because the same result is not seen in phylogenetic analyses based on rRNA sequences,[4,44] we can confidently ascribe this discrepancy to exceptionally variable rates of nucleotide sequence divergence in the homologous rRNA species of different mitochondrial lineages, leading in single-gene trees to a long-branch attraction artifact that incorrectly groups the most rapidly diverging mitochondrial sequences together with the outgroup (eubacteria, in this case).

Question 6: How Have Mitochondrial Genomes Evolved in the Different Eukaryotic Lineages?

Even in land plants, where the mitochondrial genome has expanded in size, gene loss or gene transfer have played major roles in the evolutionary reshaping of mtDNA. Complete sequencing of mitochondrial genomes provides us with complete gene inventories. By mapping these gene sets to robust phylogenetic trees, we can begin to infer what genes must have been present at each stage in the evolutionary diversification of a given set of organisms. For example, 12 protein-coding genes must have been lost in going from *R. americana* mtDNA to the common ancestor of red and green algae, with an additional 11 genes subsequently being shed in going to the common ancestor of two red algae, *Porphyra purpurea* and *Cyanidoschyzon merolae*.[43] In these sorts of comparisons, we see clearly that the same gene has been lost, independently, multiple times during mtDNA evolution. For example, within a small assemblage consisting of two red algae, two green algae, and two land plants, we can infer three independent losses each of the respiratory gene *nad10* and the ribosomal protein gene *rpl14*.[43] Multiple independent losses of both ribosomal protein and respiratory chain genes have been well documented in the case of angiosperm mtDNA.[45]

Although it is reasonable to suppose that some of the genes that have been lost from mtDNA have been transferred to the nuclear genome, from which they are now expressed (there are, in fact, clear examples of this type), it is important to appreciate that another mechanism for providing an essential mitochondrial activity is functional replacement. A striking case in point involves two ribosomal proteins in plant mitochondria.[46] In one instance (in rosids), the gene encoding mitochondrial ribosomal protein S13 has been deleted from mtDNA and functionally replaced by a divergent nuclear copy of the same gene of chloroplast origin (i.e., the original nuclear S13 gene represents a gene transfer from chloroplast DNA). In a second example (in angiosperms and gymnosperms), the mtDNA lacks a gene specifying mitochondrial ribosomal protein S8, whose function is instead provided by a divergent copy of another nuclear gene, this one encoding ribosomal protein S15A, the cytosolic counterpart of mitochondrial S8.

Question 7: Which Protist Groups Are Specifically Affiliated with the Multicellular Eukaryotic Lineages?

Because there is little evidence of evolutionary gene transfer into mtDNA, save the special case of promiscuous chloroplast DNA in the angiosperm mitochondrial genome, we can safely assume strict vertical inheritance for mtDNA and the genes it encodes. Complete sequencing of mtDNA effectively provides a set of coinherited genomic characters and protein-coding sequences that offer an alternative means of establishing eukaryotic phylogenetic relationships that have proven elusive by other approaches.[5,6] In this regard, a key goal in eukaryotic phylogeny is to define which extant unicellular groups are the closest relatives of the (primarily) multicellular kingdoms of animals, plants and fungi.

A number of organismal groupings that only a few years ago were considered to be protists (protoctists[47]) are now firmly imbedded within the multicellular kingdoms Plantae, Fungi, or Animalia, in part as a consequence of mitochondrial genome analyses. For example, the charophyte algae are now placed together with land plants in the clade Streptophyta, a sister group to the Chlorophyta, which comprises all other green algae.[48] The overall structure and organization of the mitochondrial genome of a charophyte alga, *Klebsormidium flaccidum*, provides strong molecular support for this placement (OGMP, unpublished results). Chytridiomycetes (chytrids), long noted as having some fungus-like feeding and biochemical characteristics, are now established as a basally radiating group within the monophyletic Fungi.[49]

Two unicellular groups have been proposed as nearest relatives of animals, one (Choanoflagellata) having been suggested over a hundred years ago,[50] the other (Ichthyosporea) only recently.[51] Complete sequencing of representative choanoflagellate and ichthyosporean mtDNAs has revealed strikingly different patterns of genome and gene organization, with neither mtDNA resembling the prototypical compact animal mitochondrial genome.[52] However, in phylogenetic reconstructions that include mtDNA-encoded choanoflagellate and ichthyosporean protein sequences, these two groups robustly cluster with metazoan animals, to the exclusion of fungi, in a monophyletic assemblage christened "Holozoa".[53] In this assemblage, Ichythyosporea diverges basally, with Choanoflagellata constituting the sister group to Metazoa, a relationship that could not be discerned confidently from other types of molecular data. These observations indicate that the last common ancestor of multicellular animals and their closest unicellular relatives possessed a gene-rich mtDNA, indicating that the evolutionary appearance of the compact metazoan mitochondrial genome coincided with the emergence of a multicellular body plan.

Question 8: What is the Evolutionary Origin of the Mitochondrial Proteome?

Because genes encoded by mtDNA account for only a small fraction of the mitochondrial proteome, the focus in studies of mitochondrial evolution has begun to shift to those nuclear genes that encode most of the organellar proteins. Clearly, we need in-depth and comprehensive information about nuclear DNA–encoded mitochondrial proteins to fully understand the origin and evolution of the mitochondrion as a whole, as well as the process of mitochondrial biogenesis.

This question is being approached in several ways. First, determination of complete nuclear genome sequences provides a necessary information base for eventual identification of all of the protein components of the mitochondrial proteome. Computer-assisted algorithms designed to recognize organellar targeting signals or to predict subcellular location offer the means to recover a substantial proportion of nucleus-encoded mitochondrial protein sequences from genomic information. In yeast (*Saccharomyces cerevisiae*), some 423 proteins (393 of which are specified by the nuclear genome) have been annotated as putatively encoding mitochondrial proteins.[54] Using a computational approach[55] to infer subcellular localization, Marcotte et al.[9] deduced that there are about 630 mitochondrial proteins in yeast (representing ~10% of the organism's coding capacity). In a more direct analysis using high-throughout immunolocalization of epitope-tagged proteins, Kumar et al. recently estimated that ~13% of the yeast proteome, or some 800 proteins, are mitochondrial.[56] The numbers thus obtained from independent approaches are roughly comparable, but the limitations of the computational methods employed in these studies almost certainly means that some portion of the mitochondrial proteome will remain unidentified. Of course, another limitation of the whole-genome approach is that it is applicable only to those relatively few nuclear genomes whose sequences have been or are being determined; as in the case of mitochondrial genomes, protist genomes are relatively poorly represented among current nuclear genome sequencing projects.

A comprehensive expressed sequence tag (EST) approach offers the prospect of being able to identify mitochondrial proteins across a wide range of eukaryotes. However, this approach is limited in depth of coverage, and for the most part generates partial rather than complete protein sequences. Moreover, even with high coverage of normalized libraries, organism-specific EST databases are likely to be biased toward sequences encoding highly expressed proteins. For these reasons, the EST approach may be particularly useful for obtaining a phylogenetically comprehensive look at the evolution of a selection of relatively highly expressed mitochondrial protein sequences.

A third route to investigating the composition and evolution of the mitochondrial proteome is direct analysis. Here, purified mitochondrial or submitochondrial fractions are resolved by methods such as two-dimensional gel electrophoresis followed by mass spectrometry.[57] A proteomics approach yields a large amount of peptide sequence data; however, full utilization of this information depends heavily on the availability of extensive EST or complete genome sequence information. Combined with such data, proteome analysis offers a powerful means of identifying the protein components of the mitochondrial proteome, particularly those components of currently unknown function or those without obvious organellar targeting information. This methodology has been applied to isolated mitochondria from several eukaryotes, including *Homo* (human heart[58]), *Arabidopsis*,[59,60] and *Oryza*[61] (rice), as well as to submitochondrial fractions and isolated mitochondrial complexes, such as mitochondrial ribosomes.[62–64] So far, there have been no comparable proteomics analyses in protists.

In the case of yeast, where we have the most information, what can we say about the evolution of the mitochondrial proteome? Notably, only a small proportion of mitochondrial proteins (38 in the study of Karlberg et al.[7]) can confidently be identified as originating from the α-proteobacterial symbiont that contributed the mitochondrial genome. The largest proportion of the yeast mitochondrial proteome is actually encoded by "prokaryote-like" genes: genes that have matching sequences in Bacteria (but not specifically α-Proteobacteria) or Archaea. Many of the these prokaryote-like genes have homologs in other eukaryotes and therefore may have been present in the last common ancestor of the three domains; conceivably, genes in this category were already extant in the "host" genome at the time of the symbiotic event that led to the eventual emergence of the mitochondrion.

Another substantial fraction of the yeast mitochondrial proteome consists of genes having homologs only in other eukaryotes, with no eubacterial or archaeal matches; presumably, these proteins are eukaryote-specific inventions, recruited to function in the evolving organelle. Finally, a smaller proportion of the mitochondrial proteome is represented by unique nuclear sequences, without any database matches. Either these genes encode yeast-specific mitochondrial proteins, or they are genes whose primary structure evolves so rapidly that homologs in other eukaryotes are not readily recognizable. Overall, these observations lead to the conclusion that diverse genomic sources have contributed to the yeast mitochondrial proteome in the course of its evolution.

Summary and Future Prospects

Comparative mitochondrial genomics, an approach involving systematic and comprehensive determination, analysis, and comparison of complete mtDNA sequences, has revealed much about mitochondrial genome form, size, organization, gene content, patterns of gene expression, and evolution. This approach has led to the discovery of gene-rich mtDNAs retaining considerable ancestral, eubacteria-like character, thereby giving us new insights into the nature of the primitive proto-mitochondrial genome. Mitochondrial genomic characters yield valuable information about the timing of evolutionary events and about evolutionary affiliations. Phylogenetic reconstructions based on concatenated mtDNA-encoded protein sequences are proving a particularly valuable adjunct to nuclear genomic/genetic data in helping to decipher evolutionary relationships between and among eukaryotic lineages. Complete mtDNA sequences provide inventories of the encoded genes, allowing one to make inferences about the number and timing of gene losses during mitochondrial genome evolution.

Because such a small proportion of the mitochondrial proteome is actually specified by mtDNA, attention is increasingly being directed to nucleus-encoded mitochondrial proteins and their genes. Analyses in yeast clearly point to a dual (or even multiple) evolutionary origin of the mitochondrial proteome, in contrast to a single (monophyletic) origin of the mitochondrial genome itself. This intriguing conclusion raises the question of how similar the mitochondrial proteome is in different eukaryotes. Can we identify a core of conserved mitochondrial proteins that dates to the very earliest stages of mitochondrial evolution? If so, has this core been supplemented by additional proteins in a kingdom-, phylum-, or even organism-specific fashion? How similar is the protein composition of submitochondrial complexes, such as mitochondrial ribosomes, throughout the range of eukaryotes? Answering these questions will provide much of the focus for continuing investigations of how mitochondria originated and how they have evolved. Once again, a comparative genomics/proteomics survey, this time centered on the nuclear genome, will lead the way, and once again, protists will loom large in such a survey.

Acknowledgments Continuing work in the author's laboratory on "Structure, Function, and Evolution of Mitochondrial Nucleic Acids" is supported by an operating grant (MOP-4355) from the Canadian Institutes of Health Research. Salary and interaction support from the Canada Research Chairs Program and Canadian Institute for Advanced Research (Program in Evolutionary Biology) is gratefully acknowledged.

References

1. L. Margulis, *Origin of Eukaryotic Cells* (New Haven, CT: Yale University Press, 1970).
2. M.W. Gray and W.F. Doolittle, "Has the Endosymbiont Hypothesis Been Proven?" *Microbiological Reviews* 46 (1982): 1–42.
3. M.W. Gray, "The Endosymbiont Hypothesis Revisited," *International Review of Cytology* 141 (1992): 233–357.
4. M.W. Gray, "Rickettsia, Typhus and the Mitochondrial Connection," *Nature* 396 (1998): 109–110.
5. M.W. Gray, G. Burger, and B.F. Lang, "Mitochondrial Evolution," *Science* 283 (1999): 1476–1481.
6. B.F. Lang, M.W. Gray, and G. Burger, "Mitochondrial Genome Evolution and the Origin of Eukaryotes," *Annual Review of Genetics* 33 (1999): 351–397.
7. O. Karlberg, B. Canbäck, C.G. Kurland, and S.G.E. Andersson, "The Dual Origin of the Yeast Mitochondrial Proteome," *Yeast* 17 (2000): 170–187.
8. C.G. Kurland and S.G.E. Andersson, "Origin and Evolution of the Mitochondrial Proteome," *Microbiology and Molecular Biology Reviews* 64 (2000): 786–820.
9. E.M. Marcotte, I. Xenarios, A.M. van der Bliek, and D. Eisenberg, "Localizing Proteins in the Cell from Their Phylogenetic Profiles," *Proceedings of the National Academy of Sciences USA* 97 (2000): 12115–12120.
10. M.W. Gray, G. Burger, and B.F. Lang, "The Origin and Early Evolution of Mitochondria," *Genome Biology* 2 (2001): 1018.1–1018.5.
11. S.G.E. Andersson, A. Zomorodipour, J.O. Andersson, T. Sicheritz-Pontén, U.C.M. Alsmark, R.M. Podowski, A.K. Näslund, A.-S. Eriksson, H.H. Winkler, and C.G. Kurland, "The Genome Sequence of *Rickettsia prowazekii* and the Origin of Mitochondria," *Nature* 396 (1998): 133–140.
12. B.F. Lang, G. Burger, C.J. O'Kelly, R. Cedergren, G.B. Golding, C. Lemieux, D. Sankoff, M. Turmel, and M.W. Gray, "An Ancestral Mitochondrial DNA Resembling a Eubacterial Genome in Miniature," *Nature* 387 (1997): 493–497.
13. M. Turmel, V. Côté, C. Otis, J.-P. Mercier, M.W. Gray, K.M. Lonergan, and C. Lemieux, "Evolutionary Transfer of ORF-Containing Group I Introns between Different Subcellular Compartments (Chloroplast and Mitochondrion)," *Molecular Biology and Evolution* 12 (1995): 533–545.
14. Y. Cho, Y.-L. Qiu, P. Kuhlman, and J.D. Palmer, "Explosive Invasion of Plant Mitochondria by a Group I Intron," *Proceedings of the National Academy of Sciences USA* 95 (1998): 14244–14249.
15. J. Marienfeld, M. Unseld, and A. Brennicke, "The Mitochondrial Genome of *Arabidopsis* is Composed of Both Native and Immigrant Information," *Trends in Plant Science* 4 (1999): 495–502.
16. L. Margulis, *Symbiosis in Cell Evolution* (San Francisco. W.H. Freeman, 1981).
17. C. de Duve, "The Birth of Complex Cells," *Scientific American* 274 (1996): 38–45.
18. T. Cavalier-Smith, "A 6-Kingdom Classification and a Unified Phylogeny," in W. Schwemmler and H.E.A. Schenk, eds., *Endocytobiology II* (Berlin: de Gruyter, 1983), 1027–1034.
19. T. Cavalier-Smith, "Eukaryotes with No Mitochondria," *Nature* 326 (1987): 332–333.
20. W. Martin and M. Müller, "The Hydrogen Hypothesis for the First Eukaryote," *Nature* 392 (1998): 37–41.
21. T.M. Embley and R.P. Hirt, "Early Branching Eukaryotes?" *Current Opinion in Genetics and Development* 8 (1998): 624–629.
22. V.V. Emelyanov, "Mitochondrial Connection to the Origin of the Eukaryotic Cell," *European Journal of Biochemistry* 270 (2003): 1599–1618.

23. P.J. Keeling and N.M. Fast, "Microsporidia: Biology and Evolution of Highly Reduced Intracellular Parasites," *Annual Review of Microbiology* 56 (2002): 93–116.
24. A.J. Roger, "Reconstructing Early Events in Eukaryotic Evolution," *The American Naturalist* 154 (1999): S146–S163.
25. E.T. Bui, P.J. Bradley, and P.J. Johnson, "A Common Evolutionary Origin for Mitochondria and Hydrogenosomes," *Proceedings of the National Academy of Sciences USA* 93 (1996): 9651–9656.
26. T.M. Embley, M. van der Giezen, D.S. Horner, P.L. Dyal, and P. Foster, "Mitochondria and Hydrogenosomes are Two Forms of the Same Fundamental Organelle," *Philosophical Transactions of the Royal Society of London Series B* 358 (2003): 191–203.
27. J. Tovar, A. Fischer, and C.G. Clark, "The Mitosome, a Novel Organelle Related to Mitochondria in the Amitochondriate Parasite *Entamoeba histolytica*," *Molecular Microbiology* 32 (1999): 1013–1021.
28. Z. Mai, S. Ghosh, M. Frisardi, B. Rosenthal, R. Rogers, and J. Samuelson, "Hsp60 is Targeted to a Cryptic Mitochondrion-Derived Organelle ("Crypton") in the Microaerophilic Protozoan Parasite *Entamoeba histolytica*," *Molecular and Cellular Biology* 19 (1999): 2198–2205.
29. B.A.P. Williams, R.P. Hirt, J.M. Lucocq, and T.M. Embley, "A Mitochondrial Remnant in the Microsporidian *Trachipleistophora hominis*," *Nature* 418 (2002): 865–869.
30. W. Martin, C. Rotte, M. Hoffmeister, U. Theissen, G. Gelius-Dietrich, S. Ahr, and K. Henze, "Early Cell Evolution, Eukaryotes, Anoxia, Sulfide, Oxygen, Fungi First (?), and a Tree of Genomes Revisited," *IUBMB Life* 55 (2003): 193–204.
31. S.G.E. Andersson and C.G. Kurland, "Origins of Mitochondria and Hydrogenosomes," *Current Opinion in Microbiology* 2 (1999): 535–541.
32. A.J. Roger and J.D. Silberman, "Mitochondria in Hiding," *Nature* 418 (2002): 827–829.
33. J.D. Silberman, A.G.B. Simpson, J. Kulda, I. Cepicka, V. Hampl, P.J. Johnson, and A. J. Roger, "Retortamonad Flagellates are Closely Related to Diplomonads—Implications for the History of Mitochondrial Function in Eukaryote Evolution," *Molecular Biology and Evolution* 19 (2002): 777–786.
34. G. Burger, B.F. Lang, M. Reith, and M.W. Gray, "Genes Encoding the Same Three Subunits of Respiratory Complex II are Present in the Mitochondrial DNA of Two Phylogenetically Distant Eukaryotes," *Proceedings of the National Academy of Sciences USA* 93 (1996): 2328–2332.
35. K. Henze, A. Badr, M. Wettern, R. Cerff, and W. Martin, "A Nuclear Gene of Eubacterial Origin in *Euglena gracilis* Reflects Cryptic Endosymbioses during Protist Evolution," *Proceedings of the National Academy of Sciences USA* 92 (1995): 9122–9126.
36. W.F. Doolittle, "You Are What You Eat: A Gene Transfer Ratchet Could Account for Bacterial Genes in Eukaryotic Nuclear Genomes," *Trends in Genetics* 14 (1998): 307–311.
37. M.W. Gray, B.F. Lang, R. Cedergren, G.B. Golding, C, Lemieux, D. Sankoff, M. Turmel, N. Brossard, E. Delage, T.G. Littlejohn, I. Plante, P. Rioux, D. Saint-Louis, Y. Zhu, and G. Burger, "Genome Structure and Gene Content in Protist Mitochondrial DNAs," *Nucleic Acids Research* 26 (1998): 865–878.
38. D.J. Patterson and M.L. Sogin, "Eukaryote Origins and Protistan Diversity," in H. Hartman and K. Matsuno, eds., *The Origin and Evolution of the Cell* (Singapore: World Scientific, 1992), 13–46.
39. J. Fan and R.W. Lee, "Mitochondrial Genome of the Colorless Green Alga *Polytomella parva*: Two Linear DNA Molecules with Homologous Inverted Repeat Termini," *Molecular Biology and Evolution* 19 (2002): 999–1007.
40. D.J. Oldenburg and A.J. Bendich, "Mitochondrial DNA from the Liverwort *Marchantia polymorpha*: Circularly Permuted Linear Molecules, Head-to-Tail Concatemers, and a 5' Protein," *Journal of Molecular Biology* 310 (2001) 549–562.
41. J.E. Feagin, "Mitochondrial Genome Diversity in Parasites," *International Journal of Parasitology* 30 (2000): 371–390.
42. B.L. Ward, R.S. Anderson, and A.J. Bendich, "The Mitochondrial Genome is Large and Variable in a Family of Plants (Cucurbitaceae)," *Cell* 25 (1981): 793–803.

43. M.W. Gray, "Evolution of Organellar Genomes," *Current Opinion in Genetics and Development* 9 (1999): 678–687.
44. M.W. Gray and D.F. Spencer, "Organellar Evolution," in D. McL. Roberts, P. Sharp, G. Alderson, and M. Collins, eds., *Evolution of Microbial Life* (Cambridge: Cambridge University Press, 1996), 109–126.
45. K.L. Adams, Y.-L. Qiu, M. Stoutemyer, and J.D. Palmer, "Punctuated Evolution of Mitochondrial Gene Content: High and Variable Rates of Mitochondrial Gene Loss and Transfer to the Nucleus During Angiosperm Evolution," *Proceedings of the National Academy of Sciences USA* 99 (2002): 9905–9912.
46. K.L. Adams, D.O. Daley, J. Whelan, and J.D. Palmer, "Genes for Two Mitochondrial Ribosomal Proteins in Flowering Plants are Derived from their Chloroplast or Cytosolic Counterparts," *Plant Cell* 14 (2002): 931–943.
47. L. Margulis and V.K. Schwartz, *Five Kingdoms* (New York: W.H. Freeman, 1982, 1988).
48. M.W. Gray, C. Lemieux, G. Burger, B.F. Lang, C. Otis, I. Plante, and M. Turmel, "Mitochondrial Genome Organization and Evolution Within the Green Algae and Land Plants," in I.M. Møller, P. Gardeström, K. Glimelius, and E. Glaser, eds., *Plant Mitochondria: From Gene to Function* (Leiden: Backhuys Publishers, 1998), 1–8.
49. B. Paquin, M.-J. Laforest, L. Forget, I. Roewer, Z. Wang, J. Longcore, and B.F. Lang, "The Fungal Mitochondrial Genome Project: Evolution of Fungal Mitochondrial Genomes and Their Gene Expression," *Current Genetics* 31 (1997): 380–395.
50. H. James-Clark, "Note on the Infusoria Flagellata and the Spongiae Ciliatae," *American Journal of Science* 1 (1886): 113–114.
51. M.A. Ragan, C.L. Goggin, R.J. Cawthorn, L. Cerenius, A.V.C. Jamieson, S.M. Plourde, T.G. Rand, K. Söderhäll, and R.R. Gutell, "A Novel Clade of Protistan Parasites Near the Animal-Fungal Divergence," *Proceedings of the National Academy of Sciences USA* 93 (1996): 11907–11912.
52. G. Burger, L. Forget, Y. Zhu, M.W. Gray and B.F. Lang, "Unique Mitochondrial Genome Architecture in Unicellular Relatives of Animals," *Proceedings of the National Academy of Sciences USA* 100 (2003): 892–897.
53. B.F. Lang, C. O'Kelly, T. Nerad, M.W. Gray, and G. Burger, "The Closest Unicellular Relatives of Animals," *Current Biology* 12 (2002): 1773–1778.
54. P.E. Hodges, A.H.Z. McKee, B.P. Davis, W.E. Payne, and J.I. Garrels, "The Yeast Proteome Database (YPD): A Model for the Organization and Presentation of Genome-Wide Functional Data," *Nucleic Acids Research* 27 (1999): 69–73.
55. E.M. Marcotte, "Computational Genetics: Finding Protein Function by Nonhomology Methods," *Current Opinion in Structural Biology* 10 (2000): 359–365.
56. A. Kumar, S. Agarwal, J.A. Heyman, S. Matson, M. Heidtman, S. Piccirillo, L. Umansky, A. Drawid, R. Jansen, Y. Liu, K.-H. Cheung, P. Miller, M. Gerstein, G.S. Roeder, and M. Snyder, "Subcellular Localization of the Yeast Proteome," *Genes and Development* 16 (2002): 707–719.
57. T.J. Griffin and R. Aebersold, "Advances in Proteome Analysis by Mass Spectrometry," *Journal of Biological Chemistry* 276 (2001): 45497–45500.
58. S.W. Taylor, E. Fahy, B. Zhang, G.M. Glenn, D.E. Warnock, S. Wiley, A.N. Murphy, S.P. Gaucher, R.A. Capaldi, B.W. Gibson, and S.S. Ghosh, "Characterization of the Human Heart Mitochondrial Proteome," *Nature Biotechnology* 21 (2003): 281–286.
59. V. Kruft, H. Eubel, L. Jänsch, W. Werhahn, and H.-P. Braun, "Proteomic Approach to Identify Novel Mitochondrial Proteins in Arabidopsis," *Plant Physiology* 127 (2001): 1694–1710.
60. A.H. Millar, L.J. Sweetlove, P. Giegé, and C.J. Leaver, "Analysis of the Arabidopsis Mitochondrial Proteome," *Plant Physiology* 127 (2001): 1711–1727.
61. J.L. Heazlewood, K.A. Howell, J. Whelan, and A.H. Millar, "Towards an Analysis of the Rice Mitochondrial Proteome," *Plant Physiology* 132 (2003): 230–242.
62. E.C. Koc, W. Burkhart, K. Blackburn, A. Moseley, and L.L. Spremulli, "The Small Subunit of the Mammalian Mitochondrial Ribosome. Identification of the Full Complement of Ribosomal Proteins Present," *Journal of Biological Chemistry* 276 (2001): 19363–19374.
63. T. Suzuki, M. Terasaki, C. Takemoto-Hori, T. Hanada, T. Ueda, A. Wada, and K. Watanabe, "Proteomic Analysis of the Mammalian Mitochondrial Ribosome. Identification of Protein

Components in the 28 S Small Subunit," *Journal of Biological Chemistry* 276 (2001): 33181–33195.

64. E.C. Koc, W. Burkhart, K. Blackburn, M.B. Moyer, D.M. Schlatzer, A. Moseley, and L.L. Spremulli, "The Large Subunit of the Mammalian Mitochondrial Ribosome. Analysis of the Complement of Ribosomal Proteins Present," *Journal of Biological Chemistry* 276 (2001): 43958–43969.

12

On the Origin and Evolution of Plastids

JOHN M. ARCHIBALD
PATRICK J. KEELING

It is difficult to overstate the role of endosymbiosis in the evolution of eukaryotic cells. The endosymbiosis that gave rise to plastids (chloroplasts), the light-harvesting organelles of plants and algae, had an enormous effect on the course of eukaryotic evolution and, consequently, the evolution of the Earth's biosphere. The great majority of the planet's primary producers are not prokaryotic photosynthesizers but are eukaryotic phototrophs, organisms that owe their photosynthetic capabilities to an ancient endosymbiosis between a heterotrophic eukaryote and a photosynthetic cyanobacterium. It is widely believed that this process, known as "primary endosymbiosis," occurred only once, and that all plastids descend from a single common ancestor.

As important as the primary endosymbiosis was to the evolution of eukaryotes, the evolutionary history of plastids has proven to be exceedingly complex, much more so than can be accounted for by a single endosymbiotic event. Photosynthesis has also spread horizontally among unrelated eukaryotic groups by endosymbioses involving two eukaryotic cells. This process, referred to as "secondary endosymbiosis," has produced some of the most complex cells known, with an elaborate internal membrane structure surrounding the endosymbiont compartment, a sophisticated protein targeting machinery, and four distinct genomes (two nuclear genomes [host and endosymbiont], a mitochondrial genome, and a plastid genome). Yet another layer of complexity has been revealed with the discovery that some algae have replaced their ancestral secondary plastids with that of an unrelated alga. These so-called "tertiary" endosymbioses have occurred numerous times between dinoflagellate algae and a variety of secondary plastid-containing endosymbionts.

Primary, secondary, and tertiary plastid-containing organisms inhabit a wide range of aquatic and terrestrial ecosystems. Many are macroscopic, such as the trees and plants that inhabit dry land and the giant kelps that cling to the sea floor. Many more are benthic or planktonic microorganisms that exhibit a bewildering array of molecular, biochemical, and cell biological diversity. In large part, it is this diversity that has made the task of understanding the evolutionary history of photosynthetic eukaryotes and their plastids so difficult. The realization that secondary endosymbiosis has been responsible for the spread of primary plastids across the eukaryotic tree has greatly improved our understanding of plastid evolution; however, it has also generated as many questions as it has answered. How many times has secondary endosymbiosis occurred during the evolution of eukaryotes? What was the nature of the host and endosymbiont involved in these mergers? To what extent has tertiary endosymbiosis played a role in the evolution of plastids? In this chapter we discuss the origin and evolution of primary, secondary, and tertiary plastids, focusing on what has been established as fact, what remains controversial, and the kinds of data necessary to resolve some of the more contentious issues in this fast-paced field.

Primary Endosymbiosis

Ultimately, all plastids trace back to an endosymbiotic event with a cyanobacterium: The evidence for this is so diverse and so strong that it is now undisputed. In addition to their shared and unique form of photosynthesis that uses two photosystems and cleaves water to generate oxygen, plastids and cyanobacteria share a wealth of features at practically all levels of organization. Indeed, the gross similarities observed between cyanobacteria and chloroplasts led to the first theories of an endosymbiotic origin for the organelle in the early twentieth century.[1,2] Today, evidence from drug sensitivities, biochemistry, and, perhaps most powerfully, molecular biology and phylogeny have demonstrated beyond any doubt that plastids are derived from cyanobacteria by endosymbiosis.

This process is thought to have taken place as depicted in figure 12.1. A free-living cyanobacterium was engulfed by a phagotrophic eukaryote (fig. 12.1a) and, rather than being digested as food, was retained in the cytoplasm. If a eukaryote engulfed a gram-negative cyanobacterium by phagocytosis, the resulting structure would be bound by three membranes, but modern primary plastids possess only two membranes, so it has been proposed that the phagosomal membrane surrounding the plastid was lost. It is most likely that this endosymbiosis did not occur quickly but, rather, was a gradual coadaptation, where a eukaryote repeatedly engulfed and transiently retained endosymbionts, perhaps for progressively longer periods. In any case, once established inside the host, the endosymbiont was increasingly reduced so that most of its genome was lost, and many genes were transferred to the host nucleus. These genes were expressed by the host transcription and translation apparatus, and their protein products targeted back to the organelle using a transit peptide, a short amino-terminal leader that is recognized by the plastid and used to direct proteins across the two plastid membranes.[3] Such plastids, derived directly from endosymbiosis with a cyanobacterium, are called "primary plastids" (fig. 12.1b) and are found in three lineages: glaucophytes, red algae, and green algae along with their land plant relatives. Although this simplified scheme is thought to represent the origin of these plastids in a general way, a number of important aspects of this process and its history remain debated. We will discuss four of these: the kind of cyanobacterium that was engulfed, the number of times plastids originated,

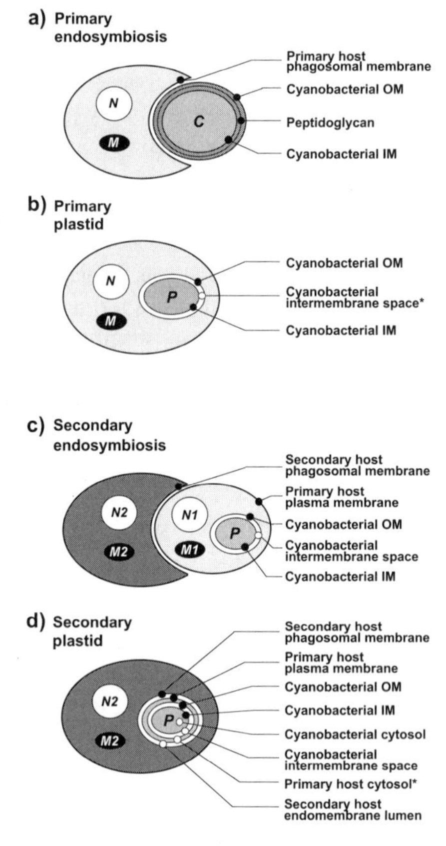

Figure 12.1. Diagram depicting the general features and end products of primary and secondary endosymbiosis. (a) Primary endosymbiosis involves the uptake of a photosynthetic cyanobacterium by a nonphotosynthetic, heterotrophic eukaryote. (b) A primary plastid-containing organism with two membranes surrounding its organelle, derived from the outer and inner membranes of the gram-negative prokaryotic cell. In glaucophyte algae, the intermembrane space still contains the layer of peptidoglycan present in the cyanobacterial endosymbiont. Primary plastids reside in the cytosol of the host. (c) Secondary endosymbiosis occurs when a nonphotosynthetic, heterotrophic eukaryote engulfs a primary plastid-containing eukaryote and retains its photosynthetic apparatus. (d) Secondary plastids are surrounded by one or more additional membranes and, in contrast to primary plastids, reside within the lumen of the host cell's endomembrane system. In two secondary plastid-containing groups, the chlorarachniophytes and cryptomonads, the nucleus of the algal endosymbiont (called the nucleomorph) persists in the space between the inner and outer pairs of plastid membranes (the primary host cytosol). Abbreviations: N, nucleus; M, mitochondrion; C, cyanobacterium; P, plastid; OM, outer membrane; IM, inner membrane.

the relationships among primary plastid-bearing eukaryotes, and the extent of the primary endosymbiont's contribution to the host cell at the molecular level.

The strongest evidence for a cyanobacterial origin for plastids comes from molecular phylogenetic analyses of plastid encoded genes, which consistently show a relationship between plastid and cyanobacterial homologues, typically with high statistical support. Curiously, however, this approach has failed to identify a single kind of cyanobacterium that is more like plastids than others. Cyanobacteria are a very diverse group of prokaryotes at the morphological and molecular level.[4,5] Several large studies based on the molecular phylogeny of cyanobacterial small-subunit ribosomal RNA (SSU rRNA) genes have identified a number of major subdivisions within the group, but these typically only partly coincide with the divisions expected based on morphology or pigmentation.[4,6] More interestingly still, although plastids form a strongly supported lineage within the cyanobacteria in these studies, their position within the group is totally undefined, and no particular cyanobacterium has been observed to be particularly closely related to plastids.[4,6,7] Accordingly, at this time it is impossible to say what kind of cyanobacterium gave rise to plastids. It is possible that the current sampling of cyanobacteria, although very broad, has missed the key lineage that is closely related to the progenitor of plastids. Alternatively, plastids may have originated very soon after the origin of cyanobacteria, and the major cyanobacterial lineages diverged at about the same time, making the phylogeny difficult to resolve, or sampling additional genes may reveal the true cyanobacterial sister group of plastids. Either way, the exact nature of the plastid progenitor remains elusive.

Although the position of the plastids within cyanobacterial SSU rRNA phylogeny remains unresolved, plastid sequences do form a unified group, indicating that all plastids are the product of a single common endosymbiosis. [4,6,7] Plastid genomes also share a variety of characters, including their characteristic rRNA operon-encoding inverted repeat structure and other unique gene orders.[8,9] In addition, all photosynthetic plastids use members of a particular family of light-harvesting antenna proteins that are unique to plastids and not found in cyanobacteria.[10,11] Nevertheless, there has been some dissent against the common view that all primary plastids are derived from a single endosymbiosis, and these come from phylogenetic trees inferred from nucleus-encoded genes.

Though phylogenies based on plastid genes typically show a very strong plastid clade, nuclear genes often do not. The first genes characterized from the nuclei of red, green, and glaucophyte algae were SSU rRNA and the cytoskeletal proteins actin, alpha-tubulin, and beta-tubulin. Analyses of these genes showed the three primary plastid-containing groups branching in (sometimes) three different places in the eukaryotic tree, although not with any support.[12–14] This raised some doubts about the monophyly of plastids,[12] but it could also be interpreted as nothing more than a lack of phylogenetic resolution. The true dissent came from analysis of the largest subunit of RNA polymerase II (RPB1), which showed a relatively strong separation between red and green algae (no glaucophyte gene has been characterized).[15,16]

To account for this phylogeny and the apparent close relationship between all plastids in phylogenies of plastid-encoded genes, it was proposed that the red and green algae were not, in fact, closely related, but that one of the two lineages had acquired its plastid by primary endosymbiosis and subsequently passed it to the other.[15] However, analysis of other nucleus-encoded genes showed either weak (enolase, heat shock protein 90) or strong (elongation factor-2 [EF2]) relationships between red and green algae,[17–19] indicating that the RPB1 phylogeny might be misleading. Subsequent analysis of RPB1 also showed that support for separating red and green algae was relatively weak[19] and that, in some analyses, they even branched together, as expected if there were a single origin of primary plastids (D. Longet, J.M. Archibald, P.J. Keeling, and J. Pawlowski, unpublished data). Other nucleus-encoded genes do not show a specific relationship between glaucophytes, red algae, and green algae; in many instances, this likely reflects a lack of sampling or insufficient information present in the molecule to adequately reconstruct such ancient events.

The number of characters shared by plastids, together with the strong support for the monophyly of plastid-encoded genes and the weaker but growing support for the monophyly of their host lineages, all support the simple explanation that a single endosymbiotic event led to all primary plastids. There are several qualifications to this, however, including the lack of data from glaucophytes, the possibility that plastids are derived from multiple independent endosymbioses involving closely related cyanobacteria,[4] and the possibility that plastids have been passed from one eukaryote to another without detection. However, without data specifically supporting such a complex story, the simple hypothesis that plastids originated once must be favored.

If the plastids of glaucophytes, red algae, and green algae all originated from a single primary endosymbiotic event, it is important to determine the relationship between these three lineages. This will allow the reconstruction of several aspects of plastid evolution, as well as overall tendencies, such as patterns of gene transfer and loss. However, there is currently so little molecular data available from glaucophytes that this question has remained somewhat controversial. Glaucophytes are a small group of algae, consisting of only three recognized genera that are not particularly common in nature.[20] Their chief claims to fame are that they are one of the three lineages of primary plastid-containing algae and that they alone have retained the peptidoglycan wall of the gram-negative cyanobacterial endosymbiont (fig. 12.1). For this reason, it has been suggested that glaucophytes may have been the first of the three primary lineages to diverge.[21] However, all possible alternative relationships between the three groups have been proposed at one time or another, based on different kinds of evidence.[22–26]

The paucity of molecular data from glaucophytes, together with the problem that several of the first gene trees examined failed to recover a distinct primary algal lineage, has meant that molecular phylogenetic analyses have led to no strong conclusion. Nuclear actin, tubulin, and SSU rRNA trees fail to unite the primary algae,[12–14] whereas plastid-encoded elongation factor-Tu trees show glaucophytes branching within the red algae.[23] In phylogenies of ribulose-1,5-bisphosphate carboxylase/oxygenase (RuBisCO), red algae possess one type of enzyme and glaucophytes and green algae possess another.[26,27] Plastid-encoded SSU rRNA is the gene most often used to analyze these relationships, and in these phylogenies the glaucophytes are typically found to be the most basal of the primary algal lineages, although sometimes with poor statistical support.[6,7] Recently, large-scale analyses of plastid-encoded protein genes were conducted, and in these trees, the glaucophytes are also the most basal of the three lineages, although once again alternative topologies could not be confidently rejected.[25,28] The data thus far appear to be converging on this topology for the tree of primary algae, but a firm conclusion awaits further analysis and more data.

One last issue surrounding the origin of primary plastids concerns a fundamental aspect of endosymbiont and host coevolution: What is the extent of gene transfer from the cyanobacterial endosymbiont to the host genome? Even the largest plastid genomes are mere shadows of their cyanobacterial progenitors and encode only a fraction of the proteins needed for plastid function. Most of the genes for plastid proteins have moved to the host nuclear genome. This special case of lateral gene transfer was hypothesized long ago, and current molecular data strongly support its occurrence, but was this the only molecular contribution of the endosymbiont? A few studies[29] have suggested that endosymbiont genes could have invaded the host genome and taken over the function of resident host genes in a process called "endosymbiotic gene replacement." Although several examples of this phenomenon have been well documented, they are relatively rare. However, recent analysis of the complete genome sequence of the flowering plant *Arabidopsis thaliana* has indicated that the endosymbiont contributed a great deal more.

A comparison of all conserved *Arabidopsis* proteins with those encoded in other finished genomes revealed that a large proportion (18%) were more similar to cyanobacterial homologs than to any other known sequences.[28] Extrapolating to the genome as a whole suggested that ~4,500 *Arabidopsis* nuclear genes are derived from the endosymbiont, more than reside in many modern-day cyanobacterial genomes. Surprisingly, a large fraction of the proteins encoded by these genes are not predicted to be targeted to the plastid, indicating that they have assumed some function in the host cell.[28] This is not to say that each and every one of these endosymbiont genes is derived from a unique cyanobacterial homolog: Many are certainly members of gene families that multiplied and diversified after their transfer to the nuclear genome. It does, however, indicate that the effect of primary endosymbiosis on the evolution of the host cell was more profound than previously considered.[28,30]

Second-Hand Plastids

A dramatic twist in the story of plastid evolution has come with the realization that although all plastids share a common ancestor, the same cannot be said for their hosts. Plastids have moved horizontally between unrelated eukaryotic lineages in a process called "secondary endosymbiosis" (fig. 12.1c). This involves the uptake and retention of a primary plastid-containing alga by a nonphotosynthetic, heterotrophic eukaryote.

The basic structure and cell biology of secondary plastids (fig. 12.1d) is distinguished from primary plastids in two important and related respects. First, compared to their primary counterparts, secondary plastids are surrounded by one or more additional membranes. This is a natural consequence of eukaryotic phagocytosis and means that unlike primary plastids, which reside in the cytosol of the host cell (fig. 12.1b), secondary plastids exist in the lumen of the endomembrane system (fig. 12.1d). As in primary plastids, the inner two plastid membranes appear to be derived from the inner and outer membranes of the cyanobacterial endosymbiont, and the third (if present) and fourth (outermost) membranes correspond to the primary host's plasma membrane and phagosomal membrane, respectively.[3] Second, the mechanism of targeting proteins to secondary plastids is more complex than for primary plastids. As mentioned above, most plastid proteins are nucleus encoded, translated on cytosolic ribosomes, and postranslationally targeted to the organelle with the use of an amino-terminal transit peptide. Although this is true of both primary and secondary plastids, the extra membranes surrounding secondary plastids means that an extra step in the targeting pathway is required. Secondary plastid-containing algae take advantage of the host cell's signal peptide secretion system and use bipartite amino-terminal leaders on their plastid proteins. In addition to the amino-terminal transit peptide, proteins targeted to secondary plastids contain signal peptides that direct them to the endomembrane system of the host. From here, proteins are localized to the plastid (by an as yet unidentified mechanism) and sent across the inner two plastid membranes by their transit peptides, as occurs in primary plastid-containing algae.[3]

By virtue of their size, trees and plants, which are descendents of primary plastid-containing green algae, are perhaps the most obvious and well-known eukaryotic photosynthesizers. However, in terms of sheer variety and abundance, secondary plastid-containing algae constitute a far greater fraction of the Earth's photosynthetic diversity. These organisms can be divided into two groups: those with plastids derived from green algae, and those with red algal plastids. Two distinct eukaryotic lineages, the euglenids and chlorarachniophytes, possess green algal secondary plastids. The euglenids are an abundant and speciose group of unicellular marine and freshwater flagellates that include a variety of photosynthetic and nonphotosynthetic genera. Chlorarachniophytes are a comparatively rare group of marine pseudopod-forming amoeboflagellates and flagellates, composed of only a handful of recognized genera. In the context of secondary endosymbiosis, the chlorarachniophytes are of particular interest: Chlorarachniophytes and cryptomonads (below) are the only two groups of secondary algae that have retained the nucleus of their secondary endosymbiont. This nucleus is referred to as the "nucleomorph" and resides between the second and third plastid membranes (fig. 12.1d). The chlorarachniophyte nucleomorph was first identified in the 1980s[31,32] and has since been shown to contain a bona fide eukaryotic nuclear genome, composed of a small number of AT-rich genes partitioned among three short chromosomes.[33–35]

The evidence for the green algal origin of the euglenid and chlorarachniophyte plastids is strong. Preliminary examinations revealed that although their plastids are bound by three and four membranes, respectively, both possess a combination of photosynthetic pigments characteristic of plants and green algae, chlorophylls *a* and *b*.[31,36] Molecular phylogenies inferred from plastid- and (in the case of chlorarachniophytes) nucleomorph-encoded genes have since confirmed that both plastids are green algal in nature.[10,37–41] Pinpointing the exact source of the euglenid and chlorarachniophyte plastids within the green algal/plant lineage has, however, proven difficult.

Early molecular phylogenies of plastid-encoded SSU rRNA and the large subunit of RuBisCO placed the chlorarachniophytes as a sister group to the green algae (chlorophytes) and land plants (streptophytes).[39] In contrast, nucleomorph-encoded SSU rRNA phylogenies suggested that the endosymbiont was derived from within the chlorophyte green algae, possibly from a trebouxiophyte[41] or an ulvophyte[38] green alga. Nucleomorph-encoded genes tend to be AT rich and divergent[17,35] and are, thus, difficult to place accurately on phylogenetic trees, so these results should be interpreted with caution. Nevertheless, analyses of plastid-encoded elongation factor Tu (EF-Tu) are also in line with an ulvophyte origin for the endosymbiont.[37] The situation has recently become even more confusing, as a nucleus-encoded enolase gene with significant similarity to streptophyte enolases has been characterized in chlorarachniophytes.[18] With respect to the precise origin of the euglenid plastid, molecular data has been similarly ambiguous. Although the plastid genome of *Euglena gracilis*[42] shares many features with other green algal genomes, phylogenetic analyses have been largely uninformative. They have, however, singled out the chlorophyte green algae as a possible source of the euglenid endosymbiont.[40]

A much greater array of eukaryotes possess secondary plastids derived from red algae. Three prominent groups with secondary red algal plastids are the heterokonts, haptophytes, and cryptomonads. The heterokonts (or stramenopiles) are a diverse and abundant group of algae that, in addition to having large numbers of unicellular forms, also include the giant kelps that carpet the coastlines of many marine habitats. The haptophytes are an equally abundant and ecologically significant algal lineage, known for the calcarious tests possessed by many of its members. Such tests are the major component of chalk sediments worldwide, including the white cliffs of Dover. Finally, the cryptomonads are a relatively common algal group, best known because, together with the chlorarachniophytes, they have retained a nucleomorph. The nucleomorph genome of the cryptomonad *Guillardia theta* has recently been completely sequenced[43] and has shed considerable light on the process of genome and endosymbiont reduction.[35,43] The plastids of heterokonts, haptophytes, and cryptomonads are all surrounded by four membranes and contain chlorophylls *a* and *c*. They are also unusual in that the outermost membrane of their plastid is continuous with the outer membrane of the host nuclear envelope and endoplasmic reticulum.[44]

As was the case for the green algal plastids of chlorarachniophytes and euglenids, a red algal origin for the heterokont, haptophyte, and cryptomonad plastids has been proven convincingly through consideration of molecular data. Plastid genomes have been completely sequenced from heterokonts and cryptomonads,[45,46] and both possess many features in common with red algal plastid genomes. All three groups have also been shown to possess a highly unusual proteobacterial form of RuBisCO, similar to that found in red algae, that appears to have been acquired by lateral gene transfer. This is in contrast to the cyanobacterial-type RuBisCO found in glaucophytes, green algae, euglenids, and chlorarachniophytes.[27,47] Furthermore, molecular phylogenies of a variety of plastid- and nucleomorph-encoded genes reveal a close relationship between heterokont, haptophyte, and cryptomonad sequences and those of red algae.[17,48–51]

In addition to the heterokonts, haptophytes, and cryptomonads, two other eukaryotic groups, the dinoflagellate algae and apicomplexan parasites, possess red algal secondary plastids. However, as we shall see, the history of their plastids is much less straightforward and is, in fact, extremely controversial. The dinoflagellates are an abundant, diverse and ecologically important algal group, best known as the causative agents of "red tides" and toxic shellfish poisoning. Most dinoflagellates possess three-membrane plastids containing chlorophylls *a* and *c*, as well as peridinin. The story of the origin or origins of the dinoflagellate plastid is intimately tied to that of the plastid in apicomplexans, a highly derived group of nonphotosynthetic intracellular parasites that, not surprisingly, have traditionally figured little in discussions of plastid origins. Apicomplexa and dinoflagellates are closely related and, together with ciliates, form a monophyletic group referred to as alveolates.[52] Virtually all apicomplexa are obligate intracellular parasites, so the discovery of a relict plastid in apicomplexans[53,54] came as a shock, and the origin and evolution of the organelle has been extraordinarily difficult to elucidate.

Although the plastid is clearly of secondary endosymbiotic origin, some lines of evidence appear to support a green origin for it, whereas others are more in line with a red origin (discussed in detail below). Dinoflagellate plastids contain chlorophyll *c*, which is otherwise only found in the red secondary plastids of heterokonts, haptophytes, and cryptomonads (as well as a few cyanobacteria[55]). This is consistent with the view that the dinoflagellate plastid is derived from a red alga and is perhaps related to these other, secondary plastids. Support for this idea has come from recent analyses of dinoflagellate plastid genes[56,57] which, curiously, reside on small single-gene minicircles.[57] These analyses not only indicated a red algal origin for the dinoflagellate plastid[56,57] but also hinted at a specific relationship between the dinoflagellate and apicomplexan plastids.[58] The interpretation of this data, however, is hindered by the fact that both the dinoflagellate and apicomplexan plastid-encoded genes are AT rich and extremely fast evolving—it is possible that their affinity for one another in plastid phylogenies is a result of methodological artifact rather than common ancestry.

How Many Secondary Endosymbioses?

Chlorarachniophyte and euglenid plastids are obviously derived from green algae, whereas the dinoflagellate, heterokont, haptophyte, and cryptomonad plastids clearly evolved from a red alga. Accordingly, at least two secondary endosymbiotic events are necessary to account for this diversity.[22,59,60] However, several authors have argued for many more events, in some cases suggesting that each secondary plastid-containing lineage acquired its plastid in a separate endosymbiosis.[61,62] The huge diversity of plastid types and the even greater diversity of the host cells in which they reside have made for a lengthy and engaging debate. Fortunately, recent molecular phylogenetic data has gone a long way toward settling some of the major questions.

With respect to the chlorarachniophytes and euglenids, there are (basically) two evolutionary scenarios that could explain the presence of photosynthesis in both groups. Either their plastids are the result of separate secondary endosymbiotic events involving two different hosts and two different green algae, or they are the products of a single endosymbiosis in their common ancestor. Distinguishing between these two possibilities requires one not only to examine the origins of the chlorarachniophyte and euglenid endosymbionts but also to consider the evolutionary affinities of their respective hosts. The host component of euglenids belongs to the Euglenozoa, a large, predominantly nonphotosynthetic protist assemblage that, in addition to euglenids, includes the diplonemids and kinetoplastids. The chlorarachniophytes, in contrast, are members of the Cercozoa. This morphologically diverse collection of amoeboid, amoeboflagellate, and flagellated eukaryotes has only recently been recognized as a monophyletic group on the basis of molecular data.[12,63–67] Significantly, broad-scale phylogenetic analyses of eukaryotic nuclear genes have failed to show a specific relationship between the Euglenozoa and Cercozoa (e.g., Bhattacharya et al. and Keeling[12,67]). In addition, chlorarachniophyte and euglenid plastid sequences, although clearly green algal in nature, appear unrelated to one another in molecular phylogenies.[37,39] Together with a distinct lack of morphological similarity shared between the chlorarachniophyte and euglenid hosts,[31] these facts have led to the widely held belief that their plastids are of independent endosymbiotic origin.[47,62,68]

Although the consensus view in the field is that the chlorarachniophyte and euglenid plastids evolved separately, not everyone agrees. Cavalier-Smith, who has argued strongly that secondary endosymbiosis is extremely difficult and should be invoked sparingly, has proposed a single origin of green secondary plastids.[59,60] One apparent drawback of this idea is that both the Cercozoa and Euglenozoa are composed almost entirely of nonphotosynthetic members and that, within Euglenozoa, phylogenetic analyses place the photosynthetic euglenids as a highly derived lineage, nested within nonphotosynthetic, heterotrophic groups (e.g., Preisfeld et al., and Leander et al.[69–71]). A single, ancient origin for the chlorarachniophyte and euglenid plastids demands extensive secondary loss of plastids within Euglenozoa, Cercozoa, and other eukaryotic lineages. This is at odds with the general observation that plastids have never been observed in nonphotosynthetic members of Euglenozoa or Cercozoa (except in a few instances where nonphotosynthetic plastids are known to exist). Moreover, analysis of cytoskeletal data indicated that phototrophy evolved relatively recently within the evolution of euglenids.[70,71] That said, several "plant-like" metabolic enzymes have recently been characterized from two nonphotosynthetic kinetoplastids, *Trypanosoma* and *Leishmania*.[72] Although the taxonomic sampling necessary to convincingly demonstrate the algal origin of these genes is unavailable at this time, if they can be proven to be plastid derived, then the Euglenozoa may have evolved from a photosynthetic, plastid-bearing ancestor.[72,73] Whether this also applies to the common ancestor of Cercozoa and Euglenozoa remains to be seen. On balance, current data provide little support for the idea.

The question of single or multiple origins for the red algal plastids of heterokonts, haptophytes, cryptomonads, and dinoflagellates has been an even thornier issue. Before the availability of molecular sequences, a variety of morphological and biochemical data were brought to bear on the question. As mentioned above, the topology of the membranes surrounding the plastids of heterokonts, haptophytes, and cryptomonads is highly unusual. The outermost plastid membrane in these organisms is fused to the outer membrane of the nuclear envelope and endomembrane system,[44] and as a result, their photosynthetic machinery resides within the lumen of the host endoplasmic reticulum. The possibility that this situation evolved more than once seems unlikely, and the plastids present in the three groups have thus been suggested to be derived from a single endosymbiosis in their common ancestor.[22,74,75] A variety of other cellular and biochemical features are also consistent with a single origin. In addition to their shared presence of chlorophyll *c*, both heterokonts and cryptomonads possess flagella with unusual tubular hairs called mastigonemes,[75] and haptophytes and heterokonts also share fucoxanthin and chrysolaminaran as well as three thylakoids per stack in their plastids.

If the heterokont, haptophyte, and cryptomonad plastids do share a common evolutionary origin, plastid and nuclear gene sequences from the three groups should form a monophyletic group in molecular trees. Surprisingly, preliminary analyses did not show this to be the case. In plastid SSU rRNA and RuBisCO phylogenies, heterokont, haptophyte, and cryptomonad sequences branched as separate lineages within the red algae,[48,76–78] and nuclear SSU rRNA trees also failed to unite the three groups.[79] Although these initial results were interpreted as evidence that these plastids were acquired from different red algae in three separate secondary endosymbioses, a more comprehensive recent analysis of a multigene plastid dataset paints a different picture that is more in line with the biochemical and ultrastructural data. In phylogenetic trees constructed from concatenated *psa*A, *psb*A, *tuf*A, *rbc*L, and SSU rRNA coding sequences, heterokonts, haptophytes, and cryptomonads form a well-supported monophyletic group.[51] This result indicates that rather than three separate endosymbioses, a single endosymbiotic event occurred in the common ancestor of the three groups.

Recent analyses of the metabolic enzyme glyceraldehyde-3-phosphate dehydrogenase (GAPDH) have lent further support to the notion of a common evolutionary origin for the heterokont, haptophyte, and cryptomonad plastids and, importantly, has brought the dinoflagellates and apicomplexans firmly into the mix. Plants and algae possess both cytosolic and plastidic isoforms of GAPDH, both of which are nucleus encoded. In red and green algae and euglenids, the plastid-targeted GAPDH homolog is, as expected, closely related to GAPDH in cyanobacteria. However, in apicomplexa, dinoflagellates, heterokonts, and cryptomonads, the plastid-targeted GAPDH is not cyanobacterial but is, instead, derived from the eukaryotic cytosolic isoform.[80]

This has now also been confirmed for haptophytes,[81] the only other group with a red secondary plastid. At some point during the early evolution of these organisms, the cytosolic GAPDH gene was duplicated and one of the duplicates acquired signal and transit peptide coding information. As a consequence, its protein product was targeted to the plastid and took over the role of plastid GAPDH, replacing the cyanobacterial-derived protein. This is an example of "reverse endosymbiotic gene replacement," the opposite of the situation discussed earlier in regard to the extent to which eukaryotic nuclear genes have been replaced by cyanobacterial homologs.[82] The picture is complicated by the fact that, in addition to possessing the cytosolic-derived plastid GAPDH, at least one dinoflagellate (*Pyrocystis*) also possesses a cyanobacterial-like plastid homolog that is closely related to that of *Euglena*.[83] The origin of this gene is unclear. Nevertheless, taken as a whole, the data indicate that the GAPDH endosymbiotic replacement took place only once, in the common ancestor of heterokonts, haptophytes, cryptomonads, dinoflagellates, and apicomplexa,[80] and that their plastids are the product of a single secondary endosymbiosis.

Independent corroborating evidence for this idea is emerging from phylogenies of the host lineages. Analyses of nucleus-encoded rRNA genes have revealed that the alveolates (dinoflagellates, apicomplexans, and ciliates) and heterokonts appear to represent a monophyletic group on the global tree of eukaryotes,[49,84] and concatenated protein-coding genes also support this conclusion.[85] Although haptophyte and cryptomonad nuclear sequences are presently scarce, if one considers the nuclear rRNA and concatenated protein phylogenies in conjunction with the GAPDH data and the results of recent plastid gene trees,[51] all five of the putative secondary red algal–containing groups are accounted for. This adds up to what appears to be a single, ancient origin for the alveolate, haptophyte, heterokont, and cryptomonad plastids. Together, these organisms make up a eukaryotic "supergroup" referred to as the chromalveolates.[59]

Alveolate Plastids: In One Vacuole, out the Other?

A global scheme for the evolution of plastids by primary and secondary endosymbiosis is presented in figure 12.2. Following a single endosymbiosis between a eukaryote and a cyanobacterium, three distinct primary plastid-containing lineages evolved: glaucophytes, red algae, and green algae. The latter two groups were subsequently involved in additional mergers: Two secondary endosymbiotic events gave rise to the green algal plastids of chlorarachniophytes and euglenids, and the plastids of chromalveolates are derived from a single secondary endosymbiosis with a red alga. This scenario represents a consensus view of the broad strokes of plastid evolution, but at a fine scale the situation is much more complex. For example, although the common ancestor of all alveolates appears to have had a red algal plastid (fig. 12.2), the dinoflagellates are known to have repeatedly "swapped" their ancestral organelle for plastids derived from other algae. As well, recent molecular data has renewed the debate over whether the apicomplexan plastid is truly red algal–derived. The twists and turns in our understanding of alveolate plastid evolution provides an excellent example of how our hypotheses have continued to evolve in response to new data and how absence of evidence should not be taken as evidence of absence.

The dinoflagellates are, without a doubt, the algal masters at plastid acquisition and exchange. Approximately half of known dinoflagellate species are photosynthetic,[86] and of those that are, most possess a three-membrane plastid containing chlorophyll *a+c* and peridinin. However, this plastid is by no means ubiquitous: Several lineages have substituted their peridinin-containing plastid with that of another alga by "tertiary endosymbiosis" or "serial secondary endosymbiosis." The extent to which these "replacement" endosymbionts and plastids have integrated with their dinoflagellate hosts varies greatly, ranging from transient endosymbioses[87,88] to fully integrated, heritable organelles.[47,89] Perhaps the best-documented examples of true tertiary endosymbiosis are in the dinoflagellates: *Karlodinium micrum*, *Karenia brevis*, and *Karenia mikimotoi*.[90,91]

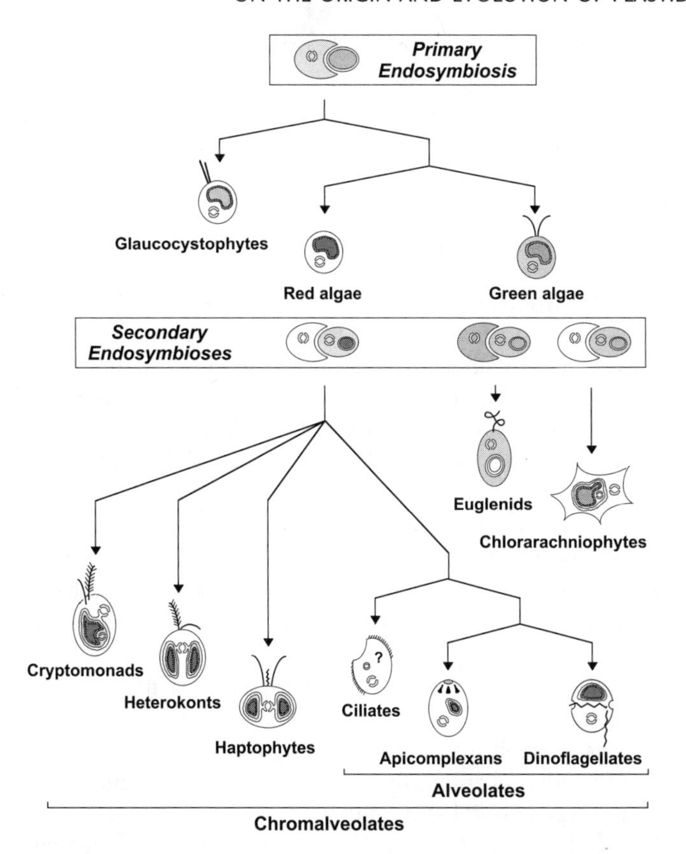

Figure 12.2. A possible scheme for the evolutionary history of photosynthetic eukaryotes and their plastids. (Top) A single primary endosymbiosis between a nonphotosynthetic, heterotrophic eukaryote and a photosynthetic cyanobacterium led to three groups of primary plastid-containing algae, glaucophytes, red algae, and green algae (and their land plant relatives). (Bottom) After the three groups diversified, a single secondary endosymbiosis between a eukaryotic heterotroph and a primary plastid-containing red alga, produced a lineage that ultimately gave rise to the cryptomonads, heterokonts, and haptophytes, as well as the alveolates (ciliates, apicomplexans, and dinoflagellates), collectively referred to as chromalveolates. In addition, two separate secondary endosymbioses involving different eukaryotic hosts and different green algae gave rise to the secondary plastids found in euglenids and chlorarachniophytes. Ciliates, apicomplexans, and many dinoflagellates have lost photosynthesis, and numerous dinoflagellates have replaced their ancestral plastids with ones derived from green algae, cryptomonads, heterokonts, and haptophytes (see text). The possible presence of a cryptic plastid in ciliates is indicated by a (?).

These species have been shown to possess a red algal plastid derived from a haptophyte.[47,92] Other examples are members of the genus *Dinophysis* that have cryptomonad plastids[93] and *Durinskia* (*Peridinium*) *balticum* and *Krypoperidinium* (*Peridinium*) *foliacium*, which harbor relatively unreduced diatom (heterokont) endosymbionts.[94]

In each of these instances, the dinoflagellate host has replaced its ancestral, secondary red algal plastid with another secondary red algal plastid, which, according to the scenario outlined in figure 12.2, is ultimately derived from the same secondary endosymbiotic event. However, not all plastids stolen by dinoflagellates are of chromalveolate origin. *Lepidodinium viride*, for example, has been shown to possess a chlorophyll *a+b*-containing plastid derived from a green alga.[95] Because green algae have primary plastids, this is not a tertiary endosymbiosis but rather a serial secondary endosymbiosis.

The reasons why dinoflagellates so readily exchange their ancestral plastids for new ones are not entirely clear. It may have to do with the fact that they are mixotrophic organisms; that is, they are both heterotrophs and phototrophs.[96] Mixotrophy provides an alternate energy source during periods when photosynthesis is insufficient or nutrients are not available from the plastid, but it also provides a constant source of potential replacement plastids from algal prey taken from the environment. This dual mode of nutrition also explains why dinoflagellates appear to loose plastids—or at least photosynthesis—as quickly as they gain them. A recent comprehensive analysis of rRNA sequences allowed at least eight independent losses of photosynthesis to be inferred.[89]

Regardless of the reasons for the comings and goings of dinoflagellate plastids, tertiary endosymbiosis has potentially significant effects on dinoflagellate molecular biology, and in particular, on the array of nucleus-encoded, plastid-targeted proteins. During the integration of a tertiary endosymbiont, each and every one of the nucleus-encoded genes for plastid-targeted proteins has the potential to be replaced by its corresponding homolog from the incoming tertiary endosymbiont nucleus. Alternatively, plastid-targeted proteins that serviced the ancestral peridinin plastid can simply be recycled and imported into the newly acquired organelle. Unfortunately, almost none of these genes have been studied. In the one interesting exception, the gene encoding the thylakoid lumen protein PsbO has been characterized in the haptophyte plastid-containing dinoflagellate *Karenia brevis*. Here, the original nuclear-encoded PsbO from the peridinin-type plastid has been replaced by a version that came in with its tertiary haptophyte endosymbiont.[97] Whether this represents the exception or the rule is presently unknown, but it is possible that the complement of proteins in tertiary plastids is a mixture cobbled together from both old and new plastids.

If the dinoflagellates are promiscuous with respect to plastid uptake and replacement, then what can be said of their close relatives, the apicomplexa? The discovery of a cryptic plastid in these unusual organisms was arguably the biggest surprise in plastid research in the last two decades. This is because apicomplexa are a group composed almost entirely of obligate intracellular parasites. Many apicomplexa cause serious disease in humans and other animals, the most notorious being the opportunistic pathogen *Toxoplasma*, the common gastrointestinal pathogen *Cryptosporidium*, and the malaria parasite *Plasmodium*. As intracellular parasites, apicomplexa are obviously not photosynthetic and have never been thought to contain a plastid. However, in the mid-1970s an electron microscopical examination of extrachromosomal DNA in *Plasmodium* revealed a circular molecule of about 35 Kbp with a cruciform structure, characteristic of an inverted repeat.[98] Although at first assumed to be the mitochondrial genome,[99] when gene sequences from this element were characterized, they were clearly bacterial and resembled plastid homologs as much as they did mitochondrial ones.[100,101] Eventually, the characterization of several sequences led Iain Wilson and colleagues to make the bold claim that this element was a plastid genome.[102–104]

At about the same time, a second extrachromosomal element was discovered in *Plasmodium*. This 6-Kbp linear element was found to encode *cox*I, *cox*III, *cob*, and fragments of rRNA, all related to mitochondrial homologues.[105,106] This element cofractionated with the mitochondrion,[107] providing further support for the nonmitochondrial nature of the 35-Kb circular DNA. Eventually, the complete sequence of the 35-Kb circle from *Plasmodium* and *Toxoplasma* demonstrated beyond any doubt that this was a plastid genome,[54,108] and the organelle was identified by in situ hybridization.[53]

Even with the complete genome sequence, the function of a plastid in *Plasmodium* was not obvious, although knocking out plastid gene expression was shown to prevent the parasite from completing its life cycle.[109,110] Clearly the plastid is not photosynthetic, so why is it maintained? The answer lies in the metabolic diversity of plastids. Although plastids are typically thought of in the context of photosynthesis and the biosynthesis of various compounds related to photosynthesis such as chlorophyll, they are also responsible for several other important metabolic functions in plants and algae. In particular, the biosynthesis of fatty acids, isopentyl diphosphate (the primary substrate for isoprenoid synthesis), heme, and aeromatic amino acids are all carried out in plastids.[111–113] In plants and other algae in which it has been examined (which is not many), the enzymes that make up these pathways are derived from the cyanobacterial endosymbiont. However, the genes encoding these enzymes are among the many that have been transferred to the nuclear genome and express proteins that are postranslationally targeted back to the organelle.

The secret of the function of the apicomplexan plastid, therefore, lay in the nucleus. It has now been shown that enzymes for the biosynthesis of fatty acids, isopentyl diphosphate, and heme are all encoded in the *Plasmodium* nuclear genome,[114–118] and they possess bipartite leaders that direct them to the plastid.[118–120] In contrast, *Plasmodium* does not possess plastid-derived enzymes for the biosynthesis of aromatic amino acids: This pathway does exist in *Plasmodium* and other apicomplexa,[121] but it is cytosolic,[122] and the enzymes are not derived from the plastid.[123]

The unexpected discovery of a plastid in these important parasites immediately sparked a debate over their origins. The plastids are certainly a product of secondary endosymbiosis, but whether from a red or green alga has been hotly debated. The first indications from the characteristics of several genes, the gene content, and the order of certain genes in the *Plasmodium* plastid genome were that the plastid was red and not green. However, the phylogeny of the plastid EF-Tu indicated that it was green and not red,[108] although this result was extremely weakly supported. The discovery of the plastid GAPDH gene replacement (see earlier) tipped the scales decisively in favor of a red algal origin for the apicomplexan plastid by demonstrating that the organelle originated long before the apicomplexa, in the ancestor of all chromalveolates.[80] Nevertheless, some recent data have revived the debate over the nature of the apicomplexan plastid and its relationship to the plastids of other chromalveolates.

In virtually all eukaryotes, the mitochondrial protein CoxII is the product of a single open reading frame encoded in the mitochondrial genome. However, in certain green algae the gene has been split roughly in half and transferred to the nucleus, so that these organisms target two subunits of CoxII to their mitochondria.[124,125] It has now been shown that *Plasmodium* and *Toxoplasma* share this rare characteristic, and this has been argued to indicate that the plastid is derived from a green alga.[126] These data are not consistent with the GAPDH data, which indicate that all the chromalveolates shared a single common secondary endosymbiosis with a red alga. To explain both genes, it has been suggested that the apicomplexa ancestrally contained a chromalveolate-like red algal plastid, but that this plastid was lost and replaced with a second plastid of green algal origin.[127]

Such replacements have taken place repeatedly in dinoflagellates (see above), and this would explain both CoxII (which would have come in with the new plastid) and GAPDH (which would have been retained from the original plastid and retargeted to the new one). In either case, the CoxII data demand a remarkable lateral transfer of two subunits from different chromosomes. If this did in fact take place, one would expect the *Plasmodium* genome to be chock-a-block with green algal genes, but it is not.[114] More to the point, the CoxII analysis that led to this conclusion did not consider the CoxII proteins of ciliates, which are critical, as ciliates are closely related to apicomplexa.[126] When the phylogeny was reanalyzed with ciliates included, the apicomplexa and ciliates formed a clade, as expected, if no lateral gene transfer took place.[128] More interestingly still, ciliate CoxII proteins contain a very large (300 amino acid) insertion in exactly the position where the split took place.[129] This could be regarded as a prelude to the split, or at least as evidence that this region of the protein in alveolates is susceptible to change. Considering both the phylogeny and the characteristics of the CoxII proteins, it appears more likely that the split took place twice independently in green algae and apicomplexa. It does not likely represent a lateral gene transfer but, instead, could be a solution to the problem of targeting CoxII: By splitting the protein in two, the hydrophobicity is reduced, allowing the proteins to be successfully translocated across the mitochondrial membranes. At present, no dinoflagellate CoxII has been characterized, but we would predict that they may also be nucleus encoded, and if so, they are likely also split. Altogether, the CoxII data is interesting, but it does little to challenge the red algal ancestry of apicomplexan plastids.[9,80,104,130]

Interestingly, another recent study has also proposed that the apicomplexan and dinoflagellate plastids are not directly related, but in this case it is suggested that the ancestral dinoflagellate plastid was replaced.[131] This hypothesis is based on phylogenies of two photosystem proteins, PsbA and PsaA, which showed the dinoflagellates with "typical" peridinin plastids to be sisters to those that have replaced their plastid with a haptophyte plastid (see above). Together, these two lineages branched as sisters to the haptophytes themselves. The authors suggested that the peridinin plastid was ultimately derived from a haptophyte by a plastid replacement that took place early in the history of dinoflagellates.[131] Once again, the prediction for plastid replacements in dinoflagellates certainly supports this as a possibility, but more evidence will be needed to support such a remarkable conclusion. Indeed, the PsbA/PsaA phylogeny itself does not quite match this scenario, as the dinoflagellate grouping does not branch from within the haptophytes, as one would expect, but as a distant sister to the entire lineage, including *Pavlova*, which is widely recognized as the deepest known lineage among haptophytes.[132] In addition, the divergent nature of both the peridinin- and haptophyte-type dinoflagellate sequences makes accurate phylogenetic inferences difficult.

Prospectus

The last decade has seen extraordinary advances in our understanding of the evolutionary history of plastids. With these advances, however, has come the realization that both the tempo and mode of plastid evolution are far more complex than previously imagined. As in most fields of molecular evolution, the cry from workers in the field is "more data!" What is needed most is nuclear gene sequence data from a wide range of primary, secondary, and tertiary plastid-containing organisms, such that the evolutionary affinities of their respective host cells can better understood and the origin of their nucleus-encoded, plastid-targeted proteins elucidated. Fortunately, in the era of complete genome sequencing, a much more comprehensive understanding of plastid history may now be within reach.

One of the serious challenges that must be faced in coming years is developing a better understanding of the process of plastid loss. Most current models of plastid evolution support the notion that the loss of photosynthesis is a much more pervasive phenomenon than previously believed. For example, if the common ancestor of alveolates, haptophytes, heterokonts, and cryptomonads had a plastid, then the ciliates, an entirely nonphotosynthetic alveolate lineage, must have evolved from a plastid-containing ancestor. It will be important to determine which nonphotosynthetic lineages are truly descended from algal ancestors and to better understand the process by which they have discarded photosynthesis. In particular, it is generally not clear whether plastids are ever actually lost, or whether they are always retained in a cryptic form. Until the process of plastid "loss" is better understood, a full understanding of plastid origins will remain elusive.

Acknowledgments We thank B.S. Leander for helpful comments on the manuscript. This work was supported by a grant from the Canadian Institutes of Health Research (CIHR) to P.J. Keeling. J.M. Archibald is supported by postdoctoral fellowships from CIHR and the Killam Foundation (University of British Columbia). P.J. Keeling is a scholar of the Canadian Institute for Advanced Research, CIHR, and the Michael Smith Foundation for Health Research.

References

1. C. Mereschkowsky, "Über Natur Und Ursprung Der Chromatophoren Im Pflanzenreiche," *Biol. Centralbl.* 25 (1905): 593–604.
2. J. Sapp, *Evolution by Association: A History of Symbiosis* (New York: Oxford University Press, 1994).
3. G.I. McFadden, "Plastids and Protein Targeting," *Journal of Eukaryotic Microbiology* 46 (1999): 339–346.
4. S. Turner, "Molecular Systematics of Oxygenic Photosynthetic Bacteria," *Plant Systematics and Evolution* [*Suppl.*] 11 (1997): 13–52.
5. D.M. Ward, M.J. Ferris, S.C. Nold, and M.M. Bateson, "A Natural View of Microbial Biodiversity within Hot Spring Cyanobacterial Mat Communities," *Microbiology and Molecular Biology Reviews* 62 (1998): 1353–1370.
6. S. Turner, K.M. Pryer, V.P. Miao, and J.D. Palmer, "Investigating Deep Phylogenetic Relationships among Cyanobacteria and Plastids by Small Subunit rRNA Sequence Analysis," *Journal of Eukaryotic Microbiology* 46 (1999): 327–338.
7. B. Nelissen, Y. Van de Peer, A. Wilmotte, and R. De Wachter, "An Early Origin of Plastids within the Cyanobacterial Divergence Is Suggested by Evolutionary Trees Based on Complete 16S rRNA Sequences," *Molecular Biology and Evolution* 12 (1995): 1166–1173.
8. B. Stoebe and K.V. Kowallik, "Gene-Cluster Analysis in Chloroplast Genomics," *Trends in Genetics* 15 (1999): 344–347.
9. G. I. McFadden and R.F. Waller, "Plastids in Parasites of Humans," *Bioessays* 19 (1997): 1033–1040.
10. D.G. Durnford, J.A. Deane, S. Tan, G.I. McFadden, E. Gantt, and B.R. Green, "A Phylogenetic Assessment of the Eukaryotic Light-Harvesting Antenna Proteins, with Implications for Plastid Evolution," *Journal of Molecular Evolution* 48 (1999): 59–68.

11. G.R. Wolfe, F.X. Cunningham, D.G. Durnford, B.R. Green, and E. Gantt, "Evidence for a Common Origin of Chloroplasts with Light-Harvesting Complexes of Different Pigmentation," *Nature* 367 (1995): 566–568.
12. D. Bhattacharya, T. Helmchen, and M. Melkonian, "Molecular Evolutionary Analyses of Nuclear-Encoded Small Subunit Ribosomal RNA Identify an Independent Rhizopod Lineage Containing the Euglyphidae and the Chlorarachniophyta," *Journal of Eukaryotic Microbiology* 42 (1995): 64–68.
13. D. Bhattacharya and K. Weber, "The Actin Gene of the Glaucocystophyte *Cyanophora paradoxa*: Analysis of the Coding Region and Introns, and an Actin Phylogeny of Eukaryotes," *Current Genetics* 31 (1997): 439–446.
14. P.J. Keeling, J.A. Deane, C. Hink-Schauer, S.E. Douglas, U.G. Maier, and G.I. McFadden, "The Secondary Endosymbiont of the Cryptomonad *Guillardia theta* Contains Alpha-, Beta-, and Gamma-Tubulin Genes," *Molecular Biology and Evolution* 16 (1999): 1308–1313.
15. J.W. Stiller and B.D. Hall, "The Origin of Red Algae: Implications for Plastid Evolution," *Proceedings of the National Academy of Sciences USA* 94 (1997): 4520–4525.
16. J.W. Stiller, J. Riley, and B.D. Hall, "Are Red Algae Plants? A Critical Evaluation of Three Key Molecular Data Sets," *Journal of Molecular Evolution* 52 (2001): 527–539.
17. J.M. Archibald, T. Cavalier-Smith, U. Maier, and S. Douglas, "Molecular Chaperones Encoded by a Reduced Nucleus—The Cryptomonad Nucleomorph," *Journal of Molecular Evolution* 52 (2001): 490–501.
18. P.J. Keeling and J.D. Palmer, "Lateral Transfer at the Gene and Subgenic Levels in the Evolution of Eukaryotic Enolase," *Proceedings of the National Academy of Sciences USA* 98 (2001): 10745–10750.
19. D. Moreira, H. Le Guyader, and H. Phillippe, "The Origin of Red Algae and the Evolution of Chloroplasts," *Nature* 405 (2000): 69–72.
20. D. Bhattacharya and H.A. Schmidt, "Division Glaucocystophyta," in D. Bhattacharya, ed., *Origin of Algae and Their Plastids* (New York: Springer, 1997), 139–148.
21. M. Herdman and R. Stanier, "The Cyanelle: Chloroplast or Endosymbiotic Procaryote?" *FEMS Microbiological Letters* 1 (1977): 7–12.
22. T. Cavalier-Smith, "The Origins of Plastids," *Biological Journal of the Linnean Society* 17 (1982): 289–306.
23. C.F. Delwiche, M. Kuhsel, and J.D. Palmer, "Phylogenetic Analysis of *tuf*A Sequences Indicates a Cyanobacterial Origin of All Plastids," *Molecular Phylogenetics and Evolution* 4 (1995): 110–128.
24. K.V. Kowallik, "Origin and Evolution of Chloroplasts: Current Status and Future Perspectives," in H.E. Schenk, R.G. Herrmann, K.W. Jeon, N.E. Müller, and W. Schwemmler, eds., *Eukaryotism and Symbiosis: Intertaxonic Combination versus Symbiotic Adaptation* (Berlin: Springer, 1997), 3–23.
25. W. Martin, B. Stoebe, V. Goremykin, S. Hansmann, M. Hasegawa, and K.V. Kowallik, "Gene Transfer to the Nucleus and the Evolution of Chloroplasts," *Nature* 393 (1998): 162–165.
26. K. Valentin and K. Zetsche, "Nucleotide Sequence of the Gene for the Large Subunit of Rubisco from *Cyanophora paradoxa*—Phylogenetic Implications," *Current Genetics* 18 (1990): 199–202.
27. C.F. Delwiche, and J.D. Palmer, "Rampant Horizontal Transfer and Duplication of Rubisco Genes in Eubacteria and Plastids," *Molecular Biology and Evolution* 13 (1996): 873–882.
28. W. Martin, T. Rujan, E. Richly, A. Hansen, S. Cornelsen, T. Lins, D. Leister, B. Stoebe, M. Hasegawa, and D. Penny, "Evolutionary Analysis of *Arabidopsis*, Cyanobacterial, and Chloroplast Genomes Reveals Plastid Phylogeny and Thousands of Cyanobacterial Genes in the Nucleus," *Proceedings of the National Academy of Sciences USA* 99 (2002): 12246–12251.
29. H. Brinkmann and W. Martin, "Higher Plant Chloroplast and Cytosolic 3-Phosphoglycerate Kinases: A Case of Endosymbiotic Gene Replacement," *Plant Molecular Biology* 30 (1996): 65–75.
30. J.M. Archibald and P.J. Keeling, "Plant Genomes: Cyanobacterial Genes Revealed," *Heredity* 90 (2003): 2–3.
31. D.J. Hibberd and R.E. Norris, "Cytology and Ultrastructure of *Chlorarachnion Reptans*

(Chlorarachniophyta Divisio Nova, Chlorarachniophyceae Classis Nova)," *Journal of Phycology* 20 (1984): 310–330.

32. M. Ludwig and S.P. Gibbs, "Evidence That Nucleomorphs of *Chlorarachnion reptans* (Chlorarachniophyceae) Are Vestigial Nuclei: Morphology, Division and DNA-Dapi Fluorescence," *Journal of Phycology* 25 (1989): 385–394.
33. P.R. Gilson and G.I. McFadden, "The Miniaturized Nuclear Genome of a Eukaryotic Endosymbiont Contains Genes That Overlap, Genes That Are Cotranscribed, and the Smallest Known Spliceosomal Introns," *Proceedings of the National Academy of Sciences USA* 93 (1996): 7737–7742.
34. P.R. Gilson, U.G. Maier, and G.I. McFadden, "Size Isn't Everything: Lessons in Genetic Miniaturisation from Nucleomorphs," *Current Opinion in Genetics and Development* 7 (1997): 800–806.
35. P.R. Gilson and G.I. McFadden, "Jam Packed Genomes—a Preliminary, Comparative Analysis of Nucleomorphs," *Genetica* 115 (2002): 13–28.
36. S.P. Gibbs, "The Chloroplasts of *Euglena* May Have Evolved from Symbiotic Green Algae," *Canadian Journal of Botany* 56 (1978): 2883–2889.
37. K. Ishida, Y. Cao, M. Hasegawa, N. Okada, and Y. Hara, "The Origin of Chlorarachniophyte Plastids, as Inferred from Phylogenetic Comparisons of Amino Acid Sequences of EF-Tu," *Journal of Molecular Evolution* 45 (1997): 682–687.
38. K. Ishida, B.R. Green, and T. Cavalier-Smith, "Diversification of a Chimaeric Algal Group, the Chlorarachniophytes: Phylogeny of Nuclear and Nucleomorph Small-Subunit rRNA Genes," *Molecular Biology and Evolution* 16 (1999): 321–331.
39. G.I. McFadden, P.R. Gilson, and R.F. Waller, "Molecular Phylogeny of Chlorarachniophytes Based on Plastid rRNA and *rbcL* Sequences," *Archiv für Protistenkunde* 145 (1995): 231–239.
40. M. Turmel, C. Otis, and C. Lemieux, "The Complete Chloroplast DNA Sequence of the Green Alga *Nephroselmis olivacea*: Insights into the Architecture of Ancestral Chloroplast Genomes," *Proceedings of the National Academy of Sciences USA* 96 (1999): 10248–10253.
41. Y. Van de Peer, S.A. Rensing, and U.-G. Maier, "Substitution Rate Calibration of Small Subunit Ribosomal RNA Identifies Chlorarachniophyte Endosymbionts as Remnants of Green Algae," *Proceedings of the National Academy of Sciences USA* 93 (1996): 7732–7736.
42. R.B. Hallick, L. Hong, R.G. Drager, M.R. Favreau, A. Monfort, B. Orsat, A. Spielmann, and E. Stutz, "Complete Sequence of *Euglena gracilis* Chloroplast DNA," *Nucleic Acids Research* 21 (1993): 3537–3544.
43. S.E. Douglas, S. Zauner, M. Fraunholz, M. Beaton, S. Penny, L. Deng, X. Wu, M. Reith, T. Cavalier-Smith, and U.-G. Maier, "The Highly Reduced Genome of an Enslaved Algal Nucleus," *Nature* 410 (2001): 1091–1096.
44. S.P. Gibbs, "The Chloroplast Endoplasmic Reticulum: Structure, Function, and Evolutionary Significance," *International Review of Cytology* 72 (1981): 49–99.
45. K.V. Kowallik, B. Stoebe, I. Schaffran, P. Kroth-Pancic, and U. Freier, "The Chloroplast Genome of a Chlorophyll *a+c*-Containing Alga, *Odontella sinensis*," *Plant Molecular Biology Reporter* 13 (1995): 336–342.
46. S.E. Douglas and S.L. Penny, "The Plastid Genome from the Cryptomonad Alga, *Guillardia theta*: Complete Sequence and Conserved Synteny Groups Confirm Its Common Ancestry with Red Algae," *Journal of Molecular Evolution* 48 (1999): 236–244.
47. C.F. Delwiche, "Tracing the Thread of Plastid Diversity through the Tapestry of Life," *American Naturalist* [*Supplement*] 154 (1999): S164–S177.
48. N. Daugbjerg and R.A. Andersen, "Phylogenetic Analyses of the *rbcL* Sequences from Haptophytes and Heterokont Algae Suggest Their Chloroplasts Are Unrelated," *Molecular Biology and Evolution* 14 (1997): 1242–1251.
49. Y. Van de Peer and R. De Wachter, "Evolutionary Relationships among Eukaryotic Crown Taxa Taking into Account Site-to-Site Variation in 18s rRNA," *Journal of Molecular Evolution* 45 (1997): 619–630.
50. G. Van der Auwera, C.J.B. Hofmann, P. De Rijk, and R. De Wachter, "The Origin of Red Algae and Cryptomonad Nucleomorphs: A Comparative Phylogeny Based on Small and Large

Subunit rRNA Sequences of *Palmaria palmata*, *Gracilaria verrucosa*, and the *Guillardia theta* Nucleomorph," *Molecular Phylogenetics and Evolution* 10 (1998): 333–342.
51. H.S. Yoon, J.D. Hackett, G. Pinto, and D. Bhattacharya, "The Single, Ancient Origin of Chromist Plastids," *Proceedings of the National Academy of Sciences USA* 99 (2002): 15507–15512.
52. J. Wolters, "The Troublesome Parasites: Molecular and Morphological Evidence That Apicomplexa Belong to the Dinoflagellate-Ciliate Clade," *Biosystems* 25 (1991): 75–84.
53. G.I. McFadden, M. Reith, J. Munholland, and N. Lang-Unnasch, "Plastid in Human Parasites," *Nature* 381 (1996): 482.
54. R.J.M.I. Wilson, P.W. Denny, D.J. Preiser, K. Rangachari, K. Roberts, A. Roy, A. Whyte, M. Strath, D.J. Moore, P.W. Moore, and D.H. Williamson, "Complete Gene Map of the Plastid-Like DNA of the Malaria Parasite *Plasmodium falciparum*," *Journal of Molecular Biology* 261 (1996): 155–172.
55. A.W.D. Larkum, C. Scaramuzzi, G.C. Cox, R.G. Hiller, and A.G. Turner, "Light-Harvesting Chlorophyll C-Like Pigment in *Prochloron*," *Proceedings of the National Academy of Sciences USA* 91 (1994): 679–683.
56. K. Takishita and A. Uchida, "Molecular Cloning and Nucleotide Sequence Analysis of *psbA* from the Dinoflagellates: Origin of the Dinoflagellate Plastid," *Phycological Research* 47 (1999): 207–216.
57. Z. Zhang, B.R. Green, and T. Cavalier-Smith, "Single Gene Circles in Dinoflagellate Chloroplast Genomes," *Nature* 400 (1999): 155–159.
58. Z. Zhang, B.R. Green, and T. Cavalier-Smith, "Phylogeny of Ultra-Rapidly Evolving Dinoflagellate Chloroplast Genes: A Possible Common Origin for Sporozoan and Dinoflagellate Plastids," *Journal of Molecular Evolution* 51 (2000): 26–40.
59. T. Cavalier-Smith, "Principles of Protein and Lipid Targeting in Secondary Symbiogenesis: Euglenoid, Dinoflagellate, and Sporozoan Plastid Origins and the Eukaryote Family Tree," *Journal of Eukaryotic Microbiology* 46 (1999): 347–366.
60. T. Cavalier-Smith, "Membrane Heredity and Early Chloroplast Evolution," *Trends in Plant Science* 5 (2000): 174–182.
61. D. Bhattacharya and M. Melkonian, "The Phylogeny of Plastids: A Review Based on Comparisons of Small-Subunit Ribosomal RNA Coding Regions," *Journal of Phycology* 31 (1995): 489–498.
62. C.F. Delwiche and J.D. Palmer, "The Origin of Plastids and Their Spread Via Secondary Endosymbiosis," in D. Bhattacharya, ed., *Origins of Algae and Their Plastids* (New York: Springer, 1997), 53–86.
63. T. Cavalier-Smith and E.E. Chao, "Sarcomonad Ribosomal RNA Sequences, Rhizopod Phylogeny, and the Origin of Euglyphid Amoebae," *Archiv für Protistenkunde* 147 (1997): 227–236.
64. T. Cavalier-Smith, "A Revised Six-Kingdom System of Life," *Biological Reviews of the Cambridge Philosophical Society* 73 (1998): 203–266.
65. P.J. Keeling, J.A. Deane, and G.I. McFadden, "The Phylogenetic Position of Alpha- and Beta-Tubulins from the *Chlorarachnion* Host and *Cercomonas* (Cercozoa)," *Journal of Eukaryotic Microbiology* 45 (1998): 561–570.
66. J.M. Archibald, D. Longet, J. Pawlowski, and P.J. Keeling, "A Novel Polyubiquitin Structure in Cercozoa and Foraminifera: Evidence for a New Eukaryotic Supergroup," *Molecular Biology and Evolution* 20 (2003): 62–66.
67. P.J. Keeling, "Foraminifera and Cercozoa Are Related in Actin Phylogeny: Two Orphans Find a Home?" *Molecular Biology and Evolution* 18 (2001): 1551–1557.
68. J.M. Archibald and P.J. Keeling, "Recycled Plastids: A Green Movement in Eukaryotic Evolution," *Trends in Genetics* 18 (2002): 577–584.
69. A. Preisfeld, I. Busse, M. Klingberg, S. Talke, and H.G. Ruppel, "Phylogenetic Position and Inter-Relationships of the Osmotrophic Euglenids Based on SSU rDNA Data, with Emphasis on the Rhabdomonadales (Euglenozoa)," *International Journal of Systematic and Evolutionary Microbiology* 51 (2001): 751–758.
70. B.S. Leander, R.E. Triemer, and M.A. Farmer, "Character Evolution in Heterotrophic Euglenids," *European Journal of Protistology* 37 (2001): 337–356.

71. B.S. Leander, R.P. Witek, and M.A. Farmer, "Trends in the Evolution of the Euglenid Pellicle," *Evolution* 55 (2001): 2215–2235.
72. V. Hannaert, E. Saavedra, F. Duffieux, J.-P. Szikora, D.J. Rigden, P.A.M. Michels, and F.R. Opperdoes, "Plant-Like Traits Associated with Metabolism of *Trypanosoma* Parasites," *Proceedings of the National Academy of Sciences USA* 100 (2003): 1067–1071.
73. W. Martin and P. Borst, "Secondary Loss of Chloroplasts in Trypanosomes," *Proceedings of the National Academy of Sciences USA* 100 (2003): 765–767.
74. T. Cavalier-Smith, "Eukaryote Kingdoms: Seven or Nine?" *Biosystems* 14 (1981): 461–481.
75. T. Cavalier-Smith, "The Kingdom Chromista: Origin and Systematics," *Progress in Phycological Research* 4 (1986): 309–347.
76. L.K. Medlin, A. Cooper, C. Hill, S. Wrieden, and U. Wellbrock, "Phylogenetic Position of the Chromista Plastids Based on Small Subunit rRNA Coding Regions," *Current Genetics* 28 (1995): 560–565.
77. K.M. Müller, M.C. Oliveira, R.G. Sheath, and D. Bhattacharya, "Ribosomal DNA Phylogeny of the Bangiophycidae (Rhodophyta) and the Origin of Secondary Plastids," *American Journal of Botany* 88 (2001): 1390–1400.
78. M.C. Oliveira and D. Bhattacharya, "Phylogeny of the Bangiophycidae (Rhodophyta) and the Secondary Endosymbiotic Origin of Algal Plastids," *American Journal of Botany* 87 (2000): 482–492.
79. D. Bhattacharya, T. Helmchen, C. Bibeau, and M. Melkonian, "Comparison of Nuclear-Encoded Small-Subunit Ribosomal RNAs Reveal the Evolutionary Position of the Glaucocystophyta," *Molecular Biology and Evolution* 12 (1995): 415–420.
80. N.M. Fast, J.C. Kissinger, D.S. Roos, and P.J. Keeling, "Nuclear-Encoded, Plastid-Targeted Genes Suggest a Single Common Origin for Apicomplexan and Dinoflagellate Plastids," *Molecular Biology and Evolution* 18 (2001): 418–426.
81. J.T. Harper and P.J. Keeling, "Nucleus-Encoded, Plastid-Targeted Glyceraldehyde-3-Phosphate Dehydrogenase (GAPDH) Indicates a Single Origin for Chromalveolate Plastids," *Molecular Biology and Evolution* (in press).
82. W. Martin and C. Schnarrenberger, "The Evolution of the Calvin Cycle from Prokaryotic to Eukaryotic Chromosomes: A Case Study of Functional Redundancy in Ancient Pathways through Endosymbiosis," *Current Genetics* 32 (1997): 1–18.
83. T.M. Fagan and J.W. Hastings, "Phylogenetic Analysis Indicates Multiple Origins of Chloroplast Glyceraldehyde-3–Phosphate Dehydrogenase Genes in Dinoflagellates," *Molecular Biology and Evolution* 19 (2002): 1203–1207.
84. A. Ben Ali, R. De Baere, G. Van der Auwera, R. De Wachter, and Y. Van de Peer, "Phylogenetic Relationships among Algae Based on Complete Large-Subunit rRNA Sequences," *International Journal of Systematic and Evolutionary Microbiology* 51 (2001): 737–749.
85. S.L. Baldauf, A.J. Roger, I. Wenk-Siefert, and W.F. Doolittle, "A Kingdom-Level Phylogeny of Eukaryotes Based on Combined Protein Data," *Science* 290 (2000): 972–977.
86. F.J.R. Taylor, *The Biology of Dinoflagellates* (Oxford: Blackwell Scientific Publications, 1987).
87. M.A. Farmer and K.R. Roberts, "Organelle Loss in the Endosymbiont of *Gymnodinium acidotum* (Dinophyceae)," *Protoplasma* 153 (1990): 178–185.
88. L.W. Wilcox and G.J. Wedemayer, "Dinoflagellate with Blue-Green Chloroplasts Derived from an Endosymbiotic Eukaryote," *Science* 227 (1985): 192–194.
89. J.F. Saldarriaga, F.J.R. Taylor, P.J. Keeling, and T. Cavalier-Smith, "Dinoflagellate Nuclear SSU rRNA Phylogeny Suggests Multiple Plastid Losses and Replacements," *Journal of Molecular Evolution* 53 (2001): 204–213.
90. N. Daugbjerg, G. Hansen, J. Larsen, and O. Moestrup, "Phylogeny of Some of the Major Genera of Dinoflagellates Based on Ultrastructure and Partial LSU rDNA Sequence Data, Including the Erection of Three New Genera of Unarmoured Dinoflagellates," *Phycologia* 39 (2000): 302–317.
91. G. Hansen, N. Daugbjerg, and P. Henriksen, "Comparative Study of *Gymnodinium mikimotoi* and *Gymnodinium aureolum*, Comb. Nov (= *Gyrodinium aureolum*) Based on Morphology, Pigment Composition, and Molecular Data," *Journal of Phycology* 36 (2000): 394–410.

92. T. Tengs, O.J. Dahlberg, K. Shalchian-Tabrizi, D. Klaveness, K. Rudi, C.F. Delwiche, and K.S. Jakobsen, "Phylogenetic Analyses Indicate That the 19'hexanoyloxy-Fucoxanthin-Containing Dinoflagellates Have Tertiary Plastids of Haptophyte Origin," *Molecular Biology and Evolution* 17 (2000): 718–729.
93. J.D. Hackett, L. Maranda, H.S. Yoon, and D. Bhattacharya, "Phylogenetic Evidence for the Cryptophyte Origin of the Plastid of *Dinophysis* (Dinophysiales, Dinophyceae)," *Journal of Phycology* 39 (in press).
94. J.M. Chesnick, W.H. Hooistra, U. Wellbrock, and L.K. Medlin, "Ribosomal RNA Analysis Indicates a Benthic Pennate Diatom Ancestry for the Endosymbionts of the Dinoflagellates *Peridinium foliaceum* and *Peridinium balticum* (Pyrrhophyta)," *Journal of Eukaryotic Microbiology* 44 (1997): 314–320.
95. M.M. Watanabe, S. Suda, I. Inouye, I. Sawaguchi, and M. Chihara, "*Lepidodinium viride* gen et sp. nov. (Gymnodiniales, Dinophyta), a Green Dinoflagellate with a Chlorophyll *a*- and *b*-Containing Endosymbiont," *Journal of Phycology* 26 (1990): 741–751.
96. D.K. Stoecker, "Mixotrophy among Dinoflagellates," *Journal of Eukaryotic Microbiology* 46 (1999): 397–401.
97. K. Ishida and B.R. Green, "Second- and Third-Hand Chloroplasts in Dinoflagellates: Phylogeny of Oxygen-Evolving Enhancer 1 (psbO) Protein Reveals Replacement of a Nuclear-Encoded Plastid Gene by That of a Haptophyte Tertiary Endosymbiont," *Proceedings of the National Academy of Sciences USA* 99 (2002): 9294–9299.
98. A. Kilejian, "Circular Mitochondrial DNA from the Avian Malarial Parasite *Plasmodium lophurae*," *Biochimica et Biophysica Acta* 390 (1975): 267–284.
99. M.J. Gardner, P.A. Bates, I.T. Ling, D.J. Moore, S. McCready, M.B. Gunasekera, R.J. Wilson, and D.H. Williamson, "Mitochondrial DNA of the Human Malarial Parasite *Plasmodium falciparum*," *Molecular and Biochemical Parasitology* 31 (1988): 11–17.
100. M.J. Gardner, J.E. Feagin, D.J. Moore, D.F. Spencer, M.W. Gray, D.H. Williamson, and R.J. Wilson, "Organisation and Expression of Small Subunit Ribosomal RNA Genes Encoded by a 35-Kilobase Circular DNA in *Plasmodium falciparum*," *Molecular and Biochemical Parasitology* 48 (1991): 77–88.
101. M.J. Gardner, D.H. Williamson, and R.J. Wilson, "A Circular DNA in Malaria Parasites Encodes an RNA Polymerase Like That of Prokaryotes and Chloroplasts," *Molecular and Biochemical Parasitology* 44 (1991): 115–123.
102. M.J. Gardner, J.E. Feagin, D.J. Moore, K. Rangachari, D.H. Williamson, and R.J. Wilson, "Sequence and Organization of Large Subunit rRNA Genes from the Extrachromosomal 35 Kb Circular DNA of the Malaria Parasite *Plasmodium falciparum*," *Nucleic Acids Research* 21 (1993): 1067–1071.
103. M.J. Gardner, N. Goldman, P. Barnett, P.W. Moore, K. Rangachari, M. Strath, A. Whyte, D.H. Williamson, and R.J. Wilson, "Phylogenetic Analysis of the *rpo*B Gene from the Plastid-Like DNA of *Plasmodium Falciparum*," *Molecular and Biochemical Parasitology* 66 (1994): 221–231.
104. D.H. Williamson, M.J. Gardner, P. Preiser, D.J. Moore, K. Rangachari, and R.J. Wilson, "The Evolutionary Origin of the 35 Kb Circular DNA of *Plasmodium falciparum*: New Evidence Supports a Possible Rhodophyte Ancestry," *Molecular and General Genetics* 243 (1994): 249–252.
105. J.E. Feagin, M.J. Gardner, D.H. Williamson, and R.J. Wilson, "The Putative Mitochondrial Genome of *Plasmodium falciparum*," *Journal of Protozoology* 38 (1991): 243–245.
106. A.B. Vaidya, R. Akella, and K. Suplick, "Sequences Similar to Genes for Two Mitochondrial Proteins and Portions of Ribosomal RNA in Tandemly Arrayed 6–Kilobase-Pair DNA of a Malarial Parasite," *Molecular and Biochemical Parasitology* 35 (1989): 97–107.
107. R.J. Wilson, M. Fry, M.J. Gardner, J.E. Feagin, and D.H. Williamson, "Subcellular Fractionation of the Two Organelle DNAs of Malaria Parasites," *Current Genetics* 21 (1992): 405–408.
108. S. Köhler, C.F. Delwiche, P.W. Denny, L.G. Tilney, P. Webster, R.J.M. Wilson, J.D. Palmer, and D.S. Roos, "A Plastid of Probable Green Algal Origin in Apicomplexan Parasites," *Science* 275 (1997): 1485–1489.

109. M.E. Fichera and D.S. Roos, "A Plastid Organelle as a Drug Target in Apicomplexan Parasites," *Nature* 390 (1997): 407–409.
110. C.Y. He, M.K. Shaw, C.H. Pletcher, B. Striepen, L.G. Tilney, and D.S. Roos, "A Plastid Segregation Defect in the Protozoan Parasite *Toxoplasma gondii*," *EMBO Journal* 20 (2001): 330–339.
111. J.L. Harwood, "Recent Advances in the Biosynthesis of Plant Fatty Acids," *Biochimica et Biophysica Acta* 1301 (1996): 7–56.
112. F. Rohdich, K. Kis, A. Bacher, and W. Eisenreich, "The Non-Mevalonate Pathway of Isoprenoids: Genes, Enzymes and Intermediates," *Current Opinion in Chemical Biology* 5 (2001): 535–540.
113. K.M. Herrmann, "The Shikimate Pathway as an Entry to Aromatic Secondary Metabolism," *Plant Physiology* 107 (1995): 7–12.
114. M.J. Gardner, N. Hall, E. Fung, O. White, M. Berriman, R.W. Hyman, J.M. Carlton, A. Pain, K.E. Nelson, S. Bowman, I.T. Paulsen, K. James, J.A. Eisen, K. Rutherford, S.L. Salzberg, A. Craig, S. Kyes, M.S. Chan, V. Nene, S.J. Shallom, B. Suh, J. Peterson, S. Angiuoli, M. Pertea, J. Allen, J. Selengut, D. Haft, M.W. Mather, A.B. Vaidya, D.M. Martin, A.H. Fairlamb, M.J. Fraunholz, D.S. Roos, S.A. Ralph, G.I. McFadden, L.M. Cummings, G.M. Subramanian, C. Mungall, J.C. Venter, D.J. Carucci, S.L. Hoffman, C. Newbold, R.W. Davis, C.M. Fraser, and B. Barrell, "Genome Sequence of the Human Malaria Parasite *Plasmodium falciparum*," *Nature* 419 (2002): 498–511.
115. H. Jomaa, J. Wiesner, S. Sanderbrand, B. Altincicek, C. Weidemeyer, M. Hintz, I. Turbachova, M. Eberl, J. Zeidler, H.K. Lichtenthaler, D. Soldati, and E. Beck, "Inhibitors of the Nonmevalonate Pathway of Isoprenoid Biosynthesis as Antimalarial Drugs," *Science* 285 (1999): 1573–1576.
116. S. Sato and R. J. Wilson, "The Genome of *Plasmodium falciparum* Encodes an Active Delta-Aminolevulinic Acid Dehydratase," *Current Genetics* 40 (2002): 391–398.
117. S.A. Ralph, M.C. D'Ombrain, and G.I. McFadden, "The Apicoplast as an Antimalarial Drug Target," *Drug Resistance Updates* 4 (2001): 145–151.
118. R.F. Waller, P.J. Keeling, R.G. Donald, B. Striepen, E. Handman, N. Lang-Unnasch, A.F. Cowman, G.S. Besra, D.S. Roos, and G.I. McFadden, "Nuclear-Encoded Proteins Target to the Plastid in *Toxoplasma gondii* and *Plasmodium falciparum*," *Proceedings of the National Academy of Sciences USA* 95 (1998): 12352–12357.
119. G.G. van Dooren, V. Su, M.C. D'Ombrain, and G.I. McFadden, "Processing of an Apicoplast Leader Sequence in *Plasmodium falciparum* and the Identification of a Putative Leader Cleavage Enzyme," *Journal of Biological Chemistry* 277 (2002): 23612–23619.
120. R.F. Waller, M.B. Reed, A.F. Cowman, and G.I. McFadden, "Protein Trafficking to the Plastid of *Plasmodium falciparum* is via the Secretory Pathway," *EMBO Journal* 19 (2000): 1794–1802.
121. F. Roberts, C.W. Roberts, J.J. Johnson, D.E. Kyle, T. Krell, J.R. Coggins, G.H. Coombs, W.K. Milhous, S. Tzipori, D.J. Ferguson, D. Chakrabarti, and R. McLeod, "Evidence for the Shikimate Pathway in Apicomplexan Parasites," *Nature* 393 (1998): 801–805.
122. T. Fitzpatrick, S. Ricken, M. Lanzer, N. Amrhein, P. Macheroux, and B. Kappes, "Subcellular Localization and Characterization of Chorismate Synthase in the Apicomplexan *Plasmodium falciparum*," *Molecular Microbiology* 40 (2001): 65–75.
123. P.J. Keeling, J.D. Palmer, R.G. Donald, D.S. Roos, R.F. Waller, and G.I. McFadden, "Shikimate Pathway in Apicomplexan Parasites," *Nature* 397 (1999): 219–220.
124. X. Perez-Martinez, S. Funes, E. Tolkunova, E. Davidson, M.P. King, and D. Gonzalez-Halphen, "Structure of Nuclear-Localized cox3 Genes in *Chlamydomonas reinhardtii* and in Its Colorless Close Relative *Polytomella* Sp.," *Current Genetics* 40 (2002): 399–404.
125. X. Perez-Martinez, A. Antaramian, M. Vazquez-Acevedo, S. Funes, E. Tolkunova, J. d'Alayer, M.G. Claros, E. Davidson, M.P. King, and D. Gonzalez-Halphen, "Subunit II of Cytochrome *c* Oxidase in Chlamydomonad Algae Is a Heterodimer Encoded by Two Independent Nuclear Genes," *Journal of Biological Chemistry* 276 (2001): 11302–11309.
126. S. Funes, E. Davidson, A. Reyes-Prieto, S. Magallón, P. Herion, M.P. King, and D. Gonzalez-Halphen, "A Green Algal Apicoplast Ancestor," *Science* 298 (2002): 2155.

127. J.D. Palmer, "The Symbiotic Birth and Spread of Plastids: How Many Times and Whodunnit?" *Journal of Phycology* 39 (2003): 4–11.
128. R.F. Waller, P.J. Keeling, G.G. van Dooren, and G.I. McFadden, "Comment on 'a Green Algal Apicoplast Ancestor'," *Science* 301 (2003): 49a.
129. G. Burger, Y. Zhu, T.G. Littlejohn, S.J. Greenwood, M.N. Schnare, B.F. Lang, and M.W. Gray, "Complete Sequence of the Mitochondrial Genome of *Tetrahymena pyriformis* and Comparison with *Paramecium aurelia* Mitochondrial DNA," *Journal of Molecular Biology* 297 (2000): 365–380.
130. J.L. Blanchard and J.S. Hicks, "The Non-Photosynthetic Plastid in Malarial Parasites and Other Apicomplexans Is Derived from Outside the Green Plastid Lineage," *Journal of Eukaryotic Microbiology* 46 (1999): 367–375.
131. H.S. Yoon, J.D. Hackett, and D. Bhattacharya, "A Single Origin of the Peridinin- and Fucoxanthin-Containing Plastids in Dinoflagellates through Tertiary Endosymbiosis," *Proceedings of the National Academy of Sciences USA* 99 (2002): 11724–11729.
132. B. Edvardsen, W. Eikrem, J.C. Green, R. Andersen, S.Y. Moon-van der Staay, and L.K. Medlin, "Phylogenetic Reconstructions of the Haptophyta Inferred from 18S Ribosomal DNA Sequences and Available Morphological Data," *Phycologia* 39 (2000): 19–35.

13

The Karyomastigont Model of Eukaryosis

HANNAH MELNITSKY
FREDERICK A. RAINEY
LYNN MARGULIS

"Eukaryosis" refers to the origin of eukaryotic cells: The evolution of the first organisms with membrane-bounded nuclei and, presumably, associated cytoskeleton for storage and distribution of genetic material. Eukaryotes, as acritarchs and Ediacarans, abound in the late Proterozoic Eon between 1,000 and 542 mya,[1] and other fossils document eukaryosis even before 1,200 mya.[2] Molecular evidence indicates that the earliest protists may have appeared as long as 2,700 mya.[3] The status for the search of the first eukaryotes in the pre-Phanerozoic fossil record was accessibly reviewed.[4]

The origin of mitosis and of other aspects of the microtubule-based cytoskeleton remains a mystery. The two universal components of eukaryotes, the nucleus and the cytoskeletal-based motility system for segregation of nuclear DNA, irreducibly coevolved. Universally associated with nuclear division (karyokinesis) is a paradesmose or other form of mitotic spindle. The ubiquitous cytoskeleton, of which the mitotic apparatus is a part, is composed of hundreds of proteins: Its development requires a large, complex genome. Analogous data to those that established the origin of mitochondria and plastids from bacteria by symbiogenesis (cortical inheritance in ciliates, behavior of symbiotic bacteria) led us to the concept of the symbiogenetic origin of the nucleated cell with its internal motility system.[5–7]

The Karyomastigont Model

The organellar system essential to our evolutionary scheme for the origin of the nucleus was first described in 1915 by Polish parasitologist C. Janicki. The karyomastigont organellar system[8] that he described is conspicuous in many cells (fig. 13.1). Minimally,

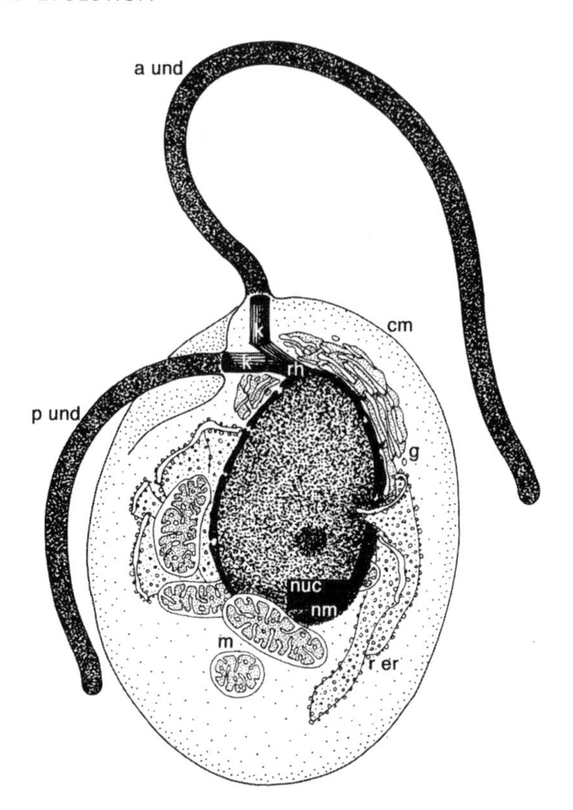

Figure 13.1. The karyomastigont. Generalized drawing of a karyomastigont in a heterokont cell, based on electron micrographs of *Proteromonas* (Phylum Zoomastigina, Class Proteromonadida), with the cell shape of a zoospore of *Ophiocytium* (a xanthophyte). Note the anterior undulipodium (**a und**) is longer than the posterior undulipodium (**p und**). The rhizoplast (or nuclear connector; **rh**) connects the kinetosomes (**k**) to the nuclear membrane (**nm**) of the nucleus (**nuc**). cm, cell membrane; **g**, Golgi; **m**, mitochondrion; **rer**, rough endoplasmic reticulum. Drawing by Kathryn Delisle.

the system consists of the nucleus, the undulipodium (cilium or "eukaryotic flagellum") with its microtubular axoneme that invariably develops from a centriole-kinetosome ("basal body"), and a proteinaceous connector, also called a rhizoplast, that joins them.

The karyomastigont represents to us a relic of the earliest nucleated cell morphology. Karyomastigonts are widespread in many protist taxa: Archaeprotista (amitochondriate protists), Chlorophyta (green algae), Chytridiomycota (i.e., *Blastocladiella*), Cryptomonadida, Dinomastigota (both biundulipodiate algae and heterotrophs), Zoomastigota (motile, mitochondriate protists), and Granuloreticulosa (in the motile foraminiferan cells).[9,10] In our view, in several lineages the tethered nucleus was released from the rest of the karyomastigont to generate both the akaryomastigonts of parabasalids such as *Calonympha* and *Snyderella* and solitary nuclei of plant, fungal, and animal cells—protoctists such as plasmodial slime molds, amoebae, and ciliates.[11,12]

We hypothesize that the eukaryotic cell evolved from a symbiotic consortium of *Spirochaeta*-like eubacteria with archaebacteria that resembled extant *Thermoplasma* (fig. 13.2).[13] The spirochetes presumably swam in search of nutrients in habitats reminiscent of intestinal fluids of extant xylophagous dictyopterans (wood-ingesting roaches and termites): sulifidic, viscous, and replete with organic compounds. Comparable modern environments support syntrophic consortia of prokaryotes analogous to those hypothesized in our model for eukaryosis. In "Thiodendron" sulfur mats, for instance, sulfidogenic bacteria *Dethiosulfovibrio*[14] join spirochetes of the genus *Spirochaeta*[15,16]

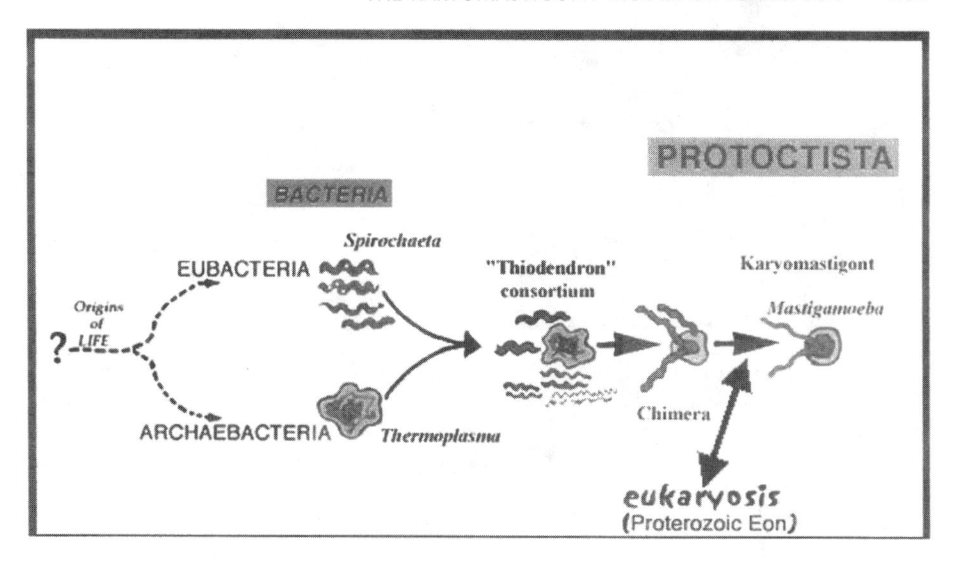

Figure 13.2. The karyomastigont model of eukaryosis. Left to right as a function of time from 4.0 until 2.0×10^9 years ago. Examples of modern codescendant genera in italics. Events depicted before origin of mitochondria and chloroplasts. See text and Margulis and Schwartz[9] for details. Drawing by Kathryn Delisle.

in stable associations, visible to the unaided eye in geochemical hot springs and marine intertidal habitats as white mats (fig. 13.3).

The 16S rRNA genes of the putative spirochete partner of three "Thiodendron" consortia organisms, isolated separately by Galina Dubinina and her colleagues, were sequenced by previously described methods.[17] These data indicate that although spirochetes are diverse with respect to habitat, optimal temperature, and pH for growth and range of substrate use, they represent a phylogenetically unified group. They form a distinct lineage at the phylum level within the prokaryotes.[18] The phylogenetic placement of these strains originating from the "Thiodendron" mat within the spirochete lineage of the bacteria and their close relationship to members of the genus *Spirochaeta* confirms the morphological observations (fig. 13.4).[15] Furthermore, the differences in 16S rRNA genes of the three "Thiodendron" isolates indicate independently evolved consortia in at least three of the six geographically isolated consortia.[16]

We hypothesize that spirochetes in syntrophic consortia attached themselves to sulfidogens comparable to *Dethiosulfovibrio* or *Thermoplasma*, on whose exudates they fed. Spirochete attachment structures evolved then as they have many times in contemporary organisms. Some spirochetes aggressively and permanently penetrated their archaebacterial consorts. Syntrophy, growth, and regulated reproduction rates were selected for as members of the consortia became progressively more integrated. Motility was retained as integration led to metabolic chimeras.[5,6] The ligation structure, at first a "spirochete attachment site,"[6] was selected to become an organellar system that ensured synchronous DNA inheritance and segregation in now-fused partners. This organellar system is still detectable in many organisms—cells and animals such as the choanoflagellate cells of sponges and the choanomastigotes.[9] Syntrophy, catabolism,

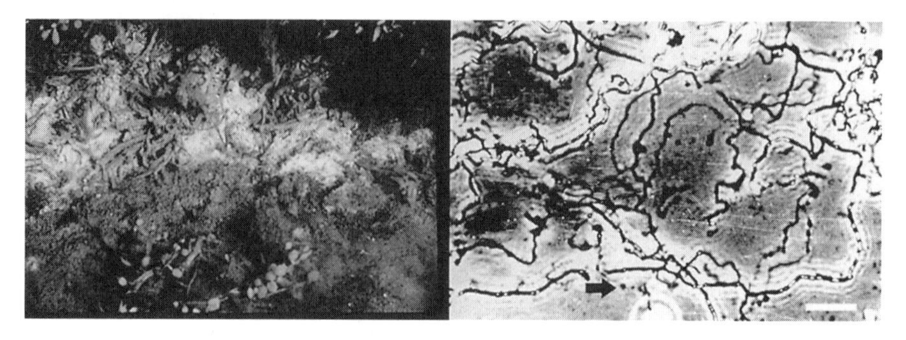

Figure 13.3. "Thiodendron." The brown alga *Fucus* and the white sulfur mat containing "Thiodendron" in the intertidal zone (left). The composition of the "Thiodendron" mat, viewed with phase contrast light microscopy (right). White globules are intracellular sulfur deposits. Small black bodies (arrow) are *Dethiosulfovibrio* cells (Surkov et al.[14]). Under strict anoxic conditions these filaments revert to the typical *Spirochaeta* morphology. Bar = 5 μm. Photos courtesy of Galina Dubinina.

and eventual anabolic-metabolic pathway fusion led to composite individuals in which protruding surface spirochetes emerged as undulipodia. The underlying kinetosomes became mitotic centrioles in response to selection pressure for equal distribution of indispensable DNA from both the archae- and the eubacterial member of the consortium. The kinetosomes, which had evolved directly from the complex spirochete attachment structures, dedifferentiated to become centrioles that were passively distributed to the consortia offspring. As the mitotic spindle evolved in numerous lineages, the former attachment sites became kinetosome-centrioles. Their evolution slowed as they became what they are still today: nucleating loci for the maturation of the organelles of motility, the undulipodia ("eukaryotic flagella"), in the next generation (fig. 13.5).

The proliferation of membrane and of cytoskeletal elements with nuclear and cytoplasmic, including spindle microtubules we trace to the archaebacterial-eubacterial fusion. Microtubule-based cytoplasmic organization is specifically of spirochaetal origin[6]; therefore, modern cytoplasmic-tubule bearing spirochetes (e.g., *Diplocalyx*, *Hollandina*, and other small pillotinas)[19] should be more closely related to the fused eubacterial endobiont than are other bacteria. Examples of the spirochete legacy include vertebrate sensory cilia of rods, cones, and auditory kinocilia; mitotic spindle fibers; Microtubule Organizational Centers (MTOCs) including fungal and plant spindle pole bodies, parabasalid atractophores, oxymonad axostyles, foraminiferal reticulopods, and actinopod axonemes; egg centrosomes; and sperm tail axonemes. The tubule proteins in *Azotobacter* or in those of the cell wall in the cyanobacterium *Synechococcus*)[20] should be less homologous to eukaryotic tubules than those in tubule-bearing spirochetes.

Thermoplasma remains the best candidate codescendant for the ancestral sulfidogenic archaebacterial component of the eukaryotic cell for several reasons outlined by Dennis Searcy. These include its histone- and actin-like proteins, tolerance of acid and heat, lack of a cell wall, and H_2S [sulfide] production.[6,13,21,22] We endorse Searcy's "sulfur hypothesis" and agree entirely with him that "the ancestor of the cytoplasm reduced S^0 [elemental sulfur] to H_2S and premitochondrial symbionts oxidized it back to elemental sulfur."[22] However, we maintain one major caveat: we identify the "premito-

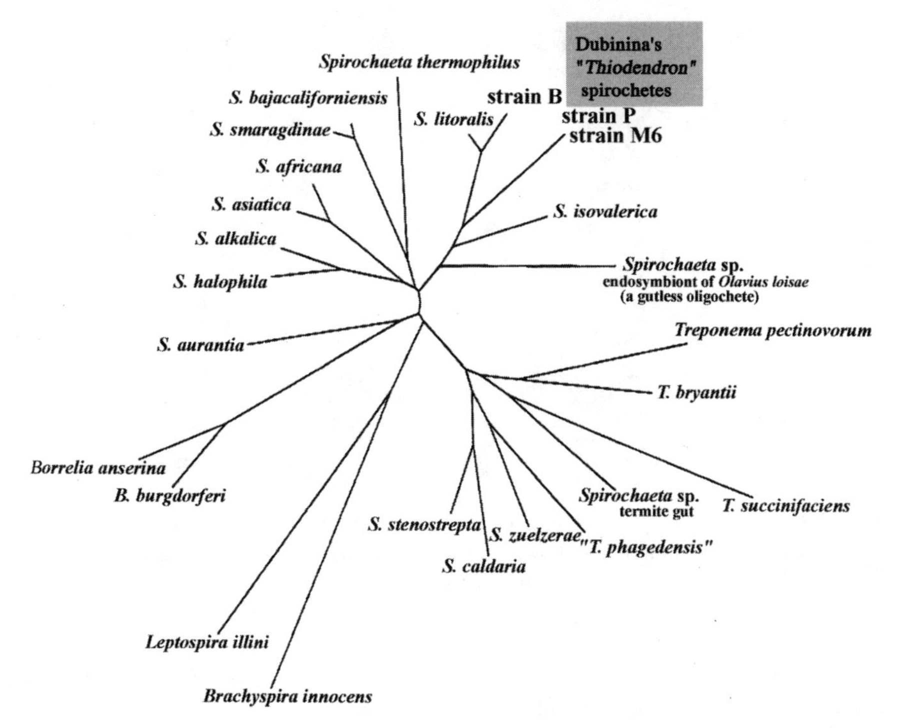

Figure 13.4. 16S rRNA gene sequence–based tree showing the relationship of the spirochetes isolated from "Thiodendron" to other members of the spirochete lineage. Almost complete 16S rRNA gene sequences comprising between 1428 and 1479 nucleotides were determined for strains P, M6, and B. The phylogenetic dendrogram was constructed using the neighbor-joining methods from distance matrices. These strains were found to cluster within the genus *Spirochaeta*, where they showed the highest similarity to *S. litoralis* and *S. isovalerica*. Strain B shows 96% 16S rRNA gene sequence similarity to *S. litoralis*.The other two, P and M6, with 99.6 sequence similarity to each other, because they have less than 90% sequence similarity to any other member of this genus, can be considered strains of a novel branch within the genus *Spirochaeta*. These uncharacterized strains require further study and description. The 16S rRNA gene sequences of strains P, M6, and B have been deposited in GenBank under the accession numbers AY337318, AY337319, and AY337320.

chondrial symbionts" in Searcy's model not as α-proteobacterial oxygen respirers but as the spirochete members of the consortium that preceded mitochondria by millions of years. The sulfide-sulfur oxidoreduction cycle evolved before mitochondrial acquisition and therefore is expected to be present in amitochondriates incapable of further mitochondrial oxidation of S^0 to sulfate or thiosulfate. Eventually, in a subsequent symbiotic event, the acquisition of the α-proteobacterial oxygen-respirers led to mitochondria. Mitochondrial integration accelerated sulfur metabolism and ensured that intracellular oxygen became the sulfide oxidizing agent. We suggest, with Searcy, that the original archaebacterial–eubacterial association was a sulfur syntrophy and involved endogenous sulfide production (by the archaebacterium), which has been measured by Searcy in dramatic experiments in selected organisms from all major

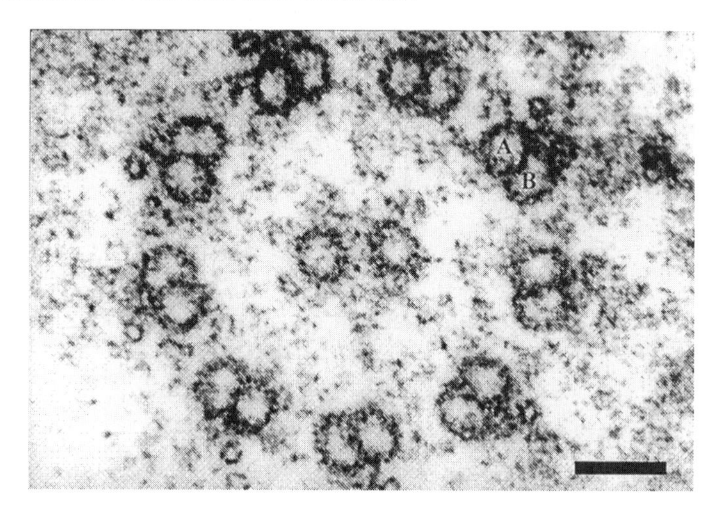

Figure 13.5. The structure of the undulipodium. A [9(2)+2] transverse axoneme section of the protist *Hexamita* sp. undulipodium showing α,β-tubulin dimers that make up walls of A, B tubules. Bar = 50 nm. Transmission electron micrograph by the late David Chase.

eukaryotic lineages. Although he hypothesizes that the original sulfide oxidation was performed by the early mitochondrion, an α-proteobacerium, we believe that the first sulfur syntrophy involved sulfide oxidation to elemental sulfur by the spirochete that became the undulipodium.

The oxidizing agent was ambient O_2 in variable supply—ultimately, of course, a product of cyanobacterial oxygenic photosynthesis. Our hypothesis is analogous to the sulfur syntrophy discovered in the spirochete–Thiodendron consortium by Dubinina and her colleagues.[16–18] After the much later integration, of the premitochondrion, the *Thermoplasma* (archaebacterial-derived) hydrogen sulfide was routinely and thoroughly oxidized not only to S^0 (an ability of the spirochete) but all the way to thiosulfate (HSO_4^-) or sulfate ($SO_4^=$), an exclusive metabolic virtuosity of the oxygen-respiring mitochondrial symbiont.

Our testable karyomastigont model offers a parsimonious explanation for the origin of interdependent eukaryotic organelles: the nucleus, endoplasmic reticulum (ER), Golgi, and cytoskeleton. Protists interpretable as representatives of intermediate states in the evolution of the plant–animal–fungus mitotic apparatus and cell organization have been observed.[6,10] Some termite hindgut protist species have a single nucleus, others have hundreds; in some cells the nuclei are exclusively associated with karyomastigonts, and in others akaryomastigonts abound. Still other protists have both akaryomastigonts and karyomastigonts.[23]

Studies of *Staurojoenina assimilis*[24,25] and *Mixotricha paradoxa*,[26] hindgut parabasalid protists, are especially relevant as modern examples of processes we propose. Cell membranes in these archaeprotists are sites of bacterial attachment.[27] Most notably, the surface of *Mixotricha*, each cell of which harbors 250,000 treponeme spirochetes, 200 larger *Canaleparolina darwiniensis* spirochetes,[28] and a *Borrelia*-like spirochete in the ingestive zone, is an excellent analog system to understand integration of free-living microbes in the origin of complex individuality on the cell level (fig. 13.6). The attached surface spirochetes provide motility for *Mixotricha*.[26] The spirochetes absorb protist waste products as food products of cellulose digestion.[28] Some termite spirochetes fix nitrogen, and many catabolize cellobiose to produce acetate, which flows through the intestinal chitin layers to animal mitochondria.[29] The microbial community therefore provides a major source of nutrition for the insect.[30]

Wood-ingesting roaches and termites harbor complex, predictably present communities of anaerobic protists that tend to form associations like those we hypothesize were important in the origin of nucleated cells.[31] The hindgut of *Cryptotermes cavifrons* was used in the studies reported here. Some of the most conspicuous of which are the devescovinids *Foaina reflexa* and *Caduceia versatilis*,[32] the monocercomonad *Tricercomitus divergens*, and the calonymphids *Snyderella tabogae* and *Stephanonympha* sp.[33,34] The hindguts of *Reticulitermes flavipes* and *R. tibialis* house similar protists, most notably the oxymonads *Pyrsonympha* and *Dinenympha* and the parabasalid *Trichonympha*.[34] These microbial community protists were used in studies described here. Many types of prokaryotes, over 50% of which are spirochetes,[29] also live in the intestines of these termites. The frequency of heterotrophic motility and syntrophic associations in anoxic carbon-rich environments permit identification of extant processes that represent each stage hypothesized in our karyomastigont model of eukaryosis.[11,35] Microbial symbioses analogous to those in our model have been videographed in live material.[36]

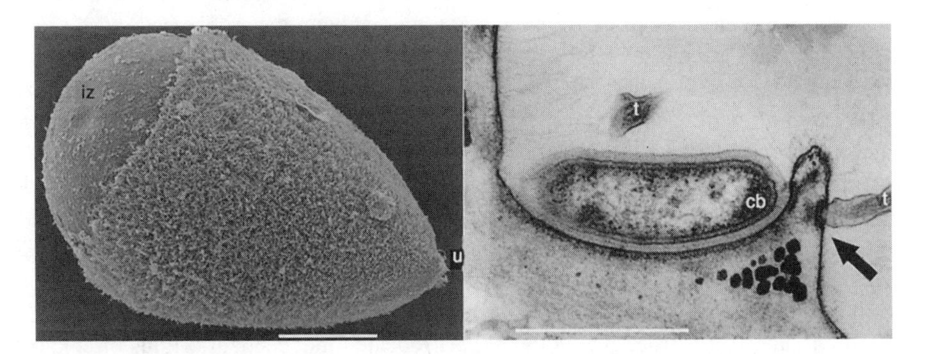

Figure 13.6. *Mixotricha paradoxa*. Protist's anterior cortex studded with about 250,000 *Treponema*-like spirochetes before division that extend to the posterior smooth ingestive zone (Iz), which sports irregularly dispersed rod bacteria and is lined with *Borrelia*-like spirochetes (b; left). *M. paradoxa* moves by the synchronous undulations of the treponemes, arranged regularly along its cortex, each of which is associated with a rod-shaped bacterium. The protist, a hypertrophied trichomonad, has four undulipodia (u; one forward, three trailing) that are located at the anterior tip of the cell. Together they act as a rudder: they control the direction of cell movement. *M. paradoxa* cortical rod bacteria (cb) are attached to a "bracket," a raised portion of the protist cortex where a treponeme spirochete (t) is inserted on the opposite side (right). The treponeme attachment site (arrow) shows that both protist and spirochete are modified. Left, scanning electron micrograph, bar = 50μm. Right, transmission electron micrograph, bar = 1.0μm. Transmission electron micrograph courtesy of A.V. Grimstone.

Alternative Models of Eukaryosis

The Hydrogen Hypothesis

The hydrogen hypothesis proposed by Martin and Müller posits eukaryosis from syntrophic bacterial associations that simultaneously produced mitochondriate cells.[37] Mitochondria and hydrogenosomes share a common ancestor: all amitochondriate eukaryotes evolved by their secondary loss. Accordingly, an anaerobic, hydrogen-dependent, autotrophic methanogen (an archaebacterium) fused with a respiring, anaerobic, heterotrophic eubacterium that released hydrogen and carbon dioxide as waste to form the first eukaryotes. A mitochondriate cell's eubacterial liposynthetic enzymes produced the nuclear membrane and the ER, suggests Martin, who rejects an additional endosymbiotic event in the origin of the nucleus.[38] The weakness of Martin's hypothesis is that it accounts for neither the ubiquity of microtubule systems nor the complete absence of methanogenetic physiology in all eukaryotes. Moreover, the ER's *N*-linked protein glycosylation pathway, nearly invariant in all eukaryotes, actually resembles that of archaebacteria, not eubacteria. The Golgi membrane synthetic system, however, appears to be of a different, most likely eubacterial, origin.[39] These data are more consistent with our karyomastigont model than with that of Martin.

The Methanogenic Syntrophy Hypothesis

An alternative syntrophy hypothesis proposed by López-García and Moreira[40] depicts an initial symbiotic association between a δ-proteobacterium, which reduced sulfate and produced hydrogen and carbon dioxide, and a methanogen that consumed these products. Symbiotic associations with a methanotrophic α-proteobacterium would have occurred at the same time or shortly thereafter. Membranous structures and a protonuclear region containing archaebacterial cytoplasm are proposed to have arisen from this association, as methanogens form nucleosome-like structures and contain DNA-binding proteins homologous to histones and eukaryotic DNA-associated enzymes.[40] Accordingly, the eubacterial genes were transferred to the archaebacterial genome, where genes for metabolism replaced those of the methanogen, and genes encoding the genetic machinery of the eubacteria were lost. Support for this syntrophy hypothesis comes from the myriad accounts of symbiotic associations between methanogens and proteobacteria in nature today. If it is true, one would expect repeated independent origins of eukaryotic cells, for which there is no evidence.[41]

The Chimera Hypothesis

On the basis of molecular sequence data, Gupta's chimera hypothesis proposes that all eukaryotes evolved from a common ancestor; the thermoacidophilic archaebacteria (eocytes) are deemed the closest extant relatives to eukaryotes.[42] An initial engulfment of the archaebacterium by an α- or a δ-proteobacterium led to formation of the first eukaryote after full integration of the two genomes.[42,43] Although Gupta does not endorse our spirochete hypothesis for the ancestral eubacterial component of eukaryotes, he does show evidence of an early eubacterial endosymbiont acquisition before the later eubacteria that then gave rise to mitochondria and plastids. However, except for the ER and the nucleus, no hypothesis for the origin of uniquely eukaryotic structures, the cytoskeleton including the mitotic apparatus, are proffered. Furthermore, any "engulfment" implies phagocytosis, which presupposes exactly the intracellular actin–tubulin cytoskeletal system, conspicuously absent in prokaryotes, for which the evolutionary scenario must account.

The Chronocyte Hypothesis

In our view, the weakest hypothesis for eukaryosis is that of the "chronocyte," an imaginary early organism without descendants that contained all eukaryotic proteins devoid of homology to any known bacterial proteins. Accordingly, the eukaryotic cell would have originated when a chronocyte engulfed archae- and eubacteria.[44] This idea is supported by "eukaryotic signature proteins," uniquely eukaryotic proteins with no homology to any bacterial proteins. The weakness of this proposal is that no known extant or fossil organism even remotely resembles the chronocyte.

Hypotheses of the Origin of Centrioles

Although several recent eukaryosis scenarios concur on the relevance of microbial symbiogenesis and syntrophy to the origin of eukaryotes, ours is the only one in which the origin of cytoskeletal motility proteins is central. Intracellular motility, and particularly the presence of the mitotic spindle and its connection to the [9(3) + 0] microtubule-arrayed structure of the centriole-kinetosome, may not be ignored in any analysis of the origin of nucleated cells.[20] That centrioles, kinetosomes, and their cilia (undulipodia) originated from bacteria by symbiosis was suggested early in the twentieth century by B.M. Kozo-Polyansky in 1924, and by I.E. Wallin in 1927.[5] In seminal work that introduced the MTOC concept, Pickett-Heaps[45] also suggested that centrioles and kinetosomes were of endogenous origin and that, after the evolution of intranuclear mitosis, MTOCs were externalized as locomotory organelles. However, the fact that undulipodia are remarkably uniform whereas mitosis varies enormously indicates that the [9(2)+2] locomotory organelle, already evolutionarily stabilized, was ancestral to the myriad mitotic descendants. Comparative protist ultrastructure confirms our notion that the primordial role of the centriole-kinetosome is to organize the undulipodium.[45]

Our karyomastigont model differs from all endogenous models also in its claim that bacterial symbiogenesis led to highly motile (swimming) protists before acquisition of mitochondria and plastids.[11] Unlike other models, ours postulates only steps that are documented biological phenomena in a sequence consistent with the fossil record of the Proterozoic Eon. No published plausible detailed alternatives that draw directly on live organisms and the Proterozoic fossil record have come to our attention for the origin of the motile nucleated cell that divides by mitosis.

Eukaryotic Motility Proteins

Some of the larger pillotinaceous spirochetes bear 24-nm cytoplasmic tubules in their protoplasmic cylinders,[47] including *Diplocalyx cryptotermitidis*, which lives in the hindgut of *C. cavifrons*,[28,47] and *Pillotina calotermitidis* of *Reticulitermes*.[19] By ultrastructural analysis, these prokaryote tubules resemble bona fide microtubules, but no tubulin or other chemistry has been reported. Both sequence (for *Prosthecobacter*) and morphological data (for another verrucomicrobial eubacterium) do, however, indicate the existence of a bacterial form of tubulin.[48,49] Bacterial tubulins **a** and **b** will replace FtsZ as the best candidates for a prokaryotic homolog to eukaryotic tubulin if lateral gene transfer from a eukaryote can be definitively ruled out as the source for the genes.[48] FtsZ, unlike the bacterial tubulins, appears to be universal in both eu- and archaebacteria.[50] These discoveries indicate that tubulin genes originated in eubacterial cells. Furthermore, the bacterial protein MreB is probably homologous to actin, a major component of the cytoskeleton.[51] Although consistent with the symbiotic theory of the origin of undulipodia, these findings do not distinguish the various models of eukaryosis.

If the cytoskeleton including mitosis evolved in part from spirochetes, highly conserved proteins, including the most conserved tubulin, should more resemble spirochete proteins than those of other prokaryotes. No cultivation in vitro of pillotinaceous spirochetes from termite hindguts has been achieved despite many attempts.[29] Failure to isolate the relevant spirochetes, those that contain conspicuous 24-nm cytoplasmic tubules, presents a daunting obstacle to identification of the motility proteins. The inferred protein sequences of the two fully sequenced spirochetes, *Treponema pallidum* and *Borrelia burgdorferi*, share surprisingly few sequences in common. No homology with the proteins BUB, cenexin, centrin, dynein, dynactin, kinesin, MAD, NuMA, and pericentrin was detected in preliminary studies.[52]

Eukaryotic motility protein candidates for homology studies are listed (table 13.1). Although our own experiments were limited to well-studied, widely distributed tubulin proteins (α, β, and γ), cenexin, and pericentrin, the number of potential protein candidates for motility sequence comparison exceeds 650.[53]

The 50-kDa proteins a-tubulin and β-tubulin, among the most highly conserved, make up the heterodimers of microtubule walls.[54] Microtubules, dynamically unstable, break down and reform in ways that ensure segregation of chromatin during meiosis and mitosis.[55] "Motor proteins" like the large ATPases dynein and kinesin travel along microtubules. Organelle movement, vesicle motility, protein transport, chromosome segregation, pigment dispersal, and much other locomotion in nucleated cells depend on the microtubule-based cytoskeleton. The formation of the cytoskeleton requires 25-nm-diameter γ-tubulin rings that nucleate microtubule assembly, possibly by transient stabilization of the minus-end of the standard microtubule.[56] Because other microtubule proteins (such as δ, ϵ-, ξ-, and η-tubulins) are limited to the kinetosome in undulipodiated eukaryotes[57] and so many eukaryotes that have microtubule organizing centers permanently lack kinetosomes, these minor tubulins are poorer candidates for homology studies.

Pericentrin, ubiquitous in eukaryotes, is a major component of the pericentriolar material of MTOCs; this "granuloreticular fuzz" often seems, in electron micrographs, to be the locus of microtubule generation. Pericentrin, therefore, is a good candidate for homology studies and forms the basis of our research plan. First discovered in patient scleroderma antisera,[58,59] it is known from insects (e.g., *Drosophila*), amphibians (e.g., *Xenopus* and its acentriolar eggs), and ciliates (*Tetrahymena*).[60] We believe that the evolutionary importance of pericentrin is underscored by its presence in the amoebal life history stage of the amoebomastigote *Naegleria*, when at the electron-microscopic level all traces of the undulipodium are absent.[61]

Table 13.1. Sequenced mitotic and other motility proteins

	Molecular Weight	Description	References
Proteins			
astrin	134 kDa	Specific association with mitotic and meiotic spindles	72
BUB[a]	140–150 kDa	Activates GTPase; mitotic checkpoint component	73, 74
cenexin	96 kDa	Acquired by immature centriole at G2-M transition; inner centriole wall component	75
CENP-A	17 kDa	Involved in mitotic kinetochore assembly, recruits other centromeric proteins	76
CENP-E	312 kDa	Requires GTP, assists in kinetochore MT binding, maintains spindle pole structure, kinesin-like	77
centrin (cyclin-dependent kinase)	20 kDa	Requires Ca^{2+}; recognized by antibodies to spasmin; localizes to centrosome; involved in MT organization, centriole replication	78, 79
dynein	>1,000 kDa	Minus-end directed dogbone structure ATPase involved in mitotic spindle organization; forms "arms" of axonemes	81, 81, 82
dynactin	varies	Binds to dynein; binds membrane, NuMA, kinetochore	83, 84
kinesin	120 kDa (varies)	Usually plus-end directed ATPase; particle transport along MTs; found in axonemes; involved in anaphase chromosome segregation	85, 86
kinectin	160 kDa	Integral kinesin-binding ER membrane protein; moves endosomes (vesicles)	87, 88
MAD[a]	25 kDa	Mitotic checkpoint component	89, 73
NuMA	240 kDa	Involved in vertebrate spindle attachment, MT organization; similar protein found in yeast polar bodies	90, 91
Pericentrin	220 kDa	PCM component; lattice in centriole rings of γ-tubulin; centrosome and mitotic spindle formation and function; MT nucleation	58, 92
Tubulin proteins			
α-tubulin	50 kDa	Requires GTP, Ca^{2+}; forms walls of MTs	54
β-tubulin	50 kDa	Requires GTP, Ca^{2+}; forms walls of MTs	54
γ-tubulin	50 kDa	Defines MT polarity; nucleating agent for centriolar/kinetosomal replication	56, 69
δ-tubulin	51 kDa	Forms triplet MTs of centrioles and kinetosomes	57

[a]Tension-sensitive proteins

Preliminary Experimental Results

Methods

Hindguts of *Cryptotermes cavifrons*, *Reticulitermes tibialis*, and *R. flavipes*, removed using fine forceps, were teased open in a 5-μL drop of termite Ringer's solution. Cells were allowed to settle onto 22 × 22 mm coverslips coated with a 1 mg/mL poly-*L*-lysine hydrochloric aqueous solution. Each coverslip was dipped in a Columbia jar containing Streck Tissue Fixative (Streck Laboratories) and then stored for 5 min in a Columbia jar of cytoskeleton buffer (10 m*M* MES [Sigma], 150 m*M* NaCl, 5 m*M* EGTA, 5 m*M* $MgCl_2$, 5 m*M* glucose).[61] Before storage in CB coverslips designated for treatment with γ-tubulin or pericentrin antibodies were dipped for 5–10 s in cold acetone to perforate the membranes and thus facilitate antibody diffusion.[63]

Coverslips were inverted onto 15 μL of diluted antibody and incubated 30 min in the dark in a humidity chamber. Antibodies were diluted with Tris buffer solution (200m*M* Tris [Merck], 154 m*M* NaCl, 20 m*M* EGTA, 20m*M* $MgCl_2$, pH 7.5 at room temperature, diluted 1:9 with water)[62] to concentrations indicated (table 13.2). Control mammalian PtK cells, grown on coverslips in Dulbecco's Modified Eagle's Medium with nonessential amino acids and Earle's salts (Gibco), received the same antibody as treated hindgut cells. Coverslips incubated in antipericentrin antibodies were twice washed with cytoskeleton buffer and incubated for 30 min in 15 μL TRITC-conjugated secondary antibody. Coverslips incubated in anti-γ-tubulin or cenexin antibodies were, after two buffer washes, treated with FITC-conjugated secondary antibody. Antibodies to α-tubulin and β-tubulin were directly conjugated to FITC, thus eliminating the need for a second incubation period with a fluorophore-labeled secondary antibody. Coverslips were washed twice with cytoskeleton buffer, mounted on slides, and viewed using fluorescence light microscopy.

An Optronix camera mounted on a Nikon Optiphot microscope fitted with fluorescence, Nomarski differential interference, and phase contrast microscopy was used for videomicroscopy of stained material. The video images were stored on three-quarter-inch Sony U-matic 60-min tapes and confirmed by still photographs taken with 160ASA 35 mm Ektachrome film through the same microscope.

Table 13.2. Antibodies to eukaryotic motility proteins

Antibody (aliquot)	Dilution	Species of Origin	Immunogen	Antibody Subclass	Clonality	Source
Anti-α-tubulin	1:25	mouse	Rat brain tubulin	IgG	Monoclonal	Sigma
Anti-β- tubulin	1:50	mouse	Rat brain tubulin	IgG	Monoclonal	Sigma
Anti-γ- tubulin	1:10,000	mouse	Synthetic peptide	IgG	Monoclonal	Sigma
Anticenexin (CDIB4)	1:1,00	mouse	Denatured lamb thymus centrioles	IgM	Monoclonal	Keith Gull
Antipericentrin (UM225)	1:400	rabbit	Human centrosomes	IgG	Polyclonal	Stephen Doxsey

Antibodies to α-tubulin and β-tubulin were conjugated to FITC, whereas other antibodies required incubation with 15μL of 1:160 diluted secondary antibody conjugated to FITC (Sigma), or TRITC (Sigma) in the case of pericentrin.

Results and Discussion

Only the α-tubulin antibody gave the expected positive results in all the protists in all the preparations (several hundred cells in at least five sets of preparations). Given the highly conserved nature of this protein among eukaryotes[54] these observations were reassuring. The known microtubular structures (axonemes of the undulipodia, centriole-kinetosomes, bundled axostyles of parabasalids, and single axostyles of devescovinids) at least were stained (fig. 13.7). Because these protist structures, when treated with FITC-conjugated anti-α-tubulin, consistently fluoresced bright green, they served as excellent control for the prokaryotes in the same preparations. In one of the slides treated with the conjugated anti-α-tubulin antibody, unidentified spirochetes of at least the diameter of the undulipodia fluoresced with the same or even greater intensity than the undulipodia (fig. 13.7). Some unidentifiable rod-shaped bacteria may have also given a positive signal, but their smaller size makes interpretation difficult (fig. 13.8). We are acutely aware of the possibility of false positives in immunofluorescent cytological preparations,[64] but this preliminary observation indicates that further work, in principle, can yield robust results.

The intrinsic fluorescence of wood particles, both inside the protists and free in the intestine, tend to render interpretation more difficult, yet wood is often distinguishable from the protein fluorescence because of its dull yellow glow relative to FITC bright green. Because the only observed fluorescence in the cenexin experiments was caused by wood, we plan no further work with this protein. Unexpected and unpredictable results were seen with in the β-tubulin antibody experiments. Microtubular structures of only about 50% of all of the interpretable large protist cells treated with anti-β-tubulin fluoresced, and some cells of the same genus lacked any nonwood fluorescence. These baffling results lead us to plan no further work with this antibody. Because the PtK control cells, which no doubt contain β-tubulin, showed similar ambiguous patterns, we plan to seek a more predictable conserved tubulin system in future studies.

About 10% of the hypermastigote cells displayed uniform dots of fluorescence near the cell surface; whether or not this punctate fluorescent pattern corresponds to the locations of the surface centriole-kinetosomes needs investigation. Wood staining of protists tested for pericentrin made observation difficult, but the fact that some cells

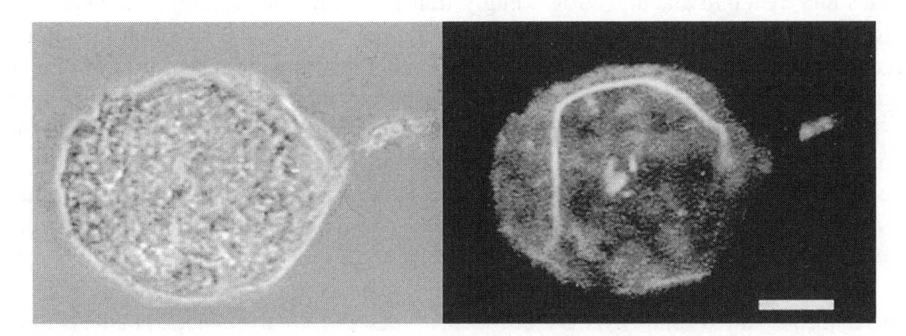

Figure 13.7. Alpha-tubulin. *Caduceia versatilis* from the hindgut of *Cryptotermes cavifrons*. Phase contrast image (left). Axostyle (right) visualized by epifluorescence microscopy, fluorescence of *C. versatilis* stained with anti-α-tubulin antibodies conjugated to FITC. Bar = 10μm.

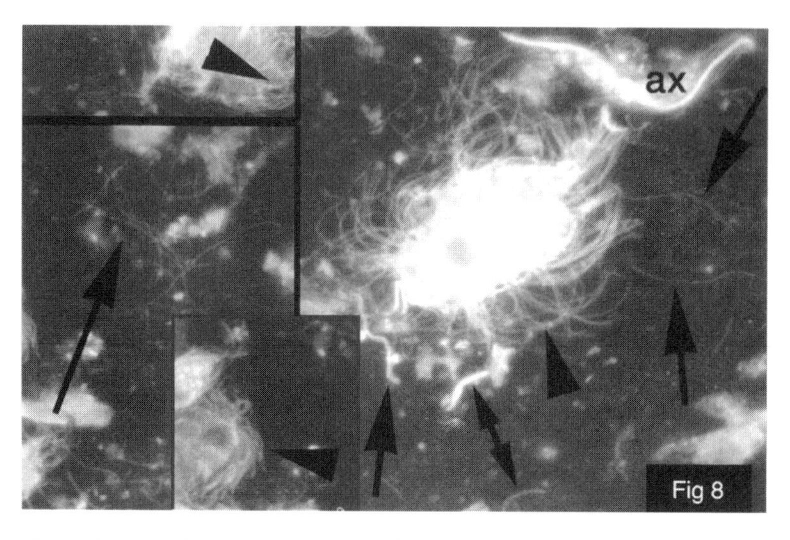

Figure 13.8. Alpha-tubulin. *Cryptotermes cavifrons* intestine stained with anti-α-tubulin antibodies conjugated to fluorescein isothiocyanate (FITC). Bar = 10 μm. Arrows to gut spirochetes, arrowheads to undulipodia of poorly preserved hypermastigotes, ax, axostyle of pyrsonymphid.

appeared to have a few very small fluorescent dots encourages us to proceed with further study.

Caduceia versatilis, another amitochondriate, is the largest parabasalid devescovinid conspicuous in the gut of *Cryptotermes cavifrons*. Force is generated for the rotation by the microtubules of the axostyle.[65] When treated with FITC-conjugated anti-γ-tubulin, a distinctive punctate pattern of fluorescence at *C. versatilis*' shear zone resulted.[63] The organism is nicknamed "*Rubberneckia*" because at the shear zone the cell membrane's break-and-reform rotation ring enables constant clockwise rotation of the "head."[32]

Further positive results with γ-tubulin antibodies in these amitochondriate termite protists have been reported.[63] Because the details of the unique rotation of the anterior of *C. versatilis* cells are not understood, the reason for specific γ-tubulin localization lacks a satisfactory explanation. Because γ-tubulin nucleates the minus end of microtubules and dynein/dynactin retrogradingly transports membrane fragments along microtubules, perhaps γ-tubulin is enriched at the point in the cell at which membrane reconstruction continuously occurs. About 25% of the termite protists and fewer mammalian cells exhibited punctate fluorescence when treated with FITC-conjugated anti-g-tubulin. The variation was made more difficult to interpret because staining was in part masked by wood. Yet the fluorescence pattern, although of course inconclusive, is consistent with γ-tubulin antibody binding to centrosomes. These preliminary results emcourage us to seek the most highly conserved centriole-kinetosome protein domains in amitochondriate protists and candidate prokaryotes.

No nucleic acids are present inside centrioles, kinetosomes, or undulipodia.[33,66] Rather, the centriole-kinetosome DNA is embedded, at least in *Chlamydomonas*, in the chromatin of the *uni* linkage group, chromosome XIX, it lies beneath the nuclear membrane.[66] The most likely reason that no ubiquitous eukaryotic motility proteins have been discovered in spirochetes or in any other prokaryotes directly by immunofluorescence is because they have not been properly sought. The banded shear zone labeling with anti-γ-tubulin in *C. versatilis* establishes a specific intracellular location and perhaps a unique role for this highly conserved motility protein. Gamma-tubulin has been reported in membranes of plants[68] and of *Tetrahymena*'s macro- and micronuclei.[69] Alpha-tubulin and β-tubulin are also found in membranes; namely, in those of mitochondria, where they are associated with voltage-gated ion channels.[70] Because of the proliferation and widespread location of MTOCs and microtubules in nucleated cells, from the cilia of sense organs to the neurotubules of brain axons and dendrites to proliferating centriole-kinetosomes of the sperm of the *Ginkgo* tree we judge that the ancestral motility proteins that originated in the prokaryotic constituents of the karyomastigont underwent a similar process of redeployment. Probably, motility protein domain shuffling began with the earliest amitochondriate eukaryotes.

Did any centromeric proteins or their homologs evolve in prokaryotes, or did their evolution occur after the origin of chromatin, the histone, protein-rich material that comprises the chromosomes of eukaryotes? The search for BUB, MAD, and other (table 13.1) motility proteins in amitochondriate termite protists and their bacteria should help constrain their time of evolutionary appearance.

Our major contribution is development of an evolutionary model that is specific enough to test in detail. Here we show the potential usefulness of the application of the protein immunofluorescence label techniques to organisms we judge relevant to the search for the earliest eukaryotes. Amitochondriate protists thrive in the anaerobic natural habitats they share with prokaryotes such that conserved motility protein domains can be sought in both groups of organisms simultaneously. Particularly advantageous is the plethora of spirochetes that naturally abide in these anoxic, organic-rich habitats. Identification of any motility proteins, such as those in table 13.1, in prokaryotes that contain 24-nm-diameter cytoplasmic tubules would be especially significant. Such results, in principle, can distinguish between current hypotheses: whether or not spirochetes, methanogens, chlorobia, α-proteobacteria, other single lineages, or symbiotic consortia of prokaryotes were direct ancestors to the first nucleated cells. We have enhanced the chances that these questions of microbial evolution can be answered by observation and direct experiment. We maintain that the karyomastigont hypothesis of the origin of the mitotic nucleus explains more facts more clearly than the published alternatives.

Because this book deals with microbes, and even their relation to the origins of species, and is about evolution, we invited Charles Darwin to have the last word: "Anyone

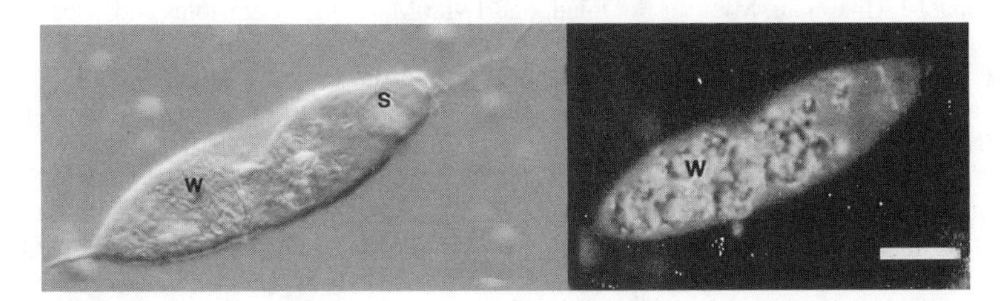

Figure 13.9. ***Caduceia versatilis* shear zone.** *C. versatilis* ("*Rubberneckia*") phase contrast image (left). The fluorescent image (right) reveals γ tubulin at the "neck" where the anterior portion of the cell rotates 360° relative to the posterior (S=shear zone) and also shows wood fluorescence (W). Bar=10m.

whose disposition leads him to attach more weight to unexplained difficulties than to the explanation of a certain number of facts will certainly reject my theory."[71]

Acknowledgments Michael Dolan, who maintained and provided termites, aided in every aspect of this work. See chapter 14, "The Missing Piece: The Microtubule Cytoskeleton and the Origin of Eukaryotes" in this volume. Michael Chapman helped in both the work and the manuscript preparation. We thank Dennis Searcy and Susan Leschine (University of Massachusetts, Amherst) for help, Stephen Doxsey (University of Massachusetts Medical School) for advice and for provision of antibodies, and Keith Gull (University of Manchester) for the cenexin antibody. Robin Kolnicki aided in literature search on kinetochore proteins, and Arturo Becerra and Antonio Lazcano provided us with unpublished data. Thanks to Patricia Wadsworth for the supply of PtK cells. We thank Celeste Asikainen and Johanna Lelke for aid in manuscript preparation. This research was supported by funds from the University of Massachusetts Commonwealth College to Hannah Melnitsky and by the Graduate School, NASA Space Sciences, Marta Norman of the Richard Lounsbery Foundation and the Alexander von Humboldt Stiftung to Lynn Margulis.

References

1. Gonzalo Vidal, "Proterozoic and Cambrian Bioevents," Homenaje al Prof. Gonzalo Vidal, *Revista Española de Paleontología*, (1998): 11–16.
2. Gonzalo Vidal and Malgorzata Moczydlowska-Vidal, "Biodiversity, Speciation, and Extinction Trends of Proterozoic and Cambrian Phytoplankton," *Paleobiology* 23 (1997): 230–246.
3. S. Blair Hedges, Hsiong Chen, Sudhir Kumar, Daniel Y.-C. Wang, Amanda S. Thompson, and Hidemi Watanabe, "A Genomic Timescale for the Origin of Eukaryotes," *BMC Evolutionary Biology* 1 (2001): 4.
4. Andrew Knoll, *Life on a Young Planet: The First Three Billion Years of Evolution on Earth* (Princeton, NJ: Princeton University Press, 2003).
5. Lynn Margulis, "Symbiosis in Evolution: Origins of Cell Motility," in S. Osawa and T. Honjo, eds., *Evolution of Life: Fossils, Molecules, and Culture* (Tokyo: Springer, 1991), 305–324.
6. Lynn Margulis, *Symbiosis in Cell Evolution*, 2nd edition (New York: W.H. Freeman, 1993).
7. Lynn Margulis, Michael Dolan, and Jessica Whiteside, "Origin of the Nucleus: Attraction Spheres and the Chimeric Karyomastigont," *Paleobiology* (in press).
8. C. Janicki, "Untersuchungen an Parasitischen Flagellaten," *Zeitschrift für Wissenschftliche Zoologie* 112 (1915): 573–691.
9. Lynn Margulis and Karlene Schwartz, *Five Kingdoms*, 3rd ed. (New York: W.H. Freeman, 1995), 5.
10. Michael J. Chapman, Michael F. Dolan, and Lynn Margulis, "Centrioles and Kinetosomes: Form, Function, and Evolution," *Quarterly Review of Biology* 75 (2000): 409–429.
11. Lynn Margulis, "Archaeal–Eubacterial Mergers in the Origin of Eukarya: Phylogenetic Classification of Life," *Proceedings of the National Academy of Sciences USA* 93 (1996): 1071–1076.
12. Michael F. Dolan, Hannah Melnitsky, Lynn Margulis, and Robin Kolnicki, "Motility Proteins and the Origin of the Nucleus," *The Anatomical Record* 268 (2002): 290–301.
13. Dennis G. Searcy, "The Archaebacterial Histone 'Hta'," *in* C.O. Gualerzi and C.L. Pon, eds., *Bacterial Chromatin* (Berlin: Springer, (1986), 175–184. Searcy, D.G., Phylogenetic and Phenotypic Relationships between the Eukaryotic Nucleocytoplasm and Thermophilic Archaebacteria. *Annals of the New York Academy of Science* 503 (1987): 168–179.
14. Alexander V. Surkov, Galina A. Dubinina, Anatoly M. Lynesko, Frank O. Glöckner, and Jan Kuever, "*Dethiosulfovibrio russensis* sp. nov., *Dethiosulfovibrio marinus* sp. nov. and *Dethiosulfovibrio acidaminovorans* sp. nov., Novel Anaerobic, Thiosulfate- and Sulfur-

Reducing Bacteria Isolated from '*Thiodendron*' Sulfur Mats in Different Saline Environments," *International Journal of Systematic and Evolutionary Microbiology* 51 (2001): 327–337.

15. Galina A. Dubinina, N.V. Leshcheva, and Margarita Y. Grabovich, "The Colorless Sulfur Bacterium *Thiodendron* is Actually a Symbiotic Association of Spirochetes and Sulfidogens," *Microbiology* 62 (1993): 432–444.
16. G. A. Dubinina, M.Y. Grabovich, and N.V. Leshcheva, "Occurrence, Structure, and Metabolic Activity of '*Thiodendron*' Sulfur Mats in Various Saltwater Environments," *Microbiology* 62 (1993): 450–456.
17. Frederick A. Rainey, Naomi Ward-Rainey, Reiner M. Kroppenstedt, and Erko Stackebrandt, "The Genus *Nocardiopsis* Represents a Phylogenetically Coherent Taxon and a Distinct Actinomycete Lineage: Proposal of Nocardiopsaceae fam. nov.," *International Journal of Systematic Bacteriology* 46 (1996): 1088–1092.
18. Frederick A. Rainey, Matthias Dorsch, Hugh W. Morgan, and Erko Stackebrandt, "16S rDNA Analysis of *Spirochaeta thermophila*: Its Phylogenetic Position and Implications for the Systematics of the Order *Spirochaetales*," *Systematic and Applied Microbiology* 15(1992): 197–202.
19. Lynn Margulis, "Spirochetes," in Joshua Lederberg, ed., *Encyclopedia of Microbiology*, 2nd ed. (New York: Academic Press, 2000), 353–363.
20. David Bermudes, Lynn Margulis, and George Tzertzinis, "Prokaryotic Origin of Undulipodia: Application of the Panda Principle to the Centriole Enigma," *Annals of the New York Academy of Sciences* 503 (1987): 187–197.
21. Dennis G. Searcy and William G. Hixon, "Cytoskeletal Origins in Sulfur-Metabolizing Archaebacteria," *Biosystems* 25 (1991): 1–11.
22. Rothsovann Yong and Dennis G. Searcy, "Sulfide Oxidation Coupled to ATP Synthesis in Chicken Liver Mitochondria," *Comparative Biochemistry and Physiology B-Biochemistry and Molecular Biology* 129 (2001): 129–137.
23. Michael F. Dolan and Harold Kirby, "*Gyronympha*, *Prosnyderella*, and *Criconympha*, Three New Genera of Calonymphids (Parabasalia: Trichomonadida) from Wood-Eating Termites," *European Journal of Protistology* 38 (2002): 73–81.
24. Michael F. Dolan and Lynn Margulis, "*Staurojoenina* and Other Symbionts in *Neotermes* from San Salvador Island, Bahamas," *Symbiosis* 22 (1997): 229–239.
25. Andrew M. Wier, Michael F. Dolan, and Lynn Margulis, "Cortical Symbionts and Hydrogenosomes of the Amitochondriate Protist, *Staurojoenina assimilis*," Symbiosis 36 (2004): 153–168.
26. L.R. Cleveland and A.V. Grimstone, "The Fine Structure of the Flagellate *Mixotricha paradoxa* and Its Associated Micro-Organisms," *Proceedings of the Royal Society Series B* 159 (1964): 668–686.
27. M. Wenzel, R. Radek, G. Brugerolle, and H. Konig, "Identification of the Ectosymbiotic Bacteria of *Mixotricha paradoxa* Involved in Movement Symbiosis," *European Journal of Protistology* 39 (2003): 11–23.
28. Andrew Wier, Jon Ashen, and Lynn Margulis, "*Canaleparolina darwiniensis*, gen. nov. sp. nov., and Other Pillotinaceous Spirochetes from Insects," *International Microbiology* 3 (2000): 213–223.
29. John A. Breznak, "Phylogenetic Diversity and Physiology of Termite Gut Spirochetes," *American Zoologist* 40 (2000): 954–955.
30. Jared R. Leadbetter, Thomas M. Schmidt, Joseph R. Graber, and John A. Breznak, "Acetogenesis from H_2 Plus CO_2 by Spirochetes from Termite Guts," *Science* 283 (1999): 686–689.
31. Andrew M. Weir, Michael F. Dolan, and Lynn Margulis, "Cortical symbionts and hydrogenosomes of the amitochondriate protist *Staurojoenina assimilis*," *Symbiosis* 36 (2004): 153–168.
32. Ugo d'Ambrosio, Michael Dolan, Andrew M. Wier, and Lynn Margulis, "Devescovinid Trichomonad with Axostyle-Based Rotary Motor ("Rubberneckia"): Taxonomic Assignment as *Caduceia versatilis* sp. nov.," *European Journal of Protistology* 35 (1999): 327–337.
33. Michael F. Dolan, "DNA Fluorescent Stain Accumulates in the Golgi but not in the Kinetosomes of Amitochondriate Protists," *International Microbiology* 3 (2000): 45–49.

34. Michael Yamin, "Flagellates of the Orders Trichomonadida Kirby, Oxymonadida Grassé, and Hypermastigida Grassi and Foà Reported from Lower Termites (Isoptera Families Mastotermitidae, Kalotermitidae, Hodotermitidae, Termopsidae, Rhinotermitidae, and Serritermitidae) and from the Wood-Eating Roach *Cryptocercus* (Dictyoptera: Cryptocercidae)," *Sociobiology* 4 (1979): 1–120.
35. Lynn Margulis, Michael F. Dolan, and Ricardo Guerrero, "The Chimeric Eukaryote: Origin of the Nucleus from the Karyomastigont in Amitochondriate Protists," *Proceedings of the National Academy of Sciences USA* 97 (2000): 6954–6959.
36. Lynn Margulis and James MacAllister, *Eukaryosis: Origin of Nucleated Cells*, 2004, 17 min video, unpublished.
37. William Martin and Miklós Müller, "The Hydrogen Hypothesis for the First Eukaryote," *Nature* 392 (1998): 37–41.
38. William Martin, "A Briefly Argued Case that Mitochondria and Plastids are Descendants of Endosymbionts, but That the Nuclear Compartment is Not," *Proceedings of the Royal Society of London* 266 (1999): 1387–1395.
39. Ari Helenius and Marcus Aebi, "Intracellular Functions of *N*-linked Glycans," *Science* 291 (2001): 2364–2369.
40. Purificación López-García and David Moreira, "Metabolic Synthesis and the Origin of Eukaryotes," *Trends in Biological Sciences* 24 (1999): 88–93.
41. Radhey S. Gupta, "Origin of Eukaryotic Cells: Was Metabolic Synthesis Based on Hydrogen as a Driving Force?" *Trends in Biological Sciences* 24 (1999): 423.
42. Radhey S. Gupta, "Protein Phylogenies and Signature Sequences: A Reappraisal of Evolutionary Relationships Among Archaebacteria, Eubacteria, and Eukaryotes," *Microbiology and Molecular Biology Reviews* 62 (1998): 1435–1491.
43. Radhey S. Gupta and G. Brian Golding, "The Origin of the Eucaryotic Cell," *Trends in Biological Sciences* 21 (1996): 166–171.
44. Hyman Hartman and Alexei Federov, "The Origin of the Eukaryotic Cell: A Genomic Investigation," *Proceedings of the National Academy of Sciences USA* 99 (2002): 1420–1425.
45. Jeremy Pickett-Heaps, "The Evolution of Mitosis and the Eukaryotic Condition," *BioSystems* 6 (1974): 37–48.
46. Jean-Pierre Mignot, "The Centrosomal Big Bang: From a Unique Central Organelle Towards a Constellation of MTOC's," *Biology of the Cell* 86 (1996): 81–91.
47. David Bermudes, Gregory Hinkle, and Lynn Margulis, "Do Prokaryotes Contain Microtubules?" *Microbiology Reviews* 58 (1994): 387–400.
48. Cheryl Jenkins, Ram Samudrala, Iain Anderson, Brian P. Hedlund, Giulio Petroni, Natasha Michailova, Nicolas Pinel, Ross Overbeek, Giovanna Rosati, and James T. Staley, "Genes for the Cytoskeletal Protein Tubulin in the Bacterial Genus *Prosthecobacter*," *Proceedings of the National Academy of Sciences USA* 99 (2002): 17049–17054.
49. Giulio Petroni, Stefan Spring, Karl-Heinz Schleifer, Franco Verni, and Giovanna Rosati, "Defensive Extrusive Ectosymbionts of *Euplotidium* (Ciliophora) that Contain Microtubule-Like Structures are Bacteria Related to *Verrucomicrobia*," *Proceedings of the National Academy of Sciences USA* 97 (2000): 1813–1817.
50. Harold P. Erickson, "FtsZ, a Tubulin Homologue in Prokaryote Cell Division," *Trends in Cell Biology* 7 (1997): 362–367.
51. Fusinita van den Ent, Linda A. Amos, and Jan Löwe, "Prokaryotic Origin of the Actin Cytoskeleton," *Nature* 413 (2001): 39–44.
52. Antonio Lazcano and Arturo Becerra, Department of Biology, Universidad Nacional Autónoma de México, Mexico City, unpublished data.
53. John L. Hall and David J.L. Luck, "Basal Body-Associated DNA: In-Situ Studies in *Chlamydomonas reinhardtii*," *Proceedings of the National Academy of Sciences USA* 92 (1995): 5129–5133.
54. Christophe Noël, Delphine Gerbod, Naomi M. Fast, René Wintjens, Pilar Delgado-Viscogliosi, W. Ford Doolittle, and Eric Viscogliosi, "Tubulins in *Trichomonas vaginalis*: Molecular Characterization of α-Tubulin Genes, Posttranslational Modifications, and Homology Modeling of the Tubulin Dimer," *Journal of Eukaryotic Microbiology* 48 (2001): 647–654.

55. Rebecca Heald and Eva Nogales, "Microtubule Dynamics," *Journal of Cell Science* 115 (2002): 3–4.
56. Michelle Moritz, Michael B. Braunfeld, John W. Sedat, Bruce Alberts, and David A. Agard, "Microtubule Nucleation by g-Tubulin-Containing Rings in the Centrosome," *Nature* 378 (1995): 638–640.
57. Susan K. Dutcher, "The Tubulin Fraternity: Alpha to Eta," *Current Opinion in Cell Biology* 13 (2001): 49–54.
58. Stephen J. Doxsey, Pascal Stein, Linda Evans, Patricia Calarco, and Marc Kirschner, "Pericentrin, a Highly Conserved Protein of Centrosomes Involved in Microtubule Organization," *Cell* 76 (1994): 639–650.
59. Irina Gavanescu, Dolores Vazquez-Abad, J. McCauley, J.-L. Senecal, and Stephen Doxsey, "Centrosome Proteins: A Major Class of Autoantigens in Scleroderma," *Journal of Clinical Immunology* 19 (1999): 166–171.
60. Tim Stearns and Marc Kirschner, "In Vitro Reconstitution of Centrosome Assembly and Function: The Central Role of γ-Tubulin," *Cell* 76 (1994): 623–637.
61. M.R. Suh, J.W. Han, Y.R. No, and J. Lee, "Transient Concentration of a Gamma-Tubulin-Related Protein with a Pericentrin-Related Protein in the Formation of Basal Bodies and Flagella during the Differentiation of *Naegleria gruberi*," *Cell Motility and the Cytoskeleton* 52 (2002): 66–81.
62. Brigitte Mies, Klemens Rottner, and J. Victor Small, "Multiple Immunofluorescence Microscopy of the Cytoskeleton," in Julio E. Celis, ed., *Cell Biology: A Laboratory Handbook*, 2nd ed. (New York: Academic Press, 1998), 469–476.
63. H. Melnitsky and L. Margulis, "Centrosomal proteins in terminte symbionts: Gamma-tubulin and a Scleroderma Antibodies bind the rotation zone of *Caduceia versatilis*," *Symbiosis* 37 (2004): 5–15.
64. D. Munson, R. Obar, G. Tzertzinis, and L. Margulis, "The 'Tubulin-like' S1 Protein of *Spirochaeta* is a Member of the HSP-65 Stress Protein Family," *BioSystems* 31 (1993): 161–167.
65. Sid L. Tamm, "Membrane Movements and Fluidity During the Rotational Motility of a Termite Flagellate," *The Journal of Cell Biology* 80 (1979):141–149.
66. E. Szathmáry, "Early Evolution of Microtubules and Undulipodia," *Biosystems* 20 (1987): 115–131.
67. Ursula W. Goodenough, "Basal Body Chromosomes?" *Cell* 59 (1989): 1–3.
68. Denisa Dryková, Vera Cenklová, Vadym Sulimenko, Jindrich Volc, Pavel Dráber, and Pavla Binarová, "Plant γ-Tubulin Interacts with αβ-Tubulin Dimers and Forms Membrane-Associated Complexes," *The Plant Cell* 15 (2003): 465–480.
69. Yuhua Shang, Bing Li, and Martin A. Gorovsky, "*Tetrahymena thermophila* Contains a Conventional γ-Tubulin That Is Differentially Required for the Maintenance of Different Microtubule-Organizing Centers," *The Journal of Cell Biology* 158 (2002): 1195–1206.
70. Manon Carré, Nicolas André, Gérard Carles, Hélène Borghi, Laetitia Brichese, Claudette Briand, and Diane Braguer, "Tubulin is an Inherent Component of Mitochondrial Membranes That Interacts with the Voltage-Dependent Anion Channel," *The Journal of Biological Chemistry* 277 (2002): 33664–33669.
71. Charles Darwin, *Origin of Species* (Penguin Edition 1859, 1957): 463.
72. Gary J. Mack and Duane A. Compton, "Analysis of Mitotic Microtubule-Associated Proteins Using Mass Spectrometry Identifies Astrin, a Spindle-Associated Protein," *Proceedings of the National Academy of Sciences USA* 98 (2001): 14434–14439.
73. Dmitrios A. Skoufias, Paul R. Andreassen, Françoise B. Lacroix, Leslie Wilson, and Robert L. Margolis, "Mammalian Mad2 and Bub1/BubR1 Recognize Distinct Spindle-Attachment and Kinetochore-Tension Checkpoints," *Proceedings of the National Academy of Sciences USA* 98 (2001): 4492–4497.
74. Hilary E. Sharp-Baker and Rey-Huei Chen, "Spindle Checkpoint Protein Bub1 Is Required for Kinetochore Localization of Mad1, Mad2, Bub3, and CENP-E, Independently of Its Kinase Activity," *The Journal of Cell Biology* 153 (2001): 1239–1249.
75. Bodo M.H. Lange and Keith Gull, "A Molecular Marker for Centriole Maturation in the Mammalian Cell Cycle," *The Journal of Cell Biology* 130 (1995): 919–927.

76. Aaron A. Van Hooser, Ilia I. Ouspenski, Heather C. Gregson, Daniel A. Starr, Tim J. Yen, Michael L. Goldberg, Kyoko Yokomori, William C. Earnshaw, Kevin F. Sullivan, and B.R. Brinkley, "Specification of Kinetochore-Forming Chromatin by the Histone H3 Variant CENP-A," *Journal of Cell Science* 114 (2001): 3529–3542.
77. Linda Wordeman, William C. Earnshaw, and Rebecca L. Bernat, "Disruption of CENP Antigen Function Perturbs Dynein Anchoring to the Mitotic Kinetochore," *Chromosoma* 104 (1996): 551–560.
78. W. Nagl, "Cdc2–Kinases, Cyclins, and the Switch from Proliferation to Polyploidization," *Protoplasma* 188 (1995): 143–150.
79. Jeffrey L. Salisbury, Kelly M. Suino, Robert Busby, and Margaret Springett, "Centrin-2 is Required for Centriole Duplication in Mammalian Cells," *Current Biology* 12 (2002): 1287–1292.
80. Tirso A. Gaglio, Mary A. Dionne, and Duane A. Compton, "Mitotic Spindle Poles are Organized by Structural and Motor Proteins in Addition to Centrosomes," *Journal of Cell Biology* 138 (1997): 1055–1066.
81. Aaron Young, Jason B. Dictenberg, Aruna Purohit, Richard Tuft, and Stephen J. Doxsey, "Cytoplasmic Dynein-Mediated Assembly of Pericentrin and γ-Tubulin onto Centrosomes," *The American Society for Cell Biology* 11 (2000): 2047–2056.
82. J. Zhou, H.-B. Shu, and Harish C. Joshi, "Regulation of Tubulin Synthesis and Cell Cycle Progression in Mammalian Cells by γ-Tubulin-Mediated Microtubule Nucleation," *Journal of Cellular Biochemistry* 84 (2002): 472–483.
83. Cristophe J. Echeverri, Bryce M. Paschal, Kevin T. Vaughan, and Richard B. Vallee, "Molecular Characterization of the 50-kD Subunit of Dynactin Reveals Function for the Complex in Chromosome Alignment and Spindle Organization During Mitosis," *Journal of Cell Biology* 132 (1996): 617–633.
84. Wendy Zimmerman and Stephen J. Doxsey, "Construction of Centrosomes and Spindle Poles by Molecular Motor-Driven Assembly of Protein Particles," *Traffic* 1 (2000): 927–934.
85. Douglas G. Cole, Dennis R. Diener, A. Himelblau, Peter L. Beech, J. C. Fuster, and Joel L. Rosenbaum, "*Chlamydomonas* Kinesin-II-Dependent Intraflagellar Transport (IFT): IFT Particles Contain Proteins Required for Ciliary Assembly in *Caenorhabditis elegans* Sensory Neuron," *The Journal of Cell Biology* 141 (1998): 993–1008.
86. Todd Maney, Andrew W. Hunter, Mike Wagenbach, and Linda Wordeman, "Mitotic Centromere-Associated Kinesin is Important for Anaphase Chromosome Segregation," *The Journal of Cell Biology* 142 (1998): 787–801.
87. Itaru Toyoshima, Hanry Yu, Eric R. Steuer, and Michael P. Sheetz, "Kinectin, a Major Kinesin-Binding Protein on ER," *The Journal of Cell Biology* 118 (1992): 1121–1131.
88. Janardan Kumar, Hanry Yu, and Michael P. Sheetz, "Kinectin, an Essential Anchor for Kinesin-Driven Vesicle Motility," *Science* 267 (1995): 1834–1837.
89. Rong Li and Andrew W. Murray, "Feedback Control of Mitosis in Budding Yeast," *Cell* 66 (1991): 519–531.
90. Alexey Khodjakov, Richard W. Cole, Berl R. Oakley, Conly L. Rieder, "Centrosome-Independent Mitotic Spindle Formation in Vertebrates," *Current Biology* 10 (2000): 59–67.
91. Laurence Haren and Andreas Merdes, "Direct Binding of NuMA to Tubulin is Mediated by a Novel Sequence Motif in the Tail Domain That Bundles and Stabilizes Microtubules," *Journal of Cell Science* 115 (2002): 1815–1824.
92. Alexey Khodjakov and Conly L. Rieder, "The Sudden Recruitment of Gamma-Tubulin to the Centrosome at the Onset of Mitosis and its Dynamic Exchange Throughout the Cell Cycle Do Not Require Microtubules," *The Journal of Cell Biology* 146 (1999): 585–596.

14

The Missing Piece: The Microtubule Cytoskeleton and the Origin of Eukaryotes

MICHAEL F. DOLAN

Eukaryotes are characterized by a membrane-bounded nucleus and, among other attributes, a microtubule cytoskeleton that is used to separate the chromosomes in mitosis. The recent molecular- and biochemical-based hypotheses on the origin of eukaryotes fail to adequately address the evolutionary origin of microtubules. This stems in part from the replacement of morphological- and organism-based approaches to cell evolution with molecular- and biochemical-based ones. Morphological, natural historical approaches look for tubules or tubule-organizing centers in bacteria, or for simplified microtubule structures in protists, and then consider the biochemical components involved. Proponents of such approaches usually insist that the clues to the earliest lineages can be found among extant taxa. Molecular and biochemical approaches emphasize gene or amino acid sequences and protein chemistry and consider the organismic biology secondarily, as in the case of FtsZ (filament temperature-sensitive protein Z), the putative bacterial tubulin ancestor or homolog. Advocates of these molecular approaches are not troubled by premitochondrial eukaryotes, or the premicrotubule-containing eukaryotes, which are purported to be evolutionary intermediates but that have left no descendants. A hypothesis on the origin of microtubules that synthesizes both approaches is lacking.

The Unexplained

Despite recent advances in molecular phylogenetics, which allow for the better testing of hypotheses on microbial evolution, there are many aspects of the origin of eukaryotic cells that remain unexplained. There are many aspects of the eukaryotic cells that appear to have no counterparts in bacteria. One of these is the microtubular cytoskeleton, which includes the mitotic spindle, the eukaryotic flagellum or undulipodium, and other persistent cytoskeletal elements. Accounting for the origin of the microtubular cytoskeleton is a requisite of any theory for the origin of eukaryotes because the mitotic spindle is used universally in eukaryotes in the separation of the chromosomes in nuclear division. There are other aspects that require evolutionary explanation: the motor proteins that move along the tubules, the kinetochores that form the junction of the tubules and the chromosomes, and the spindle pole bodies that nucleate and anchor the tubules. Here I consider only the microtubules themselves.

Questioning recent hypotheses of eukaryotic evolution from this perspective, I begin with an overview of the makeup of microtubules, the protein tubulin, and microtubular dynamics. The different hypotheses of eukaryosis are then perused. I then examine tubulin's origin and the bacterial protein FtsZ, which is widely accepted as a homolog of tubulin. I also point to some confusion in concepts and terminology when comparing bacterial and eukaryotic cell division, particularly in the areas of growth and motility. Donna Kubai's comment on the erroneous finding of mitosis in cyanobacteria rings true for contemporary theories of eukaryote origins: "Minimization of the significance of any differences noted in unusual modes of nuclear division could surely lead one to formulate procrustean analogies with the events of standard mitosis. As an example of this sort of error, we recall the attempts made to describe the distribution of the bacterial genophore in the terminology of typical mitosis."[1]

Tubulin and Microtubules

Microtubules are polymers of the α- and β-tubulin heterodimers. They spontaneously form polymers, or protofilaments, in the cell, and thirteen of these protofilaments align in parallel to form a cylinder with an outer diameter of 25 nm, called the microtubule.[2] Often these microtubules are anchored at one end, the minus end, at "a microtubule organizing center" (MTOC), such as the centrosome in animal cells or the spindle pole body in yeasts.[3] The tubules grow at the plus end by the continued addition of dimers. There is a continuous growth and contraction, a polymerization and breakdown of the microtubules from the MTOC. This is how the microtubules grow and attach to other cell components such as the chromosomes at mitosis.

The polymerization of microtubules is associated with the GTPase activity of the β-tubulin component of the tubulin heterodimer. As tubulin is added to the growing end of the microtubule, the GTP is hydrolyzed to GDP. A growing microtubule has a "GTP cap" in which the GTP has not hydrolyzed to GDP. If GTP hydrolysis occurs all the way to the growing end of the tubule, the microtubule becomes unstable and depolymerizes.[2]

Microtubules are generally nucleated from certain sites in the cell, for example, the spindle pole, the nuclear envelope, or the basal body (i.e., the kinetosome at the base of the flagellum, sometimes referred to as an "undulipodium"). This later structure is a highly complex organelle, composed of hundreds of proteins. It has no bacterial homolog, and it undoubtedly evolved into its current form after the origin of microtubules. However, it is a remarkably standardized organelle across diverse eukaryotic taxa, which indicates that it evolved once in the early evolution of eukaryotes.

Theories of Eukaryosis

The development of molecular phylogenetics has given rise to several new hypotheses on the origin of eukaryotes. When examining them from the perspective of the microtubule cytoskeleton, we consider the last 15 years or so (since the widespread acceptance of the three domain phylogeny).[4,5] Before that, most biologists who were interested in cell evolution assumed that microtubules evolved endogenously in eukaryotes with little if any bacterial precursor structures, or ignored the question entirely—as most still do today. Margulis, representing the morphology-based or natural history approach, had long hypothesized that the eukaryotic flagellum or undulipodium and the microtubular cytoskeleton evolved symbiogeneticially from a spirochete.[6] However, the initial reports of a tubulin-like protein in *Spirocheta bajacaliforniensis*[7] were not supported by subsequent study.[8]

Reports of DNA in centriole-kinetosomes, which might be a residual organellar genome, analagous to those found in mitochondria and chloroplasts, were also not supported by subsequent studies.[9–11] Although the molecular nature of the centriole/kinetosome is not fully characterized, there is no evidence for DNA within this structure. Nor is there any evidence for spirochete genes in the nuclear genome of eukaryotes (although this aspect has not been systematically studied). Thus, there is simply no molecular or biochemical evidence to support the spirochete hypothesis. The search for in vivo tubular, cytoskeletal elements in bacterial cells, an approach based on electron microscopy, has been supplanted by molecular biological experimentation in recent years, although there are occasional reports of cytoskeletal structures in bacteria, as with the fibrocrystalline body of *Halobacterium salinarum*.[12,13] Recent hypotheses for the origin of eukaryotes are based on compilations of molecular and biochemical evidence and ignore or gloss over morphological concerns.

In his three-domain scheme, Carl Woese presented a radical new hypothesis for eukaryotic origins: that the eukaryotes were as old as eubacteria and archaeabacteria, and had evolved along a distinct lineage separate from these two domains.[5] After the first complete genomes were published and it was found that the three-domain phylogeny was not consistent for a variety of genes, Woese revised his hypothesis, emphasizing widespread lateral gene transfer before the hypothetical genetic "annealing" of life into three distinct domains.[14]

In contrast to the long debate over the symbiogenetic origin of mitochondria and plastids, several new models that involve a chimeric origin for the nucleus from two bacterial lineages have been proposed and quickly added to textbooks. The uniqueness of the three domains was challenged by Gupta, who also argued that eukaryotes had arisen as a chimera, a fusion of an archaebacterium and a gram-negative eubacterium, before the symbiotic origin of mitochondria.[15,16] Although Gupta developed a hypothesis to explain the endomembrane system, he did not consider the microtubule cytoskeleton.

The hydrogen hypothesis of Martin and Müller[17] also postulated a chimeric origin of eukaryotes through the fusion of a methanogen and an alpha proteobacterium. Under this hypothesis, there were no ancestrally amitochondriate eukaryotes. The hydrogenosomes found in some anaerobic, amitochondriate protists evolved from the same alpha proteobactercal symbiont. The syntrophy hypothesis of Moreira and Lopez-Garcia[18] similarly postulated a primordial fusion of a methanogen and a proteobacterium, but argued that this occurred before the mitochondrial symbiosis. In neither case did the original hypothesis include a consideration of the microtubule cytoskeleton. Updates of both hypotheses have attempted to explain the origin of tubulin by the prokaroyotic protein FtsZ.[19,20]

The protein FtsZ has become widely accepted as the "ancestor of tubulin." It was adopted by Rizzoti in his peduncle hypothesis, as a response to Margulis's spirochete origin of microtubules, and by Cavalier-Smith in his latest phagotrophy hypothesis.[21,22] To hypothesize that FtsZ is the origin of tubulin, one must explain how this protein, which is involved in septum formation in dividing prokaryotic cells, became the basis for structures involved in the movement of chromosomes. As Cavalier-Smith has recently put it, "A key feature of the origin of the eukaryotic cell from a bacterial ancestor, sadly often ignored, is how bacterial mechanisms for DNA replication and segregation, cell division, and cell-cycle controls have been converted into eukaryotic ones."[22]

FtsZ as the Prokaryotic Homolog of Tubulin

FtsZ is a crucial structural component in bacterial cell division.[23] As the name indicates, mutants lacking the functional protein do not divide but continue to elongate into filaments. The ring of molecules it forms at the cell surface, the Z ring, is part of a complex structure called variously the "divisome," the "septalsome," or the "septator" that forms across the inner membrane, the periplasm, and cell wall at the site of cell division. The septator functions to direct the division process in ways that are not clear.

FtsZ has been put forward as the tubulin homolog because it is a GTPase, polymerizes like tubulin, has a similar tertiary structure, and is involved with cell division. Purified FtsZ binds to and hydrolyzes GTP. The purified protein also polymerizes to form long tubules in a GTP-dependent manner. These structures closely resemble tubulin protofilaments. FtsZ's C terminus is not needed for polymerization, as is the case with tubulin.[24] The crystal structures of α- and β-tubulin and FtsZ revealed that FtsZ has the same type of unusual GTP-binding domain as tubulin.[25,26]

While Harold Erickson had originally concluded in 1995 that FtsZ was the bacterial homolog of tubulin, he questioned its structural similarity to tubulin ("The tubular structures formed by FtsZ seem substantially different from MTs [microtubules], and it is questionable even whether these FtsZ tubules exist in vivo").[27] Within a year, however, his views had radically changed: "We conclude that the functional FtsZ polymer in vivo must be some form of protofilament sheet, probably close to the long, narrow sheets or bundles (observed in vitro). . . . The geometry of a protein subunit that can assemble into a protofilament is so complex and precise that this assembly could not exist in vitro unless it was constantly selected for, and hence functionally important in vivo."[28] He then endorsed the "FtsZ as ancestor scenario," "A scenario for the evolution of the FtsZ/tubulin cytoskeleton should postulate FtsZ as the primordial protein, since it is found in all bacteria examined. . . . Curiously, eukaryotic cells have replaced the FtsZ system with the actin-based cytokinetic ring and use microtubules for very different functions," a view later echoed by Cavalier-Smith in his revised phagotrophy hypothesis.[22]

FtsZ has become widely accepted as the molecule from which microtubules evolved. This conclusion is crucially important because it bears on the classical distinction between bacteria and eukaryotes as representing the most profound evolutionary leap and being structurally worlds apart. Moreover, implicitly, it offers a smooth transition of eukaryotes from bacteria, one that would not require that eukaryotes were fundamentally chimeric beings resulting from some kind of symbiosis. Thus, Lutkenhaus has remarked that the bacterial cytoskeleton has been recognized, "with the confirmation that bacteria not only have FtsZ, the ancestral homolog of tubulin, but also have MreB and ParM, which are ancestral homologs of actin, involved in two fundamental aspects of cell growth—cell shape and chromosome segregation. Thus, one of the criteria often used by cell biologists to differentiate prokaryotic cells from eukaryotic cells can now be discarded, along with our naiveté about the simplicity of bacterial cells."[28]

Questioning the FtsZ Origin of Tubulin

The three characteristics of classical explanation of cell evolution criticized by Woese in his paper, "On the Evolution of Cells,"[29] are the invocation of fully evolved cells, rather than protocells, as the earliest ancestors; the evolution of eukaryotes after prokaryotes; and the difference in character of prokaryotic and eukaryotic evolution.[29] Two of these criticisms can also be applied to the FtsZ–tubulin question. Proponents of FtsZ invoke proteins that are already fully evolved. FtsZ is regularly referred to as the "ancestor" of tubulin, or the "ancestral homolog" of tubulin.[28,30] In addition, the eukaryotic cell, in this case the microtubular cytoskeleton, is seen as evolving after its prokaryotic counterpart. Hence, FtsZ's in vitro characteristics are preadaptations for microtubular structure. However, Woese's third component of the classical explanation of cell evolution, that prokaryotic and eukaryotic evolution are different in character, is accurate, stands up to his criticism, and can be applied here to the FtsZ origin of tubulin. Woese does not appreciate that eukaryotic cells can evolve by engulfing and consuming other cells and that eukaryotic lineages can evolve through the incorporation of whole genomes. This has occurred with the symbiotic origin of mitochondria and plastids as well as with, for example, the whole-cell incorporation of red algal cells in the evolution of several phyla of algae.

Despite the widespread acceptance of FtsZ as the origin of tubulin, the reasoning underlying this argument can be criticized from several perspectives. These include using the contemporary protein as an ancestor of tubulin, ignoring functional differences of the two proteins, using in vitro data to explain in vivo phenomena, using eukaryotic terminology to describe prokaryotic phenomena, and ignoring the "wild card" of eukaryotic evolution—whole-genome incorporation.

Though evidence from crystal structures and from in vitro biochemical tests indicates that FtsZ and tubulin had a common ancestor protein, this is different from saying that one contemporary protein is the ancestor of another contemporary protein. Both proteins have become firmly established in their respective structures, FtsZ in the Z-ring and tubulin in microtubules. Over billions of years of evolution, they have both no doubt changed to fit into their respective structures. Because they have very little gene sequence in common, are functionally different, and presumably diverged from their common ancestor long ago, a comparison of these two proteins cannot provide much information on the genetic basis of microtubules. To say that these two proteins derived from a common gene product provides virtually no insight into the evolutionary origin of the microtubules.

Doubts can be raised about the FtsZ ancestry of tubulin because of the functional differences in cell division between the two proteins. Microtubules form the mitotic spindle that runs from pole to pole and pole to chromosome. They are anchored at a nucleation site and rapidly polymerize and depolymerize. The motor proteins dynein and kinesin move along the tubules. The Z-ring forms between the poles of a dividing bacterial cell. FtsZ forms a scaffold for the numerous other proteins that make up the Z-ring. In early studies, the Z-ring's role was compared to that of actin because FtsZ forms a ring in the same position in the cell as does the actin-based cleavage furrow of animal cells.

To bridge the functional cell biological gap between the two proteins, the results of in vitro FtsZ studies are often used to explain in vivo phenomena even though FtsZ does not form structures in the cell as it does when purified. FtsZ does not appear to polymerize in vivo. It forms a ring that is bound to the cell membrane, but it does not form filaments in the normal cell. "Spiral tubules" have been induced in vivo when the protein is overexpressed, but it is hard to say the structures in this case were tubules because the study was done using light microscopy.[31] There have been no reports using electron microscopy to show FtsZ structures in cells in contrast to those that reveal the easily seen tubulin microtubules. FtsZ only displays in vitro the tubulin-like properties that tubulin displays in vivo.

Analysis of the evolutionary origin of tubulin is further complicated, particularly in the bacteriological literature, which is supplying the FtsZ data, by imprecise language that evokes eukaryotic cytoskeletal features and eukaryotic-like motility phenomena for the Z-ring and other aspects of bacterial cell division. It is crucial to distinguish the two types of motility involved in this discussion: motion produced by growth, and that produced by motile proteins. Bacterial cells reproduce very quickly by adding proteins and other compounds to their cell walls, and so "move" across the medium. Motile proteins, such as myosin or dynein, in contrast, are attached to a structure and undergo a conformational change that generates motion. After it was shown with immunoelectron microscopy that the Z-ring is constricted or reduced in diameter as cell division progresses, the Z-ring was then said to "contract" (i.e., display the attributes of, or be acted upon by) a motor protein. The ring was then dubbed a "cytokinetic ring," which again evokes cytokinesis as in animal cells. However, the Z-ring is attached to the cell wall, so it is hard to see how it could contract. Furthermore there are no known motor proteins in *Escherichia coli*.[31]

It seems more likely that the constriction is caused by growth rather than by motor proteins. Lutkenhaus tries to get around this problem by writing, "the Z-ring contracts at the leading edge of the invagination, suggesting it is a dynamic structure that is undergoing remodeling."[28] Contraction (i.e., motor protein-based motility) and remodeling are two different phenomena. Because it has been found that the Z-ring has a half-life of less than a minute, it seems likely that the reduction of the ring as the septum is formed is caused by remodeling (i.e., growth), rather than contraction.[33] Although Stricker et al. admit, "the substructure of the Z-ring is not known," and others concur that "the physiologically relevant, assembled form of FtsZ in the Z ring is not known," it is still assumed, based on in vitro studies, that FtsZ forms protofilaments in the cell.[34] This uncorroborated model is then further elaborated to invoke force generation of the Z-ring based on annealing of the hypothesized protofilaments.[35] Neither protofilaments nor force generation have been observed in cells. The in vitro properties of FtsZ are often reported without an indication that they are found in vitro not in live cells.[36]

The extension of eukaryotic cell division terminology to bacteria has become widespread, sometimes in an apparent effort to bridge the gap between the two groups. Although disagreement over the use of the eukaryotic term "chromosome" in bacteria is well known, other problematic terms including "bacterial mitosis," "spindle," "tubules," "contraction," and "leading edge" are used without noting what we take to be the profound differences between bacterial and eukaryotic cell division. The recent discovery of active chromosome segregation in prokaryotes has brought forth images of "mitotic-like machinery" and an "analogue of the eukaryotic spindle."[38] However, these processes of moving DNA molecules very short distances seem a far cry from "the action of a spindle apparatus and 'motor' proteins."[37] Mitosis is characteristic of eukaryotes. It is defined as condensed chromosomes moved on a microtubule spindle. It should not be used synonymously with DNA segregation.

Proponents of the FtsZ origin of tubulin ignore one other crucial factor that should be considered in questioning the direct filiation, FtsZ-became-tubulin argument: the eukaryotic cell's propensity for whole-genome acquisition. Nucleated cells' ability to engulf and retain another cell, whether bacterial or eukaryotic, has never been found in bacteria, which are apparently blocked by their cell walls and lack of motor proteins from carrying out this process. We know from the cases of mitochondria and chloroplasts that genes have been transferred between the organelle and the nucleus and that enzymes can contain subunits from different organismic sources (e.g., one subunit from the chloroplast, one from the nucleus). A possible chimeric origin of tubulin has been put forth by Gupta, who has found that glyceraldehye-3-phosphate dehydrogenase (GAPDH) shares interesting attributes with tubulin.[38] FtsZ slight homology with tubulin could then reflect a chimeric origin of the protein, with sequence derived from two organismic lineages, rather than by direct filiation from a common FtsZ-tubulin ancestor. Thus, the evolutionary origin of microtubules remains unclear. The possibilities include direct filiation of tubulin from a common ancestor with FtsZ, a fusion product of the FtsZ ancestor and that of another protein (e.g., GAPDH), an as-yet-to-be described protein that perhaps forms cytoplasmic tubules, or some combination of these. There is no molecular or biochemical evidence to support the addition of microtubules to the eukaryotic nucleocytoplasm from a distinct symbiont. The prospect of concluding whether tubulin evolved from a direct filiation within the nucleocytoplasm or from a symbiotic addition to the ancient eukaryotic ancestor rests on our ability to distinguish symbiont-dervied bacterial genes in eukaryotes from those in the genetic soup of our genetically not-yet-annealed ancestors.[39]

References

1. Donna F. Kubai, "The Evolution of the Mitotic Spindle," *International Review of Cytology* 43 (1975): 167–227, 168.
2. Dennis Bray, *Cell Movements* (New York: Garland Publishing, 1992), 204.
3. Jeremy Pickett-Heaps, "The Evolution of Mitosis and the Eukaryotic Condition," *BioSystems* 6 (1974): 37–48.
4. Carl R. Woese, "Bacterial Evolution," *Microbiological Reviews* 51 (1987): 221–271.
5. C.R. Woese, O. Kandler, and M.L. Wheelis, "Towards a Natural System of Organisms: Proposal for the Domains Arachaea, Bacteria, and Eucarya," *Proceedings of the National Academy of Sciences USA* 87 (1990): 4576–4579.
6. Lynn Margulis, *Symbiosis in Cell Evolution* (New York: H. Freeman, 1981). See also

L. Margulis, M.F. Dolan, and R. Guerrero, "The Chimeric Eukaryote: Origin of the Nucleus from the Karyomastigont in Amitochondriate Protists," *Proceedings of the National Academy of Sciences USA* 97 (2000): 6954–6959.

7. D. Bermudes, S.P. Fracek Jr., R.A. Laursen, L. Margulis, R. Obar, and G. Tzertzinis, "Tubulin-Like Protein from *Spirochaeta bajacaliforniensis*," *Annals of the New York Academy of Sciences* 503 (1987): 1987–1997.
8. D. Munson, R. Obar, G. Tzertzinis, and L. Margulis, "The 'Tubulin-like' S1 Protein of *Spirochaeta* is a Member of the hsp65 Stress Protein Family," *BioSystems* 31 (1993):161–167.
9. J.L. Hall, Z. Ramanis, and D.J.L. Luck, "Basal Body Centriolar DNA—Molecular genetic Studies in *Chlamydomonas*," Cell 59 (1989): 121–132.
10. J.L. Hall and D.J.L. Luck, "Basal Body-Associated DNA—*In Situ* Studies in *Chlamydomonas reinhardtii*," *Proceedings of the National Academy of Sciences USA* 92 (1995): 5129–5133.
11. Jan Sapp, "Freewheeling Centrioles," *History and Philosophy of the Life Sciences* 20 (1998): 255–290.
12. D. Bermudes, G. Hinkle, and L. Margulis, "Do Prokaryotes Contain Microtubules?" *Microbiological Reviews* 58 (1994): 387–400.
13. I. Alba, M. Torreblanca, M. Sánchez, M.F. Colom, and I. Meseguer, "Isolation of the Fibrocrystalline Body, a Structure Present in Haloarchaeal Species, from *Halobacterium salinarum*," Extemophiles 5 (2001): 169–175.
14. C. Woese, "The Universal Ancestor," *Proceedings of the National Academy of Sciences USA* 95 (1998): 6854–6859.
15. R.S. Gupta, K. Aitken, M. Falah, and B. Singh, "Cloning of *Giardia lamblia* Heat Shock Potein HSP70 Homologs: Implications Regarding Origin of Eukaryotic Cells and of Endoplasmic Reticulum," *Proceedings of the National Academy of Sciences USA* 91 (1994): 2895–2899.
16. R.S. Gupta, "Protein Phylogenies and Signature Sequences: A Reappraisal of Evolutionary Relationships among Archaebacteria, Eubacteria and Eukaryotes," *Microbiology and Molecular Biology Reviews* 62 (1998): 1435–1491.
17. W. Martin and M. Müller, "The Hydrogen Hypothesis for the First Eukaryote," *Nature* 392 (1998): 37–41.
18. D. Moreira and P. López-García, "Symbiosis Between Methanogenic Archaea and Delta-Proteobacteria as the Origin of Eukaryotes; the Syntrophic Hypothesis," *Journal of Molecular Evolution* 47 (1998): 517–530.
19. W. Martin, M. Hoffmeister, C. Rotte, and K. Henze, "An Overview of Endosymbiotic Models for the Origins of Eukaryotes, Their ATP-producing Organelles (Mitochondria and Hydrogenosomes) and Their Heterotrophic Lifestyle," *Biol. Chem.* 382 (2001): 1521–1539.
20. P. López-García and D. Moreira, "The Syntrophy Hypothesis for the Origin of Eukaryotes," in J. Seckbach, ed., *Symbiosis: Mechanisms and Model Systems* (Boston: Kluwer, 2002), 133–146.
21. M. Rizzotti, "Cilium: Origin and 9-Fold Symmetry," *Acta Biotheoretica* 43 (1995): 227–240.
22. T. Cavalier-Smith, "The Phagotrophic Origin of Eukaryotes and Phylogenetic Classification of Protozoa," *International Journal of Systematic and Evolutionary Microbiology* 52 (2002): 297–354.
23. J.F. Lutkenhaus, H.Wolf-Watz, and W.D. Donachie, "Organization of Genes in the ftsA-envA Region of the *Escherichia coli* Genetic Map and Identification of a New Fts Locus (FtsZ)," *Journal of Bacteriology* 142 (1980): 615–620.
24. H.P. Erickson, "FtsZ, A Prokaryotic Homolog of Tubulin?" *Cell* 80 (1995): 367–370.
25. E. Nogales, S.G. Wolf, and K.H. Downing, "Structure of the Tubulin Dimer by Electron Crystallography," *Nature* 391 (1998): 199–203.
26. J. Löwe and L.A. Amos, "Crystal Structure of the Bacterial Cell-Division Protein FtsZ," *Nature* 391 (1998): 203–206.
27. H.P. Erickson, D.W. Taylor, K.A. Taylor, and D. Bramhill, "Bacterial Cell Division Protein FtsZ Assembles into Protofilament Sheets and Mini-Rings, Structural Homologs of Tubulin Polymers," *Proceedings of the National Academy of Sciences USA* 93 (1996): 519–523.

28. J. Lutkenhaus, "Dynamic Proteins in Bacteria," *Current Opinion in Microbiology* 5 (2002): 548–552.
29. C.R. Woese, "On the Evolution of Cells," *Proceedings of the National Academy of Sciences USA* 99 (2002): 8742–8747.
30. S.G. Addinall and B. Holland, "The Tubulin Ancestor, FtsZ, Draughtsman, Designer and Driving Force for Bacterial Cytokinesis," *Journal of Molecular Biology* 318 (2002): 219–236.
31. X. Ma, D.W. Ehrhardt, and W. Margolin, "Co-Localization of Cell Division Proteins FtsZ and FtsA to Cytoskeletal Structures in Living *Escherichia coli* Cells Using Green Fluorescent Protein," *Proceedings of the National Academy of Sciences USA* 93 (1996): 12998–13003.
32. N. Nanninga, "Morphogenesis of *Escherichia coli*," *Microbiology and Molecular Biology Reviews* 62 (1998): 110.
33. J. Stricker, P. Maddox, E.D. Salmon, and H.P. Erickson, "Rapid Assembly Dynamics of the *Escherichia coli* FtsZ-ring Demonstrated by Fluorescence Recovery after Photobleaching," *Proceedings of the National Academy of Sciences USA* 99 (2002): 3171–3175.
34. W. Margolin, "Bacterial Division: The Fellowship of the Ring," *Current Biology* 13 (2003): R16–R18.
35. J. Stricker, P. Maddox, E.D. Salmon and H.P. Erickson, "Rapid Assembly Dynamics of the *Escherichia coli* FtsZ-Ring Demonstrated by Fluorescence Recovery after Photobleaching," *Proceedings of the National Academy of Sciences USA* 99 (2002): 3171–3175.
36. N. Buddelmeijer and J. Beckwith, "Assembly of Cell Division Proteins at the *E. coli* Cell Center," *Current Opinion in Microbiology* 5 (2003): 553–557.
37. M.E. Sharpe and J. Errington, "Upheaval in the Bacterial Nucleoid," *Trends in Genetics* 15 (1999): 70–74.
38. R.S. Gupta and B.J. Soltys, "Prokaryotic Homolog of Tubulin? Consideration of FtsZ and Glyceraldehyde 3-Phosphate Dehydrogenase as Probable Candidates," *Biochemistry and Molecular Biology International* 38 (1996): 1211–1221.
39. C. Woese, "The Universal Ancestor," *Proceedings of the National Academy of Sciences USA* 95 (1998): 6854–6859.

15

Heritable Microorganisms and Reproductive Parasitism

JOHN H. WERREN

Some of the most important evolutionary transitions of life have involved symbiotic relationships. For example, evidence is now overwhelming that mitochondria, a vital eukaryotic organelle, originally evolved from an intracellular bacterium present in a primitive eukaryote.[1–5] The very greening of our planet is the result of a second symbiont–host relationship; chroloplasts originally evolved from a cyanobacterial symbiont present in the ancestors of algae and plants.[1–3,6] However, symbioses are not just a thing of the past. Intimate associations between microbial symbionts and eukaryotic hosts are incredibly widespread in nature and may be an important force in evolution.[7–13]

Symbiotic relationships with hosts range from mutualistic to parasitic, and modes of symbiont transmission range along a continuum from completely horizontal (infectious) to completely vertical (inherited).[14–17] Examples of horizontally transmitted symbionts include the nitrogen-fixing bacteria of plants, algal-bearing symbionts of corals, and a variety of pathogenic microorganisms, including some human disease agents. Other symbionts are transmitted vertically, being passed from hosts to their offspring. Some "heritable symbionts" reside within the cells of their hosts and are transmitted within eggs, whereas others are transmitted via the female's reproductive tract or get passed from parent to offspring by other mechanisms.[8] Some microorganisms employ a mixed vertical and horizontal transmission strategy.

The mechanism of transmission between hosts is a key feature shaping symbiont relationships with their hosts.[15, 17–20] It is generally recognized that symbionts with horizontal transmission can readily evolve either parasitic or mutualistic host associations. In contrast, there was a long-held view that heritable symbionts must inevitably evolve mutualistic associations with their hosts.[18–20] The idea is simple: because heritable symbionts are transmitted via the reproduction of their hosts, any harmful effects on their hosts should be selected against and beneficial ones favored. However, we now know that this view is overly simplistic. There is an alternative set of adaptations available to heritable symbionts that can best be described as "reproductive parasitism."[15,21] Reproductive parasites alter the reproduction of hosts in ways that are advantageous to the symbiont, but potentially harmful to the host. Evidence is accumulating that microbial exploitation of host reproduction is widespread and common. Examples of microbial alterations of host reproduction include male killing, feminization, parthenogenesis induction (reproduction without mating), and induction of sperm–egg incompatibilities.[12,21–27] Reproductive parasites include fungi, bacteria, protozoans, and viruses, and they have been found infecting all the major animal and plant taxa.[12,26] Reproductive parasitism can involve both vertically and horizontally transmitted microbes, although the dynamics and adaptations selected for by these two transmission modes differ. For example, some horizontally transmitted microbes also infect the reproductive tracts of their hosts and increase their transmission directly via host sexual behavior (e.g., sexual transmission) or indirectly by reproductive castration of the host.[28–31] Attention in this chapter will be focused on reproductive parasites that are transmitted primarily via a vertical transmission mode.

Reproductive parasites may be important in shaping the evolution of eukaryotic hosts. Here, I explore the basic biology of reproductive parasitism. Principles are then illustrated with a more detailed discussion of *Wolbachia*, a widespread group of intracellular alpha proteobacteria that induce a number of reproductive alterations in hosts. The generality of these principles to all microorganisms with a vertical transmission pattern is emphasized.

Reproductive Parasitism as an Adaptive Strategy

Vertically transmitted microorganisms are common in nature, and exclusive or nearly exclusive vertical transmission has evolved independently many times and by different microorganisms in different hosts. Examples include the independent evolution of inherited γ-proteobacterial endosymbionts in different insects, heritable symbioses by some dinoflagellate associates of coral, eubacteria in deep-sea vent-worms, microsporidia associated with various arthropods, and parasitic fungi of plants.[25–40]

The factors that cause microorganisms to evolve from horizontal to predominantly vertical transmission are still not completely understood, although presumably they arise from microorganisms with mixed (vertical and horizontal) transmission strategies and subsequent selection favoring enhancement of the vertical and discouragement of the horizontal mode. A likely scenario would be a change in ecological circumstances that reduce the efficiency of horizontal transmission (e.g., a reduction in population density), thus enhancing the relative value of the heritable transmission mode.[41] Once predominantly vertical transmission has evolved, the adaptive "options" available to these "inherited" microorganisms become more restricted. For example, killing the infected host no longer makes adaptive sense, as this would lead to a reduction in transmission of the parasite. As previously mentioned, this observation led to the widely held belief that vertically transmitted symbionts evolve only mutualistic associations with their hosts. However, this is not correct. It is important to also consider the opportunities for manipulation of host reproduction that are available to inherited microbes.

The idea is most readily visualized by considering the evolution of sex ratio–distorting microbes. An important feature of vertical transmission, particularly in organisms with anisogamous gametes (gametes of unequal size such as sperm and eggs or pollen and ovules), is that vertical transmission is asymmetric. Inherited microorganisms are transmitted to offspring through eggs, but they typically have low or no transmission through sperm. This asymmetric inheritance creates a divergence in "genetic interests" between the symbiont and host[15] that is overlooked in some theoretical treatments of this question.[17–19] The asymmetry in inheritance creates strong selection on heritable microbes to increase the number, survival, and fitness of infected females (the transmitting sex), regardless of their negative fitness consequences to males (the nontransmitting sex).[15,42] Therefore, inherited microbes are selected to distort sex ratios toward female production, and such sex ratio distorters are widespread in nature.[22–25]

We can divide reproductive parasitism into five broad categories based on the phenotypic effect on hosts. Each involves an alteration of the host's reproduction in ways that enhance transmission of the microorganism. They are sex ratio distortion, cytoplasmic incompatibility, postsegregation distortion, germline enhancement, and reproductive castration.

Werren and O'Neill[15] developed a simple framework that illustrates the spectrum of adaptive options available to strictly vertically transmitted symbionts. For such a symbiont to become established in a host population, an infected female must produce, on average, more infected daughters (weighted by their survival and fecundity) than an uninfected female produces uninfected daughters. Definition of the following basic terms allows the construction of a formula for the statement above. The symbiont is transmitted vertically to "α" proportion of the daughters, fitness (survival and fecundity) of infected females is W_i and of uninfected females is W_u, and the proportion of female progeny (the primary sex ratio) produced by infected and uninfected females is x_i and x_u, respectively. The symbiont will be maintained in the population so long as

$$\alpha x_i W_i / x_u W_u > 1.$$

In words, the infection increases in frequency when the "effective" number of infected daughters produced by an average infected female is greater than the effective number of uninfected daughters produced by an average uninfected female. It should be noted that fitness of male progeny is irrelevant to the increase of a cytoplasmically inherited symbiont unless the females preferentially mate with their siblings.

The "classic" option available to an inherited symbiont is mutualism—the symbiont increases in frequency by enhancing fitness of infected females (W_i). Mutualistic endosymbioses are widespread in nature, and many of these involve heritable microorganisms. Many fascinating examples are described by Buchner[8] and have been reviewed by others.[13,41,43] Therefore, I will not consider them further except to point out that mutualistic symbionts will also be selected for features of reproductive parasitism when such features enhance their transmission, just as reproductive parasites can be selected for mutualistic effects. That is, mutualism and reproductive parasitism are not mutually exclusive strategies.[12]

Besides mutualism, the following modes of reproductive parasitism are available to heritable microbes, and examples of most of these have been found in nature.

Sex Ratio Distortion

Sex ratio–distorting microbes are incredibly common in nature. Four basic types of sex ratio distorters have been described; *feminizers* (which convert genetic males into functional females), *parthenogenesis inducers* (which induce production of female offspring by unmated females), *sex ratio biasers* (which alter the primary sex ratio among infected mother's offspring), and *male killers* (which induce death of male offspring but not female offspring of infected mothers).

Feminizers have been described in a number of species. For example, *Wolbachia* convert genetic males into female in the isopod *Armadillidium vulgare*[44,45] and some moths.[46,47] Similarly, microsporidia induce feminization in the intertidal shrimp *Gammarus duebeni*.[48,49] With reference to the formula above, feminizing microbes increase in frequency not by increasing the fitness of infected individuals but by increasing the production of female offspring (x_i), the sex that transmits the microbe. In a large, random-mating population, sex ratio distorters can in principle increase to fixation (100%), thus driving the population to extinction.[42,50,51] The actual scarcity of males will not prevent increase of the sex ratio distorter unless infected females are more likely to go unmated than uninfected females. However, various factors can restrain their spread, including selection for host suppression and deme (group) level selection between local populations.

Male-killing microbes have been reviewed in several recent treatments.[52–54] A diverse set of microorganisms have been found to induce male killing in different hosts. They include γ-proteobacteria, α-proteobacteria, spiroplasms, and microsporidea.[55–61] This indicates that male killing can evolve readily and that the ecological circumstances that favor its evolution occur frequently in nature. Male killing by a heritable microorganism is selectively favored when elimination of the male enhances the fitness of infected female siblings. Thus, male killers do not increase by directly altering the primary sex ratio of their hosts (x_i), but, indirectly, by increasing the fitness of infected females (W_i) via killing of male siblings. The conditions favoring male killing can occur commonly when siblings compete for limited resources or when male killing reduces sibling mating and therefore inbreeding depression.[42] Male killing in animals is selectively analogous to mitochondrial-induced cytoplasmic male sterility (CMS) in plants.[62] In the latter case, mitochondrial variants in a wide range of plant species inhibit development of anthers, thus converting bisexual plants into females. Because mitochondria, like heritable microorganisms, are typically inherited through the matriline in plants but not through the patriline, CMS is selectively favored by mitochondria so long as the fitness through ovules is increased.

An interesting form of male killing occurs in the mosquito *Culex salinarius*. Some members of this species are infected with the microsporidean *Amblyospora californica*, which is transmitted through the eggs of infected females. Microsporidea occurring in male offspring proliferate in the larvae, typically inducing death. The resulting spores are not capable of directly infecting mosquitoes, but infect an intermediate copepod host, which in turn produces spores that infect filter-feeding mosquito larvae.[56] Thus, *A. californica* uses females for vertical transmission and males for horizontal transmission.

In principle, sex ratio distorters also can increase in frequency by decreasing the sex ratio of uninfected individuals (x_u). To date, only one unequivocal case of this has been found: *Wolbachia*-infected males in the haplodiploid insect *Nasonia vitripennis* induce a sperm–egg incompatibility that converts fertilized eggs of uninfected females (which would normally develop into female progeny) into males.[63] It is also possible that symbionts, by imprinting modifications of paternally derived sex determination genes, could manipulate sex ratio in uninfected females; however, there is currently no clear-cut evidence to support this proposition.

Sex ratio distorters have important implications for the evolution of sex determination[63] and mating systems,[64–66] which will be discussed in more detail later.

Cytoplasmic Incompatibility

Cytoplasmic incompatibility (CI) is an incompatibility between the sperm of infected males and the eggs of uninfected females (or females infected with a different symbiont strain) that typically results in death of the zygote (fig. 15.1). CI is the mirror-image alternative to mutualism. Whereas mutualistic symbionts increase in frequency in host populations by increasing the survival and reproduction of infected females (W_i), CI-inducing symbionts indirectly increase in frequency by decreasing the fitness of uninfected females (W_u). Many strains of the widespread bacterial group *Wolbachia* are known to induce CI.[67–69] Recently, a strain of Citophaga-Like-Organisms (or CLOs), another widespread group of arthropod endosymbionts,[70–74] has been found to cause CI.[71] The cytological mechanisms of CI have been described for *Wolbachia*, where incompatible crosses typically result in defects in maturation of the male pronucleus, and disrupted

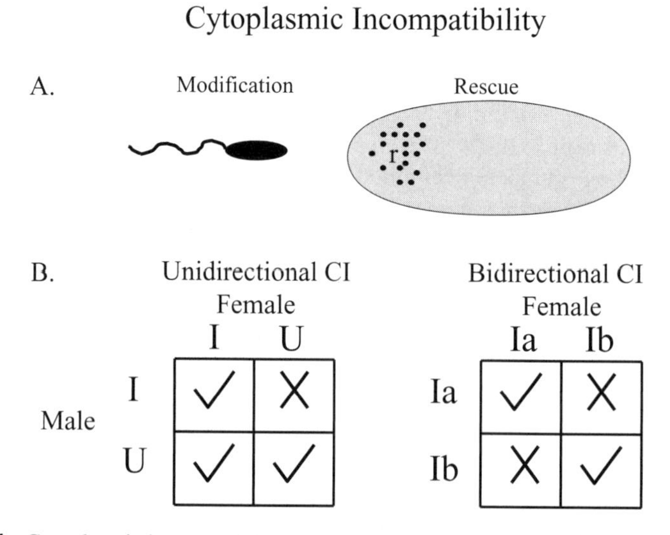

Figure 15.1. Cytoplasmic incompatibility (CI). (A) CI involves a modification of the sperm by the symbiont and a rescue in the eggs. The biochemical mechanisms remain unclear. (B) There are two forms of CI. Unidirectional CI typically occurs when a male is infected and a female is uninfected. Offspring of such crosses suffer increased mortality resulting from destruction of the paternal chromosomes. All other cross combinations are compatible. Bidirectional CI typically occurs in crosses between individuals infected with different Wolbachia that are not able to rescue the other's modification. As a result, incompatibility occurs in both reciprocal crosses.

early mitotic divisions,[75–78] typically leading to embryonic death. These observations have led to a model that *Wolbachia* modifies sperm and that this modification can be rescued when the same *Wolbachia* strain is also present in the egg (figure 15.1).[27] Consistent with sperm modification, *Wolbachia* apparently need to be present in the developing sperm cysts if they are to induce CI when the sperm fertilizes the egg, although *Wolbachia* themselves are not present in mature sperm.[79,80] However, the biochemical mechanisms of CI remain obscure.

There are two types of CI: unidirectional and bidirectional (fig. 15.1). Unidirectional incompatibility typically occurs between infected males and uninfected females; the sperm from males of infected strains is incompatible with uninfected eggs, whereas sperm from infected eggs is compatible with either infected or uninfected eggs—hence the name unidirectional incompatibility. It can also occur between two infections if one is incapable of rescuing the modification of the other. Bidirectional incompatibility (BiCI) occurs when the male and female carry different strains of symbiont that are mutually incapable of rescue of modifications induced by the alternate strain. BiCI is found both within and between species in nature.[81–83] There is widespread interest in the possibility that CI promotes divergence and speciation in arthropods, and this point will be discussed further later. Although so far only two groups of bacteria have been found to induce CI, the phenotype can be selectively favored whenever microorganisms are predominantly or exclusively inherited through host eggs.[15,84,85]

Post Segregation Killing and Maiming

Postsegregation killing is a mechanism employed by some genetic elements that have incomplete transmission to progeny. Postsegregation killing involves two components: a modification in the parent ("toxin") and a rescue in the offspring ("antidote"). The modification persists in progeny, and therefore the genetic element must also be transmitted to rescue, or the progeny dies. The classic examples are killer-plasmids in bacteria and spore-killers in fungi,[86] although even restriction–modification systems of many bacteria may have evolved because of the advantages of postsegregation killing.[87] The advantages of postsegregation killing to the genetic factor is that it increases the frequency of individuals within a population that carry the genetic element. However, this advantage only occurs when the infected and uninfected progeny of a parent compete with each other for resources. Under these circumstances, the infected progeny gain in fitness from the death of their uninfected siblings, favoring spread of the phenotype. The phenotype is especially likely to be important during the initial phases of invasion of heritable symbionts into a population, when microbial densities within hosts are likely to be lower and, therefore, transmission rates are also lower.

An explicit case of postsegregation killing has not yet been described for a heritable symbiont. Demonstrating its existence requires documenting preferential mortality of embryos that do not receive the symbiont, and also documenting that this is not the result of mutualistic benefits of the symbiont but only occurs when the parent is infected and the progeny looses the infection. Some cases of mutualism that have been inferred from a reduction in fitness following antibiotic curing of symbionts may be caused by postsegregation killing.

It should be emphasized that killing is just an end-point along the continuum. Any interacting set of gene products from the symbiont that has the property of reducing the fitness of progeny that lose the symbiont can be selectively favored. These basic properties are persistence of the modification effect in the egg, and loss of the rescue effect such that the symbiont must also be present in the progeny for rescue to occur. Later, I also describe a case in which postsegregation killing of germ cells can be favored during germ cell proliferation.

Germline Manipulation

Heritable microorganisms will be under strong selection to increase their transmission through the germline. This can be accomplished by a number of different mechanisms. First, vertically transmitted microorganisms are under intense selection to localize to the ovaries to enhance transmission. Buchner[8] describes a number of mechanisms by which this is achieved in heritable symbionts of insects. For organisms that have a discrete region of the egg that develops into the germline and associated biochemical signals (e.g., the polar granules of *Drosophila*), heritable microorganisms will localize to the germinal pole, as is observed in many cases.[8,63,88] The mechanisms by which this is achieved are still not completely clear. However, eggs that contain germ-cell determinants are found in many species of protostomes and deuterostomes,[89] and we therefore expect symbiotic microorganisms to adapt to these signals for orientation to the germline in eggs.

More subtle effects, although not described, are expected to occur. For example, microorganisms in the general region of the germ-pole should be selected to produce inputs that enhance the probability that the cells in which they occur develop into germ cells. This could be achieved by localizing polar granules to their vicinity or by production of products that enhance the cellular pathways leading to germline determination. These scenarios are not as far-fetched as they might seem. In *Drosophila*, the polar granules include a nontranscribed mitochondrial ribosomal RNA that is required for germ-cell determination, and translation of other polar RNA's may require this mitochondrial ribosomal RNA machinery.[90,91] Indeed, this may reflect an ancestral mechanism to ensure that mitochondria, a highly derived endosymbiont, localize to the germ-line. It is easy to imagine that localization of this or other germ particle determinants to the surface of endosymbiotic microbes would enhance their probability of occurring within germ cells.

A second mechanism of ensuring representation in eggs could occur during proliferation of primordial germ cells. During cell proliferation, endosymbionts must also proliferate and segregate to daughter cells. During this process, it is likely that stochastic processes will lead to loss of endosymbionts in some daughter cells, and such loss will be most likely when microbial densities are low during the initial stages of germ-cell proliferation (see fig. 15.2). This is precisely the kind of scenario that can favor postsegregation killing by endosymbionts. Because germ cells effectively compete for transmission to the next generation (a female makes many more primordial cells than will become mature eggs), postsegregation killing of germ cells that have lost the symbiont will be advantageous to the microorganism and will increase the effective transmission rate. Such an effect may be occurring in the wasp *Asobara tabida*, where antibiotic treatment of larvae results in failure of ovaries to mature.[92] This may be the result of a mutualistic effect of the endosymbiotic bacterium but could also be caused by the death of germ-cells that have lost the bacterium. More detailed cytological stud-

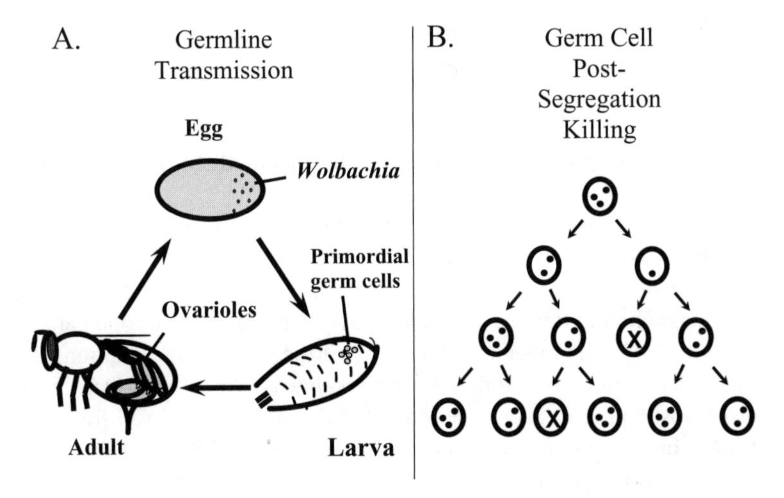

Figure 15.2. (A) Germ-line transmission of *Wolbachia*. *Wolbachia* within the egg localize to the egg germ-pole and, as a result, infect primordial germ cells and end up in the ovarioles, where they are again transmitted to eggs. *Wolbachia* vary in the extent to which they localize to the egg germ-pole in different host species. The strong germ-pole localization pattern observed in the parasitic wasp *Nasonia vitripennis* is shown here. (B) Hypothesized germ cell postsegregation killing by reproductive parasites such as *Wolbachia*. By a modification-rescue system (modification in parent cell and rescue in the daughter cell), those daughter cells that do not receive the symbiont (e.g., due to stochastic loss) will die or fail to replicate. This mechanism enhances germ-line transmission of the bacterium, particularly when loss of uninfected germ cells is compensated for during development or the host is not egg-limited during reproduction in nature.

ies, and experiments using subcuring doses of bacteria, would help resolve whether this is the case. The role of heritable microorganisms in germ-line manipulation has been an understudied topic. However, focus on this area is likely to yield a number of interesting discoveries.

Parasitic Castration

A number of microorganisms have been described that sterilize their hosts.[25,92] The selective advantages of parasitic castration are variable. In some cases, the reproductive organs are converted into "factories" to produce parasites, which then horizontally infect new hosts. In other cases, castration prolongs the life of the host, resulting in greater opportunities for infectious transmission. Parasitic castration is also relevant to reproductive parasites with a vertical transmission mode. In hosts with both sexual and asexual reproduction modes, castration of the sexual organs can lead to an increase in asexual reproduction, and hence to an increase in vertical transmission of the infection. For example, species of the grass *Danthonia* produce both open sexual flowers that reproduce by outcrossing and closed flowers that reproduce by selfing and develop next to the parent. *Danthonia* are often infected with a parasitic fungus (*Atkinsonella*) that castrates the sexual flowers and infects ovules in the selfing flowers, resulting in effective vertical transmission.[25] Many endophytic fungi employ mixed vertical and horizontal transmission modes, with vertical transmission occurring through tillers or seeds.[25,92]

The Diversity and Abundance of Reproductive Parasites

A surprisingly diverse assemblage of heritable microorganisms induce reproductive alterations in their hosts. For example, male-killing microorganisms include γ-proteobacteria such as *Arsenophonus nasoniae*, α-proteobacteria such as *Rickettsia* and *Wolbachia*, other bacterial groups including flavobacteria and spiroplasms, and eukaryotic parasites such as microsporidea.[52–61] This diversity indicates that the ecological conditions that favor male killing occur commonly and that it is relatively easy for microorganisms to detect the sex of the host and to induce male lethality. The pattern observed for cytoplasmic incompatibility inducing microorganisms is quite different. Until recently, *Wolbachia* was the only bacterial group shown to induce CI, but recent studies indicate that relatives of the bacterial group *Citophaga* can also cause CI in some hosts.[70] The pattern indicates that the modification of sperm and rescue in eggs required of CI is relatively difficult to evolve. Feminizing microorganisms have been found in the *Wolbachia* and microsporidea, and *Rickettsia* have also been implicated as feminizers.[44–49] Parthenogenesis-inducing microorganisms include the *Wolbachia* and *Citophaga*-like bacteria.[23,71–74] Parasitic fungi in plants, and *Wolbachia* in insects and nematodes, have been found to induce host sterility under some conditions.[31,92]

Microbiologists almost certainly vastly underestimate the distribution and abundance of reproductive parasites. For example, the *Rickettsia* are generally recognized as a group of intracellular bacteria that are vectored by arthropods and that induce various pathologies in vertebrates (e.g., Rocky Mountain Spotted Fever, Scrub Typhus, and Murine Typhus). However, recent studies have revealed a number of *Rickettsia* that induce male killing in their arthropod hosts that have no known vertebrate pathology.[57,60] Given the ascertain bias here (human or vertebrate disease agents were much more likely to be detected historically) and the incredible abundance of arthropods, a reasonable prediction is that that the vast majority of *Rickettsia* will turn out to be reproductive parasites of arthropods, with only a small minority inducing vertebrate disease. Indeed, it may be that most vertebrate pathogens from this genus evolve from reproductive parasites of arthropods. It remains to be seen whether most *Rickettsia* are traditional infectious pathogens of arthropods or more usually reproductive parasites.

Additional evidence points to the underdetection of reproductive parasites. Many of these microbes are difficult to culture, and thus therefore have gone largely unnoticed by microbiologists until the advent of molecular detection methods such as polymerase chain reaction (PCR). For example, the *Wolbachia* were considered to be a rather obscure and inconsequential bacterial group until the use of PCR methods revealed their incredible abundance in arthropods and filarial nematodes.[93–98] *Arsenophonus nasoniae*, a relative of *E. coli*, was originally detected because of its induction of male killing in the parasitic wasp *Nasonia vitripennis*, but recent studies have uncovered close relatives of this bacterium (based on 16S rDNA sequence) in other insects and ticks.[99] The bacterial group is likely to be widespread. That *Citophaga*-like bacteria induce reproductive alterations has only recently been discovered, and evidence indicates that these, too, are widespread in arthropods.[71–74]

Similar surveys in vertebrates have not yet been done. It is possible that some vertebrate groups also harbor heritable microorganisms, and these would be selected to manipulate reproduction in their hosts. Finally, the large number of mutualistic symbionts found in invertebrates and plants are also subject to the same selective pressures favoring reproductive manipulations described above—mutualism may not be the only quiver in their bow.

Wolbachia: Master Manipulators of Eukaryotic Reproduction

Among the diverse array of reproductive parasites, the currently undisputed "master manipulator" of host reproduction is the genus *Wolbachia*. *Wolbachia* are a widespread group of cytoplasmically inherited intracellular bacteria, found in 20%–75% of insect species and also commonly in arachnids, crustaceans, and nematodes.[93–98]

Wolbachia cause a number of reproductive alterations in their eukaryotic hosts, including sperm–egg incompatibility (CI), feminization, male killing, and (MK) parthenogenesis induction and are required for ovarian development in some hosts.[23,24,26,27,100] Although the common mode of transmission of *Wolbachia* within host species is cytoplasmic (vertical), these bacteria are also transmitted horizontally between arthropod species.[101,102] The mechanisms of horizontal movement of *Wolbachia* between arthropods remain unknown, but a consequence is that *Wolbachia* are among the more abundant symbiotic bacteria in terrestrial ecosystems.[93–98] Arguably, the spread of *Wolbachia* represents one of the great pandemics of life on this planet. Nevertheless, the importance of *Wolbachia* to arthropod ecology and evolution remains controversial and the topic of empirical and theoretical study.[103–110]

Wolbachia are intracellular symbionts of invertebrates that occur in both reproductive and somatic tissues[63,69]; are widespread and common in arthropods and nematodes; are maternally transmitted within a host species, but readily move between host species by unknown mechanisms; cause various reproductive alterations in their hosts; and may have important consequences for arthropod ecology and evolution. Some relevant details are provided here.

Wolbachia Phylogeny and Ecology

Wolbachia are obligatory intracellular alpha proteobacteria in the order Rickettsiales. Their closest known relatives are Rickettsiales that are arthropod-vectored pathogens of vertebrates.[111,112] To date, *Wolbachia* have only been found to infect invertebrates. A note on nomenclature: the type species of *Wolbachia* is *W. pipientis*, described from the mosquito *Culex pipiens*. Throughout this chapter I simply refer to "*Wolbachia*" without a species designation. The reason for this is that *Wolbachia* are a diverse and abundant bacterial group almost certainly involving more than one species, but for which species designations have not been resolved.

There are two major subdivisions of *Wolbachia* infecting arthropods (A and B groups), estimated to have diverged about 60 million years ago based on 16S and protein gene synonymous substitution rate estimates for bacteria,[101,102,113] and two additional subdivisions have been added for divergent *Wolbachia* found in termites and collembola.[114]

Wolbachia in filarial nematodes comprise two other subgroups (C and D), and all six subgroups are monophyletic with respect to other *Rickettsia* and related genera.[95,111–114] Phylogenetic studies of *Wolbachia* have been performed using 16S rDNA, the cell-cycle gene FtsZ, and the outer surface protein wsp. In general, the trees based on different genes are consistent with each other, although there is some evidence of recombination within clades and possible recombination between subgroups.[115,116] Phylogenetic studies also indicate that parthenogenesis induction and male killing have evolved multiple times independently in both the A and B groups.[101] However, an alternative explanation is that recombination events (or transfer of genes via commonly associated phage) have introduced the machinery for parthenogenesis induction (and male killing) into different *Wolbachia*.

Molecular phylogenetic studies reveal that closely related *Wolbachia* can be found in very different host species.[101,102] Movement of *Wolbachia* across insect orders appears to be common. However, documenting particular patterns of horizontal movement is currently risky because the data are based on only one to two protein coding genes per strain, and therefore, gene phylogenies can be misinterpreted as actual bacterial phylogenies.

The mechanisms and patterns of *Wolbachia* movement between host species remains unclear, although a few cases of natural transfection within a species have been observed.[117] In addition, infections of individuals in a species with multiple *Wolbachia* strains (typically double infections) are not uncommon,[93,118,119] further indicating that relatively frequent horizontal transfers between species occurs; however, transmission within species appears to be predominantly vertical as evidenced by tight associations of *Wolbachia* types with mitochondrial haplotypes.[120]

Surveys of *Wolbachia* within insects have been conducted in Panama, Indiana, and Britain, revealing similar frequencies of infected species in all three regions.[93,94,97] However, the surveys are typically based on one to two individuals per species, and therefore they almost certainly represent an underestimate.[121] A few attempts have been made to document movement of *Wolbachia* within insect communities.[94] However, these have been limited in scale and lacking the genomic precision to clearly define recent horizontal transfers. These studies do confirm the general finding that distantly related *Wolbachia* can be found in closely related hosts, as well as many cases in which one insect species is infected and a close relative is not.

The patterns indicate that *Wolbachia* are acquired and lost within species,[97] but the mechanisms of loss are also poorly understood; a number of theories have been proposed.[97] One possibility is that selection for genotypes resistant to *Wolbachia* infection could explain their loss. Resistance to parthenogenesis induction and male-killing *Wolbachia* is expected, introducing the possibility of "Red-Queen" dynamics in these systems. However, unlike normal infections, females infected with CI *Wolbachia* will generally not be selected to "resist" the infection because their eggs will become incompatible with infected males in the population. Host genotypes that suppress CI *Wolbachia* modification in males or that mimic rescue in females can be selected for. Studies of host genetic effects can help reveal possible trajectories of *Wolbachia* infections within species.

Wolbachia may be undergoing rapid expansion within insects. Further sampling is needed to determine distribution of different *Wolbachia* within ecologically, taxonomically, and geographically associated insects. More detailed studies are also needed to determine whether some strains of *Wolbachia* are undergoing rapid host range expansions within insects or are moving globally within insect communities. Such studies need more precise genomic tools for characterization of *Wolbachia* strains and should not be based on single-gene phylogenies.

Wolbachia have been experimentally transferred using embryonic microinjection between species.[122–129] One interfamilial transfection has been done[123] (*Culex* to *D. simulans*) and one interorder (Hymenoptera to Diptera).[127] In the latter example, however, the hymenopteran *Wolbachia* infection was not stably maintained in *Drosophila*, indicating potential host range limitations. In general, bacteria cause similar phenotypes when transfected into a new host. In contrast, a recent interspecific transinfection resulted in a change of the *Wolbachia* reproductive phenotype from a feminizer to a male killer.[129] Transfected *Wolbachia* express CI strongly in *D. simulans* but weakly in its natural host *D. melanogaster*, indicating host effects on CI level.[126] However, systematic experimental studies of host range and effects of hosts on *Wolbachia* expression have not yet been done.

Phenotypic Effects on Hosts and Mechanisms of Action

Wolbachia cause an amazing range of phenotypic effects upon their hosts, including manipulation of sex determination (parthenogenesis, feminization, and male killing), sperm–egg incompatibilities, and alteration of oogenesis and embryogenesis. Many of these effects involve developmental and cellular functions of the host, and they probably reflect the long and intimate evolutionary association of these bacteria and their progenitors with eukaryotic cells; *Wolbachia* and related genera such as *Rickettsia*, *Anaplasma*, and *Erhlichia* are all obligatory intracellular bacteria.[111,112] This long and intimate association of *Wolbachia* with host cells and reproductive tissues is probably why *Wolbachia* have succeeded in evolving such a diverse set of mechanisms for manipulation of host reproduction.

Biology of CI

Wolbachia-induced CI arises when sperm from a male infected with *Wolbachia* fertilizes an uninfected egg or an egg infected with a different *Wolbachia* strain. Sperm from an infected male are compatible with eggs infected with the same strain of *Wolbachia* (see fig. 15.1). Incompatible crosses typically result in defects in maturation of the male pronucleus and disrupted early mitotic divisions. This leads to either embryonic death or (in haplodiploids) in conversion of the embryo to a male when the sperm chromosomes are completely eliminated. These observations have led to a model that *Wolbachia* modify sperm and that this modification is rescued when the same *Wolbachia* strain is also present in the egg.[27] BiCI occurs when the male and female carry different strains of *Wolbachia* that are not capable of mutual rescue. BiCI is found both within and between species in nature.[63,81,82,130]

Cytological analysis reveals that CI is the result of the paternal chromosome complement failing to properly condense and align on the metaphase plate during the first mitosis.[75–78] As a consequence, only the maternal chromosome complement segregates normally. These studies indicate two models for *Wolbachia* action; either *Wolbachia* block steps involved in chromosome condensation or alter timing of cellular events that occur immediately following fertilization, resulting in asynchrony of the pronuclei and disruption of paternal pronuclear processing.[78] In preparation for the first mitosis, the sperm is converted to a pronucleus that is capable of participating in development. This process involves removal of the sperm nuclear envelope; decondensation of the sperm chromatin; replacement of sperm chromosomal proteins with maternally supplied histone; assembly of a nuclear envelope, lamina, and matrix; and chromosome replication and condensation.[131] *Wolbachia* may block a distinct step in this process; for example, by targeting proteins of the condensin complex, such as topoisomerase II, or histones H1 and H3. Alternatively, *Wolbachia* may act by disrupting the rate at which the sperm is converted to a male pronucleus, leading to the male and female pronuclei entering mitosis asynchronously. Under this model, *Wolbachia* present in the egg would rescue the effect by similarly altering the rate at which the female pronucleus is prepared. Thus, infected sperm and egg would be compatible. A recent study shows that timing of pronuclear envelope breakdown in *Nasonia* is altered in incompatible crosses, supporting the timing model.[78] However, it is not yet known whether this is the primary lesion or a consequence of disruption of an earlier step.

Parthenogenesis Induction, Feminization, and Male-Killing

Wolbachia alter host sex determination and sex ratios of hosts by different mechanisms, including parthenogenesis induction, feminization, and male-killing. Some strains of *Wolbachia* induce female parthenogenetic development (females developing from unfertilized eggs) in a number of wasp species and may be involved in parthenogenetic development in others.[23] Stouthamer and Kazmer[132] demonstrated in *Trichogramma* that the first mitotic anaphase is aborted in parthenogenesis induction–infected, unfertilized eggs. The two mitotic sets of chromosomes then form a single diploid nucleus, and subsequent mitotic divisions are normal. The egg, therefore, develops as a diploid completely homozygous infected female. In some cases, gynandromorphs are produced. A similar pattern occurs in *Muscidifurax uniraptor*.[133] Interestingly, in *Trichogramma*, fertilization of an infected egg suppresses the endoduplication event, resulting in a sexually produced offspring. The cytological mechanisms for these two processes are unclear. It is obvious that we do not yet have a detailed understanding of the cellular mechanism through which *Wolbachia* induces parthenogenesis, nor whether parthenogenesis induction in different taxa of *Wolbachia* has evolved the same or divergent mechanisms.

The selective advantages of parthenogenesis induction are clear. *Wolbachia* that induce this effect are converting unfertilized eggs into female progeny, which are capable of transmitting the maternally inherited bacteria. In some cases, parthenogenesis inducing *Wolbachia* have gone to fixation, resulting in a "parthenogenetic species." Interestingly, in some cases, antibiotic curing results in complete reversion to sexual reproduction, whereas in others the resulting males are defective in some aspect of sexuality or females are unwilling or incapable of mating with males or of using sperm, thus resulting in irreversible parthenogenesis.[134–135] This is presumably caused by the accumulation of mutations affecting sexual reproduction because of the absence of natural selection to maintain it.

Feminizing *Wolbachia* have so far been described in some isopods and two moths.[42–47] Again, the adaptive significance for the bacterium is clear: conversion of genetic males to females increases the frequency of infected females in the population, giving maternally inherited bacteria that cause such an effect an advantage. Additional study is likely to reveal feminizing *Wolbachia* to be much more widespread in arthropods than currently observed. However, there may be particular aspects of host sex determination that enhance the chance that feminizing *Wolbachia* (or other heritable microbes) evolve.[24] For example, in isopods, an "androgenic gland" physiologically determines sex; manipulation of this gland by *Wolbachia* can cause feminization, resulting in functional females that are genetically male. *Wolbachia* may be particularly effective in manipulation sex determination in those taxa with systemic determination of sex rather than cell autonomous sex determination. Host taxa with female heterogamety (heteromorphic sex chromosomes in females, homomorphic sex chromosomes in males) may also be prone to bacterial manipulation of sex determination. The reason is that bacteria within the egg have the opportunity to affect which chromosome moves to the polar body versus the functional pole, thus determining sex of the offspring. In arthropods, female heterogamety is found in butterflies, moths, and some flies; in vertebrates it is found in snakes, birds, and some fish.[136]

Male-killing *Wolbachia* have been discovered in several different insects, including butterflies, beetles, and flies.[54,58,59] Given that this is a recent discovery, male killing *Wolbachia* may be far more widespread among insects.[121] Male-killing *Wolbachia* induce death of male offspring, whereas the female offspring develop normally. Females carry the infection and transfer it cytoplasmically to their female progeny, thus perpetuating the infection. Male killers are selectively favored when death of male siblings increases the fitness of infected females and can spread to moderate to high frequencies in host populations, depending on the selective circumstances. Sometimes they can become so common that they alter the mating system of their hosts.[42] The molecular and cellular mechanisms of male killing are unknown. Because male-killing bacteria have been found in male heterogametic, male homogametic, and haplodiploid species, it is likely that there is a variety of mechanisms by which the bacteria induce male death. Basic studies of the mechanisms of male killing are needed, including determining whether male-killing bacteria can cause the same phenotype when transferred to different host taxa. Interestingly, the reverse has been found where transfection of *Wolbachia* from a moth, causing feminization, resulted in the bacteria inducing male killing in its new host moth species.[47]

Other Wolbachia Phenotypes

In filarial nematodes, *Wolbachia* are believed to have evolved a mutualistic symbiosis with their hosts.[95] Antibiotic curing of *Wolbachia* results in degeneration of the host's ovaries, and antibiotic curing seems to interfere with proper molting of developing worms, indicating a mutualistic relationship.[137,138] In the parasitoid *Asobara tabida*, antibiotic curing of wasps results in inhibition of ovarian development.[92] In a startling recent finding, it was shown with certain germ-line sterile mutants in *Drosophila* that *Wolbachia* can partially rescue the sterility by inducing ovarian development.[139] These findings are consistent with the view that these endosymbionts have been selected to manipulate germline development. *Wolbachia* have been found in some species that can rescue but do not induce CI, whereas other *Wolbachia* have no discernable phenotypic effects.[68] A *Wolbachia* has also been found in laboratory cultures of *D. melanogaster* that causes premature death of adult flies.[140] Thus, effects of *Wolbachia* on hosts can range from reproductive parasitism to mutualism to no discernible effect.

Wolbachia Genome Evolution

Recent studies have begun to elucidate the genome structure of *Wolbachia*. *Wolbachia* have relatively small genome sizes (~1.2 Mb for insect *Wolbachia*), but they nevertheless contain repetitive and phage-like sequences.[141–146] The complete genome of the *D. melanogaster* Wolbachia (*w*Mel-A group) has recently been sequenced,[146] and data are available on the TIGR Web site (www.tigr.org) and the NCBI bacterial genomes site. Overall, more than 10% of the genome is composed of repeat elements. A prophage-like element designated WO has been detected, and virus-like particles have been detected that may also be WO.[144] Clearly, the *Wolbachia* genome differs from those of many other endosymbiotic bacteria that are characterized by small genomes and little repetitive DNA.[5,13] The large number of repeat elements (including phage) could play a role in intragenomic rearrangements as well as in recombination between strains of *Wolbachia*. Although recombination has been documented in a few instances,[116,117] the extent and significance of recombination in *Wolbachia* is unknown. In particular, we do not know at what rates recombination occurs within *Wolbachia* genomes, whether insertional elements serve as sites for recombination, or whether recombination and genetic exchange among *Wolbachia* is associated with double infections, host shifts, or *Wolbachia* phenotypic shifts (e.g., shifts from CI to MK). Additional genome projects are underway for nematode and insect *Wolbachia*. Considerable progress is expected over the next several years as researchers exploit the information emerging from the *Wolbachia* genome projects.[145]

Evolutionary Consequences

Given the widespread occurrence of these bacteria and their diverse effects on hosts, it is possible that *Wolbachia* have had major effects on the evolution of arthropods and filarial nematodes. Interest has focused on several aspects of the evolutionary consequences of *Wolbachia*, including host genome evolution, sex determination and mating system evolution, and host speciation and extinction.

Overall, these bacteria could have profound effects on their hosts. Alternatively, although *Wolbachia* are widespread, they may nevertheless be relatively inconsequential to arthropods. Proponents of this view argue that levels of CI are insufficient in nature to be an important component in insect speciation,[107,110] that *Wolbachia* infections are transient in host populations and unlikely to cause significant or long-lasting genetic changes in hosts, and that these bacteria probably have little effect on competition among species. The importance of *Wolbachia* remains controversial and is a topic open to empirical and theoretical investigation. Here I discuss the current state of knowledge on this topic.

Host Genetic Structure and Genome Evolution

When *Wolbachia* enter a species by horizontal transfer, the bacterium becomes associated with a particular mitochondrial haplotype. Because both are inherited maternally (cytoplasmically), a *Wolbachia* sweep through the population is expected to cause an associated mitochondrial sweep.[68,120] Such mitochondrial hitchhiking was observed during a *Wolbachia* sweep in *D. simulans*.[120] The *Wolbachia* need not go to fixation to cause a mitochondrial replacement; because of incomplete *Wolbachia* transmission, the unassociated haplotypes will eventually be replaced. A consequence of this is that, in the absence of horizontal *Wolbachia* transmission, mitochondrial variation will be eliminated following a *Wolbachia* sweep. This may allow an estimation of the relative timing of *Wolbachia* sweeps based on levels of mitochondrial variation. Alternatively, *Wolbachia* can colonize a new species via interspecific hybridization, in which case they will also introduce the heterospecific mitochondrial haplotype, which will increase along with the *Wolbachia*. There is some evidence that *Wolbachia*-mediated transfer of mitochondria between host species has occurred,[147] although the frequency of this process is unknown.

Given the ubiquity of these parasites, selection for host modifications of *Wolbachia* may be important in arthropod evolution. However, whereas hosts are expected to evolve resistance to parasites, it is not so straightforward when CI *Wolbachia* are involved. For instance, females resistant to CI *Wolbachia* could suffer severe reductions in offspring survival because of incompatibility with infected males in the population. In contrast, resistant males gain a fitness advantage because of compatibility with both infected and uninfected females. Furthermore, female genotypes that rescue CI are expected to evolve. Resistant genotypes also are expected to evolve against male-killer and parthenogenesis induction *Wolbachia*. There is some evidence of host effects on *Wolbachia* expression. For instance, *Wolbachia* that are weak CI expressers in *D. melanogaster* are stronger expressers when transfected into *D. simulans*, indicating host effects.[126] Recent cytological studies indicate that *Wolbachia* in *D. melanogaster* are not maintained in the developing spermatocyte cells, possibly explaining this effect.[79,80] When a *Wolbachia* that induces feminization in one moth (*Ostrinia scapulalis*) is transfected into another (*Ephestia kuehniella*), it induces male killing.[129]

In *Nasonia* wasps, CI shows very different patterns in closely related species.[106,131] In *Nasonia vitripennis*, it results in complete paternal genome loss and conversion of fertilized diploid eggs into haploid males (males are haploid in ants, bees, and wasps). In the sibling species *Nasonia giraulti* and *Nasonia longicornis*, CI results in embryonic lethality. Recent work shows that this is because of host genetic effects, not differences in the resident *Wolbachia*. One can postulate that this is because of changes in genes involved in early fertilization and sperm processing. The conversion of embryonic lethality to male production was possibly selected for in populations with frequent CI, as females with the conversion genotype would produce sons and therefore have increased fitness.

In general, because of the abundance of these parasites and their intimate association with hosts cells and germ-line tissues, we expect that presence of *Wolbachia* will lead to more rapid evolution of host genes interacting with these bacteria, particularly those associated with spermatogenesis and oogenesis. However, this proposition has not yet been rigorously tested.

Perhaps the most intriguing discovery is the occurrence of a gene transfer from *Wolbachia* to the genome of an insect host, the beetle *Callosobruchus chinensis*.[148] This was discovered because of a chance amplification using primers for the *Wolbachia* surface protein wsp, with subsequent confirmation of insertion of a region containing multiple *Wolbachia* genes.[149] Given that only a few *Wolbachia* genes are typically used to screen for these bacteria and that a genomic transfer has been detected, it is likely that such transfers occur relatively often in arthropods. The important question is, What are the fates of such transfers? Do they result in degeneration of the bacterial genes, or are they sometimes "recruited" for host functions? The latter would indicate a possibly important process for introducing evolutionary innovations into arthropod genomes.[9]

Sex Determination and Mating System Evolution

Heritable microorganisms (and cytoplasmically inherited organelles such as mitochondria) are expected to evolve adaptations to manipulate sex determination of hosts to produce more females, which is the sex that transmits them. It has been proposed that the "genetic conflict" resulting from differential selection on heritable microbes and host genomes may be a "motor" for rapid evolution of sex determining mechanisms.[64] One intriguing example comes from feminizing *Wolbachia* that occur in the isopod *Armadillidium vulgare*.[44,45,121] This species normally has female heterogamety (ZW sex chromosomes in females and ZZ sex chromosomes in males). However, empirical and theoretical studies indicate that presence of feminizing *Wolbachia* in some populations may lead to the elimination of the W female-determining chromosome, because the strongly female-biased sex ratios resulting from feminized ZZ males reduces the fitness of ZW females, leading to their elimination. In some cases, *Wolbachia* suppressor genes are favored at autosomal loci, which could result in the "neo-evolution" of male heterogamety subsequent to *Wolbachia*-induced loss of the W chromosome. The dynamics are complex, and although the appropriate genotypes are present, it remains to be determined how often sex determination shifts of the expected pattern occur in nature.

Feminizing *Wolbachia* may be very widespread. The strong female-biased populations that result from these *Wolbachia* place a premium on genotypes that can produce rare males and also on genotypes that can circumvent the bacterial manipulations that induce feminization. Both processes are expected to lead to more rapid evolution in sex determining genes. Similar arguments are made for male-killing *Wolbachia*, particularly with regard to genetic modifications that thwart early male killing. Studies over the next several years will reveal whether *Wolbachia*, and other inherited microorganisms, actually do lead to accelerated evolution of sex determination.

Recent studies indicate that sex ratio-distorting *Wolbachia* may also lead to rapid evolutionary changes in mating systems.[65,66,150,151] The best example of this comes from the butterfly *Acraea encedana*.[65,66] In this species, male-killing *Wolbachia* can reach very high frequencies, which results in many females going unmated. This leads to a shift in the mating system to a lek mating structure, where females gather and release pheromones that attract males and females solicit mating from males. This sex role reversal (in most animals males form leks to attract females) is not observed in related species that do not have highly female-biased sex ratios. Such sex ratio–distorter effects on mating systems may be common but have not been widely explored.

Speciation and Extinction

Recent reviews have considered the possible role of *Wolbachia* in host speciation.[104,152] The possibilities fall into three broad categories: CI and other *Wolbachia* may promote genetic divergence and speciation in insects by causing reproductive incompatibility between incipient species; CI between populations may select for premating isolation, thus accelerating the speciation process; or *Wolbachia* may accelerate rates of evolution in the host genome, particularly in genes involved in reproduction, thus causing nuclear gene incompatibilities between diverging populations.

Bidirectional incompatibility between populations is one mechanism that could reduce gene flow between populations and species. Studies in the parasitic wasp *Nasonia* show that bidirectional incompatibility occurs between three closely related species because of different bidirectionally incompatible *Wolbachia*.[63,106] This involves both a younger and an older species pair. A recent study shows that BiCI has evolved between the younger species pair before other isolating mechanisms, such as hybrid lethality and hybrid incompatibility, indicating that *Wolbachia*-induced isolation can arise early in the speciation process.[106] However, the two species occur in different geographic regions, so *Wolbachia* are not actively maintaining isolation between them, although recent field studies indicate that they may come in contact with each other in the Midwest.

Unidirectional incompatibility seems less likely to promote speciation, as gene flow is only reduced in one direction but not the other. However, in combination with other isolating factors in the other direction, unidirectional CI could be a significant component. This may be the case in some mushroom-feeding *Drosophila* species. [105] The potential importance of unidirectional CI is increased by the fact that it is likely to be more common in nature than BiCI.

A number of arguments have been made against a role of CI in insect speciation. Among them are that BiCI between populations does not occur frequently enough in nature to be of importance, that BiCI differences between populations are unstable, and that natural CI levels are insufficient to permit genetic divergence or select for premating isolation.[107,110]

The relevant issues are amenable to empirical and theoretical study. Experimental transfer of CI *Wolbachia* into a common host background will allow determination of the minimum number of bidirectional CI types, which is important to assess the possibilities of BiCI in nature. Geographic studies will reveal how frequently geographic populations (and closely related species) harbor different *Wolbachia*. Recent theoretical work shows that the presence of BiCI between populations subject to migration can greatly increase divergence at a selected locus.[153,154] Associations develop between the resident *Wolbachia* type and selected alleles, which increases local frequencies of both in the face of gene flow. Models also indicate that CI will select for premating isolation, an effect that could accelerate and stabilize genetic isolation.[155] Clearly, more theoretical and empirical study is needed to assess the role of *Wolbachia* in the evolution of arthropod species.

The studies described above indicate that reproductive alterations induced by these bacteria and the intimate association of *Wolbachia* with host cells may create selective pressures that result in more rapid evolution of host genes involved in Wolbachia–host interactions. If correct, then *Wolbachia* infections could accelerate host evolution, which could secondarily accelerate host divergence and speciation, particularly if such evolution results in genetic incompatibilities between populations in genes involved in reproduction and cellular biology. Genetic exchange between *Wolbachia* and hosts is also a potential means by which these symbiotic bacteria could promote host evolution and speciation.

Extinction is a difficult process to study. However, basic models indicate that sex ratio–distorting microorganisms can increase to high frequencies, possibly driving host populations to extinction.[42,50] How often this occurs in nature is difficult to assess, because the infected species we observe tend to be those that have features leading to more stable associations. Nevertheless, very high frequencies of male killers have been observed in some species, which results in significant proportions of females going unmated.[65] More subtly, the presence of sex ratio–distorting or CI *Wolbachia* can impose a cost to a population that would reduce its competitive ability. For instance, a CI *Wolbachia* at a frequency of 50% in a host population would result in 25% of the matings being incompatible. Infection polymorphisms are not uncommon in some species.[26,68] Detailed field and experimental studies would be needed to test the possible role of *Wolbachia* in population and species extinction. Given their abundance, it is not an unreasonable consideration.

Conclusion

Heritable microorganisms are extremely common in nature and are likely to be more pervasive than previously appreciated. A common strategy for heritable microorganisms is to manipulate host reproduction in ways that enhance transmission of the parasite. Reproductive parasitism includes manipulation of sex determination and germ-line development, induction of sperm–egg incompatibility, host castration, and postsegregation killing. A growing number of empirical and theoretical studies indicate that reproductive parasites could be important in host evolution, particularly in genome evolution, sex determination, germ-line development, and mating systems of invertebrates. Research over the next decade may well resolve whether reproductive parasites are an important motor shaping invertebrate evolution.

Acknowledgments I thank B.J. Velthuis for valuable comments on a draft of this chapter. I also thank J. Sapp for organizing the symposium on microbial evolution and the participants for stimulating discussions. The U.S. National Science Foundation is thanked for support of this work (DEB 9981634 and EF-0328363).

References

1. L. Margulis, *Symbiosis in Cell Evolution* (New York: Freeman, 1993).
2. M.W. Gray, "The Endosymbiont Hypothesis Revisited," *International Review of Cytology* 141 (1992): 233–357.
3. M.W. Gray, "Origin and Evolution of Organelle Genomes," *Current Opinions in Genetic Development* 3 (1993): 884–890.
4. P. Lopez-Garcia and D. Moreira, "Symbiosis between Methanogenic Archaea and Delta-Proteobacteria as the Origin of Eukaryotes: The Syntrophic Hypothesis," *Journal of Molecular Evolution* 47 (1998): 517–530.
5. S.G. Andersson, A. Zomorodipour, J.O. Andersson, T. Sicheritz-Ponten, U.C. Alsmark, R.M. Podowski, A.K. Naslund, A.S. Eriksson, H.H. Winkler, and C.G. Kurland, "The Genome Sequence of *Rickettsia prowazeki* and the Origin of Mitochondria," *Nature* 396 (1998): 133–140.
6. R.K. Trench, "*Cyanophora paradoxa* Korschikoff and the origins of chloroplasts," in

L. Margulis, and R. Fester, eds., *Symbiosis as a Source of Evolutionary Innovation* (Cambridge, MA: The MIT Press, 1991).
7. I.E. Wallin, *Symbionticism and the Origin of Species* (Baltimore, MD: Williams and Wilkins, 1927).
8. P. Buchner, *Endosymbiosis of Animals with Plant Microorganisms* (New York: Wiley Interscience, 1965).
9. L. Margulis and R. Fester, eds. *Symbiosis as a Source of Evolutionary Innovation* (Cambridge, MA: The MIT Press, 1991).
10. J. Maynard Smith, "A Darwinian View of Symbiosis," in L. Margulis, and R. Fester, eds., *Symbiosis as a Source of Evolutionary Innovation* (Cambridge, MA: The MIT Press, 1991).
11. Jan Sapp, *Evolution by Association. A History of Symbiosis.* (New York: Oxford University Press, 1994).
12. S.L. O'Neill, A.A. Hoffman, and J.H. Werren, eds., *Influential Passengers: Inherited Microorganisms and Arthropod Reproduction* (New York: Oxford University Press, 1997).
13. N.A. Moran and P. Baumann, "Bacterial Edosymbionts in Animals," *Current Opinions in Microbiology* 3 (2000): 270–275.
14. N.A. Moran and J.J. Wernegreen, "Are Mutualism and Parasitism Irreversible Evolutionary Alternatives for Endosymbiotic Bacteria? Insights from Molecular Phylogenetics and Genomics." *Trends in Ecology and Evolution* 15 (2000): 321–326.
15. J.H. Werren and S. O'Neill, "The Evolution of Heritable Symbionts," in S.L. O'Neill, A.A. Hoffmann, and J.H. Werren, eds., *Influential Passengers: Inherited Microorganisms and Arthropod Reproduction* (New York: Oxford University Press, 1997).
16. J. Korb and D.K. Aanen, "The Evolution of Uniparental Transmission of Fungal Symbionts in Fungus-growing Termites (Macrotermitinae)," *Behav. Ecol. & Sociobiol.* 53 (2003): 65–71.
17. P.W. Ewald, "Transmission Modes and Evolution of Parasitism–Mutualism Continuum," *Annals of the New York Academy of Science* 503 (1987):295–306.
18. P.E.M. Fine, "Vectors and Vertical Transmission: An Epidemiologic Perspective," *Annals of the New York Academy of Science* 266 (1975): 173–194.
19. M. Lipsitch, M.A. Nowak, D. Ebert, and R.M. May, "The Population Dynamics of Vertically and Horizontally Transmitted Parasites," *Proceedings of the Royal Society London B* 260 (1995): 321–327.
20. S.A. Frank, "Kin Selection and Virulence in the Evolution of Protocells and Parasites," *Proceedings of the Royal Society London B* 258 (1994): 153–161.
21. C. Bandi, A.M. Dunn, G.D.D. Hurst, and T. Rigaud, "Inherited Microorganisms, Sex-Specific Virulence and Reproductive Parasitism," *Trends in Parasitology* 17 (2001): 88–94.
22. G.D.D. Hurst, L.D. Hurst, and M.E.N. Majerus, "Cytoplasmic Sex-Ratio Distorters," in S.L. O'Neill, A.A. Hoffmann, and J.H. Werren, eds., *Influential Passengers: Inherited Microorganisms and Arthropod Reproduction* (New York: Oxford University Press, 1997).
23. R. Stouthamer, "Wolbachia-Induced Parthenogenesis," in S.L. O'Neill, A.A. Hoffmann, and J.H. Werren, eds., *Influential Passengers: Inherited Microorganisms and Arthropod Reproduction*, 102–124. (New York: Oxford University Press, 1997).
24. T. Rigaud, "Inherited Microorganisms and Sex Determination of Arthropod Hosts," in S.L. O'Neill, A.A. Hoffmann, and J.H. Werren, eds., *Influential Passengers: Inherited Microorganisms and Arthropod Reproduction*, 155–175. (New York: Oxford University Press, 1997).
25. K. Clay and C. L. Schardl. "Evolutionary Origins and Ecological Consequences of Endophyte Symbiosis with Grasses," *American Naturalist* 160 (2002): S99–S127.
26. K. Bourtzis and T.A. Miller, *Insect Symbiosis* (Boca Raton, FL: CRC Press, 2003).
27. J.H. Werren, "Biology of Wolbachia," *Annual Review of Entomology* 42 (1997): 587–609.
28. P.H. Thrall, J. Antonovics, and J. Bever, "Sexual Transmission of Disease and Host Mating Systems. I. Within Season Reproductive Success," *American Naturalist* 149 (1996): 485–506.
29. P.H. Thrall and J. Antonovics, "Polymorphism in Sexual vs. Non-Sexual Transmission," *Proceedings of the Royal Society of London Series B.* 264: 581–587.
30. A.B. Lockhart, P.H. Thrall, and J. Antonovics, "The Distribution and Characteristics of

Sexually Transmitted Diseases in Animals: Ecological and Evolutionary Implications," *Biological Reviews of the Cambridge Philosophical Society* 71 (1997): 415–471.
31. K. Clay, "Parasitic Castration of Plants by Fungi," *Trends in Ecology and Evolution* 6 (1991): 162–166.
32. N.A. Moran and P. Baumann, "Phylogenetics of Cytoplasmically Inherited Microorganisms of Arthropod," *Trends in Ecology and Evolution* 9 (1994): 15–20.
33. P. Baumann, N.A. Moran and L. Baumann. "Bacteriocyte-Associated Endosymbionts of Insects," in M. Dworkin, ed., *The Prokaryotes, Third Edition, A Handbook on the Biology of Bacteria: Ecophysiology, Isolation, Identification, Applications* (New York: Springer, 2000).
34. K.M. Usher, J. Kuo, J. Fremont, and D.C. Sutton, "Vertical Transmission of Cyanobacterial Symbionts in the Marine Sponge *Chondrilla australiensis* (*Demospongiae*)," *Hydrobiologia* 461 (2001): 15–23.
35. J.P. Sandstrom, J.A. Russell, J.P. White, and N.A. Moran, "Independent Origins and Horizontal Transfer of Bacterial Symbionts of Aphids," *Molecular Ecology* 10 (2001): 2058–2064.
36. L.A. Hurtado, M. Mateos, R.A. Lutz, and R.C. Vrijenhoek, "Coupling of Bacterial Endosymbiont and Host Mitochondrial Genomes in the Hydrothermal Vent Clam *Calyptogena magnifica,*" *Applied and Environmental Microbiology* 69 (2003): 2058–2064.
37. J. Korb and D.K. Aanen, "The Evolution of Uniparental Transmission of Fungal Symbionts in Fungus-Growing Termites (Macrotermitinae)," *Behav. Ecol. and Sociobiol.* 53 (2003): 65–71.
38. A.R. Spie, A.E. Wilbur, and S.C. Cary, "Bacterial Symbiont Transmission in the Wood-Boring Shipworm *Bankia setacea* (Bivalvia: Teredinidae)," 66 (2000): 1685–1691.
39. A.F. Azad, J.B. Sacci Jr., W.M. Nelson, G.A. Dasch, E.T. Schmidtmann, and M. Carl, "Genetic Characterization and Transoverial Transmission of a Typhus-Like Rickettsia Found in Cat Fleas," *Proceedings of the National Academy of Sciences USA* 89 (1992): 43–46.
40. C. Bandi, M. Sironi, G. Damiani, L. Magrassi, C.A. Nalepa, U. Laudani, et al., "The Establishment of Intracellular Symbiosis in an Ancestor of Cockroaches and Termites," *Proceedings of the Royal Society of London Series B: Biological Sciences* 259 (1995): 293–299.
41. A.E. Douglas, "Host Benefit and the Evolution of Specialization in Symbiosis," *Heredity* 81 (1998): 599–603.
42. J.H. Werren, "The Coevolution of Autosomal and Cytoplasmic Sex Ratio Factors," *Journal of Theoretical Biology* 124 (1987): 317–334.
43. A.E. Douglas, "Nutritional Interactions in Insect-Microbial Symbioses: Aphids and Their Symbiotic Bacteria Buchnera," *Annual Review of Entomology* 43 (1998): 17–37.
44. T. Rigaud and P. Juchault, "Conflict between Feminizing Sex-Ratio Distorters and an Autosomal Masculinizing Gene in the Terrestrial Isopod *Armadillidium vulgare*, Latr," *Genetics* 133 (1993): 247–252.
45. T. Rigaud and P. Juchault, and J.P. Macquard, "The Evolution of Sex Determination in Isopod Crustaceans," *BioEssays* 19 (1997): 409–416.
46. M. Hiroki, Y. Kato, T. Kamito, and K. Miura. "Feminization of Genetic Males by a Symbiotic Bacterium in a Butterfly, *Eurema hecabe* (Lepidoptera : Pieridae)," *Naturwissenschaten* 89 (2002): 167–170.
47. Y. Fujii, D. Kageyama, S. Hoshizaki, H. Ishikawa, and T. Sasaki, "Transfection of Wolbachia in Lepidoptera: The Feminizer of the Adzuki Bean Borer *Ostrinia scapulalis* Causes Male Killing in the Mediterranean Flour Moth *Ephestia kuehniella,*" *Proceedings of the Royal Society of London B* (2001) 268: 855–859.
48. A.M. Dunn, J. Adams, and J.E. Smith, "Transovarial Transmission and Sex Ratio Distortion by a Microsporidian Parasite in a Shrimp," *Journal of Invertebrate Pathology* 61 (1993): 248–252.
49. A.M. Dunn, M.J. Hatcher, R.S. Terry, and C. Tofts, "Evolutionary Ecology of Vertically Transmitted Prasites: Transovarial Transmission of a Microsporidian Sex-Ratio Distorter in *Gammarus duebeni,*" *Parasitology* 111(1995): S91–109.
50. H.W. Howard, "The Genetics of *Armadillidium vulgare* Latr. II. Studies on the Inheritance of Monogeny and Amphogeny," *Journal of Genetics* 44 (1942): 143–59.

51. D. Lewis, "Male-Sterility in Natural Populations of Hermaphrodite Plants," *New Phytology* 40 (1941): 56–63.
52. L.D. Hurst, "The Incidences, Mechanisms and Evolution of Cytoplasmic Sex-Ratio Distorters in Animals." *Biol. Rev. Camb. Phil. Soc.* 68 (1983): 121–94.
53. G.D.D. Hurst and F.M. Jiggins, "Male-Killing Bacteria in Insects: Mechanisms, Incidence, and Implications," *Emerging Infectious Diseases* 6 (2000): 329–336.
54. G.D.D. Hurst, F.M. Jiggins, and M.E.N. Majerus. "Inherited Microorganisms that Selectively Kill Male Hosts: The Hidden Players of Insect Evolution?" in K. Bourtzis and T.A. Miller, eds., *Insect Symbiosis* (Boca Raton, FL: CRC Press, 2003).
55. R. Gherna, J.H. Werren, W. Weisburg, R. Cote, C.R. Woese, L. Mandelco, and R. Brenner, "*Arsenophonus nasoniae*, Genus Novel, Species Novel, Causative Agent of Sonkiller Trait in the Parasitic Wasp, *Nasonia vitripennis*," *Inter. J. Bact. Syst.* 41 (1991): 563–565.
56. J.J. Becnel, "Horizontal Transmission and Subsequent Development of *Ambyospora californica* (Microsporida: Ambryosporidae) in the Intermediate and Definitive Hosts," *Diseases of Aquatic Organisms* 13 (1992): 17–28.
57. J.H. Werren, G.D. Hurst, W. Zhang, J.A. Breeuwer, R. Stouthamer, and M.E. Majerus, "Rickettsial Relative Associated with Male Killing in the Ladybird Beetle (*Adalia bipunctata*)," *Journal of Bacteriology* 176 (1994): 388–394.
58. G.D.D. Hurst, F.M. Jiggins, J.H.G. von der Schulenburg, D. Bertrand, S.A. West, Goriacheva II, I.A. Zakharov, J.H. Werren, R. Stouthamer, and M.E.N. Majerus, "Male-Killing *Wolbachia* in Two Species of Insect," *Proceedings of the Royal Society of London Series B–Biological Sciences* 266 (1999): 735–740.
59. R.F. Fialho and L. Stevens, "Male-Killing *Wolbachia* in a Flour Beetle," *Proceedings of the Royal Society of London Series B–Biological Sciences* 267 (2000): 1469–1473.
60. E.T. Lawson, T.A. Mousseau, R. Klaper, M.D. Hunter, and J.H. Werren, "Rickettsia Associated with Male-killing in a Buprestid Beetle," *Heredity* 86 (2001): 497–505.
61. T. Fukatsu and H. Anbutsu, "Population Dynamics of Male-Killing and Non-Male-Killing Spiroplasmas in *Drosophila melanogaster*," *Applied and Environmental Microbiology* 69 (2003): 1428–1434.
62. P. Samitou-Laprade, J. Cuguen, and P. Vernet, "Cytoplasmic Male Sterility in Plants: Molecular Evidence and the Nucleocytoplasmic Conflict," *Trends in Ecology and Evolution* 99 (1994): 431–435.
63. J.A.J. Breeuwer and J.H. Werren, "Microorganisms Associated with Chromosome Destruction and Reproductive Isolation Between Two Insect Species," *Nature* 346 (1990): 558–560.
64. J.H. Werren and L. Beukeboom, "Sex Determination, Sex Ratios, and Genetic Conflict," *Annual Review of Ecology and Systematics* 29 (1998): 233–261.
65. F.M. Jiggins, G.D.D. Hurst, and M.E.N. Majerus, "Sex-Ratio-Distorting *Wolbachia* Causes Sex-Role Reversal in Its Butterfly Host," *Proceedings of the Royal Society of London B* 266 (2000a): 69–73.
66. J.P. Randerson, F.M. Jiggins, and L.D. Hurst, "Male Killing Can Select for Male Mate Choice: A Novel Solution to the Paradox of the Lek," *Proceedings of the Royal Society of London B* 267 (2000): 867–874.
67. J.H. Yen and A.R. Barr, "New Hypothesis of the Cause of Cytoplasmic Incompatibility in *Culex pipiens*," *Nature* 232 (1971): 657–658.
68. A.A. Hoffmann and M. Turelli, "Cytoplasmic Incompatibility in Insects," in S.L. O'Neill, A.A. Hoffmann, and J.H. Werren, eds., *Influential Passengers: Inherited Microorganisms and Arthropod Reproduction* (New York: Oxford University Press, 1997), 42–80.
69. K. Bourtzis, H.R. Braig, and T.L. Karr. "Cytoplasmic Incompatibility," in K. Bourtzis and T.A. Miller, eds. *Insect Symbiosis* (Boca Raton, FL: CRC Press, 2003).
70. A.R. Weeks and J.A.J. Breeuwer, "A New Bacterium from the Cytophaga-Flavobacterium-Bacteroides Phylum That Causes Sex-Ratio Distortion," in K. Bourtzis and T.A. Miller, eds., *Insect Symbiosis* (Boca Raton, FL: CRC Press, 2003).
71. Martha Hunter, personnal communication.
72. M.S. Hunter, "The Influence of Parthenogenesis-Inducing on the Oviposition Behavior and Sex-Specific Developmental Requirements of Autoparasitoid Wasps," *J. Evol. Biol.* 12 (1999): 765–741.

73. A.R. Weeks, et al., "A Mite Species that Consists Entirely of Haploid Females," *Science* 292 (2001): 2479–2482.
74. E. Zchori-Fein, et al., "A Newly Discovered Bacterium Associated with Parthenogenesis and a Change in Host Selection Behavior in Parasitoid Wasps," *Proceedings of the National Academy of Sciences USA* 98 (2001): 12555–12560.
75. K.M. Reed and J.H. Werren, "Induction of Paternal Genome Loss by the Paternal-Sex-Ratio Chromosome and Cytoplasmic Incompatibility Bacteria (*Wolbachia*): A Comparative Study of Early Embryonic Events," *Molecular Reproduction and Development* 40 (1995): 408–418.
76. C.W. Lassy and T.L. Karr, "Cytological Analysis of Fertilization and Early Embryonic Development in Incompatible Crosses of *Drosophila simulans,*" *Mechanisms of Development* 57 (1996): 47–58.
77. G. Callaini, et al. "*Wolbachia*-Induced Delay of Paternal Chromatin Condensation Does Not Prevent Maternal Chromosomes from Entering Anaphase in Incompatible Crosses of *Drosophila simulans*," *Journal of Cell Science* 110 (1997): 271–280.
78. U. Tram and W. Sullivan, "Role of Delayed Nuclear Envelope Breakdown and Mitosis in *Wolbachia*-Induced Cytoplasmic Incompatibility," *Science* 296 (2002): 1124–1126.
79. M.E. Clark, Z. Veneti, K. Bourtzis, and T.L. Karr, "*Wolbachia* Distribution and Cytoplasmic Incompatibility in *Drosophila*: The Cyst as the Basic Cellular Unit of CI Expression," *Mechanisms of Development*, 120 (2003):185–198.
80. M.W. Clark, Z. Veneti, K. Bourtzis, and T.L. Karr, "The Distribution and Proliferation of the Intracellular Bacteria *Wolbachia* During Spermatogenesis in *Drosophila*," *Mechanisms of Development*, 111 (2002): 3–15.
81. S.L. O'Neill and T.L. Karr, "Bidirectional Incompatibility between Conspecific Populations of *Drosophila simulans,*" *Nature* 348 (1990): 178–180.
82. M.J. Perrot Minnot, L.R. Guo, and J.H. Werren, "Single and Double Infections with *Wolbachia* in the Parasitic wasp *Nasonia vitripennis*: Effects on Compatibility," *Genetics* 143 (1996): 961–972.
83. R. Stouthamer, J.A.J. Breeuwer, and G.D.D. Hurst, "*Wolbachia pipientis*: Microbial Manipulator of Arthropod Reproduction," *Annual Review of Microbiology* 53 (1999): 71–102.
84. E. Caspari and G.S. Watson, "On the Evolutionary Importance of Cytoplasmic Sterility in Mosquitoes," *Evolution* 13 (1959): 568–570.
85. M. Turelli, "Evolution of Incompatibility-Inducing Microbes and their Hosts," *Evolution* 48 (1994): 1500–1513.
86. T.F. Cooper and J.A. Heinemann, "Postsegregational Killing Does Not Increase Plasmid Stability but acts to Mediate Exclusion of Competing Plasmids," *Proceedings of the National Academy of Sciences USA* 97 (2000): 12643–12848.
87. T. Naito, K. Kusano, and I. Kobayashi, "Selfish Behaviour of Restriction-Modification Dystems," *Science* 267 (1995): 897–899.
88. M.E. Clark and T.L. Karr, "Distribution of *Wolbachia* within *Drosophila* Reproductive Tissues: Implications for the Expression of Cytoplasmic Incompatibility," *Int & Comp Biol.* 42 (2002): 332–339.
89. D. Carré, C. Djediat, and C. Sardet, "Formation of a Large Vasa-Positive Germ Granule and Its Inheritance by Germ Cells in the Enigmatic Chaetognaths," *Development* 129 (2002): 661–670.
90. A. Nakamura, et al., "Requirement for a Non-Coding RNA in *Drosophila* Polar Granules for Germ Cell Establishment," *Science* 274 (1996): 2075–2079
91. R. Amikura, et al, "Presence of Mitochondria-Type Ribosomes Outside Mitochondria in Germ Plasm of *Drosophila* Embryos," *Proceedings of the National Academy of Sciences USA* 98 (2001): 9133–9138
92. F. Dedeine, F. Vavre, F. Fleury, B. Loppin, M.E. Hochberg, and M. Bouletreau, "Removing Symbiotic *Wolbachia* Bacteria Specifically Inhibits Oogenesis in a Parasitic Wasp," *Proceedings of the National Academy of Sciences USA* 98 (2001): 6247–6252.
93. J.H. Werren, D. Windsor, L. Guo, and L.R. Guo, "Distribution of *Wolbachia* among Neotropical Arthropods," *Proceedings of the Royal Society of London Series B Biological Sciences* 262 (1995): 197–204.

94. S.A. West, J.M Cook, J.H. Werren, and H.C.J. Godfray, "Wolbachia in Two Host-Parasitoid Communities," *Molecular Ecology* 7 (1998): 1457–1465.
95. C. Bandi, T.J.C. Anderson, C. Genchi, and M.L. Baxter, "Phylogeny of Wolbachia in Filarial Nematodes," *Proceedings of the Royal Society of London Series B Biological Sciences* 265 (1998): 2407–2413.
96. B. Bouchon, T. Rigaud, and P. Juchault, "Evidence for Widespread *Wolbachia* Infection in Isopod Crustaceans: Molecular Identification and Host Feminization," *Proceedings of the Royal Society of London Series B: Biological Sciences* 265 (1998): 1081–1090.
97. J.H. Werren and D.M. Windsor, "*Wolbachia* Infection Frequencies in Insects: Evidence of a Global Equilibrium?" *Proceedings of the Royal Society of London Series B–Biological Sciences* 267 (2000): 1277–1285.
98. A. Jeyaprakash and M.A. Hoy, "Long PCR Improves *Wolbachia* DNA Amplification: wsp Sequences Found in 76% of Sixty-Three Arthropod Species," *Insect Molecular Biology* 9 (2000): 393–405.
99. J.H. Werren, "Aresenophonus," in G.M. Garrity, ed., *Bergey's Manual of Systematic Bacteriology*, Vol. 2., (New York: Springer, 2003).
100. M.J. Wade and N.W. Chang, "Increased Male Fertility in *Tribolium confusum* Beetles after Infection with the Intracellular Parasite *Wolbachia*," *Nature* 373 (1995): 72–74.
101. J.H. Werren, W. Zhang and L.R. Guo. "Evolution and Phylogeny of *Wolbachia* Bacteria: Reproductive Parasites of Arthropods," *Proceedings of the Royal Society of London B* 261 (1995):55–71.
102. W.G. Zhou, F. Rousset, and S. O'Neill, "Phylogeny and PCR-Based Classification of *Wolbachia* Strains Using wsp Gene Sequences," *Proceedings of the Royal Society of London Series B–Biological Sciences* 265 (1998): 509–515.
103. H. Laven, "Speciation by Cytoplasmic Isolation in the *Culex pipiens* Complex," *Cold Spring Harbor Symposium on Quantitative Biology* 24 (1959): 166–173.
104. J.H. Werren, *Wolbachia and Speciation. Endless Forms: Species and Speciation.* (New York: Oxford University Press, 1998), 245–260.
105. D.D. Shoemaker, V. Katju, and J. Jaenike, "Wolbachia and the Evolution of Reproductive Isolation between *Drosophila recens* and *Drosophila subquinaria*," *Evolution* 53 (1999): 1157–1164.
106. S.R. Bordenstein, F.P. O'Hara, and J.H. Werren, "Wolbachia-Induced Incompatibility Precedes Other Hybrid Incompatibilities in Nasonia," *Nature* 409 (2001): 707–710.
107. G.D. Hurst and M. Schilthuizen, "Selfish Genetic Elements and Speciation," *Heredity* 80 (1998):2–8.
108. G.D.D. Hurst and J.H. Werren, "The Role of Selfish Genetic Elements in Eukaryotic Evolution," *Nature Reviews* 2 (2001): 597–606.
109. A. Rokas, "*Wolbachia* as a Speciation Agent," *Tree* 15 (2001): 44–45.
110. A.R. Weeks, K.T. Reynolds, A.A. Hoffmann, and H. Mann, "*Wolbachia* Dynamics and Host Effects: What Has (and Has Not) Been Demonstrated?" *Trends in Ecology and Evolution* 17 (2002): 257–262.
111. V.V. Emelyanov, "Evolutionary Relationship of Rickettsiae and Mitochondria," *FEBS Letters 501* (2001) (1): 11–18.
112. A.V. Taillardat-Bisch, D. Raoult, and M. Drancourt. "RNA Polymerase Beta-Subunit-Based Phylogeny of *Ehrlichia spp., Anaplasma spp., Neorickettsia spp.* and *Wolbachia pipientis*," *International Journal of Syst. and Evol. Microbiology* 53 (2003): 455–458.
113. S.L. O'Neill, R. Giordano, A.M.E. Colbert, T.L. Karr, and H.M. Robertson, "16S rRNA Phylogenetic Analysis of the Bacterial Endosymbionts Associated with Cytoplasmic Incompatibility in Insects," *Proceedings of the National Academy of Sciences USA* 89 (1992): 2699–2702.
114. N. Li, M. Casiraghi, E. Salati, C. Bazzocchi, and C. Bandi, "How Many *Wolbachia* Supergroups Exist?" *Mol. Biol. Evol.* 19 (2002): 341–346.
115. F.M. Jiggins, J.H.G. von der Schulenburg, G.D.D. Hurst, and M.E.N. Majerus, "Recombination Confounds Interpretations of *Wolbachia* Evolution," *Proceedings of the Royal Society of London Series B–Biological Sciences* 268 (2001): 1423–1427.
116. J.H. Werren and J.D. Bartos, "Recombination in *Wolbachia*," *Current Biology* 11 (2001): 431–435.

117. R. Cordaux, A. Michel Salzat, and D. Bouchon, "*Wolbachia* Infection in Crustaceans: Novel Hosts and Potential Routes for Horizontal Transmission," *Journal of Evolutionary Biology 14* (2001a):237–243.
118. S.L. Dobson, E.J. Marsland, and W. Rattanadechakul, "*Wolbachia*-Induced Cytoplasmic Incompatibility in Single and Superinfected *Aedes albopictus* (*Diptera Culicidae*)," *Journal of Medical Entomology* 38 (2001):382–387.
119. F. Rousset, H.R. Braig, and S.L. O'Neill, "A Stable Triple *Wolbachia* Infection in *Drosophila* with Nearly Additive Incompatibility Effects," *Heredity* 82 (1999):620–627.
120. M. Turelli and A.A. Hoffman, "Rapid Spread of an Inherited Incompatibility Factor in California *Drosophila*," *Nature* 353 (1991): 440–442.
121. Y. Caubet, M.J. Hatcher, J.-P. Mocquard, and T. Rigaud, "Genetic Conflict and Changes in Heterogametic Mechanims of Sex Determination," *Journal of Evolutionary Biology* 13 (2000): 766–777.
122. L. Boyle, S.L. O'Neill, H.M. Robertson, and T.L. Karr, "Interspecific and Intraspecific Horizontal Transfer of *Wolbachia* in *Drosophila*," *Science* 260 (1993): 1796–1799.
123. H.R. Braig, H. Guzman, R.B. Tesh, and S.L. O'Neill, "Replacement of the Natural *Wolbachia* Symbiont of *Drosophila simulans* with a Mosquito Counterpart," *Nature* 367(1994): 453–455.
124. N.W. Chang and M.J. Wade, "The Transfer of *Wolbachia pipientis* and Reproductive Incompatibility between Infected and Uninfected Strains of the Flour Beetle, *Tribolium confusum*, by Microinjection," *Canadian Journal of Microbiology* 40 (1994): 978–981.
125. D.J. Clancy and A.A. Hoffmann, "Behavior of *Wolbachia* Endosymbionts from *Drosophila simulans* in *Drosophila serrata*, a Novel Host," *American Naturalist* 149 (1997): 975–988.
126. D. Poinsot, K. Bourtzis, G. Markakis, C. Savakis, and H. Mercot, "*Wolbachia* Transfer from *Drosophila melanogaster* into *D. simulans*: Host Effect and Cytoplasmic Incompatibility Relationships," *Genetics* 150 (1998): 227–237.
127. M.M.M. van Meer and R. Stouthamer, "Cross-Order Transfer of *Wolbachia* from *Muscidifurax uniraptor* (*Hymenoptera: Pteromalidae*) to *Drosophila simulans* (*Diptera: Drosophilidae*)," *Heredity* 82 (1999): 163–169.
128. T. Rigaud, P.S. Pennings, and P. Juchault, "*Wolbachia* Bacteria Effects After Experimental Interspecific Transfers in Terrestrial Isopods," *Journal of Invertebrate Pathology* 77(4) (2001): 251–257.
129. Y. Fujii, D. Kageyama, S. Hoshizaki, H. Ishikawa, and T. Sasaki, "Transfection of *Wolbachia* in Lepidoptera: The Feminizer of the Adzuki Bean Borer *Ostrinia scapulalis* Causes Male Killing in the Mediterranean Flour Moth *Ephestia kuehniella*," *Proceedings of the Royal Society London B* 268 (2001): 855–859.
130. S.R. Bordenstein and J.H. Werren, "Effects of A and B Wolbachia and Host Genotype on Interspecies Cytoplasmic Incompatibility in *Nasonia*," *Genetics* 148 (1998): 1833–1844.
131. D. Poccia and P. Collas, "Transforming Sperm Nuclei into Male Pronuclei in vivo and in vitro," *Current Topics in Developmental Biology* 34 (1996): 25–88.
132. R. Stouthamer and D.J. Kazmer, "Cytogenetics of Microbe-Associated Parthenogenesis and its Consequences for Gene Flow in Trichogramma Wasps," *Heredity* 73 (1994): 317–327.
133. Y. Gottlieb, E. Zchori-Fein, J.H. Werren, and T. Karr, "Diploidy Restoration in *Wolbachia*-Infected *Muscidifurax uniraptor*. (Hymenoptera: Pteromalidae)," *Journal of Invertebrate Pathology* 81 (2002): 166–174.
134. Y. Gottlieb and E. Zchori-Fein, "Irreversible Thelytokous Reproduction in *Muscidifurax uniraptor*," *Entomol. Exp. Appl.* 100 (2001): 271–278.
135. J.W.A.M. Pijls, J.H. Van Steenbergen, and J.J.M. Van Alphen, "Asexuality Cured: The Relations and Differences Between Sexual and Asexual *Apoanagyrus diversicorni*," *Heridity* 76 (1996): 506–513.
136. J.J. Bull, *The Evolution of Sex Determing Mechanisms* (Menlo Park, CA: Benjamin/Cummings, 1983).
137. C. Bandi, "Tetracycline Treatment and Sex-Ratio Distortion: A Role for *Wolbachia* in the Moulting of Filarial Nematodes?" *Int. J. Parasit.* 32(12) (2002): 1457–1468.

138. M. Casiraghi, J.W. McCall, L. Simoncini, L.H. Kramer, L. Sacchi, C. Genchi, J.H. Werren, and C. Bandi, "Tetracycline Treatment and Sex-Ratio Distortion: A Role for *Wolbachia* in the Moulting of Filarial Nematodes?" *Int. J. Parasit.* 32 (12) (2002): 1457–1468.
139. D.J. Starr and T.W. Cline, "A Host-Parasite Interaction Rescues *Drosophila* Oogenesis Defects," *Nature* 418 (2002): 76–79.
140. K.T. Min and S. Benzer, "*Wolbachia*, Normally a Symbiont of *Drosophila*, Can Be Virulent, Causing Degeneration and Early Death," *Proceedings of the National Academy of Science USA* 94 (1997): 10792–10796.
141. L.V. Sun et al. "Determination of *Wolbachia* Genome Size by Pulsed-field Gel Electrophoresis," *Journal of Bacteriology* 183 (7) (2001): 2219–2225.
142. S. Masui, T. Kamoda, T. Sasaki, and H. Ishikawa, "The First Detection of the Insertion Sequence ISW1 in the Intracellular Reproductive Parasite *Wolbachia*," *Plasmid* 42 (1999): 13–19.
143. S. Masui, S. Kamoda, T. Sasaki, and H. Ishikawa, "Distribution and Evolution of Bacteriophage WO in *Wolbachia*, the Endosymbiont Causing Sexual Alterations in Arthropods," *Journal of Molecular Evolution* 51 (2000): 491–497.
144. S. Masui, S. Kamoda, T. Sasaki, and H. Ishikawa, "Distribution and Evolution of bacteriophage WO in *Wolbachia*, the Endosymbiont Causing Sexual Alterations in Arthropods," *Journal of Molecular Evolution* 51 (2000): 491–497.
145. B. Slatko, S.L. O'Neill, A.L. Scott, J.H. Werren, and M.L. Blaxter, "The *Wolbachia* Genome Consortium," *Microbial and Comparative Genomics* 4 (1999): 161–165.
146. M. Wu, V. Sun, J. Vamathevan, M. Riegler, R. Deboy, J.C. Brownlie, E.A. McGraw, W. Martin, C. Esser, N. Ahmadinejad, C. Wiegand, R. Madupu, M.J. Beanan, L.M. Brinkac, S.C. Daugherty, A.S. Durkin, J.F. Kolonay, W.C. Nelson, Y. Mohamoud, P. Lee, K. Berry, M.B. Young, T. Nierman, I.T. Paulsen, K.E. Nelson, H. Tettelin, S.L. O'Neill, and J.A. Eisen. "Phylogenomics of the Reproductive Parasite *Wolbachia Pipientis w*Mel: A Streamlined Genome Overrun by Mobile Genetic Elements," *PLoS* (2) (2004): 327–341.
147. F. Rousset and M. Solignac, "Evolution of Single and Double *Wolbachia* Symbioses During Speciation in the *Drosophila simulans* Complex," *Proceedings of the National Academy of Sciences USA* 92 (1995):6389–6393.
148. N. Kondo, N. Nikoh, N. Ijichi, M. Shimada, and T. Fukatsu, "Genome Fragment of *Wolbachia* Endosymbiont Transferred to X Chromosome of Host Insect," *Proceedings of the National Academy of Sciences USA* 98 (2002): 14280–14285.
149. T. Fukatsu, N. Kondo, N. Ijichi, and N. Nikoh. "Discovery of Symbiont–Host Horizontal Genome Transfer: A Beetle Carrying Two Bacterial and One Chromosomal Wolbachia Endosymbionts," in K. Bourtzis and T.A. Miller, eds. *Insect Symbiosis.* (Boca Raton, FL: CRC Press, 2003).
150. I. Silva and R. Stouthamer, "Can the Parthenogenesis-*Wolbachia* lead to Unusual Courtship Behaviour in Trichogramma?" *Proceedings of the Section Experimental and Applied Entomology of the Netherlands Entomological Society* 7 (1996): 27–31.
151. J. Moreau and T. Rigaud, "Variable Male Potential Rate of Reproduction: High Male Mating Capacity as an Adaptation to Parasite-induced Excess of Females?" *Proceedings of the Royal Society of London Series B–Biology* 270 (2003): 1535–1540.
152. S.R. Bordenstein, "Symbiosis and the Origin of Species," in K. Bourtzis and T.A. Miller, eds., *Insect Symbiosis* (Boca Raton, FL: CRC Press, 2003).
153. A. Telschow, P. Hammerstein, and J.H. Werren, "The Effect of *Wolbachia* on Genetic Divergence between Populations: Mainland-Island Model," *Integr. Comp. Biol.* 2 (2002): 340–351.
154. A. Telschow, P. Hammerstein, and J.H. Werren, "The Effect of *Wolbachia* on Genetic Divergence between Populations: Models with Two Way Migration," *American Naturalist* 160 (2002): S43–S66.

Index